S0-BGS-322

Springer Series in
MATERIALS SCIENCE 42

Springer
Berlin
Heidelberg
New York
Barcelona
Hong Kong
London
Milan
Paris
Singapore
Tokyo

Springer Series in
MATERIALS SCIENCE

Editors: R. Hull   R. M. Osgood, Jr.   H. Sakaki   A. Zunger

The Springer Series in Materials Science covers the complete spectrum of materials physics, including fundamental principles, physical properties, materials theory and design. Recognizing the increasing importance of materials science in future device technologies, the book titles in this series reflect the state-of-the-art in understanding and controlling the structure and properties of all important classes of materials.

27 **Physics of New Materials**
Editor: F. E. Fujita   2nd Edition

28 **Laser Ablation**
Principles and Applications
Editor: J. C. Miller

29 **Elements of Rapid Solidification**
Fundamentals and Applications
Editor: M. A. Otooni

30 **Process Technology for Semiconductor Lasers**
Crystal Growth and Microprocesses
By K. Iga and S. Kinoshita

31 **Nanostructures and Quantum Effects**
By H. Sakaki and H. Noge

32 **Nitride Semiconductors and Devices**
By H. Morkoç

33 **Supercarbon**
Synthesis, Properties and Applications
Editors: S. Yoshimura and R. P. H. Chang

34 **Computational Materials Design**
Editor: T. Saito

35 **Macromolecular Science and Engineering**
New Aspects
Editor: Y. Tanabe

36 **Ceramics**
Mechanical Properties, Failure Behaviour, Materials Selection
By D. Munz and T. Fett

37 **Technology and Applications of Amorphous Silicon**
Editor: R. A. Street

38 **Fullerene Polymers and Fullerene Polymer Composites**
Editors: P. C. Eklund and A. M. Rao

39 **Semiconducting Silicides**
Editor: V. E. Borisenko

40 **Reference Materials in Analytical Chemistry**
A Guide for Selection and Use
Editor: A. Zschunke

41 **Organic Electronic Materials**
Conjugated Polymers and Low-Molecular-Weight Organic Solids
Editors: G. Grosso and R. Farchioni

42 **Raman Scattering in Materials Science**
Editors: W. H. Weber and R. Merlin

43 **Fundamentals of Crystal Growth**
By B. Mutaftschiev

Series homepage – http://www.springer.de/phys/books/ssms/

Volumes 1–26 are listed at the end of the book.

Willes H. Weber Roberto Merlin (Eds.)

Raman Scattering in Materials Science

With 256 Figures

Springer

Dr. Willes H. Weber
Principal Research Scientist
Physics Department MD3028/SRL
Ford Motor Company
Dearborn, MI 48121-2053
USA

Prof. Roberto Merlin
Department of Physics
500 East University
University of Michigan
Ann Arbor, MI 48109-1120
USA

Series Editors:

Prof. Alex Zunger
NREL
National Renewable Energy Laboratory
1617 Cole Boulevard
Golden Colorado 80401-3393, USA

Prof. Robert Hull
University of Virginia
Dept. of Materials Science and Engineering
Thornton Hall
Charlottesville, VA 22903-2442, USA

Prof. R. M. Osgood, Jr.
Microelectronics Science Laboratory
Department of Electrical Engineering
Columbia University
Seeley W. Mudd Building
New York, NY 10027, USA

Prof. H. Sakaki
Institute of Industrial Science
University of Tokyo
7-22-1 Roppongi, Minato-ku
Tokyo 106, Japan

Library of Congress Cataloging-in-Publication Data

Raman scattering in materials science / Willes H. Weber, Roberto Merlin (eds.).
p. cm. -- (Springer series in materials science, ISSN 0933-033X ; 42)
Includes bibliographical references and index.
ISBN 3540672230 (alk. paper)
1. Raman spectroscopy. 2. Materials--Spectra. I. Weber, Willes H., 1942- II. Merlin, R. (Roberto), 1950- III. Springer series in materials science ; v. 42.

QC454.R36 R3385 2000
535.8'46--dc21

00-039469

ISSN 0933-033x

ISBN 3-540-67223-0 Springer-Verlag Berlin Heidelberg New York

Springer-Verlag Berlin Heidelberg New York
a member of BertelsmannSpringer Science+Business Media GmbH

Printed in Germany

Typesetting: Data conversion by LE-TEX, Leipzig
Cover concept: eStudio Calamar Steinen
Cover production: *design & production* GmbH, Heidelberg

Printed on acid-free paper SPIN: 10691138 57/3141/tr 5 4 3 2 1 0

Preface

The idea for this book began about two years ago when one of us (WHW) organized a focused session at the March Meeting of the American Physical Society on the applications of Raman scattering in materials characterization. The response to that session was overwhelming; we had twice as many papers as we anticipated, and it became clear at that time that Raman spectroscopy was becoming a standard tool for the materials scientist. Raman scattering has evolved from a technique practiced by dedicated laser physicists working in dark laboratories to a general-purpose characterization tool routinely applied to a wide range of materials science problems. With the addition of fiber optics, Raman probes are now in use for monitoring thin-film deposition processes and for such practical tasks as sorting waste polymers for recycling. There are two reasons for this evolution. First, the new generation of Raman instruments, with such improvements as array detectors, turnkey lasers, single-stage high-throughput spectrometers and holographic notch filters to reject Rayleigh light, made the implementation of Raman scattering simpler and less expensive. Second, many scientists in the field realized that the sort of detailed information obtainable from Raman scattering measurements could not be obtained as easily or at all with any other methods.

The purpose of the book is to provide a link between the fields of materials science and Raman spectroscopy. The first chapter gives an overview of the theory of Raman scattering in solids, with experimental examples drawn from well-known materials to illustrate most of the basic concepts. This chapter avoids rigorous mathematical treatments, but provides ample references to where such treatments can be found. The second chapter discusses the tremendous improvements in Raman instrumentation that have been made in the last decade and the impact this has had on the widespread use of Raman scattering. The remaining eight chapters focus on specific materials systems that comprise the mainstream of current research in materials science: bulk and alloy semiconductors; semiconductor heterostructures; high-T_c superconductors; catalysts materials; III-V nitrides; fullerenes, nanotubes, and other inorganic carbon-based materials; polymers; and manganites. These chapters form the heart of the book, and they provide extensive examples of Raman applications to current materials science problems. To augment these chapters, we have added short contributions on related subjects of historical or topical interest. These short contributions are referred to as "boxes" and are generally invited by the author(s) of the chapter after which each appears. They range from an anecdotal account of a meeting with C.V. Raman to a purely technical account of strain mapping in semiconductor devices. The resulting book is the most thorough collection of Raman applications in materials science ever assembled in one volume.

April 2000
Ann Arbor

Willes H. Weber
Roberto Merlin

Contents

List of Contributors

Sanford A. Asher
Department of Chemistry,
University of Pittsburgh
Pittsburgh, PA 15260, USA

Leah Bergman
Department of Physics
North Carolina State University
P.O. Box 8202
Raleigh, NC 27695, USA

Richard W. Bormett
Renishaw, Inc.
623 Cooper Court
Schaumburg, IL 60173, USA

Elias Burstein
Department of Physics
and Astronomy,
University of Pennsylvania
Philadelphia, PA 19104-6396, USA

Manuel Cardona
Max-Planck-Institut
für Festkörperforschung
Heisenbergstr. 1
70569 Stuttgart, Germany

Gene Dresselhaus
Francis Bitter Magnet Laboratory
Massachusetts Institute
of Technology
Cambridge, MA 02139, USA

Mildred Dresselhaus
Department of Electrical
Engineering and Computer Science
and Department of Physics
Massachusetts Institute
of Technology
Cambridge, MA 02139, USA

Mitra Dutta
US Army Research Office, US Army
Research Laboratory
P.O. Box 12211
Research Triangle Park,
NC 27709-2211, USA

P.C. Eklund
Department of Physics and
Astronomy and Center for Applied
Energy Research
University of Kentucky
Lexington, KY 40506-0055, USA

A. Fainstein
Instituto Balseiro and
Centro Atómico Bariloche
8400 S.C. de Bariloche, Argentina

Daniel Gammon
Code 6876
Naval Research Laboratory
4555 Overlook Ave. SW
Washington DC 20375, USA

Marcos Grimsditch
Materials Science Division, Bldg. 223
Argonne National Lab
Argonne, IL 60439, USA

Shaw Ling Hsu
Polymer Science & Engineering
University of Massachusetts
Conte Bldg.
Amherst, MA 01003, USA

B. Jusserand
Concepts and Devices for Photonics (CDP) Lab
France Télécom/CNET/DTD
and CNRS/URA 250
196 Avenue Henri Ravera
92220 Bagneux, France

R.S. Katiyar
Department of Physics,
University of Puerto Rico
San Juan, PR 00931, USA

Miles V. Klein
Department of Physics
University of Illinois,
Urbana-Champaign
Urbana, IL 61801, USA

Samuel Krimm
Department of Physics and
Biophysics Research Division
University of Michigan
Ann Arbor, MI 48109, USA

David J. Lockwood
Institute for
Microstructural Sciences
National Research Council
of Canada
Ottawa, ON K1A 0R6, Canada

Jose Menéndez
Department of Physics
Arizona State University
Tempe, AZ 85287-1504, USA

Roberto Merlin
Department of Physics
University of Michigan
Ann Arbor, MI 48109-1120, USA

Robert J. Nemanich
Department of Physics
North Carolina State University
Raleigh, NC 27695-8202, USA

M.A. Pimenta
Departamento de Fiscia
Universidade Federal
de Minas Gerais
Belo Horizonte, 30123-970, Brazil

Aron Pinczuk
Dept. of Physics
Dept. of Appl. Phys. & Appl. Math.
Columbia University, New York,
NY 10027
Lucent Technologies Bell
Laboratories
Murray Hill, NJ 07974-0636, USA

Vyacheslav B. Podobedov
Optical Technology Division
National Institute of Standards
and Technology
Gaithersburg, MD 20899, USA

A.K. Ramdas
Department of Physics
Purdue University
West Lafayette, IN 47907-1396, USA

Israel E. Wachs
Zettlemoyer Center for Surface
Studies and Department of Chemical
Engineering
Lehigh University
Bethlehem, PA 18015, USA

Alfons Weber
Optical Technology Division
National Institute of Standards
and Technology
Gaithersburg, MD 20899, USA

Willes H. Weber
Physics Department, MD3028/SRL
Ford Motor Company Research Lab
Dearborn, MI 48121-2053, USA

Ingrid de Wolf
IMEC
Kapeldreef 75
3001 Leuven, Belgium

1 Overview of Phonon Raman Scattering in Solids

R. Merlin, A. Pinczuk, and W.H. Weber

Abstract. This chapter provides a short review of the concepts underlying inelastic light scattering by phonons in solids. The discussion introduces the basic scattering mechanisms and the nomenclature used in the Raman community, but it avoids mathematical details. We give extensive references to the literature on topics that delve more deeply into theoretical issues. In addition, selected experimental results, obtained primarily from simple, well-known materials such as C, Si or GaAs, are shown to illustrate Raman spectroscopy applications to materials science.

The term Raman scattering is historically associated with the scattering of light by optical phonons in solids and molecular vibrations. In this book, the term refers to inelastic scattering by most elementary excitations associated with degrees of freedom of ions and electrons in crystalline and amorphous solids. The only exceptions are long wavelength acoustic phonons (sound waves) and acoustic magnons, which are identified with Brillouin scattering. Inelastic scattering processes are two-photon events that involve the simultaneous annihilation of an incident photon and the creation of a scattered photon [1]. If the frequency of the latter, ω_S, is smaller than that of the former, ω_L, a quantum of energy $\hbar(\omega_L - \omega_S)$ is added to the scattering medium and the event is referred to as a *Stokes* process (here, we use ω_L to designate the incident photon frequency, since the incident beam is invariably generated by a laser source). If, instead, $\omega_S > \omega_L$, we have an *anti-Stokes* process, where an elementary excitation of the medium is annihilated. For systems in thermal equilibrium, the anti-Stokes intensity depends strongly on temperature, since these processes can occur only when the medium is not in its ground state. The dominant form of Raman scattering, first-order scattering, involves a single quantum of excitation in the medium. However, it is not uncommon for materials to show strong higher-order processes leading to the creation or annihilation of two or more quanta.

Close to 75 years after the phenomenon was discovered [2], Raman scattering has become one of the most versatile spectroscopic tools to study the low-lying excitations of condensed matter systems. The group of excitations that can be accessed in Raman experiments is large and is growing as different areas of condensed matter science evolve. In solid-state media, it includes phonons, magnons, and impurity vibrational modes as well as the elementary excitations of bulk and low-dimensional electronic systems. While the focus of this chapter is on Raman scattering by phonons, the conceptual discussion of conservation laws and selection rules applies, with few modifications, to electronic excitations as well. For a general description of electronic Raman scattering, particularly for doped semiconductors, we refer the reader to Chap. 7 of the book by Hayes and Loudon [3], and the review by Klein [4]. Magnetic scattering is discussed by Cottam and Lockwood [5] and in Chap. 6 of [3]. Extensive information on the various forms of Raman scattering and a list of original references can be found in the proceedings of topical conferences [6–10] and the series *Light Scattering in Solids* [11–18].

The strongest inelastic light scattering processes are due to coupling of light to the electric moments of the scattering medium [3,19–22]. Light scattering processes in which light couples to magnetic moments are by far too weak to interpret experiments. Further, the intensities of scattering processes in which there is direct coupling of light to the motion of the ions are negligible [23]. At the relatively high frequency of light employed in current experiments, the dominant contributions to electric moments are due to excitation of electrons across energy bandgaps (these are the so-called *interband* transitions in solids) [19,22]. The coupling of incident and scattered light to the medium may be understood as the modulation of the electric susceptibility by elementary excitations [19]. It is well known that coupling of light to optical transitions of the scattering medium is enhanced when ω_{L} and ω_{S} are close to interband gaps. Such optical resonances result in large enhancements of Raman scattering cross-sections and intensities [24–26]. Events involving light at frequencies ω_{L} and ω_{S} that are close to interband transition energies are referred to as resonant Raman processes (see Sect. 1.2).

The reader may find excellent classical, i.e., macroscopic presentations of the derivation of the Raman cross-section intensities in [1,3]. In classical as well as in quantum-mechanical descriptions [19,22,24–26], the intensities are calculated as *differential cross-sections*, which represent the rate at which energy is removed from the incident beam. The calculations of the cross-sections incorporate parameters that describe modulations of the electric-dipole density by elementary excitations of the scattering media. In classical descriptions, these modulations are represented by derivatives of the electric susceptibility. Macroscopic theories offer relatively simple, one could say intuitive, understandings of many light scattering phenomena [27]. With the use of the methods of group theory, macroscopic theories also yield symmetry-based selection rules for the polarization of the incident and scattered light [3,21,22].

Microscopic, that is quantum-mechanical, formulations are required in any attempt to describe resonant inelastic light scattering processes [24–26]. The microscopic description incorporates explicit interactions that account for the coupling between the photons and the electronic states of the material. The quantum formulation also requires precise consideration of the interactions between electrons and the elementary excitations of the media. In the specific case of phonons, these are the diverse forms of the electron–phonon interaction. Inelastic light scattering resonances in bulk semiconductors have been investigated in considerable detail [25,26,28]. In particular, resonant scattering by optical phonons and by electronic excitations are reviewed in [26], and specific examples are given in several of the following chapters.

The study of resonant effects continues to be of great interest because enhanced cross-sections enable Raman scattering observations of otherwise weak processes. Early research demonstrated resonant enhancement in the spectra of optical phonons [29–32] and excitations of an electron gas at semiconductor surfaces [33]. In the latter case, light scattering occurs within ultrathin layers of thickness that in some instances are 10 nm or even smaller. The enhanced cross-sections revealed in this research played a key role in seminal light scattering studies of vibrational modes in semiconductor superlattices and quantum wells [34–37], as discussed in Chap. 3. Resonances are also crucial in studies of *low*-dimensional electron systems in semiconductor heterostructures and field-effect-devices [38–41]. Low dimensional electron systems that reside in semiconductor quantum structures are of great interest in contemporary materials science, device applications and fundamental physics.

1.1 Light Scattering Mechanisms and Selection Rules

1.1.1 Conservation Laws

Consider a monochromatic light beam of frequency ω_{L}. The magnitude of the propagation vector, $\boldsymbol{k}_{\mathrm{L}}$, is $|\boldsymbol{k}_{\mathrm{L}}| = \omega_{\mathrm{L}}\eta(\omega_{\mathrm{L}})/c$ where $\eta(\omega_{\mathrm{L}})$ is the refractive index. Due to the scattering, a fraction of the incident photons are annihilated with creation of a scattered field in which the photons have frequency ω_{S}. The propagation vector of the scattered light is $\boldsymbol{k}_{\mathrm{S}}$, with $|\boldsymbol{k}_{\mathrm{S}}| = \omega_{\mathrm{S}}\eta(\omega_{\mathrm{S}})/c$. We define the *scattering frequency* as

$$\omega = \omega_{\mathrm{L}} - \omega_{\mathrm{S}} \,, \tag{1.1}$$

and the *scattering wave vector* as

$$\boldsymbol{k} = \boldsymbol{k}_{\mathrm{L}} - \boldsymbol{k}_{\mathrm{S}} \,. \tag{1.2}$$

Inelastic scattering must satisfy conservation of energy and momentum. To consider these conservation laws we recall that in perfect crystals, that is, in idealized materials that display perfect translation symmetry, the elementary excitations can be labeled by the wave vector $\boldsymbol{q}$, also known as the crystal momentum [42]. These modes are represented by a dispersion relation that specifies a frequency ω_q for each value of $\boldsymbol{q}$ [42]. In first-order processes, only a single elementary excitation participates. In such situations, momentum conservation translates into the requirement that the scattering wave vector equal the wave vector of the excitation:

$$\boldsymbol{k} = \boldsymbol{q} \; . \tag{1.3}$$

Similarly, conservation of energy leads to

$$\omega = \omega_q . \tag{1.4}$$

In higher-order scattering processes ω becomes the sum of the frequencies of two or more quanta for which the total wave vector is $\boldsymbol{k}$ (see Sect. 1.3).

1.1.2 Kinematics: Wave Vector Conservation

Processes that conserve crystal momentum obey (1.3). The magnitude and orientation of the scattering wave vector are determined by the geometry of the scattering experiment. Figure 1.1 shows three standard arrangements for the propagation of incident and scattered beams. The smallest and the largest scattering wave vector are obtained, respectively, in the forward ($\theta = 0°$) and the backscattering ($\theta = 180°$) geometries. In the forward case, the magnitude of the scattering wave vector is

$$|\boldsymbol{k}_{\min}| = [\eta(\omega_{\mathrm{L}})\omega_{\mathrm{L}} - \eta(\omega_{\mathrm{S}})\omega_{\mathrm{S}}]\,(1/c) \; , \tag{1.5}$$

whereas, in the backscattering configuration, we have

$$|\boldsymbol{k}_{\max}| = [\eta(\omega_{\mathrm{L}})\omega_{\mathrm{L}} + \eta(\omega_{\mathrm{S}})\omega_{\mathrm{S}}]\,(1/c) \; . \tag{1.6}$$

It follows from (1.6) that for typical experiments, in the visible and near infrared, $\boldsymbol{k}_{\max} \leq 10^6\,\mathrm{cm}^{-1}$. This wave vector is much smaller, by about two orders of magnitude, than the value corresponding to the Brillouin zone boundary

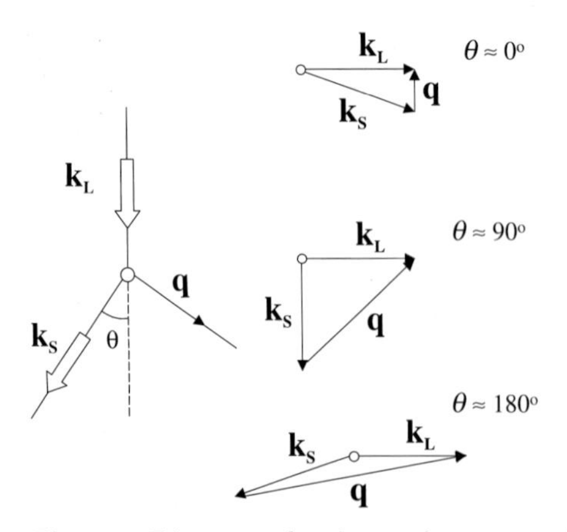

Fig. 1.1. Diagrams showing various scattering geometries

of typical crystals [28]. Thus, first-order processes that conserve wave vector access only elementary excitations at or near the center of the Brillouin zone. For higher-order processes, wave vector conservation is written as

$$\boldsymbol{k} = \sum_j \boldsymbol{q}_j \,, \tag{1.7}$$

where the summation is carried out over all the elementary excitations that take part in the scattering. In the vector sum of (1.7), the individual excitation wave vectors, $\boldsymbol{q}_j$, can range from zero to the values at the Brillouin zone boundary. Inelastic light scattering spectra obtained under conditions of wave vector conservation are thus of two kinds. In first-order, the spectra display a discrete set of peaks that are associated with elementary excitations at the center of the zone, whereas higher-order spectra give continua that probe modes of wave vectors that span the whole Brillouin zone of the crystal. In second- and higher-order processes, the prominent spectral features are related to structure in the density of states of the respective modes. These are the so-called critical points, that is, points at which $(\partial\omega_q/\partial\boldsymbol{q}) = 0$ [42].

1.1.3 Kinematics: Breakdown of Wave Vector Conservation

Wave vector conservation as represented by (1.3) and (1.7) breaks down when the medium has no translation symmetry (linear momentum is, of course, always conserved). In the language of quantum mechanics, we say that the Bloch theorem does not apply and, hence, that wave vectors are not good quantum numbers for labeling modes [42]. This applies to mildly imperfect crystals with a small concentration of defects as well as to solid solutions, alloys and amorphous solids (glasses). In all these cases, the Raman spectra are expected to display features reflecting the density of states of the particular excitation (see Sects. 1.4–1.6). We note that inelastic light scattering studies of imperfect or non-crystalline samples play a fundamental role in the characterization and materials science of such systems.

The condition of conservation of wave vector has to be considered with some care in the case of quantum structures such as artificial multilayer systems that display size quantization along one or more directions [43,44]. Size quantization, or confinement, occurs when there is a small characteristic length, so that along that direction the motion is quantized into distinct energy levels, as in a quantum well. For electrons, typical lengths are in the nanometer range, and these structures are often referred to as *nanostructures*. In quantum structures, the crystal momentum may remain as a good quantum number for components along directions orthogonal to that of the size quantization. For example, when size quantization occurs along a single direction, the wave vector is a good quantum number for the other two (orthogonal) directions and the system displays two-dimensional behavior. Confinement may also involve two directions in space and, in this case, the quantum structures known as quantum-wires exhibit one-dimensional behavior. Finally, in the so-called zero-dimensional case of a quantum-dot, wave vectors cannot be used to represent states.

Wave vector conservation as described by (1.3) also breaks down in strongly absorbing media. In the presence of optical absorption, which necessarily occurs in experiments carried out under resonance conditions, the refractive indexes $\eta(\omega_\mathrm{L})$ and $\eta(\omega_\mathrm{S})$ are complex. The wave vectors $\boldsymbol{k}_\mathrm{L}$ and $\boldsymbol{k}_\mathrm{S}$ are also complex, having imaginary parts given by $\mathrm{Im}(\boldsymbol{k}_{\mathrm{L,S}}) = \omega_{\mathrm{L,S}}\kappa(\omega_{\mathrm{L,S}})/c$, in which $\kappa(\omega_\mathrm{L})$ and $\kappa(\omega_\mathrm{S})$ are the extinction coefficients at the two frequencies [42]. We recall that in solids with translation symmetry the wave vector $\boldsymbol{q}$ is a continuous, real variable. Thus, the condition of wave vector conservation (1.3) cannot be exactly fulfilled in absorbing media. Instead, a range $\Delta\boldsymbol{q}$ of phonon wave vectors can be excited. The range is defined by the magnitude of the extinction coefficients [19], and we may write:

$$\Delta\boldsymbol{q} \approx \mathrm{Im}(\boldsymbol{k}_\mathrm{L}) + \mathrm{Im}(\boldsymbol{k}_\mathrm{S})\ . \tag{1.8}$$

The inelastic light scattering intensities will still have a major peak from the zone-center phonons, but there will be additional contributions from modes extending over a range of wave vectors defined by (1.8). A similar picture of the breakdown in wave vector conservation applies to partially disordered, or microcrystalline, materials in which there is absence of long-range order. In this case one defines a correlation length (or average crystallite size) ℓ, over which the material shows good crystalline order. Thus, the range of phonon wave vectors that contribute to the Raman line shape is determined by the crystallite size and is given by $\Delta\boldsymbol{q} = 2\pi/\ell$ [45].

1.1.4 Light Scattering Susceptibilities

Stokes and anti-Stokes radiation are created by a fluctuating electric-dipole set up in the scattering medium by the simultaneous action of the incident light beam and the elementary excitations of the solid [3,19,21,22]. The effect may be understood as that from an induced polarization $\boldsymbol{P}$ (the dipole moment per unit volume) that oscillates at the frequency ω_S. We mentioned earlier that this polarization is represented by a modulation of the electric susceptibility of the medium induced by elementary excitations of frequency ω_q. It is then natural to introduce a modulated Raman susceptibility $\delta\chi_{ij}$, defined by [3,19]

$$P_i(\omega_\mathrm{S}) = \sum_j \delta\chi_{ij} E_j(\omega_\mathrm{L})\ , \tag{1.9}$$

where $\boldsymbol{E}(\omega_\mathrm{L})$ is the electric field of the incident beam. In its most general form this susceptibility is a function of ω_L, ω_S, and also of the wave vectors of the light and the elementary excitations. The modulated susceptibility is a second rank tensor with non-zero components determined by the symmetries of the scattering medium and of the elementary excitations [3,19]. The scattering intensity, proportional to $|\boldsymbol{P}(\omega_\mathrm{S})|^2$ [46], becomes

$$I(\omega_\mathrm{L}, \omega_\mathrm{S}, \boldsymbol{k}) \propto \left|\hat{\boldsymbol{e}}_\mathrm{S} \cdot \delta\chi_{ij} \cdot \hat{\boldsymbol{e}}_\mathrm{L}\right|^2 , \tag{1.10}$$

where $\hat{\boldsymbol{e}}_\mathrm{S}$ and $\hat{\boldsymbol{e}}_\mathrm{L}$ are the unit polarization vectors of the scattered and incident beams, respectively. It is remarkable that an expression as simple as this one can be used to gain significant insights into inelastic light scattering phenomena using a purely phenomenological framework. Here, $\delta\chi_{ij}$ is written as a series expansion in powers of coordinates that represent the elementary excitations of the scattering medium, and (1.10) is used to determine the selection rules for the polarization of the incident and scattered light. As an example we consider optical vibrations. In infinite, perfect, crystalline media these modes are represented by plane waves that are the Fourier components of the motion of the ions. Explicitly, the displacement $\boldsymbol{U}$ of a particular ion in the unit cell at $\boldsymbol{r}$ is given by [1]

$$U_\mathrm{S}(\boldsymbol{r}, t) \propto Q_m(\omega, \boldsymbol{q})\mathrm{e}^{\mathrm{i}(\boldsymbol{q}\cdot\boldsymbol{r}-\omega t)} , \tag{1.11}$$

where $Q_m(\omega, \boldsymbol{q})$ is the phonon coordinate or, alternatively, the amplitude of the mode of frequency ω and wave vector $\boldsymbol{q}$ belonging to the mth-branch. In a quantum mechanical description, phonons are the quanta of the displacement fields associated with the harmonic oscillations. While the analysis of phonon symmetries generally requires the full space group of the crystal [47–49], the prescription for enumerating Raman-active modes and their associated scattering tensors is considerably simpler for it involves only the representations of the point group [3]. For instance, for the centrosymmetric silicon with the diamond structure, the group is O_h and the optical phonons at the center of the Brillouin zone are triply degenerate and transform like the *gerade* representation F_{2g}. We will use Schoenfliess' notation for the point group and Mulliken's for the representation symmetry, except that F will be used for triply degenerate modes, instead of T. This minor change in notation is now widely used since it is consistent with the sequence E, F, G, and H being used to identify 2-fold, 3-fold, 4-fold, and 5-fold degenerate modes, respectively. In the case of zincblende crystals such as GaAs, which belong to the point group T_d with no center of inversion, the long-wavelength optical modes are triply degenerate (this, ignoring the splitting between transverse and longitudinal modes due to long ranges electrostatic forces; see later) and transform like a vector corresponding to the F_2 representation. A full group theoretical treatment is beyond the scope of this chapter. The interested reader may find some guidance in [3,47–49].

In the phenomenological description, the modulated susceptibility is written as

$$\delta\chi_{ij} = \sum_m R_{ij}^{(m)} Q_m(\boldsymbol{0}) + \sum_{mn,\boldsymbol{qp}} R_{ij}^{(mn)} Q_m(\boldsymbol{q}) Q_n(\boldsymbol{p}) + \dots , \tag{1.12}$$

where the *second-rank* tensors $R_{ij}^{(m)}$ and $R_{ij}^{(mn)}$ are referred to as the first- and second-order Raman tensors, respectively. Here, the first term in the right hand side has a single phonon coordinate at $\boldsymbol{q} \approx 0$ and represents first-order scattering. The next term has two phonon coordinates and represents second-order scattering (from (1.7), we have that $\boldsymbol{q} + \boldsymbol{p} \approx 0$). In this fashion we generate expressions that describe further higher-order scattering events. The expansion of the modulated susceptibility is not limited to lattice vibrations (for plasma waves, which are the collective excitations of an electron gas, see [43,50]) or to the wave vector representation. In fact, there are situations where it is more convenient to consider expansions in terms of the ion displacements $\boldsymbol{U}$ or other local variables such as the ion spin. The latter is widely used to describe light scattering by magnons [3,5].

Equations (1.10) and (1.12) are employed in calculations of inelastic light scattering intensities. If the first term in the right side of (1.12) does not vanish for a particular mode, the mode is, by definition, Raman *active* or Raman *allowed.* Equation (1.10) is then used to obtain selection rules for the polarization of the incident and scattered light. Because Raman tensors are of second rank (like the linear dielectric tensor), they transform like the product of two vectors. Their non-vanishing components have been obtained for all the crystal classes long ago [21]. A list of all the Raman tensors and their symmetries is given, e.g., in [3,26]. Raman-allowed modes are those that transform according to one of the symmetries of second-rank tensors and scattering processes that obey the proper selection rules are referred to as *allowed* scattering. These principles apply to all elementary excitations of ions and of electrons. In addition, and as for the dielectric tensor, Raman tensors of vibrational modes are symmetric, i.e., $R_{ij} = R_{ji}$, except near resonances where they can have anti-symmetric components [3]. Hence, strictly antisymmetric phonons are forbidden in off-resonance scattering. This does not apply to excitations that are not invariant under time-reversal, such as magnons and spin-flip transitions, for which the Raman tensors are predominantly antisymmetric [3,5].

An important case is that of materials possessing a center of inversion. Here, the Raman tensor is of *even* symmetry and, accordingly, *odd* symmetry modes are Raman inactive. This applies in particular to the so-called *infrared-active* modes, which carry a dipole moment and couple directly to light. Infrared-active modes are *odd* because the dipole moment, being a vector, changes sign under inversion. It follows that the Raman allowed modes of centrosymmetric materials are infrared inactive and vice versa. Depending on the crystal structure, materials with a center of inversion may exhibit only Raman-active or infrared-active phonons, a combination of both as well as modes that are *silent*: neither Raman nor infrared active (e.g., the A_{2g} and B_{1u} optical phonons in the rutile structure; see Sect. 1.1.5). For instance, the triply degenerate optical mode of the diamond structure is infrared-forbidden while rocksalt crystals (e.g., NaCl) show only infrared-active modes [47].

To continue with examples of first-order Raman scattering we return to the case of the zincblende structure [47], where the optical modes at $\boldsymbol{q} = 0$ are described by the components of a vector. Consider now the mode for which the displacements are along the z direction. The table of [3] indicates that the corresponding first-order Raman tensor is

$$R_{ij}^{z} = \begin{bmatrix} 0 & d & 0 \\ d & 0 & 0 \\ 0 & 0 & 0 \end{bmatrix} . \tag{1.13}$$

This expression offers an example of the determination of polarization selection rules. The tensor specifies that in a coordinate system aligned with the crystal axes, Raman scattering occurs when the incident and scattered polarization are orthogonal to the phonon displacement coordinate, perpendicular to each other and lie in the x-y plane (note that, for these polarizations, the phonon displacements along the x and y directions do not contribute to the scattering). The short-hand notation for indicating this geometry for, say, backscattering from a (001) surface is $z(xy)\bar{z}$, following the Porto convention [51]. The first and last symbols indicate the propagation directions and the second and third symbols indicate the polarization directions of the incident and scattered beams, respectively.

Using polarized measurements from oriented single crystals it is usually possible to identify unequivocally the irreducible representations for all allowed Raman modes of a material. In powders, polycrystalline films, and glassy polymers polarization measurements can also give information about the properties of the Raman tensors, but the information is much more limited. In these cases one measures the depolarization ratio ϱ

$$\varrho = \frac{\boldsymbol{I}(\hat{\mathbf{e}}_{\mathrm{S}} \perp \hat{\mathbf{e}}_{\mathrm{L}})}{\boldsymbol{I}(\hat{\mathbf{e}}_{\mathrm{S}} \| \hat{\mathbf{e}}_{\mathrm{L}})} , \tag{1.14}$$

which is the intensity ratio obtained when the scattered light polarization is alternatively perpendicular and parallel to the incident polarization. Totally symmetric vibrational modes, i.e., those that maintain the symmetry of the crystal and whose scattering tensors have only diagonal elements, have smaller depolarization ratios than the other modes. A strongly *polarized* mode is one with a small value of ϱ, a strongly *depolarized* mode has a large value of ϱ. In general, $0 \leq \varrho < 3/4$, with the maximum value being obtained for those modes with only off-diagonal matrix elements [52].

1.1.5 Enumeration of Raman Active Modes

Before attempting to analyze Raman spectra from a material, one would first want to identify the number of Raman-allowed modes and their symmetries (i.e., their scattering tensors). This is a straightforward task, if the complete crystal structure of the material is known. The procedure we have found most useful for accomplishing this task is called the *nuclear site group analysis* method, as discussed in detail by Rousseau et al. [53], who also give extensive tables used to carry out the procedure. The first step is to identify the crystal structure, i.e., the *space group*, and the precise location of each atom in the unit cell. Complete descriptions of all space groups and the allowed atomic positions within the unit cell can be found in the *International Tables for Crystallography* [54]. This volume also defines the Wyckoff notation [55], used to designate the various possible atomic sites, and the site symmetries as given in Tables A of [53]. The second step is to determine the irreducible representations of the zone-center phonons that result from each of the non-equivalent, occupied sites in the unit cell (Tables B of [53]). The third and final step is to identify and determine the scattering tensors for the irreducible representations that are Raman allowed (Tables E of [53]). Two simple examples serve to demonstrate the method.

Consider again semiconductors such as Si or Ge that crystallize in the diamond structure, of space group O_h^7. There are eight equivalent atoms in the cubic (non-primitive) unit cell, and they occupy the $8a$ sites in Wyckoff's notation [55]. According to Table 32A of [53], these have T_d site symmetry. From Table 32B of [53], the irreducible representations of the zone-center phonons (at the Γ-point of the Brillouin zone) resulting from atoms occupying the $8a$ sites are

$$\Gamma = F_{1u} + F_{2g} \; . \tag{1.15}$$

From Table 32E of [53], the F_{2g} mode is Raman active. The table also shows that translations transform like F_{1u}. Therefore, this symmetry corresponds to acoustic modes. Both representations are triply degenerate, and together they account for all six degrees of freedom required for a structure with two atoms in the primitive unit cell. Furthermore, the three Raman tensors for the F_{2g} modes are symmetric matrices with only xy, xz, or yz components, respectively, all of which are equal.

As a second example, consider the rutile structure, which is found for a number of metal dioxides (e.g., TiO_2, RuO_2, IrO_2, RhO_2, OsO_2) and difluorides (e.g., CoF_2, MgF_2, NiF_2, MgF_2). The space group for the rutile structure is D_{4h}^{14}, and there are two formula units per primitive unit cell. The cations occupy $2a$ sites and the anions $4f$ sites [55], which (from Table 15A of [53]) have site symmetries of D'_{2h} and C'_{2v}, respectively. From Table 15B of [53] the irreducible representations are $A_{2u} + B_{1u} + 2E_u$ and, $A_{1g} + A_{2g} + A_{2u} + B_{1g} + B_{1u} + B_{2g} + E_g + 2E_u$, for the $2a$ and $4f$ sites, respectively, giving the result

$$\Gamma = A_{1g} + A_{2g} + 2A_{2u} + B_{1g} + 2B_{1u} + B_{2g} + E_g + 4E_u \; . \tag{1.16}$$

Finally from Table 15E [53] we identify the four Raman-active modes whose scattering matrices have the forms:

$$A_{1g}:\begin{bmatrix} a & 0 & 0 \\ 0 & a & 0 \\ 0 & 0 & b \end{bmatrix}, \quad B_{1g}:\begin{bmatrix} c & 0 & 0 \\ 0 & -c & 0 \\ 0 & 0 & 0 \end{bmatrix}, \quad B_{2g}:\begin{bmatrix} 0 & d & 0 \\ d & 0 & 0 \\ 0 & 0 & 0 \end{bmatrix}, \tag{1.17}$$

and

$$E_g:\begin{bmatrix} 0 & 0 & e \\ 0 & 0 & 0 \\ e & 0 & 0 \end{bmatrix}, \begin{bmatrix} 0 & 0 & 0 \\ 0 & 0 & e \\ 0 & e & 0 \end{bmatrix}.$$

Note that there are 18 total zone-center vibrational modes in (1.16), counting the doubly degenerate E modes twice, as required for 6 atoms per primitive unit cell. Table 15E also indicates that translations and, hence, acoustic modes transform like $A_{2u} + E_u$ and that the optical modes of symmetries A_{2g} and B_{1u} are silent.

There are two fairly common situations in which the selection rules as derived above for first-order scattering may go wrong. The first of these occurs when a Raman-allowed mode is also infrared active, which can only happen in crystals that lack inversion symmetry. Here, the problem is not the selection rules themselves but the fact that, due to the long-range nature of the Coulomb field associated with infrared-active vibrations, the frequency of a given mode depends on the direction of $\boldsymbol{q}$ [3]. In zincblende materials, this leads to a splitting of the triply-degenerate optical mode at $\boldsymbol{q} \approx 0$ into a transverse doublet (TO) and a longitudinal singlet (LO) (in non-cubic structures, the dependence of the frequency on the wave vector is complicated by the competition between electrostatic forces and crystal anisotropy; see [3]). Given that the Raman tensor does not depend on $\boldsymbol{q}$ a particular scattering geometry may give either TO or LO modes. This is discussed in detail in [3]. The second situation in which the simple selection rules may break down is under resonant conditions. This type of scattering is known as *forbidden* Raman scattering. Resonant *forbidden* Raman scattering can be much stronger than resonant *allowed* scattering. We consider a particular example, namely, forbidden LO scattering in Sect. 1.2.

1.1.6 Stokes and Anti-Stokes Scattering Intensities

Equations (1.10) and (1.12) enable simple evaluations of Raman scattering intensities. We consider the case of first-order Stokes scattering, for which the scattered intensity can be written as

$$I_{ij} \propto \left| R_{ij}^{(m)} \right|^2 \left| \langle n+1 | Q_m | n \rangle \right|^2 , \tag{1.18}$$

where the Bose factor,

$$n(\omega, T) = 1/\left[\exp(\hbar\omega/k_{\mathrm{B}}T) - 1\right] , \tag{1.19}$$

gives the equilibrium population of phonons at temperature T [42] and $\langle n+1|Q_m|n\rangle$ is the matrix element for adding a phonon to the crystal. In the case of anti-Stokes processes, the corresponding matrix element is $\langle n-1|Q_m|n\rangle$. In the harmonic approximation, we have [56]:

$$\langle n|Q_m|n+1\rangle \sim (n+1)^{1/2} , \tag{1.20a}$$

and

$$\langle n|Q_m|n-1\rangle \sim n^{1/2} . \tag{1.20b}$$

These equations are used to obtain the ratio between Stokes and anti-Stokes intensities:

$$(I_{\mathrm{S}}/I_{\mathrm{AS}}) = \exp(\hbar\omega/k_{\mathrm{B}}T) . \tag{1.21}$$

The same expression applies, of course, to any (bosonic) excitation that can be described as a harmonic oscillator (e.g., plasmons) as well as to scattering by elementary excitations of an electron gas [3]. More generally, (1.21) applies to excitations for which the Stokes and anti-Stokes processes are related by time reversal. This is not the case of scattering by magnetic excitations such as magnons [5,57]. Departures from the behavior predicted by (1.21) also occur in resonant experiments when the separation between the Stokes and anti-Stokes frequencies is comparable to the width of the electronic resonance [22].

The thermal factors that enter in higher-order scattering processes require some careful consideration. Consider as an example the case of second-order Raman scattering by phonons. Here two modes of frequencies ω_1 and ω_2 are involved, with thermal occupation factors n_1 and n_2. Generally, there are two kinds of two-phonon processes: *sum* and *difference*. In sum-processes, the two modes are created in the Stokes- and both are annihilated in the anti-Stokes component. In the difference-case, one mode is created and the other one is annihilated for both Stokes and anti-Stokes. As a result, in sum processes the Stokes intensity is proportional to $(n_1+1)(n_2+1)$ while the anti-Stokes intensity is proportional to $n_1 n_2$. For a difference process with $\omega_1 > \omega_2$, the Stokes and anti-Stokes intensity are, respectively, proportional to $(n_1+1)n_2$, and $n_1(n_2+1)$. We leave it as an exercise for the reader to prove that in either case (sum or difference), the intensity ratio is still given by (1.21).

1.2 Resonant Light Scattering and Forbidden Effects

Resonant effects in inelastic light scattering represent a major field of study (see, e.g., [26]). Measurements of the (laser) frequency dependence of the Raman cross-section offer valuable insights into the physics of interband optical excitations of the material. Also, large resonant enhancements of the cross sections enable the observations of modes that may have very weak signals under non-resonant conditions. It is clear that understanding resonant enhancements is crucial for the design of experiments. Resonant cross sections can be calculated using the methods of time-dependent perturbation theory of quantum mechanics. In this section we present some of the basic ideas involved in the description of resonant processes.

We consider as an example the case of long wavelength optical phonons in a bulk semiconductor such as GaAs. First-order Raman scattering takes place as three distinct quantum events [19,21,22,24–26]: (i) The incident photon is annihilated with creation of an interband electron-hole pair, (ii) the interband excitation is scattered by the optical phonons and (iii) the scattered photon is created with annihilation of the interband electron-hole pair. This sequence of events forms the basis for calculating the modulation of the electric susceptibility of the scattering medium by the optical phonon. The events involving photons take place through the coupling of electrons with the radiation field while those that involve phonons rely on the electron-phonon interaction [3,21,22,24–26]. Figure 1.2 shows Feynman diagrams describing terms that enter in the calculation of the first-order Raman tensor by third-order time-dependent perturbation theory [3,21,22,24–26]. These band-diagrams and the associated Feynman diagrams represent the three quantum events that contribute to the Raman tensor of a semiconductor that has no electrons (holes) in the conduction (valence) band. The diagrams are useful in the discussion of resonant processes because they highlight the electron transitions in the three events. The numbers indicate the time-sequence of the events. In the context of time-dependent perturbation theory, there are contributions of other terms with a different time sequence. However, the diagrams shown are the ones that lead to the largest resonant enhancement.

In the diagrams of Fig. 1.2, the electron interband transitions involve states of energies E_α and E_β. These transitions are associated with photon

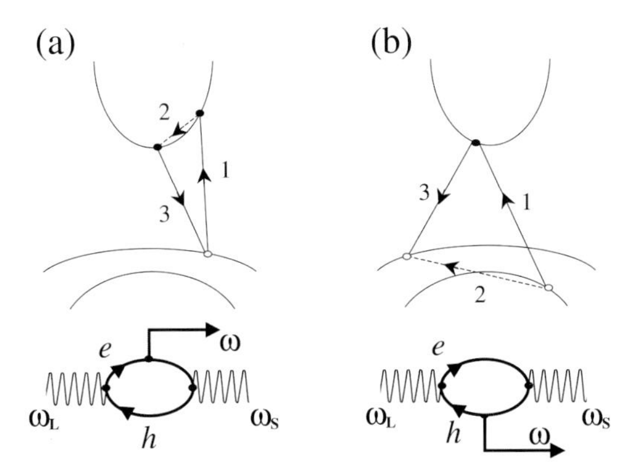

Fig. 1.2. Energy band diagrams and associated Feynman diagrams for light scattering processes (see text). Parts (**a**) and (**b**) show the electron and hole contribution to the scattering. The numbers indicate the order of the electronic transitions; e and h denote, respectively, electron and hole states

annihilation and creation. In Fig. 1.2a the electron-phonon coupling results in an *intraband* transition, while in Fig. 1.2b the *hole* undergoes an *interband* transition. The Raman tensor associated with these diagrams may be written as [3,21,22]

$$R_{ij}^{(m)} \approx \frac{C_{ij,m}}{(\hbar\omega_{\mathrm{S}} - E_\beta)(\hbar\omega_{\mathrm{L}} - E_\alpha)} \,. \tag{1.22}$$

The coefficient $C_{ij,m}$ involves the product of three matrix elements. Two are due to the electron-radiation Hamiltonian $\hat{H}_{\mathrm{R}}$ and the third one to the electron-phonon interaction $\hat{V}_{\mathrm{L}}$. The diagrams of Fig. 1.2a involve only two bands, and are called *two-band terms.* Those in Fig. 1.2b involve three bands, and are called *three-band terms.* Given that phonon energies are much smaller than those of interband transitions, that is $E_{\alpha,\beta} \gg \hbar\omega$, the two-band diagrams in Fig. 1.2a give the largest enhancement when the photon energy is in the vicinity of the gap [21,22,24–26].

The expansion (1.12) of the susceptibility in terms of the phonon coordinates suggests a relationship between resonant Raman scattering and modulation spectroscopy in that

$$R_{ij}^{(m)} \approx \partial\chi_{ij}/\partial Q_m \approx \frac{\partial\chi_{ij}}{\partial\omega} \times \frac{\partial\omega}{\partial Q_m} \,. \tag{1.23}$$

Within this approximation, the scattering resonances follow simply the frequency-dependence of the linear susceptibility probing the modulation of the electronic band structure by the atomic displacements [26]. A more thorough, but still simplified treatment shows that there is an additional contribution that reflects phonon-induced shifts of the optical oscillator strength (it gives a weaker resonance) and that $\partial\chi_{ij}/\partial\omega$ should be replaced by the finite difference $[\chi_{ij}(\omega_{\mathrm{L}}) - \chi_{ij}(\omega_{\mathrm{S}})]/\omega$. This latter expression often provides a reliable order-of-magnitude estimate of the resonant enhancement [24–26].

As mentioned earlier, larger cross sections are not the only outcome of resonant scattering; resonances often lead to the appearance of forbidden effects. A classic example is given by the work of Leite et al. [58] and Klein and Porto [59] on CdS. When the laser frequency is in the vicinity of the bandgap of CdS, and for $\hat{\mathbf{e}}_{\mathrm{S}}$and $\hat{\mathbf{e}}_{\mathrm{L}}$parallel to each other, a series of sharp Raman lines is observed, corresponding to multiple excitations of near zone-center LO phonons. As shown in Fig. 1.3, overtones of the 304 cm^{-1} LO phonon are seen out to the ninth order. If the laser frequency is slightly lower, only a broad luminescence is observed. Experimentally, the first-order LO scattering associated with the overtones cannot be accounted for by the expansion in (1.12). Theory shows that it is a *forbidden* effect that depends both on the phonon displacement and the finite scattering wave vector [60]. Overtones are observed for LO but not TO phonons because the longitudinal modes carry an electric field and, thus, they interact strongly with the carriers through the Fröhlich interaction [60]. Various models have been used to explain the

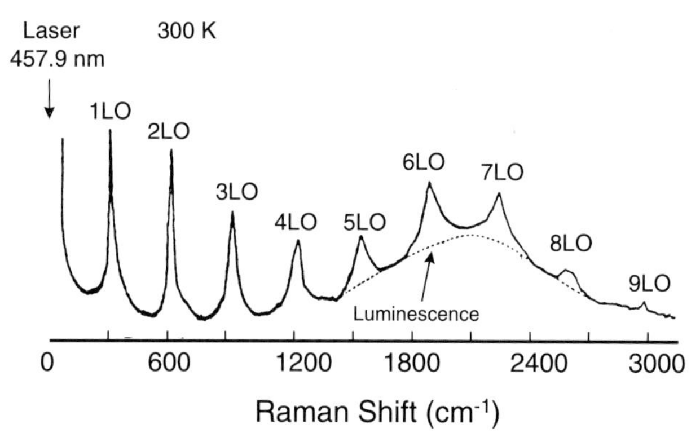

Fig. 1.3. Resonant Raman scattering showing multiple LO phonon excitation from CdS when excited with a 457.9-nm laser at room temperature. Adapted from [58]

large number of overtones including the 'cascade model where electrons cascade down the conduction band as they emit LO phonons [61] and the solid-state analog of the configuration coordinate model which provides a simple, molecular-like explanation for the occurrence of multiple scattering [62].

1.3 Two-Phonon Scattering

The energy and momentum conservation rules that apply to a two-phonon scattering event, given in (1.3) and (1.4), are quite simple. In contrast, the full theoretical treatment of the symmetry-based selection rules, which also apply, is quite complicated. One must first determine the symmetry of each phonon branch in terms of the irreducible representations of the group of the wave vector $\boldsymbol{q}$. These are then combined with the symmetries of the phonons at $-\boldsymbol{q}$ to obtain the irreducible representations for all possible two-phonon states. Only those two-phonon states whose symmetry correlates with at least one of the Raman-active symmetries for that point group will be allowed. For example, crystals belonging to the cubic O_h point group (e.g., diamond or fluorite structures) have Raman-active symmetries A_{1g}, E_g, and F_{2g}, and only those two-phonon states that belong to one of these will be allowed. The general procedure for solving this problem is given by Turrell [63], and several specific examples are discussed by others [47,64–66].

In practice the interpretation of two-phonon spectra is relatively simple. As pointed out by Cardona [26], the scattering is typically dominated by those phonon sum combinations whose irreducible representations contain the identity representation. Since all overtones (phonons from the same branch, but from opposite sides of the Brillouin zone) fall into this category, it is often reasonable to approximate the two-phonon Raman spectrum by the one-phonon density of states (DOS) with the frequency axis multiplied by two. We refer to this quantity as the 2ω-DOS. This is a rather severe approximation, but it tends to work well for many materials.

An example of the similarity between the second-order Raman spectrum and the calculated 2ω-DOS [67] is shown in Fig. 1.4 for a Si crystal. There is a weak feature observed near $620\,\mathrm{cm}^{-1}$, arising from a combination, that does not match structure in the 2ω-DOS, but the overall agreement is quite good. The frequency positions of all of the sharp structure in the spectrum associated with critical points in the phonon dispersion curves are given accurately by the 2ω-DOS, although the relative intensities of different spectral regions are not. Thus, the second-order spectrum is quite sensitive to the phonon dispersion curves throughout the Brillouin zone. Although it is not possible to determine the full dispersion curves from the second-order Raman spectrum, one can sometimes check the accuracy of various calculations of these curves, as has been done for example in the case of SiC [68]. This is a particularly useful application of Raman scattering, if the material being studied is not available in large enough single crystals to perform inelastic neutron scattering measurements of the dispersion curves.

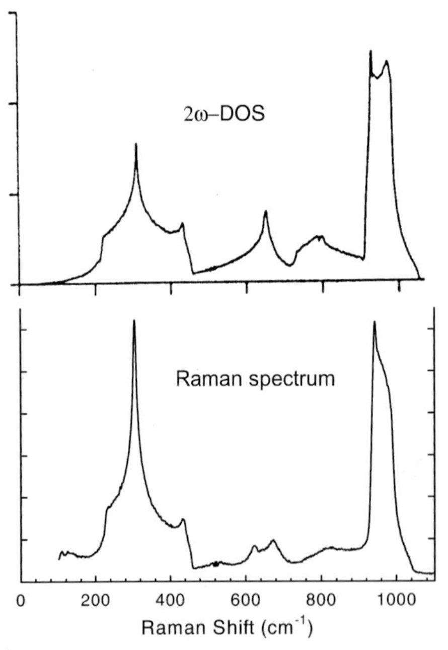

Fig. 1.4. Comparison between the Si 2ω-DOS, upper panel, as calculated by Weber [67] and the Raman spectrum of Si recorded in the $z(xx)\bar{z}$ configuration with 633-nm radiation from a (001) Si surface. The first-order mode at $521\,\mathrm{cm}^{-1}$, which is forbidden in this geometry but was observed due to polarization leakage, was artificially removed for clarity

1.4 Phonons in Semiconductor Alloys

The development of semiconductor alloy materials has led to a wide assortment of device applications that would have been much more limited, if not impossible, in pure semiconductor systems. The range of devices includes detectors, light emitting diodes and lasers, and high-speed transistors. Raman scattering has played a central role in the characterization of many of these material systems.

As an example we consider the $Si_{1-x}Ge_x$ alloy, which was one of the first and most thoroughly studied alloy system. These group IV elements have the same (diamond) crystal structure, and they are totally miscible, forming a random alloy over the full compositional range. The Raman spectrum is

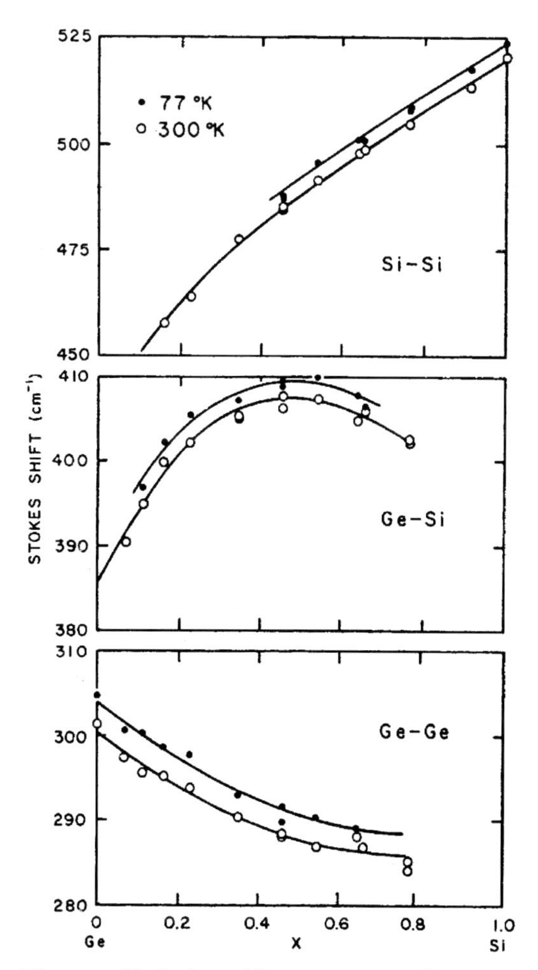

Fig. 1.5. Variation with composition of the Raman lines from bulk alloys of Si and Ge. From Renucci et al. [69]

dominated by three lines that can be loosely identified as Si–Si, Si–Ge, and Ge–Ge modes. The frequencies and intensities of these lines vary in a systematic way with composition, as shown in Fig. 1.5 [69]. The Si–Si and Ge–Ge lines approach the frequencies and intensities of the pure crystals at $x = 0$ ($520\,\mathrm{cm}^{-1}$) and $x = 1$ ($303\,\mathrm{cm}^{-1}$), respectively, while the Si–Ge line falls between these limits and has maximum intensity near the middle of the compositional range.

Another interesting aspect of Raman scattering in alloys is revealed by the results on the system $Al_xGa_{1-x}As$ reproduced in Figs. 1.6 [70] and 1.7 [71].

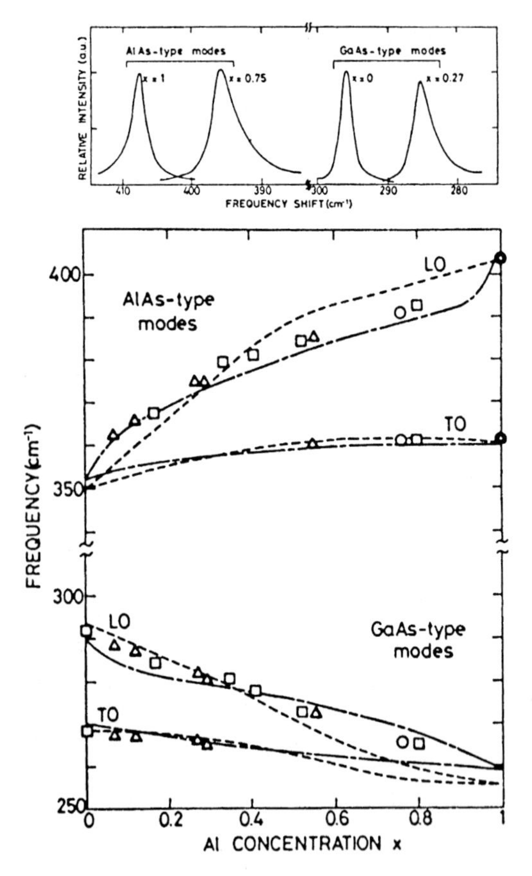

Fig. 1.6. *Upper panel*: Raman spectra showing two-mode behavior of LO phonons in $Al_{1-x}Ga_xAs$. *Lower panel*: LO and TO frequency vs. Al composition. The *triangles*, *squares* and *circles* are for samples grown by molecular-beam, liquid-phase and vapor-phase epitaxy, respectively. Early Raman measurements from R. Tsu, H. Kawamura and L. Esaki [*Proc. 11th Int. Conf. Physics Semicond.*, ed. by M. Masiek (Polish Scientific, Warsaw 1972), p. 1135] are represented by *dashed lines.* The *dashed-dotted curves* are theoretical results. Adapted from [70]

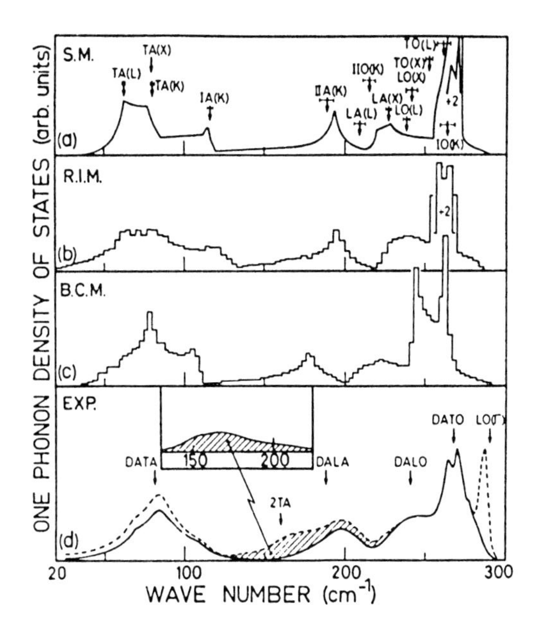

Fig. 1.7. Comparison between the calculated phonon DOS in GaAs (labels SM, RI and BC refer to the shell-, rigid-ion- and bond-charge models) and the experimental disorder-induced scattering from the alloy $Al_{0.19}Ga_{0.81}As$. The *inset* shows the estimated second-order scattering contribution. From [71]

Other than the dominant AlAs- and GaAs-like zone center TO and LO modes shown in Fig. 1.6, the spectrum of Fig. 1.7 exhibits a disorder-activated first-order continuum, which follows closely the one-phonon DOS [71]. The appearance of TO and LO pairs is known as two-mode behavior [72]. With the possible exception of $In_xGa_{1-x}P$, all the III-V binary systems that have been studied with Raman scattering show two-mode behavior [72]. The less common one-mode behavior (i.e., a single TO or a single LO over some compositional range) is observed in a handful of II-VI mixed crystals including $Zn_xCd_{1-x}S$ [73]. In either case, phonon frequency variations are usually large enough for Raman scattering to be the preferred local probe to provide a simple and accurate determination of the alloy composition.

1.5 Impurity Centers and Other Defects

At the ends of the compositional ranges for a binary alloy system it is more appropriate to view the trace constituent as an impurity in an otherwise perfect host crystal. These dilute impurities have their own local vibrational modes, which can often be observed in Raman scattering. Momentum conservation is not required in the scattering process, since crystal momentum is not a valid quantum number for the localized mode. For substitutional impurities having the same valence state as the host crystal atoms, but a different mass, the sign of the mass difference determines the nature of the local mode. If the impurity is lighter than the host, the local mode frequency is usually well above the highest frequency phonon in the host. The mode tends to be highly localized and long-lived (narrow in the frequency domain), since it cannot couple directly to the continuum of one-phonon states in the host. In an atomistic picture, the heavier host atoms cannot follow the fast motion of the light impurity. In the opposite limit of a heavier impurity, the mode is not truly "local", since it couples to the host continuum. In this case we would expect a much broader Raman line.

As examples we consider group IV substitutional impurities in a Ge lattice. For the lighter impurities, C and Si, the local modes are observed at 530 and 395 cm^{-1} [69,74], respectively. Both frequencies are well above the top of the Ge phonon bands ($\approx$ 300 cm^{-1}), and consequently the modes are highly localized and quite narrow. In contrast, the local mode for the heavier element Sn in Ge is at 263 cm^{-1}, which falls in the middle of the Ge optical phonon branch [74]. The relative width ($\Delta\omega/\omega$) of the Sn mode is larger by

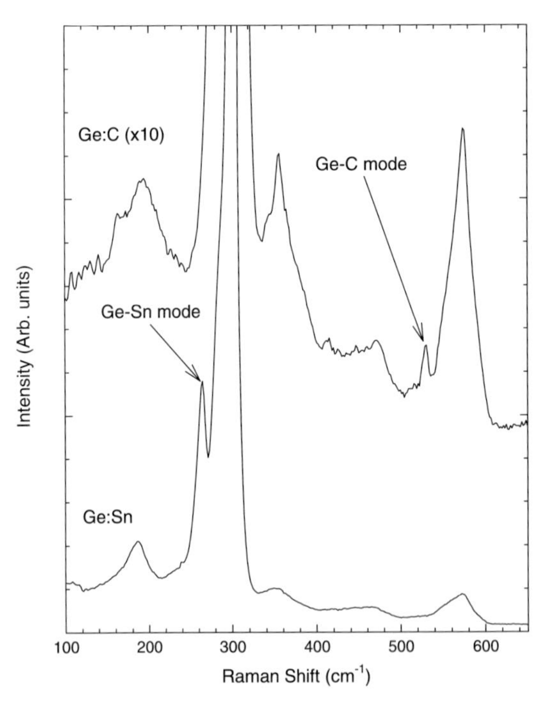

Fig. 1.8. Raman spectra obtained with 633-nm radiation in the $z(xy)\bar{z}$ scattering geometry of 50-nm thick C-doped (*upper trace*) and Sn-doped (*lower trace*) epitaxial layers grown on Ge(001) at 200 °C. Adapted from [74]. The Sn concentration is a few %; the C concentration is much lower

about a factor of three compared with the C or Si local modes. Typical Raman spectra illustrating these results for the impurity modes of C and Sn in Ge are shown in Fig. 1.8.

Since impurity centers and defects break the translational symmetry of the crystal, thereby relaxing the conservation of wave vector (1.3), they can sometimes lead to scattering by phonons in the host material that have wave vectors far away from the zone center (as for $Al_xGa_{1-x}As$ [71]; see Fig. 1.7). A classic example of a material showing such defect-induced Raman lines is given in Fig. 1.9 for graphite. Graphite has two allowed Raman modes, both of E_{2g} symmetry, whose frequencies are 42 and 1582 cm^{-1} [75]. In the lower trace of Fig. 1.9, corresponding to a nearly perfect single crystal, only the high-frequency, Raman-allowed line at 1582 cm^{-1} is seen. The upper trace shows a spectrum from highly defective graphite, which contains in addition to the allowed Raman line a strong defect-induced band near 1340 cm^{-1} and a weaker band near 1620 cm^{-1}.

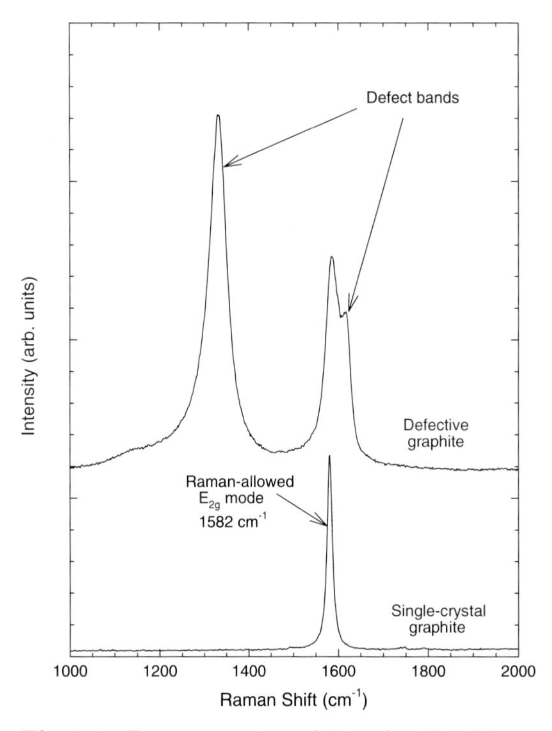

Fig. 1.9. Raman spectra obtained with 633-nm radiation showing the Raman-allowed mode at 1582 cm^{-1} in a high-quality graphite single crystal (*lower trace*) and the additional defect bands at 1340 and 1620 cm^{-1} that appear in defective graphite (*upper trace*)

1.6 Phonons in Amorphous Materials

The Raman spectrum from an amorphous material generally consists of a few broad bands, with maxima at roughly the frequencies corresponding to peaks in the broadened phonon DOS for the crystalline phase [76]. In Si, for example, as shown in Fig. 1.10 there are two maxima, one near 500 cm^{-1} and another near 150 cm^{-1}, arising from the optic and acoustic branches, respectively [77]. In an oxide glass, such as SiO_2, the spectrum is more complex, since a wider variety of vibrational excitations are possible. An example is shown in Fig. 1.11 for fused silica (amorphous SiO_2). The relatively rich spectrum from SiO_2 has been discussed by several authors [78,79]. The strong band peaking at 450 cm^{-1} is attributed to the symmetric stretching of the bridging oxygens; the structure at 490 and 670 cm^{-1} to specific types of defects; the band at 820 cm^{-1} to an unresolved TO-LO phonon pair; and the band at 1060 cm^{-1} to a TO phonon whose LO partner is the band at 1190 cm^{-1}.

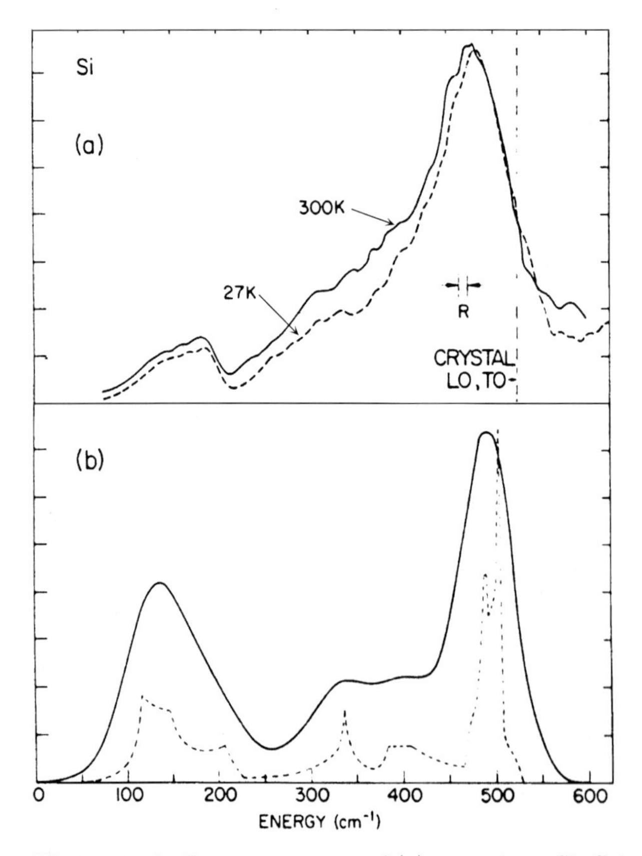

Fig. 1.10a,b. Raman spectrum of (**a**) amorphous Si. Calculated (*dashed line*) and broadened (*solid line*) DOS (**b**) for crystalline Si. From [77]

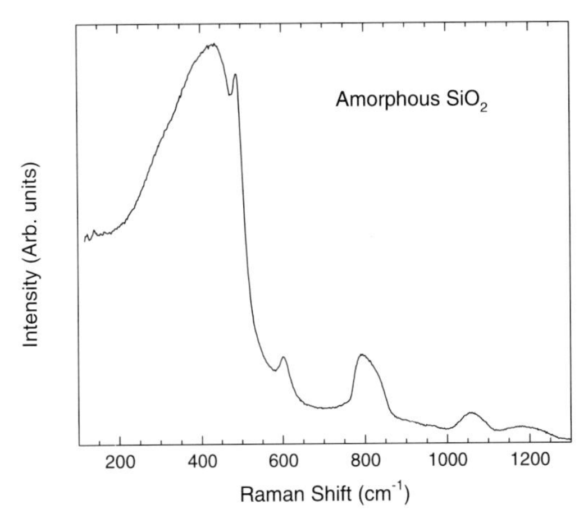

Fig. 1.11. Raman spectrum obtained with 633-nm radiation from fused silica

1.7 Structural Phase Transitions: Effects of Temperature, Pressure and Composition

Raman scattering has proven to be a valuable tool in the study of structural phase transitions [80,81]. In most experiments these transitions are induced by changes in one of the thermodynamic variables temperature or pressure. The effects of increased temperature are usually opposite those of increased pressure, since the former tends to increase the inter-atomic spacings, via thermal expansion, whereas the latter decreases inter-atomic spacings. Normally, the phonon frequencies increase slowly when the lattice contracts, and the relative changes in frequency are small, i.e., on the order of the relative changes in the lattice constants. Deviations from this "normal" behavior often signal a phase transition.

When the crystal structure changes in a discontinuous manner at the transition point, the transition is termed *first order*. An example of such a transition is the transformation of carbon from graphite to diamond. When there is a continuous change in the crystal structure through the transition point, as can be effected with infinitesimal displacements of the atomic positions, the transition is termed *second order*. Second-order transitions are generally reversible, whereas first-order transitions are not. The most interesting cases for Raman studies are those second order transitions in which a Raman-active mode has the correct symmetry to induce the transition from one phase to the other. The mode in question is then termed a *soft mode*. According to Landau's theory of second-order phase transitions, in the vicinity of the transition the soft mode frequency should behave as

$$\omega^2 \propto |T - T_c| \text{ (or } |P - P_c| \text{ with a pressure-induce transition)} , \qquad (1.24)$$

where $T_c(P_c)$ is the critical transition temperature (pressure) [82].

An example of a Raman study showing soft mode behavior is the temperature-induced, orthorhombic-tetragonal transition in $CaCl_2$, which is termed a *ferroelastic* transition. At low temperature $CaCl_2$ has the orthorhombic D_{2h}^{12} structure, which transforms into the tetragonal rutile structure (D_{4h}^{14}) at 491 K [83]. The atomic displacements that take one structure into the other involve a rotation about the c axis of the Cl octahedra surrounding each Ca. These displacements describe the eigenvector for the Raman-active $A_g(B_{1g})$ mode in the orthorhombic (tetragonal) phase, thus, suggesting soft mode behavior. Such behavior for this mode has been observed by Unruh et al. [83], as shown in Fig. 1.12. Note that in the vicinity of the transition temperature, the mode frequency varies as predicted by (1.24), but its frequency never completely vanishes, reaching a minimum of $14\,\mathrm{cm}^{-1}$ at the transition. Also note that the transition temperature appears different from above and below, an indication that the transformation is not of second-order but only approximately so.

Soft mode behavior is not the only Raman signature of a second-order (or close to second-order) phase transition. For such transitions the lower-symmetry phase must be a subgroup of the higher-symmetry phase, and it is often the case that a degenerate mode of the higher-symmetry phase splits under the lower symmetry. An example of this splitting is shown in Fig. 1.13 for a pressure-induced transition in RuO_2 obtained in a diamond-anvil cell [84]. This is also a ferroelastic transition, and the two crystal structures involved here are the same as those in Fig. 1.12. The transition, however, proceeds in the opposite direction. At ambient pressure RuO_2 has the rutile structure, which transforms into the $CaCl_2$ (ambient) structure at 11.8 GPa. A group theoretic analysis of this transition indicates that the doubly degenerate E_g mode of the low-pressure tetragonal phase splits into a non-degenerate B_{2g},

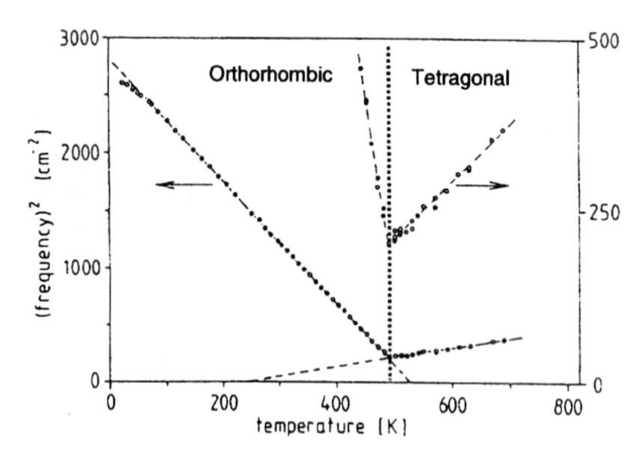

Fig. 1.12. Temperature dependence of the frequency squared for the soft mode in $CaCl_2$ as it undergoes a nearly second-order phase transition. The *dotted vertical line* indicates the transition temperature, with the D_{2h}^{12} phase on the left and the D_{4h}^{14} phase on the right. Adapted from [83]

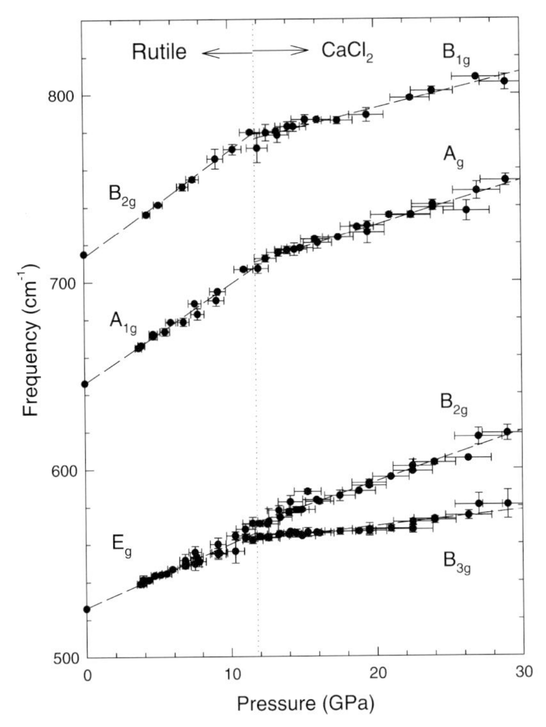

Fig. 1.13. Pressure dependence, obtained with 633-nm radiation in a diamond anvil cell, of the Raman lines in RuO_2 as it undergoes a second-order transition from the D_{4h}^{14} phase on the left to the D_{2h}^{12} phase on the right. The *dashed vertical line* indicates the transition pressure. From [84]

B_{3g} pair in the orthorhombic phase, as is observed in the experiment of Fig. 1.13. As also shown in the figure, the rates of change of the different lines with pressure are approximately straight lines. The slopes of these lines are usually characterized in terms of the *Grüneisen parameters*, γ_i, which measure the relative change of the mode frequencies with volume [85]:

$$\gamma_i = -\frac{V}{\omega_i}\frac{\partial \omega_i}{\partial V} = \frac{B}{\omega_i}\frac{\partial \omega_i}{\partial P} , \tag{1.25}$$

where V is the volume, P is the hydrostatic pressure, and $B \equiv -V(\partial P/\partial V)$ is the isothermal bulk modulus. Knowledge of the bulk modulus is needed to convert pressure slopes, as in Fig. 1.13, into Grüneisen parameters. Typical Grüneisen parameters are of order unity [86].

Owing to their device applications, there is growing interest in solid solution systems that display structural ferroelectric phase transformations as a function of composition. Raman spectra of the $Pb(Ti_{1-x}Zr_x)O_3$ system, known as PZT, were investigated in the seventies [87,88] when it was proposed that the structural instability is in some way related to a softening of an optical vibrational mode [89]. It is now recognized that solid solutions such as PZT represent a distinct class of ferroelectric materials known as relaxor ferroelectrics, which are characterized by a strong frequency dispersion of the dielectric permittivity [90]; see the box by R. Katiyar in this book. These materials often lack electric polarization and anisotropy on a macroscopic scale, but the application of an electric field induces the formation of well-defined ferroelectric states. There are numerous recent Raman investigations of relaxor ferroelectrics and, in particular studies of low-frequency vibrational modes [91–93].

1.8 Conclusions

This chapter offers a conceptual introduction to the Raman scattering investigations that are described in this book. We focused on optical vibrational modes, but in many instances the concepts also apply to inelastic light scattering by elementary excitations of electrons. The examples cited here are only a sampling of the types of material properties that can be addressed with Raman scattering. Our aim is to convince the reader that Raman applications have a very broad scope. Many of these applications will be revisited in more detail in later chapters, as they relate to specific material systems. Additional applications, not mentioned here, will also be discussed, some examples of which are strain effects, free-carrier effects in semiconductors, magnetic interactions, effects due to variations in stoichiometry, and a variety of effects that relate to the structure of polymer chains.

References

1. M. Born, K. Huang: *Dynamical Theory of Crystal Lattices* (Clarendon Press, Oxford 1954) p. 367
2. C.V. Raman, K.S. Krishnan: Nature **121**, 501 (1928)
3. W. Hayes, R. Loudon: *Scattering of Light by Crystals* (Wiley, New York 1978)
4. M.V. Klein: In *Light Scattering in Solids I*, Topics Appl. Phys., Vol. 8, ed. by M. Cardona (Springer, Berlin 1975) Chap. 4
5. M.G. Cottam, D.J. Lockwood: *Light scattering in magnetic solids* (Wiley, New York 1986)
6. G.B. Wright (Ed.): *Light Scattering Spectra of Solids* (Springer, Berlin 1969)
7. M. Balkanski (Ed.): *Light Scattering in Solids* (Flammarion, Paris 1971); M. Balkanski, R.C.C. Leite, S.P.S. Porto (Eds.): *Light Scattering in Solids* (Flammarion, Paris 1975)
8. D.J. Lockwood, J.F. Young (Eds.): *Light Scattering in Semiconductor Structures and Superlattices*, NATO ASI Series B: Physics, Vol. 273 (Plenum, New York 1991)
9. M. Agranovich, J.L. Birman (Eds.): *The Theory of Light Scattering in Solids* (Nauka, Moscow 1976); J.L. Birman, H.Z. Cummins, K.K. Rebane (Eds.): *Light Scattering in Solids* (Plenum, New York 1979)

10. S.A. Asher, P.B. Stein (Eds.): Proc. XVth Intern. Conf. on Raman Spectroscopy (Wiley, New York 1996)
11. M. Cardona (Ed.): *Light Scattering in Solids*, 1st ed.; *Light Scattering in Solids I: Introductory Concepts*, 2nd ed.; Topics Appl. Phys., Vol. 8 (Springer, Berlin 1975 and 1982)
12. M. Cardona, G. Güntherodt (Eds.): *Light Scattering in Solids II: Basic Concepts and Instrumentation*, Topics Appl. Phys., Vol. 50 (Springer, Berlin 1982)
13. M. Cardona, G. Güntherodt (Eds.): *Light Scattering in Solids III: Recent Results*, Topics Appl. Phys., Vol. 51 (Springer, Berlin 1982)
14. M. Cardona, G. Güntherodt (Eds.): *Light Scattering in Solids IV: Electronic Scattering, Spin Effects, SERS, and Morphic Effects*, Topics Appl. Phys., Vol. 54 (Springer, Berlin 1984)
15. M. Cardona, G. Güntherodt (Eds.): *Light Scattering in Solids V: Superlattices and Other Microstructures*, Topics Appl. Phys., Vol. 66 (Springer, Berlin 1989)
16. M. Cardona, G. Güntherodt (Eds.): *Light Scattering in Solids VI: Recent Results, Including High-Tc Superconductivity*, Topics Appl. Phys., Vol. 68 (Springer, Berlin 1991)
17. M. Cardona, G. Güntherodt (Eds.): *Light Scattering in Solids VII: Crystal-Field and Magnetic Excitations*, Topics Appl. Phys., Vol. 75 (Springer, Berlin 2000)
18. M. Cardona, G. Güntherodt (Eds.): *Light Scattering in Solids VIII: Fullerenes, Semiconductor Surfaces, Coherent Phonons*, Topics Appl. Phys., Vol. 76 (Springer, Berlin 2000)
19. A. Pinczuk, E. Burstein: In [11], Chap. 2
20. W. Heitler: *The Quantum Theory of Radiation* (Clarendon Press, Oxford 1954)
21. R. Loudon: Proc. Roy. Soc. A **275**, 218 (1963)
22. R. Loudon: Adv. Phys. **13**, 423 (1964)
23. A.A. Maradudin, R.F. Wallis: Phys. Rev. B **3**, 2063 (1971)
24. R.M. Martin, L.M. Falicov: In [11], Chap. 3
25. W. Richter: In: *Springer Tracts in Modern Physics*, Vol. 78 (Springer, Berlin 1976) p. 121
26. M. Cardona: In [12], Chap. 2
27. E. Burstein, A. Pinczuk: *The Physics of Optoelectronic Materials*, ed. by W.A. Albers (Plenum Press, New York 1971) p. 33
28. G. Abstreiter, M. Cardona, A. Pinczuk: In [14], Chap. 2
29. A. Pinczuk, E. Burstein: Phys. Rev. Lett. **21**, 1073 (1968)
30. R.C.C. Leite, J.F. Scott: Phys. Rev. Lett. **22**, 130 (1969)
31. F. Cerdeira, W. Dreybrodt, M. Cardona: Solid State Commun. **10**, 591 (1972)
32. P.Y. Yu, Y.R. Shen: Phys. Rev. Lett. **29**, 478 (1972)
33. A. Pinczuk, L. Brillson, E. Burstein, E. Anastassakis: Phys. Rev. Lett. **27**, 317 (1971)
34. C. Colvard, R. Merlin, M.V. Klein, A.C. Gossard: Phys. Rev. Lett. **45**, 298 (1980)
35. R. Merlin, C. Colvard, M.V. Klein, H. Morkoc, A.Y. Cho, A.C. Gossard: Appl. Phys. Lett. **36**, 43 (1980)
36. J.E. Zucker, A. Pinczuk, D.S. Chemla, A. Gossard, W. Wiegmann: Phys. Rev. Lett. **51**, 1293 (1983)
37. A.K. Sood, J. Menéndez, M. Cardona, K. Ploog: Phys. Rev. Lett. **54**, 2111 (1985)

38. E. Burstein, A. Pinczuk, S. Buchner: In *Physics of Semiconductors 1978*, ed. by B.L.H. Wilson (The Institute of Physics, London 1979) p. 1231
39. G. Abstreiter, K. Ploog: Phys. Rev. Lett. **42**, 1308 (1979)
40. A. Pinczuk, H.L. Störmer, R. Dingle, J.M. Worlock, W. Wiegmann, A.C. Gossard: Solid State Commun. **32**, 1001 (1979)
41. E. Burstein, A. Pinczuk, D.L. Mills: Surf. Sci. **98**, 451 (1980)
42. C. Kittel: *Introduction to Solid State Physics* (Wiley, New York 1996)
43. A. Pinczuk, G. Abstreiter: In [15], Chap. 4
44. B. Jusserand, M. Cardona: In [15], Chap. 3
45. F.H. Pollak: In *Analytical Raman Spectroscopy*, ed. by J.G. Grasselli, B.J. Bulkin (Wiley, New York 1991) Chap. 6, p. 137
46. J.D. Jackson: *Classical Electrodynamics* (Wiley, New York 1996)
47. J.L. Birman: *Theory of Crystal Space Groups and Lattice Dynamics* (Springer, Berlin 1984)
48. G. Burns, A.M. Glazer: *Space Groups for Solid State Scientists* (Academic Press, New York 1978)
49. M. Tinkham: *Group Theory and Quantum Mechanics* (McGraw-Hill, New York 1964)
50. P.M. Platzman, P.A. Wolff: *Waves and Interactions in Solid State Plasmas*, ed. by H. Ehrenreich, F. Seitz, D. Turnbull, Solid State Physics Suppl. 13 (Academic Press, New York 1973)
51. T.C. Damen, S.P.S. Porto, B. Tell: Phys. Rev. **142**, 570 (1966)
52. G. Herzberg: *Infrared and Raman Spectra of Polyatomic Molecules* (Van Nostrand Reinhold, New York 1945) p. 246
53. D.L. Rousseau, R.P. Baumann, S.P.S. Porto: J. Raman Spectrosc. **10**, 253 (1981)
54. T. Hahn, (Ed.): *International Tables for Crystallography*, Vol. A, 2nd ed. (Kluwer Academic Press, London 1989)
55. R.W.G. Wyckoff: *Crystal Structures*, 2nd ed., Vol. 1 (Wiley, New York 1964)
56. R.L. Liboff: *Introductory Quantum Mechanics* (Holden-Day, San Francisco 1980)
57. W. Wettling, M.G. Cottam, J.R. Sandercock: J. Phys. C **8**, 211 (1975)
58. R.C.C. Leite, J.F. Scott, T.C. Damen: Phys. Rev. Lett. **22**, 780 (1969)
59. M.V. Klein, S.P.S. Porto: Phys. Rev. Lett. **22**, 782 (1969)
60. D.C. Hamilton: Phys. Rev. **188**, 1221 (1969)
61. R.M. Martin, C.M. Varma: Phys. Rev. Lett. **26**, 1241 (1971)
62. R. Merlin, G. Güntherodt, R. Humphreys, M. Cardona, R. Suryanarayanan, F. Holtzberg: Phys. Rev. B **17**, 4951 (1978)
63. G. Turrell: *Infrared and Raman Spectra of Crystals* (Academic Press, New York 1972) p. 297
64. R. Loudon: Proc. Phys. Soc. London **84**, 379 (1964)
65. P.A. Temple, C.E. Hathaway: Phys. Rev. B **7**, 3685 (1973)
66. W.H. Weber, K.C. Hass, J.R. McBride: Phys. Rev. B **48**, 178 (1993)
67. W. Weber: Phys. Rev. B **15**, 4789 (1977)
68. W. Windl, K. Karch, P. Pavone, O. Schütt, D. Strauch, W.H. Weber, K.C. Hass, L. Rimai: Phys. Rev. B **49**, 8764 (1994)
69. M.A. Renucci, J.B. Renucci, M. Cardona: In *Light Scattering in Solids*, ed. M. Balkanski (Flamarion, Paris 1971) p. 326
70. B. Jusserand, J. Sapriel: Phys. Rev. B **24**, 7194 (1981)
71. R. Carles, A. Zwick, M.A. Renucci, J.B. Renucci: Solid State Commun. **41**, 557 (1982)

72. G.P. Srivastava: *The Physics of Phonons* (Adam Hilger, Bristol 1990) Chap. 9
73. G. Lucovsky, A. Mooradian, W. Taylor, G.B. Wright, R.C. Keezer: Solid State Commun. **5**, 113 (1967)
74. W.H. Weber, B.-K. Yang, M. Krishnamurthy: Appl. Phys. Lett. **73**, 626 (1998)
75. M.S. Dresselhaus, G. Dresselhaus: Advan. Phys. **30**, 139 (1981)
76. M.H. Brodsky: In *Light Scattering in Solids I*, Topics Appl. Phys., Vol. 8, ed. by M. Cardona (Springer, Berlin, Heidelberg, New York 1975) p. 205
77. J.E. Smith, Jr., M.H. Brodsky, B.L. Crowder, M.I. Nathan: J. Non-Cryst. Solids **8–10**, 179 (1972)
78. S.K. Sharma, D.W. Matson, J.A. Philpotts, T.L. Roush: J. Non-Cryst. Solids **68**, 99 (1984)
79. F.L. Galeener: Phys. Rev. B **19**, 4292 (1979)
80. J.F. Scott: Rev. Mod. Phys. **46**, 83 (1974)
81. P.A. Fleury, K. Lyons: In *Structural Phase Transitions I*, ed. by K.A. Müller, H. Thomas (Springer, Berlin 1981) p. 9
82. L.D. Landau, E.M. Lifshitz: *Statistical Physics*, 3rd ed., ed. by E.M. Lifshitz, L.P. Pitaevskii (Pergamon, New York 1980)
83. H.-G. Unruh, D. Mühlenberg, Ch. Hahn: Z. Phys. B **86**, 133 (1992)
84. S.S. Rosenblum, W.H. Weber, B.L. Chamberland: Phys. Rev. B **56**, 529 (1997)
85. N.W. Ashcroft, N.D. Mermin: *Solid State Physics* (Holt, Rinehart and Winston, New York 1976) p. 493
86. B.A. Weinstein, R. Zallen: In [14], Chap. 8
87. G. Burns, B.A. Scott: Phys. Rev. Lett. **25**, 1191 (1970)
88. R. Merlin, A. Pinczuk: Ferroelectrics **7**, 275 (1974); R. Merlin, J.A. Sanjurjo, A. Pinczuk: Solid State Commun. **16**, 931 (1975)
89. A. Pinczuk: Solid State Commun. **12**, 1035 (1973)
90. L.E. Cross: Ferroelectrics **151**, 305 (1994)
91. I.G. Siny, S.G. Lushnikov, R.S. Katiyar, E.A. Rogacheva: Phys. Rev. B **56**, 7962 (1997)
92. R. Farhi, M. El Marssi, J. -L. Dellis, M.D. Glinchuk, V.A. Stephanovich: Europhys. Lett. **46**, 351 (1999)
93. I. Dujovne, T.-J. Koo, A. Pinczuk, S.-W. Cheong, B.S. Dennis: Bull. Am. Phys. Soc. **45**, 677 (2000)

I The Effect of a Surface Space-Charge Electric Field on Raman Scattering by Optical Phonons

Elias Burstein

The changes in the macroscopic properties that result from the modification of the symmetry of a crystal by an external generalized force (e.g., an electric field) are called *morphic* effects, a term first coined by Hans Mueller [1] who was my mentor when I was a graduate student at MIT in the early 1940s. In optical phenomena, morphic effects manifest themselves as changes in the absorption and scattering of electromagnetic radiation by the elementary excitations of the medium [2]. The effects are particularly striking when normally forbidden optical processes become allowed. In this article, we focus our attention on the Raman scattering by LO phonons of opaque semiconductors in which a surface space-charge electric field opens a new channel for scattering.

In 1963 Loudon [3] showed that, in the limit where the scattering wave vector $\boldsymbol{q} = 0$, the electron and hole contributions to the matrix elements for the Raman scattering by LO phonons have the same magnitude but opposite signs and therefore cancel (this is for two-band processes in which the second step involves the intraband *Fröhlich* scattering of the excited electron and hole in the intermediate state by the Coulomb field of the LO phonons). The electron and hole contributions do not cancel when $\boldsymbol{q}$ is finite and $m_e \neq m_h$ [4]. Under resonance conditions, this leads to a sizable wavevector-dependent contribution to the scattering intensity. This also applies to situations in which the intermediate states involve the intraband Fröhlich scattering of Coulomb-correlated e–h pairs (i.e., excitons) by LO phonons [5]. Since the intraband scattering by the Coulomb field of the LO phonons does not change the orbital parts of the excited electron and hole wavefunctions, the $\boldsymbol{q}$-dependent terms are observed in configurations where the polarization of the incident light, $\hat{\boldsymbol{e}}_i$, is parallel to that of the scattered light, $\hat{\boldsymbol{e}}_s$.

In 1968, Pinczuk and I [6] reported an investigation of Te-doped n-type InSb (using back-scattering from air-cleaved (110) surfaces at 80 K and the 633 nm radiation of a HeNe laser) to study coupled LO phonon–plasmon modes. These modes had earlier been observed in n-type GaAs by Mooradian [7] and discussed by Burstein et al. [8]. Surprisingly, the back-scattering spectra for $\hat{\boldsymbol{e}}_i \parallel \hat{\boldsymbol{e}}_s$ exhibited a *forbidden* LO phonon peak at its *unscreened* frequency, in addition to the allowed TO mode. The intensity of the forbidden LO phonon peak, which was much weaker than the allowed TO peak in the sample with the lowest carrier density, increased with increasing carrier density, becoming larger than that of the TO mode (which was relatively in-

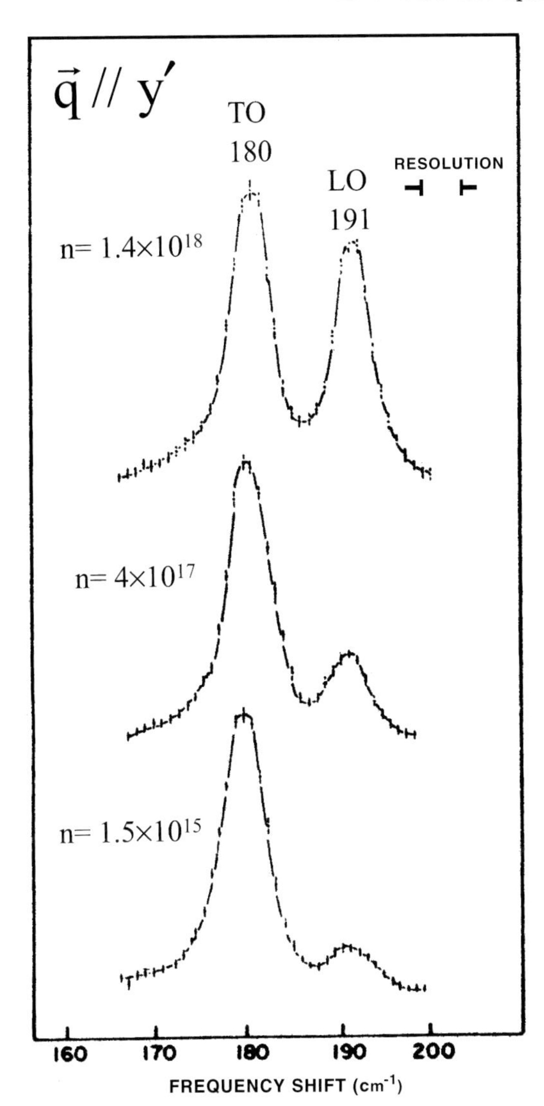

Fig. I.1. Back scattering Raman spectra of LO and TO phonons from (110) surfaces of n-InSb at 80 K [10,12]

sensitive to the carrier density) before leveling off; see Fig. I.1. For all carrier densities, the forbidden LO peak appeared at the unscreened frequency, indicating an absence of coupling with plasmons. Similar results were obtained for p-type InSb and n-type InAs [9].

The forbidden unscreened LO phonon peak and its strength dependence on the density of free carriers, was attributed to the *morphic* effect of a surface space charge electric field $\boldsymbol{E}_{\mathrm{SC}}$. The surface field is due to the pinning of the

Fermi level within the energy gap, which results in a depletion layer at the surface and absence of free carriers within the skin depth of the incident radiation. In these experiments, Raman scattering is resonantly enhanced because the incident photon frequency is close to the E_1 gap of InSb (1.96 eV). The space-charge field manifests itself as an upwards (downwards) band-bending at the surface of n-type (p-type) InSb. In the case of n-type InAs, the absence of screening cannot be attributed to the absence of free carriers, since there is an accumulation layer at the surface. It was therefore suggested that the LO phonons that take part within the narrow accumulation layer have wave vectors that are appreciably larger than the Fermi–Thomas screening wave vector and are therefore uncoupled to free carriers in the accumulation layer. Since the wave vector of the LO phonons involved is rather large, the observed Raman scattering involves both $\boldsymbol{E}_{\mathrm{SC}}$-dependent and $\boldsymbol{q}$-dependent contributions.

As a consequence of band bending, the envelope parts of the wave functions of the electronic states in the valence and conduction bands are modified from *Bloch*- to *Airy*-type. Moreover, the interband optical transitions in the first and third steps of the Raman scattering process correspond to *Franz–Keldysh* type (tunneling-assisted) optical interband transitions that create or annihilate *spatially separated* electron–hole pairs. It was conjectured by Pinczuk et al. [10] that the electric-field-induced scattering by LO phonons was due to the spatial separation (i.e., *polarization*) of the excited electron–hole pairs in the intermediate state of two-band processes and to the consequent non-cancellation of the electron and hole intraband Fröhlich interaction matrix elements. In effect, $\boldsymbol{E}_{\mathrm{SC}}$ and its associated band-bending open up a new, normally forbidden two-band channel for the Raman scattering by LO phonons, which is particularly strong under resonance conditions.

An electric-field-induced Raman scattering by LO phonons was subsequently observed by Brillson et al. [11] in the Raman inactive IV–VI compounds PbTe and SnTe, which have the NaCl structure. Use was made of the fact that the position of the Fermi level at the surface of ionic crystals is not strongly pinned, and that the energy bands can be strongly bent by the presence of metal films on the crystal surface. The bent bands and the associated $\boldsymbol{E}_{\mathrm{SC}}$ modify the Raman scattering selection rules by lowering the symmetry of the crystal and its atomic and electronic excitations. Using cleaved [100] surfaces of p-type samples coated with transparent films of Pb, Raman scattering peaks were observed, whose frequencies corresponded to those of unscreened LO phonons. Moreover, scattering was observed only for $\hat{\boldsymbol{e}}_i \parallel \hat{\boldsymbol{e}}_s$, and the observed peaks exhibited a resonant enhancement at the E_2 gap. These peaks did not appear in the absence of the Pb films. Here again, the Raman scattering by LO phonons involves the Franz–Keldysh and $\boldsymbol{q}$-dependent mechanisms.

A perturbation calculation of the contribution to the field-induced scattering by LO phonons via the Franz–Keldysh mechanism for two-band Fröhlich-scattering processes involving exciton intermediate states was carried out by Gay et al. in 1971 [12]. They found that the Raman scattering tensor is proportional to the field-induced separation of the electron and hole in the ground state of the exciton. They also carried out a computer calculation of the contribution from two-band processes involving continuum electron and hole intermediate states, which confirmed the conjecture of Pinczuk et al. that the electric field induces a spatial separation of the electron and hole in the intermediate state, leading to a non-cancellation of the corresponding contributions. Subsequently, Brillson et al. [13] carried out a theoretical calculation of the electric-field-induced scattering by LO phonons in crossed electric and magnetic fields where the electron and hole in the intermediate state have well-defined, bounded wave functions. This yields a relatively simple analytic expression for the electron and hole terms in the intraband Fröhlich interaction matrix elements. They found that, to first order in the electric field, the resultant wave-vector-independent contribution to the Raman tensor is proportional to the electric-field-induced separation of the cyclotron-orbit centers of the excited electron and hole in the intermediate state.

Buchner et al. [15] used surface space-charge electric field induced Raman scattering by LO phonons to probe the differences in the character of the $A(111)$ and $B(\bar{1}\bar{1}\bar{1})$ surfaces of n- and p-InAs. The measurements, which were carried out on air-exposed surfaces, showed that, in the case of p-InAs, the space-charge electric field of the A surface is considerably smaller than that of the B surface, whereas, in the case of n-InAs, the space charge electric fields at both surfaces are quite large. The marked difference in the forbidden scattering by LO phonons at the A and B surfaces of n- and p-InAs reflect differences in the pinning of the Fermi levels at the two surfaces. Both the (111) and $(\bar{1}\bar{1}\bar{1})$ surfaces of n-InAs were shown, by the use of externally applied electric fields, to be depleted of free carriers, in contrast to the (100) surfaces which are accumulated.

These early investigations demonstrated that Raman scattering by LO phonons in opaque crystals is a highly useful probe of energy band-bending at semiconductor surfaces, and it has continued to be of value to the present day.

References

1. H. Mueller: Phys. Rev. **58**, 805 (1940)
2. E. Burstein, A.A. Maradudin, E. Anastasakis, A. Pinczuk: Helv. Phys. Acta **41**, 730 (1966)
3. R. Loudon: Proc. Phys. Soc. (London) **A 275**, 218 (1963)
4. D.C. Hamilton: Phys. Rev. **188**, 1221 (1968)
5. E. Burstein, A. Pinczuk: In: *Physics of Optoelectronic Materials*, ed. by W.A. Albers, (Plenum, New York 1971) p. 33
6. A. Pinczuk, E. Burstein: Phys. Rev. Lett. **21**, 1073 (1968)
7. A. Mooradian, G.G. Wright: Phys. Rev. Lett. **16**, 999 (1966)

8. E. Burstein, A. Pinczuk, S. Iwasa: Phys. Rev. **157**, 60 (1967)
9. A. Pinczuk, E. Burstein: In: *Light Scattering in Solids*, ed. by M. Balkanski, (Flamarian, Paris 1971) p. 429
10. A. Pinczuk, E. Burstein: In: *Proc. Tenth Int. Conf. on the Physics of Semiconductors* (U. S. Atomic Energy Commission, Washington DC 1970) p. 727
11. L. Brillson, E. Burstein: Phys. Rev. Lett. **27**, 808 (1971); *Light Scattering in Solids*, ed. by M. Balkanski, (Flamarian, Paris 1971) p. 429
12. J.G. Gay, J.D. Dow, E. Burstein, A. Pinczuk: In: *Light Scattering in Solids*, ed. by M. Balkanski, (Flamarian, Paris 1971) p. 33
13. L. Brillson, E. Burstein: Phys. Rev. B **5**, 2973 (1972)
14. P. Cordon, A. Pinczuk, E. Burstein: In: *Proc. Tenth Int. Conf. on the Physics of Semiconductors*, (U. S. Atomic Energy Commission, Washington DC 1970) p. 734
15. S. Buchner, L.Y. Ching, E. Burstein: Phys. Rev. B **14**, 4559 (1976)

2 Raman Instrumentation

Sanford A. Asher and Richard Bormett

Abstract. The last decade has seen major advances in the applications of Raman spectroscopy to materials science. We review here the dramatic improvements in Raman instrumentation that have enabled these incisive studies. This chapter separately discusses advances in lasers, spectrometers, optics and detectors and illustrates some of these advances in a few examples of new applications and instrumentation used in materials science research.

The rapidly increasing use of Raman spectroscopy as a routine method of materials characterization is a direct result of the recent dramatic advances in lasers, detectors and spectroscopic instrumentation [1–3]. Raman instrument companies have rapidly commercialized many of these advances. Some of these advances have made it possible to commercialize new relatively inexpensive instruments ($<$ \$100 K including the laser). This has significantly increased the number of Raman instruments worldwide [4]. The new generation of low cost, high performance Raman instrumentation now makes Raman spectral measurements easy, even for nonexperts. Industrial Raman instruments are now available for real time chemical process monitoring, manufacturing quality control and routine analytical investigations. Many of these industrially rugged instruments utilize fiber optic probes that dramatically simplify measurements making them routine and easy for nonexperts [5].

The Raman instrument advances have also affected the research grade Raman instruments. Research grade instruments are available with extended wavelength coverage from the UV into the near IR spectral regions. A common feature of most of these new research-grade Raman instruments is the incorporation of a microscope. These new Raman microscopes permit spectral imaging of samples with spatial resolutions$<$ 1 mm. The high spatial resolution allows rapid chemical speciation of spatially inhomogeneous samples.

Most Raman instrumentation is fabricated by selecting the appropriate lasers, optics, spectrometers, and detectors necessary to optimize the Raman spectrometer for the required Raman spectral measurements. Thus, it is natural to organize this chapter into separate sections that separately discuss these individual components. The last section will give a few examples of use of new Raman instrumentation in materials science.

2.1 Raman Measurement Regime

2.1.1 Spontaneous, Non-resonance Raman Spectral Measurements

Spontaneous, non-resonance ("normal") Raman measurements require excitation in a spectral region, which contains no sample absorption bands [6,7]. For most samples, this limits excitation wavelengths to either the visible or near IR spectral regions. Excitation close to an absorption band results in the selective preresonance Raman enhancement of vibrations of atoms localized within the chromophoric molecular segment [8]. This will increase the relative contribution of these vibrations to the Raman spectrum. If these vibrations contain the information content desired, then preresonance enhancement will be beneficial to the measurement. In contrast, this preresonance enhancement may be detrimental since molecular information from the non-enhanced bands becomes less accessible as the smaller set of preresonance Raman enhanced bands begins to dominate the spectrum. Preresonance enhancement is commonly observed with UV excitation.

If there are no absorption bands in the UV, visible and near IR spectral regions, the choice of laser excitation wavelength for these samples depends upon other experimental requirements. Most often the experimental requirement is optimization of the spectral signal-to-noise-ratio, which requires a minimization of sample fluorescence [3,9,10]. Whether the fluorescence is intrinsic to the sample or the result of trace impurities, fluorescence can significantly degrade the Raman signal-to-noise ratios. Fluorescence typically contributes broad spectral features with intensities that can be greater than the Raman intensities of even high concentration analytes. The high intensity fluorescence background results in significant shot-noise, which dominates the Raman signal-to-noise ratios. Anecdotal evidence from our laboratory suggests that fluorescence interference is a maximum for excitation in the $\sim 300-600$ nm spectral region. Thus, both UV (< 260 nm) and near IR Raman excitation can help avoid or minimize fluorescence interference [9,10].

Unfortunately near-IR Raman instruments have several disadvantages. The Raman cross section decreases rapidly as the excitation wavelength decreases, and the most common Raman detectors (charge-coupled-detector) have lower quantum efficiencies in the near-IR. The Raman cross sections decrease because of the v^4 dependence of the Raman scattering efficiencies [6–8]. The CCD detector efficiency decreases because the detector becomes transparent to near-IR wavelength at > 1050 nm. Near-IR Raman measurements can also show interference from blackbody radiation. Thus, near-IR excitation is typically not suitable for Raman spectral measurements at elevated temperatures.

Near-IR excitation remains attractive because the diode laser sources and the diode pumped YAG lasers are relatively inexpensive. Unfortunately, the near-IR Raman band frequencies occur in regions where CCD cameras and photomultipliers are blind. These measurements then must utilize non-shot noise limited detectors and often require the use of interferometer multiplex advantages to obtain acceptable Raman signal-to-noise ratios [9].

UV excitation below 260 nm also avoids fluorescence interference and occurs in a spectral region with dramatically increased Raman cross sections [8,11,12]. Molecules with their first singlet state electronic transition below 260 nm are highly flexible. Thus, they have highly efficient nonradiative relaxation processes, dramatically decreasing their excited state lifetimes. Thus, they show negligibly small fluorescent quantum yields in condensed phase samples [10]. Although molecular species with their first singlet states at higher energy may fluoresce upon UV excitation, this fluorescence will occur in a spectral region red shifted from the $\lambda < 260$ nm Raman spectral window.

UV-Raman measurements show improved spectral signal to noise ratios due to the increased Raman cross section from the v^4 dependence of Raman cross sections. Thus, a > 250-fold increase in the Raman cross section occurs for 250 nm excitation compared to 1.06 μm excitation.

UV-Raman instruments have the disadvantage that the instruments are specialized and require special optics, and the UV laser sources are somewhat more complex and expensive [11,12]. These limitations may significantly decrease in the near future upon commercialization of UV hollow cathode sputtering lasers, as discussed below.

2.1.2 Spontaneous, Resonance Raman Spectral Measurements

Resonance Raman spectroscopy has the advantage of high selectivity and high sensitivity [12]. Excitation within the absorption band of an analyte results in the selective enhancement of those vibrational modes of the analyte that selectively couple to the oscillating dipole moment induced by the excitation electric field [7]. The intensities of the resonance Raman enhanced bands can increase by as much as 10^8-fold. Thus, it becomes possible to selectively study the vibrational spectra of dilute analytes, or chromophores in macromolecules by choosing excitation wavelengths in resonance within a particular analyte chromophore.

For resonance Raman measurements it is essential to choose a laser with an excitation wavelength within the analyte absorption bands. The first laser resonance Raman measurements were obtained from polyenes such as β-carotene, and from porphyrins and heme proteins with excitation in the visible spectral region [14]. More recently resonance Raman measurements have been extended into the UV spectral region where most species show absorption bands [11,12].

2.1.3 Nonlinear Raman Measurements

Raman scattering is a two-photon process [15]. In the case of spontaneous Raman scattering the two photons include the laser excitation photon and a photon from the vacuum field [15]. In most cases CW laser excitation produces only a small number of Raman photons, and this number is negligible compared to the vacuum field photon density. Thus, the Raman intensity is observed to increase linearly with the incident excitation beam intensity. However, if the photon density at the Raman shifted wavelength approaches or exceeds the vacuum field photon density, stimulated Raman scattering will occur. This process shows a higher intensity dependence on the incident beam intensity. This is a rich area of research with numerous important applications in its future. Stimulated Raman scattering as well as the numerous other nonlinear Raman scattering processes will not be discussed here, except to note that high peak-power laser sources are required. These methodologies generally utilize ns to ps pulsed lasers, which give high peak powers at modest pulse energies.

2.2 Choice of Raman Excitation Wavelength

As a high intensity single frequency light source, the laser is an ideal Raman excitation source. The laser excitation frequency is the major determinant of the information content of a Raman spectral measurement. The laser frequency determines the spectral range of the measurement; the operating mode (pulsed or continuous) determines the excitation photon flux, as well as its temporal characteristics. Essentially all Raman measurements are photon limited. Thus, it would appear to be desirable to utilize as bright an excitation source as possible. However, in reality the incident laser power must be constrained such that the focused energy density is below the level that causes sample photochemical or thermal degradation and below the powers that cause nonlinear optical phenomena [16,17]. This represents a major limitation for the use low duty cycle pulsed laser sources. Although they may have an average power identical to that of a CW laser, their much higher peak powers can result in significant nonlinear optical phenomena as well as increased photochemical and photothermal sample damage.

Although the first laser Raman measurements were obtained by using a pulsed Ruby laser in the near IR at 694 nm [6], pulsed laser sources generally have significant pulse-to-pulse energy fluctuations. For scanning Raman instruments this results in low spectral S/N ratios, which are dominated by the standard deviation in the pulse energies.

2.2.1 CW Lasers

The red 632.8 nm CW He–Ne laser was the first laser to be incorporated in commercial Raman instruments [6]. This laser was quickly replaced by the higher power CW Ar^+ and Kr^{++} lasers that have numerous excitation lines in the 350–700 nm region. These CW Ar^+ and Kr^{++} lasers are available with very high powers in many spectral lines (1–5 W) in the visible spectral region. In addition, these lasers can be used to pump jet stream dye lasers to obtain relatively high power excitation continuously tunable throughout the visible spectral region. These dye lasers were mainly utilized for resonance Raman measurements that required excitation wavelength tunability.

More recently near-IR diode lasers [18] around 780 nm as well as diode pumped YAG lasers have become popular because of their low cost, small size, low power consumption and reliability. As discussed below these lasers are being used as the excitation sources for a new generation of inexpensive, small Raman spectrometers that are now being commercialized.

These CW laser sources are ideal for nonabsorbing samples because the laser beam can be focused to a small beam waist in the sample. The small beam waist is ideal for efficient light collection and efficient coupling of the Raman scattered light into the spectrometer. It is possible to use very high irradiances with CW lasers to obtain very high spectral S/N ratios. The maximum irradiance possible is limited by the sample damage threshold and the threshold for the onset of nonlinear optical phenomena. For example, 1 W of CW excitation focused onto a $10\,\mu m^2$ area results in a fluence of $10\,MW/cm^2$ which can be sufficient to induce nonlinear optical processes, such as two-photon absorption followed by heating and sample degradation. Thus, a practical limit exists for the fluence of a CW laser that can be practically used for Raman spectral measurements for "nonabsorbing" samples. For absorbing samples, laser heating causes degradation at much lower incident intensities.

2.2.2 Pulsed Lasers

The increased peak powers of pulsed lasers results in the occurrence of nonlinear optical processes at much smaller average incident laser powers than for CW lasers [16,17]. Thus, pulsed laser sources are generally avoided whenever a CW laser source is available for nonresonance, spontaneous Raman scattering experiments [19].

Until very recently pulsed laser sources were the only means of extending Raman excitation into the UV [19,20]. Nonlinear optical processes such as frequency doubling and mixing in nonlinear crystals such as β-barium borate and KDP generated UV light from dye lasers pumped by Q switched YAG lasers and excimer lasers. For example, the YAG fundamental at 1.06 µm can be frequency doubled to 532 nm or tripled to 355 nm to pump a dye laser. This dye laser can be frequency doubled and mixed with the 1.06 µm YAG fundamental to generate tunable UV excitation from 196 nm to the near IR. Alternatively a XeCl excimer laser [19] at 308 nm can be used to pump a dye laser. The dye laser light can be similarly frequency doubled to provide UV excitation from 196–350 nm. Using this approach, it is possible to generate light of any wavelength in this UV spectral region.

An alternative approach to obtain UV excitation uses stimulated Raman scattering to shift the laser wavelength [21]. Most often the Raman shifting material used is H_2 gas which shows a $\sim 4200\,\mathrm{cm}^{-1}$ Raman transition. It is easy to achieve numerous excitation wavelengths by utilizing a number of anti-Stokes Raman shifts in H_2. For example, 204 nm excitation can be easily obtained from 5 H_2 anti-Stokes Raman shifts of the tripled YAG at 355 nm [21,22]. Numerous additional Raman shifted lines in the UV can also be obtained from the quadrupled YAG line at 266 nm.

A much more convenient CW UV source is obtained by frequency doubling Ar^+ and Kr^{++} lasers. For example, $> 200\,\mathrm{mW}$ of 244 and 257 nm light can be obtained from intracavity doubled Ar^+ lasers [23], while $\sim 30\,\mathrm{mW}$ can be obtained at 229 nm, with additional lines occurring between 229 and 257 nm. The intracavity frequency doubled Kr^{++} laser [24] produces a few mW of 206 nm light, which is adequate for most UV Raman measurements. A new hollow cathode UV laser is now emerging [13] that utilizes as the gain medium excited states of metal ions sputtered into the gas phase. These lasers are small, energy efficient and inexpensive. Prototypes of these hollow cathode lasers using Cu and Ag have demonstrated laser excitation at 248 and 224 nm. These lasers operate quasi-CW at $\sim 10\,\mathrm{kHz}$ repetition rates with average powers of $\sim 1\,\mathrm{mW}$, which is adequate for UV-Raman measurements.

2.3 Optical Methods for Rayleigh Rejection

The purpose of the Raman spectrometer is to reject the intense Rayleigh scattered light and to disperse the Raman scattered light into its component frequencies for detection. The relative intensity ratio of Rayleigh to the Raman scattered light is often $> 10^9$ [6]. With such a large disparity between the Raman and Rayleigh intensities, the Rayleigh light must be greatly attenuated before the spectrograph section of the Raman spectrometer. If the Rayleigh light is allowed to enter the spectrograph unattenuated, it will generate sufficient stray light to obscure all or part of the much weaker Raman spectrum. Preventing intense Rayleigh scattered light levels from entering the spectrograph stage of the Raman spectrometer is the most challenging task for the Raman spectrometer.

A modern Raman spectrometer can utilize a number of different technologies to attenuate the relatively intense Rayleigh scattered light such as holographic notch filters [25], crystalline colloidal array Bragg diffraction filters [26], dielectric filters and multi-stage spectrometers [27]. Instruments that measure Raman bands lying close ($< 150\,\mathrm{cm}^{-1}$) to the Rayleigh line, utilize more costly, complex and inefficient Rayleigh rejection devices.

A band pass or band reject filter serves as the simplest Rayleigh rejection device. Holographic notch filters are available for the visible and near IR spectral regions; these filters typically permit Raman spectral measurements of frequency shifts greater than $\sim 100\,\mathrm{cm}^{-1}$. Low cost dielectric [28] filters suitable for Rayleigh rejection are available from the UV $\sim 230\,\mathrm{nm}$ [29] through the near IR spectral regions. The hard oxide dielectric filters are suitable for UV Raman spectral measurements within $400\,\mathrm{cm}^{-1}$ of the Rayleigh line. However, multistage Raman spectrometers are typically required to routinely measure Raman shifts within $50\,\mathrm{cm}^{-1}$ of the Rayleigh line in the visible, and within $200\,\mathrm{cm}^{-1}$ with UV excitation.

2.3.1 Holographic Notch Filter

The holographic notch filter [25] selectively rejects (through Bragg diffraction) a narrow band of light, while passing light outside of the band rejection region. The notch filter is constructed in a photosensitive medium, dichromated gelatin, by exposing it to interfering laser beams, which creates a periodic modulation in the refractive index. This periodicity produces a strong 3D Bragg reflection that can efficiently ($> 99.9\%$) diffract away the Rayleigh line, while transmitting adjacent wavelengths with $> 90\%$ efficiency. These holographic notch filters are manufactured to have the center (or near center) diffraction wavelength at the Raman excitation frequency. The sharp transition from high diffraction efficiency to high transmission makes the holographic notch filters a nearly ideal Rayleigh rejection filter for Raman measurements close to the Rayleigh line ($100\,\mathrm{cm}^{-1}$). Unfortunately holographic notch filters are not yet available for excitation wavelengths shorter than 350 nm. Since each holographic notch filter efficiently operates over only a small wavelength range ($\sim 40\,\mathrm{nm}$) numerous filters would be required for Raman measurements throughout the visible and near IR spectral region. However, holographic notch filters are ideal for Raman instruments that operate with only a few laser frequencies.

2.3.2 Dielectric Edge Filters

Raman edge filters [28,29] (typically a dielectric stack or rugate) are a low cost alternative to holographic notch filters for excitation wavelengths from the NIR to the UV. The rugate filter [30] is the dielectric equivalent of the holographic notch filters; the diffracting element is a stack of dielectric coatings with spacings and refractive index modulations sufficient to diffract the desired wavelength. Unfortunately, the current Raman edge filter performance (blocking bandwidth and edge steepness) is still inferior to that of holographic notch filters. The increased blocking bandwidth of the dielectric filters typically obscures Raman bands closer than $\sim 200\,\mathrm{cm}^{-1}$ to the Rayleigh line in the visible/near-IR and $\sim 400\,\mathrm{cm}^{-1}$ in the UV. In addition, to the increased rejection bandwidth, dielectric filters typically show a $100-200\,\mathrm{cm}^{-1}$ periodic ripple in transmittance. The transmittance variation typically requires the use of an instrument intensity correction function for the Raman spectral measurement. However, the rugate filter technology is still emerging, and it is likely that the rugate filter performance will dramatically improve in the near future.

2.3.3 Pre-monochromator Rayleigh Rejection

Single or double pre-monochromators are the method of last resort for Rayleigh light rejection [27]. These pre-monochromators utilize multiple dispersive elements (gratings) and spatial filters (slits) to reduce the amount of Rayleigh light that reaches successive spectrometer stages. The stray light from the relatively intense Rayleigh scattering is attenuated by 10^3 to 10^5 per monochromator stage. For example if a single stage of a monochromator reduces the stray light by 10^4 then a double monochromator will have a stray light background decrease of 10^8, and a triple monochromator will have a stray light background decrease of 10^{12}. Only double and triple monochromators permit Raman measurements below $50\,\mathrm{cm}^{-1}$.

Unfortunately, the high Rayleigh rejection efficiency of the pre-monochromator stages is accompanied by a loss in light throughput. A triple monochromator will typically have an optical throughput of 3–10% compared to 30–50% for a single monochromator. These inefficient multistage Raman spectrometers survive because they can be used over a broad range of wavelengths and because they can uniquely measure bands close to the laser line.

2.4 Raman Spectrometers

The availability of holographic notch filters and dielectric filters for Rayleigh rejection has allowed the development of simple dispersive and non-dispersive multichannel Raman spectrometers that utilize CCD detectors. The non-dispersive Raman spectrometers separate the Raman scattered light into its component frequencies through electronically or mechanically tunable bandpass filters [31]. Alternatively, an interferometer can be used to construct a Fourier transform Raman spectrometer [9]. This FT-Raman instrument utilizes the same interferometer technology as FT-IR spectrometers. However, for the typical near IR Raman excitation ($\lambda > 900\,\mathrm{nm}$) they utilize a high purity Ge or a GaAs detector. The most common and still most versatile Raman spectrometers still utilize holographic dispersive gratings and CCD multichannel detectors. These spectrometers are useful from the UV to the near IR spectral region.

2.4.1 Dispersive Raman Spectrometers

Dispersive Raman instruments are typically characterized by their optical design. Their figures of merit are given by their light collecting power ($f/\#$), their dispersion and their optical focal lengths [3]. Dispersive spectrometers may, for example, utilize on axis transmission optics or off axis reflecting optics [2,3]. The focal length of the spectrometer and the ruled line density of the grating determine the ultimate resolution of a dispersive instrument. The most useful spectrometer designs permit easy grating changes with easy wavelength calibration and rescaling of the Raman frequency scale when changing resolution and excitation wavelengths.

Dispersive spectrometers that scan the grating angle to pass different wavelength regions across a slit use a single channel detector, such as a photomultiplier. High precision and high accuracy cosecant scanning drives are required if the spectrometer is to scan linearly in cm^{-1}. Simpler grating drives can be suitable for spectrometers that utilize multichannel detectors if the spectrometer is calibrated at each grating setting; the entire Raman spectrum is acquired without moving the grating. However, high precision grating motion is still required if the spectrometer must be repeatedly reset to a precise orientation in order to repeatedly set a particular wavelength onto a particular pixel of the multichannel detector.

Practical optical considerations typically result in spectrometers with focal lengths between 0.25 m and 1.5 m and grating groove densities between 600 g/mm and 3600 g/mm. The dispersion of the Raman spectrometer is increased by either increasing the spectrometer focal length or increasing the number of lines per millimeter of the grating. Spectrometer resolution is generally increased by increasing the grating groove density rather than increasing the focal length.

Multichannel Raman spectrometers typically sacrifice high resolution for a large spectral window. Table 2.1 shows the typical best resolution of a commercially available 0.25 m focal length multichannel Raman spectrometer for typical laser and grating combinations with a 578 channel detector (resolution is typically limited to 2.5 pixels). Multichannel spectrometers are also available that simultaneously utilize two gratings and a large multichannel detector (1024 elements) to acquire the entire Raman spectrum (100–3500 cm^{-1}) in a single scan. However, the loss of effective instrument throughput and/or loss of resolution in these instruments may not be acceptable for some applications.

Table 2.1. Typical instrument resolutions (in cm^{-1}) utilizing various grating and laser combinations with a 250 mm focal length spectrometer

Grating (g/mm) Wavelength (nm)	2400	1800	1200
780	NA	0.6	1.6
633	0.52	1.5	2.9
514	1.6	2.8	4.9
457	2.4	3.8	6.5

Optimal dispersive Raman instrument performance requires that the spectrometer and Raman collection optics meet three conditions: (1) The spectrometer operate at a slit setting that matches the detector channel size to the spectrometer resolution. (2) The Raman light collection optics magnify the laser spot size at the sample to the entrance slit size (avoid overfilling the slit). (3) The Raman light collection optics match the spectrometer f-number (avoid overfilling the spectrometer optics). The spectrometer optical efficiency and resolution can also be degraded by optical imperfections such as chromatic aberration, astigmatism, and coma. While transmission optics typically minimize astigmatism and coma, chromatic aberration can become a serious problem in the red and blue. Likewise reflective optics avoid chromatic aberrations, but can introduce astigmatism and often coma.

2.4.2 FT-Raman Spectrometers

As the Raman excitation wavelength increases beyond $\sim 850\,\text{nm}$, CCD detectors and detectors, which utilize photocathodes (such as photomultipliers, and intensified CCD and Reticons) become inefficient. Unfortunately, the only useful near IR detectors have high background noise levels [32]. Thus, it is necessary to utilize multiplex techniques to obtain acceptable spectral signal-to-noise ratios. FT-Raman spectrometers are ideally suited for use with diode pumped Nd:YAG lasers operating at 1064 nm. While FT Raman spectrometers would show no increase in S/N for shot noise limited detectors, significant S/N increases occur with the noisy detectors used with near IR Raman excitation. The use of near IR excitation often has the crucial advantage of reducing fluorescence interference. In addition, FT Raman spectrometers have both high spectral resolution and high frequency precision. However, near IR Raman measurements are disadvantaged by the smaller Raman scattering cross sections in the near IR and the poorer performance of the near IR detectors. Table 2.2 compares the main advantages and disadvantages of near IR FT-Raman measurements versus typical visible Raman measurements using dispersive Raman instruments.

2.4.3 Detectors

Until recently photomultipliers were the standard detectors used for Raman spectral measurements. The entire UV, visible to near IR spectral region ($< 900\,\text{nm}$) is well covered with an AlGaAs photocathode which has quantum yields $> 10\%$ over this entire spectral range [32]. When cooled to $-40\,^\circ\text{C}$ and used with photon counting detection, these detectors (such as the RCA 1420A-02 PMT) are almost ideal detectors with only a few counts per second of background and a linear dynamic range of $> 10^6$. In fact, PMT detectors are still used for high resolution Raman measurements because they can be masked by a final slit which can be as narrow as a few µm (in contrast to the $\sim 25\,\mu\text{m}$ limits associated with the pixels of CCD and Reticon detectors).

Table 2.2. Relative performance characteristics of dispersive and FT-Raman instruments

Feature	Dispersive Raman	FT-Raman
Available wavelengths	< 200 nm to 850 nm	1064 nm (99% systems)
Detector	CCD, shot noise limited	Ge or InGaAs, detector noise limited
Best spectral resolution	typically 1–4 cm^{-1}	$\sim 0.5\,cm^{-1}$
Fluorescence suppression	moderate at 785 nm, poor at 514 nm, good at 244 nm	Excellent
Operation at elevated temperatures	excellent, $> 1000\,°C$	poor, $< 250\,°C$
Relative v^4 advantage (from 1064)	@ 785 nm: 3.38 @ 514 nm: 18.3 @ 244 nm: 362	1

Multichannel detectors [1,3] are far superior for lower resolution studies ($> 1\,cm^{-1}$) because their multiplex advantage increases the spectral signal-to-noise ratios by the square root of the number of resolution elements over a shot-noise limited detector such as a photomultiplier. The selection of a specific CCD camera for Raman spectroscopy requires careful consideration. The wide selection of chip types, pixel sizes and operating temperatures must be considered for background dark count rate, quantum efficiency and read out noise. The scientific CCD detector with a quantum efficiency approaching 70% and very low dark current in the visible is almost ideal for most NIR and visible wavelength Raman studies. With UV Raman the detector quantum efficiency must be enhanced either by depositing UV fluorophores on the CCD surface, or by utilizing a backthinned CCD. However at the lowest light levels, where detector read noise or detector background dominates the S/N ratio, image intensifiers significantly improve the detectivities of CCD arrays.

With UV excitation the advantages of utilizing an intensified CCD typically far outweigh the disadvantages of a decreased detector dynamic range and the increased statistical variance associated with the distribution of gain in the intensifier. Most UV Raman spectral measurements involve a low photon flux. Intensified CCDs are especially helpful to align and set up a Raman measurement; the alignment is guided by observing the real time Raman spectrum. We have found that an intensified CCD array gives significantly higher spectral signal-to-noise values in the UV compared to any unintensified CCD detector.

2.4.4 Imaging Raman Spectrometers

Raman spectroscopy with a microscope is rapidly becoming the method of first choice for Raman analysis. The use of a microscope operating in a 180°

backscattering geometry eliminates the need to continually adjust the laser onto the sample and to focus the scattered light onto the spectrometer. Raman microspectrophotometers utilize research grade microscopes to focus the excitation onto the sample and to collect and transfer the Raman scattered light into the Raman spectrometer. High numerical aperture microscope objectives greatly enhance the spatial resolution and the optical collection power of the Raman instrument. Once aligned, the user is only required to place a sample under the microscope and adjust it for best optical focus. These Raman microscopes are easy to use and are capable of analyzing small areas ($\sim 1\,\mu m^2$) in order to determine the spatial distributions of chemical species.

Traditional Raman images are obtained by translating the sample across the microscopic objective focus with a motorized stage (Raman image mapping) and constructing images from data extracted from each spectrum. However new generations of Raman microscopes are emerging that utilize novel tunable filters to obtain Raman spectral images by illuminating a large field of view, typically 20–200 μm, and analyzing it for one or more specific Raman wavelengths (global Raman imaging) [31].

Commercial global Raman imaging spectrographs are available that use dielectric filters, acousto-optic tunable filters (AOTF) or liquid crystal tunable filters (LCTF). The electronic wavelength tuning ability and high image quality and high spectral resolution ($7-9\,cm^{-1}$) of LCTF devices make them the preferred global imaging dispersing elements, in spite of their lower optical efficiencies ($< 20\%$). Although, dielectric filters have a large throughput (60%) they have a lower resolution ($\sim 15\,cm^{-1}$). Raman imaging spectrometers equipped with liquid crystal tunable filters have been demonstrated to give 250 nm spatial resolution and $7\,cm^{-1}$ spectral resolution.

Still higher spatial resolution (~ 100 nm) is available with the use of a Raman scanning near field optical microscope (RSNOM) [33]. The RSNOM uses a special fiber optic tapered to an aperture less than the optical wavelength to couple the laser light to the sample. The spatial resolution is limited by the size of the aperture, the proximity of the tip to the sample and the excitation wavelength. Coupling the Raman excitation light out of the SNOM tip remains a very inefficient process, and extremely long integration times are required to obtain moderate S/N from very strong Raman scatterers. The best results have been obtained where the RSNOM excitation wavelength is in resonance with an absorption band of the sample producing a resonance Raman spectrum [33].

2.5 Examples of New Raman Instruments for Materials Characterization

In our laboratory we are using UV Raman spectroscopy to investigate the growth and structure of CVD grown diamond [34–36]. One objective is to obtain additional insight into the diamond growth mechanism, in order to optimize the process to increase the rate of diamond growth and to improve CVD diamond quality. With this in mind we built two instruments to examine CVD diamond films. For *ex situ* CVD diamond film studies we built a UV Raman microscope [37] in order to spatially examine the diamond bands to monitor stress, diamond crystal size, and to examine the structure and location of non-diamond impurities within the CVD diamond films. In addition, we built a separate UV Raman instrument to examine *in situ* growing diamond films within a CVD plasma reactor [38].

2.5.1 UV Raman Microspectrometer for CVD Diamond Studies

Figure 2.1 illustrates the optical layout of the microspectrometer. We utilize a modified Olympus U-RLA microscope with an epi-illuminator and a universal lamp housing. The excitation beam is introduced to the sample independently of the light collection optics, as opposed to the more typical epi-illumination. Our design has the advantage that the beam focal spot size and position are adjustable independently of the focusing conditions for collecting the Raman scattered light. In addition, since we use a Cassegrain objective, the prism in front of the objective does not obscure either the excitation beam or collected scattered light.

We utilize an intracavity frequency doubled argon ion laser to excite the UV Raman scattering [23]. For many of these experiments we utilized 244 and 229 nm excitation. The CW UV Raman laser beam is expanded to a ~ 10 mm

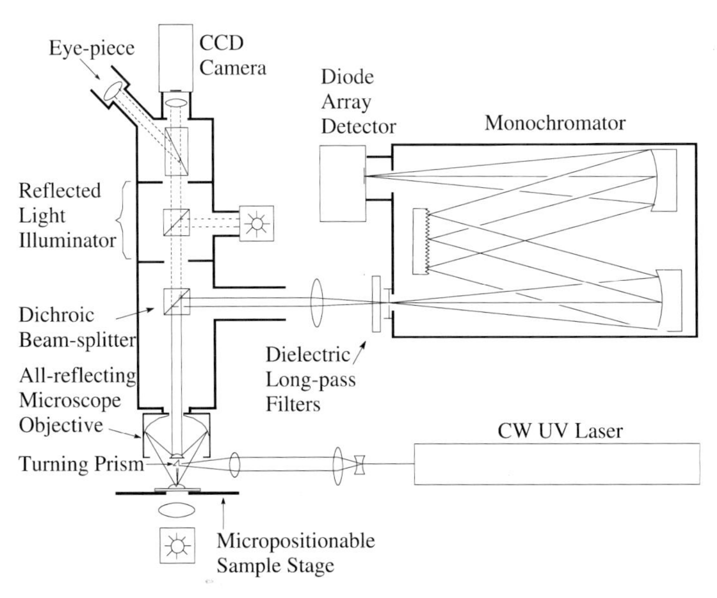

Fig. 2.1. Schematic showing the optical layout of the UV Raman microspectrometer

diameter and focused using a 5–10 cm focal-length lens onto the sample via a Suprasil 90° turning prism mounted directly below the Cassegrain objective. The beam can be focused to a spot size of 5–25 μm. An Opticon Corp. 36× all-reflective Cassegrain microscope objective with a back focus of 160 mm, a working distance of 10.5 mm and a numerical aperture (N.A.) of 0.5 is used to collect the backscattered light. The 0.5 N.A. objective enables collection of scattered radiation over a large solid angle (half angle = 30°). The objective has a dielectric over-coated aluminum coating (Al/MgF_2). This Cassegrain objective serves as a highly efficient collection optic for the scattered radiation: Collecting at $f/1$, the sample spot size of 5–10 μm can be imaged efficiently into the entrance slit (100–200 μm) of the monochromator ($f/6.8$).

An Omega Optical Inc. 290 DCLPO2, UV dichroic beam splitter was used to reflect > 90% of the scattered UV light between 230 and 265 nm towards the collecting optics of the monochromator and to transmit light between 300 and 2000 nm to the microscope trinocular eyepiece. The dichroic beam splitter is mounted in a fluorescence cube module, housed in the epi-illuminator turret that was modified to enable the efficient coupling of the scattered UV radiation into the spectrograph.

A 0.75 m single monochromator ($f/6.8$) was used to disperse the scattered light. We utilized two dielectric longpass filters, custom constructed by Omega Optical Inc. with a transmittance of 0.01% at 244 nm and 65–80% between 252 nm and 262 nm to reject the Rayleigh scattered light.

We earlier demonstrated that diamond Raman spectra excited within or close to the diamond bandgap have dramatically improved S/N ratios, due to the lack of interfering fluorescence signals [34–36]. This allowed us to monitor the spectral differences between different non-diamond carbon species. We were also able to observe for the first time the carbon-hydrogen (C – H) stretching vibrations of the non-diamond components of CVD diamond films and to examine the intensity and frequency of the third-order phonon bands of diamond. Furthermore, we were able to detect and quantify different non-diamond carbon species in the CVD diamond films.

Figure 2.2 shows the surface of the (100) face of a diamond crystallite which occurs on the surface of a CVD diamond film, while Fig. 2.3 shows the UV Raman spectra of this CVD diamond film excited at 244 nm (~ 1.5 mW). The diamond UV Raman spectra were recorded with the laser spot centered on the (100) face of this single diamond crystallite (Fig. 2.3a), or at the grain boundaries between diamond crystallites (Fig. 2.3b).

The absolute intensity of the diamond first order phonon band at 1332 cm^{-1} was approximately the same at the (100) face (Fig 2.3a)and at the grain boundaries (Fig 2.3b). However, the UV Raman spectrum taken from the grain boundaries showed a broad band at ~ 1600 cm^{-1}, assignable to non-diamond carbon impurities. This band was not present in the spectrum from the (100) crystallite face. These results demonstrate the ability to determine the spatial distribution of non-diamond impurities in CVD diamond films.

Fig. 2.2. CVD diamond film surface viewed with visible light epi-illumination through the microscope attachment of the UV Raman microspectrometer, showing the (100) faces of single crystallites

Our previous study of the oxidative degradation of CVD diamond films showed that upon oxidation the intensity of the broad non-diamond carbon

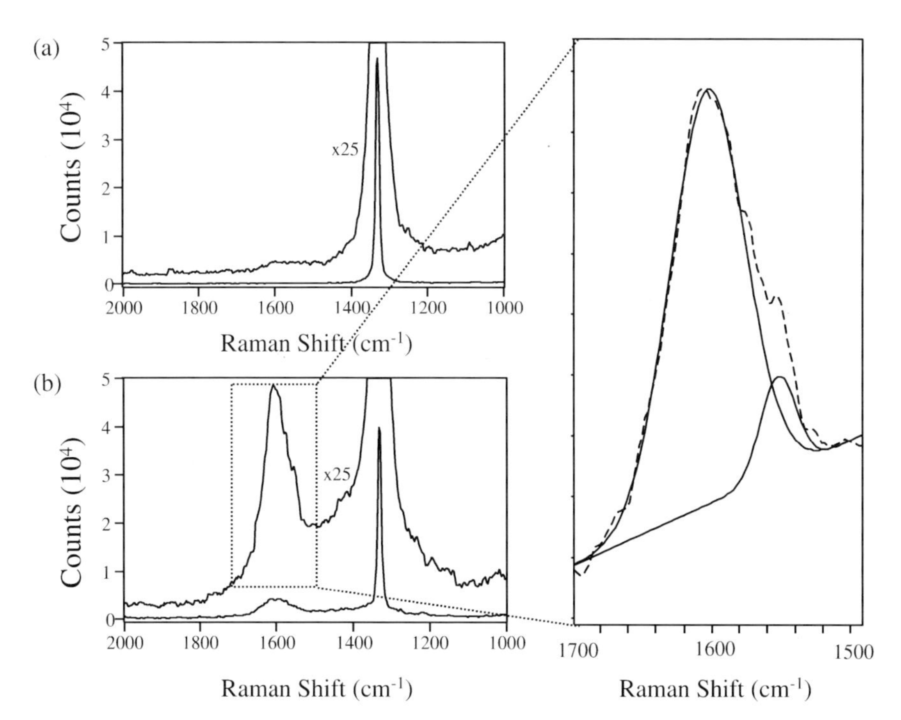

Fig. 2.3. (**a**) Raman spectra from CVD diamond film of the (100) face of a single crystallite, and (**b**) at the interstices between diamond crystallites (244 nm: ~ 1.5 mW average power: 10 s total accumulation time: 100 μm entrance slit)

band at $\sim 1550\,\text{cm}^{-1}$ and the C – H stretching band of the non-diamond components at $2930\,\text{cm}^{-1}$ decreased with respect to the diamond first order phonon band, but that the initial rate of decrease was significantly greater for the $1550\,\text{cm}^{-1}$ band than for the $2930\,\text{cm}^{-1}$ band [34–36]. These results indicate that non-diamond carbon species are oxidized in preference to diamond. They also suggest that more than one non-diamond carbon impurity is present in CVD diamond films.

Figure 2.3b illustrates that by using UV Raman microspectrometry we can resolve underlying components of the non-diamond carbon band. In this instance, a sharp low energy feature, fitted to the $\sim 1553\,\text{cm}^{-1}$ amorphous carbon band, is resolved from the broad $\sim 1603\,\text{cm}^{-1}$ non-diamond carbon band. In other instances, the sharp $\sim 1580\,\text{cm}^{-1}$ graphite band dominates the non-diamond carbon band. The limited spatial area probed enables us to speciate the different non-diamond carbon species that make up the normally broad non-diamond carbon band.

2.5.2 UV Raman Instrument for in situ Studies of CVD Diamond Growth

A UV Raman spectrometer [38] was constructed to examine *in situ* the growth of diamond in a microwave plasma CVD diamond reactor (Fig. 2.4). We utilized a 1.5 kW ASTeX microwave plasma reactor that was modified with silica viewports for spectroscopic access to the growing films during deposition. The *in situ* UV Raman spectra were excited with CW 244 nm light from a Coherent Innova 300 intracavity frequency doubled argon ion laser. The 244 nm output was expanded and focused through a silica viewport and onto the growth substrate/sample inside the reactor.

The scattered light was collimated by using a 90° off-axis parabolic mirror. Two 244 nm dielectric stack filters were used to reject Rayleigh scattering. The filtered Raman scattered light was reimaged onto the slit of a modified Spex 1701, 0.75 m single monochromator (f/6.8) equipped with a 2400 groove/mm holographic grating, and an EG&G PARC 1456 blue intensified photodiode array optical multichannel analyzer. A spatial filter (600 µm aperture) was incorporated into the collection optical train directly behind the spectrograph entrance slit to limit the measured sample volume by approximating confocal imaging. Approximate confocal imaging was used to help minimize interference from the plasma emission in the reactor. Although the plasma emission intensity in the visible is sufficiently high to prevent visible wavelength Raman measurements, it decreases dramatically in the UV spectral region below ~ 260 nm.

Figure 2.5 shows the *in situ* temperature dependence of the first order Raman band of the growing diamond films ($\sim 1332\,\text{cm}^{-1}$ at room temperature). We see very high S/N spectra for relatively short ~ 10 min spectral accumulation times. The independently measured plasma emission (measured in the absence of the excitation beam) is easily subtracted off. We calculate

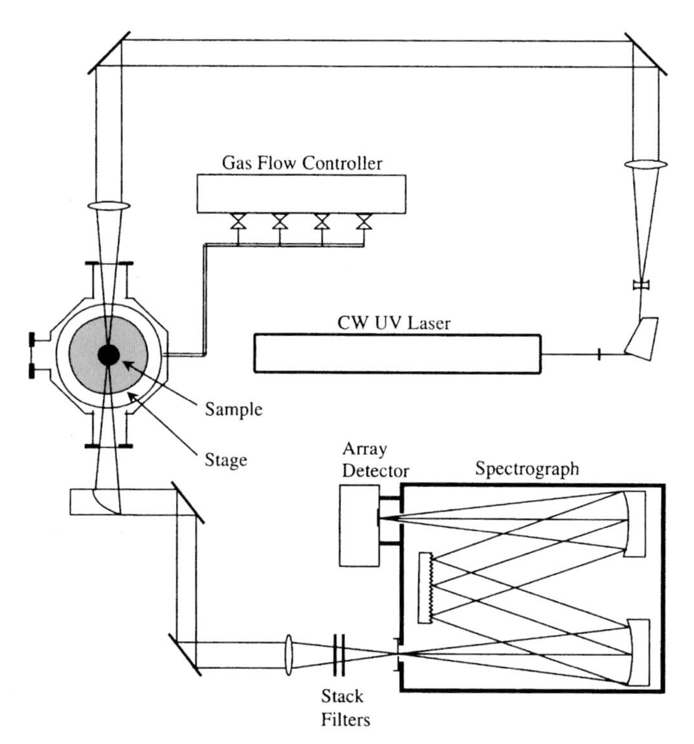

Fig. 2.4. Schematic diagram showing the UV Raman spectrometer coupled to the ASTex plasma CVD diamond growth chamber

an ∼ 8 Å thickness *in situ* detection limit for the growing CVD diamond films within the plasma reactor.

The spectra display the well known frequency decrease of the first order phonon band with temperature. This band frequency can be independently used for determining temperature. We were surprised not to observe the expected nondiamond carbon impurity bands in these spectra, since they were clearly observed in spectra of these same CVD films when they were cooled down to room temperature. To our surprise we discovered that these nondiamond carbon bands are enhanced by a narrow electronic resonance at ∼ 244 nm whose frequency is temperature dependent. This electronic resonance shifts away from this 244 nm excitation wavelength at elevated temperatures such that the Raman intensity is significantly decreased. Future experiments will use adjacent UV excitations such as 229 or 238 nm to examine these nondiamond carbon bands at the elevated temperatures required for CVD diamond growth.

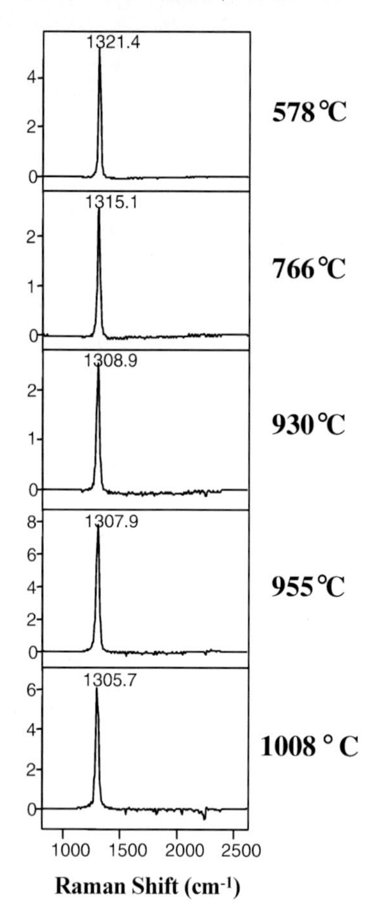

Fig. 2.5. Temperature dependence of the in situ 244 nm UV Raman spectra of CVD diamond films grown within the plasma reactor

2.6 Conclusions

These are only two examples of the development of Raman instrumentation for material science applications. These examples, show that Raman spectroscopy is a uniquely powerful probe of the underlying properties of materials and molecular structure of the constituents. The technique measures both vibrational frequencies and intensities. The frequencies report on chemical bonding, molecular structure and environment. The intensities monitor the strength of coupling between vibrational and electronic motion in the material. The spectra obviously have an extraordinarily high information content. Future instrument improvements will further enable the growth and adaptations of Raman spectroscopy to a variety of new applications. The new generations of Raman spectrometers are compact, highly efficient and highly adaptable instruments. This is a very exciting time to be working in the field of Raman spectroscopy.

References

1. L.A. Lyon, C.D. Keating, A.P. Fox, B.E. Baker, L. He, S.R. Nicewarner, S.P. Mulvaney, M.J. Natan: Anal. Chem. **70**, 341R (1998)
2. M. Delhaye, J. Barbillat, J. Aubard, M. Bridoux, E. Da Silva: in *Raman Microscopy: Developments and Applications*, ed. by G. Turrell, J. Corset (Academic, London 1996)
3. R.L. McCreery: *Modern Techniques in Raman Spectroscopy, Instrumentation for Dispersive Raman Spectroscopy*, ed. by J.J. Laserna (Wiley, New York 1996). pp. 41–72
4. Photonics Spectra **66** (December 1998)
5. (a) J. Ma, Y.S. Li: Appl. Opt. **35**, 2527 (1996) (b) S.E. Nave: ISA, **96**, 453 (1996)
6. D.A. Long: *Raman Spectroscopy* (McGraw-Hill, New York 1977); D.P. Strommen, K. Nakamoto:*Laboratory Raman Spectroscopy* (Wiley, New York 1984)
7. S.A. Asher: Ann. Rev. Phys. Chem. **39**, 537 (1988)
8. J.M. Dudik, C.R. Johnson, S.A. Asher: J. Chem. Phys. **82**, 1732 (1985)
9. D.B. Chase, J.F. Rabolt:*Fourier Transform Raman Spectroscopy* (Academic, New York 1994)
10. S.A. Asher, C.R. Johnson: Science **225**, 311 (1984)
11. S.A. Asher, C.H. Munro, Z. Chi: Laser Focus World, 99 (1997)
12. (a) S.A. Asher: Anal. Chem. **65**, 59A-66A (1993) (b) S.A. Asher: Anal. Chem. **65**, 201A (1993)
13. M. Sparrow, W. Hug, S.A. Asher: Appl. Spectrosc., in preparation (1999)
14. T.C. Strekas, T.G. Spiro: Biocim. Biophys. Acta **263**, 830 (1972)
15. M.D. Levinson: *Introduction to Non Linear Laser Spectroscopy* (Academic Press, New York (1982)
16. J. Teraoka, P.A. Harmon, S.A. Asher: J. Am. Chem. Soc. **112**, 2892 (1990)
17. P.A. Harmon, J. Teraoka, S.A. Asher: J. Am. Chem. Soc. **112**, 8789 (1990)
18. T.F. Cooney, H.T. Skinner, S.M. Angel: Appl. Spectrosc. **49**, 1846 (1995)
19. C.M. Jones, V.L. Devito, P.A. Harmon, S.A. Asher: Appl. Spectrosc. **41**, 1268 (1987)
20. S.A. Asher, C.R. Johnson, J. Murtaugh: Rev. Sci. Instr. **54**, 1657 (1983)
21. B. Hudson, R.J. Sension, R.J. Brudzynski, S. Li:*Proceedings of the XI International Conference on Raman Spectroscopy*, ed. by R.H. Clark, D.A. Long (Wiley, New York 1988)
22. (a) S.A. Asher, Z. Chi, J.S.W. Holtz, I.K. Lednev, A.S. Karnoup, M.C. Sparrow: J. Am. Chem. Soc. **121**, 4076 (1999) ; (b) I.K. Lednev, A.S. Karnoup, M.C. Sparrow, S.A. Asher: J. Am. Chem. Soc. **121**, 8074 (1999)
23. S.A. Asher, R.W. Bormett, X.G. Chen, D.H. Lemmon N. Cho, P. Peterson, M. Arrigoni, L. Spinelli, J. Cannon: Appl. Spectrosc. **47**, 628 (1993)
24. J.S.W. Holtz, R.W. Bormett, Z. Chi, N. Cho, X.G. Chen, V. Pajcini, S.A. Asher, L. Spinelli, P. Owen, M. Arrigoni: Appl. Spectrosc. **50**, 1459 (1996)
25. C.L. Schoen, S.K. Sharma, C.E. Helsley, H. Owen: Appl. Spectrosc **47**, 305 (1993)
26. P.L. Flaugh, S.E. O'Donnell, S.A. Asher: Appl. Spectrosc. **38**, 847 (1984)
27. See e.g. [3] or [6]
28. M. Futamata: Appl. Spectrosc. **50**, 199 (1996)
29. C.H. Munro, V. Pajcini, S.A. Asher: Appl. Spectrosc. **51**, 1722 (1997)

30. B.G. Bovard: Appl. Opt. **32**, 5427 (1993)
31. H.R. Morris, C.C. Hoyt, P. Miller, P.J. Treado: Appl. Spectrosc **50**, 805 (1996)
32. B. Chase: Appl. Spectrosc. 48, 14A (1994)
33. Y. Narita, T. Tadokoro, T. Ikeda, T. Saiki, S. Mononobe, M. Ohtsu: Appl. Spectrosc. **52**, 1141 (1998)
34. C.D. Zuiker, A.R. Krauss, D.M. Gruen, J.A. Carlisle, L.J. Terminello, S.A. Asher, R.W. Bormett: Mat. Res. Soc. Symp. Proc. **437**, 211 (1996)
35. R.W. Bormett, S.A. Asher, R.E. Witkowski, W.D. Partlow, R. Lizewski, F. Pettit: J. App. Phys **77**, 5916 (1995)
36. V.I. Merkulov, J.S. Lannin, C.H. Munro, S.A. Asher, V.S. Veerasamy, W.I. Milne: Phys. Rev. Lett. **78**, 4869 (1997)
37. V. Pajcini, C.H. Munro, R.W. Bormett, R.E. Witkowski, S.A. Asher: Appl. Spectrosc. **51**, 81 (1997)
38. J.C. Worthington, R.W. Bormett, C.H. Munro, R.E. Witkowski, S.A. Asher: in *Proc. XV International Conf. on Raman Spectrosc.*, ed. by S.A. Asher, P. Stein (John Wiley, 1996) p. 1218

3 Characterization of Bulk Semiconductors Using Raman Spectroscopy

J. Menéndez

Abstract. This chapter provides a brief overview of Raman scattering in semiconductors. The large amount of experimental information on these materials is reviewed from the perspective of the latest theoretical advances, and detailed expressions are given that should allow the reader to compute Raman spectra and to compare theoretical cross sections with experimental photon counts. Applications of Raman spectroscopy to measurements of crystal orientation, temperature, stress, impurities and alloying are succinctly reviewed, with emphasis on the principles underlying these applications and on the scope and limitations of the information that can be obtained with Raman spectroscopy.

The fabrication of the first transistor in 1948 represents one of the greatest triumphs of the quantum theory of materials. This historic event paved the way for the spectacular technological developments of the second half of the twentieth century. The semiconductor materials that enabled the new technologies became the target of intense research, and as a result of this effort their electronic and vibrational properties are known today with unparalleled detail, both from the experimental and theoretical points of view [1,2].

The large amount of experimental information available on semiconductors, combined with our theoretical understanding of these data, bode well for the spectroscopic characterization of these materials, since the effect of external perturbations on their electronic and vibrational spectra can be reliably predicted. In this chapter, we review some of the applications of Raman spectroscopy to the characterization of semiconductors. Inelastic light scattering – and laser spectroscopy in general – is very attractive as a characterization tool due to its contactless and non-destructive nature. Typical applications to be discussed in this chapter are the use of Raman spectroscopy for the determination of crystalline orientations, the measurement of temperature and stress, the characterization of doping levels, and the study of alloy semiconductors. Most of these applications emphasize vibrational (phonon) Raman spectroscopy, but it is important to keep in mind that one of the most significant strengths of Raman spectroscopy is its ability to probe the electronic structure of semiconductors as well. This is accomplished by detecting inelastic light scattering by electronic excitations [3,4] (instead of phonons) and also by measuring the intensity of Raman scattering by phonons as a function of the incident laser wavelength. This intensity undergoes resonant enhancements when the density of allowed interband optical transitions has a singularity at the laser photon energy. Thus the measurement of "excitation" profiles (Raman intensity versus laser wavelength), usually referred to as Resonance Raman Spectroscopy (RRS), is a powerful tool for the determination of direct band gaps and other above-gap direct transitions. RRS has been extensively reviewed in the literature [5–8], since in addition to providing information on interband transition energies it is the main experimental technique for the determination of the deformation potential parameters that characterize the electron-phonon interaction. A combination of electronic Raman scattering and RRS has been shown to be a very convenient technique for the measurement of band offsets in semiconductor heterostructures [9].

Our selection of topics for this chapter is by no means comprehensive, and even within this selection we will make no attempt at enumerating all published studies. As opposed to a standard review article, this chapter is written with those industrial and academic researchers in mind who would like to gain an initial understanding of the capabilities of Raman scattering as a characterization tool for semiconductors. Turnkey Raman systems targeted to general users (rather than expert practitioners) are becoming increasingly popular, and this chapter also provides some key formulas and experimental data to help these users make the best of their new equipment. Readers in search of comprehensive accounts are referred to the monumental *Landolt-Börstein* tables [10,11] for raw experimental data, to several excellent textbooks [1,2,12,13] for a reasoned and detailed exposure of semiconductor theory, and to a number of very instructive books and reviews on related topics [7,14–17]. Powerful search engines are now available on the Internet and they should facilitate the identification of the vast majority of the very valuable work not mentioned in our sampling of results.

Our chapter starts with an introductory section, mainly theoretical, which defines the basic concepts and explains the latest advances in our understanding of lattice dynamics and Raman spectra in semiconductors. This is followed by sections describing each of the selected applications areas. The box by Marcos Grimsditch on Brillouin scattering illustrates the applications of this "sister technique", and the box by Ingrid de Wolf describes how stress characterization is carried out in practice.

3.1 Inelastic Light Scattering by Phonons in Semiconductors

A general introduction to lattice dynamics and light scattering has been given in Chap. 1. Here we concentrate on those aspects of the theory that are particularly relevant to semiconductors. We focus on group IV elemental semiconductors and III-V or II-VI compounds which crystallize in the diamond or zincblende structures. In the remainder of this chapter, the word "semiconductors" will be meant to refer to these cubic tetrahedral semiconductors and not necessarily to other semiconducting systems. Table 3.1 shows some characteristic data for these materials. It should be mentioned that some tetrahedral semiconductors – including the nitride family, which has recently attracted much attention for their remarkable optoelectronic properties – crystallize in the wurtzite structure. This structure will not be discussed any further here, but nitride semiconductors are reviewed in Chap. 7.

Table 3.1. Selected structural, electronic, and vibrational properties of cubic semiconductors at room temperature. The last two columns show the frequencies of the longitudinal and transverse optic phonons at $\boldsymbol{q} \cong 0$. These are the only allowed Raman phonons in diamond and zincblende semiconductors

Material	Lattice parameter a (in Å)	Indirect gap (eV)	Direct gap (eV)	$\omega_{\mathrm{LO}}(\mathrm{cm}^{-1})$	$\omega_{\mathrm{TO}}(\mathrm{cm}^{-1})$
Diamond	3.56683	5.4	6.5		1332
Si	5.43086	1.12	4.25		521
Ge	5.65748	0.66	0.81		303
α-Sn	6.489		0		196
AlP	5.4635	2.41	3.62	499	440
AlAs	5.660	2.14	3.03	403	362
AlSb	6.136	1.63	2.30	340	319
GaP	5.4505	2.26	2.78	403	367
GaAs	5.6534		1.429	292	269
GaSb	6.0958		0.75	237	227
InP	5.8687		1.34	345	304
InAs	6.0584		0.356	243	218
InSb	6.47877		0.18	193	180
ZnS	5.4102		3.69	352	271
ZnSe	5.6676		2.69	250	205
ZnTe	6.1037		2.25	205	177
CdTe	6.482		1.49	168	139
HgTe	6.453		0	137	118

3.1.1 Phonons in Semiconductors

Within the harmonic approximation, vibrations are solutions to the eigenvalue problem [18]

$$\left(\boldsymbol{\Phi} - \omega_f^2 \boldsymbol{M}\right) \varepsilon(f) = 0. \tag{3.1}$$

Here $\boldsymbol{\Phi}$ is a force-constants matrix whose elements are given by $\Phi_{ij}(l\kappa, l'\kappa') = \partial^2 V(u)/(\partial u_i(l\kappa)\partial u_j(l'\kappa'))$, where $V(\boldsymbol{u})$ is the crystalline potential and the atomic displacements are indicated collectively by $\boldsymbol{u}$, which is a $3nN$-dimensional vector formed by combining the 3-dimensional displacement vectors for each of the n atoms in the N unit cells of the system. The displacement in the i-direction of the κth atom in the l-th unit cell is denoted by $u_i(l\kappa)$. Therefore, $\Phi_{ij}(l\kappa, l'\kappa')$ is equal to minus the force in the j-direction on atom $(l'\kappa')$ for a unit displacement of atom $(l\kappa)$ in the i-direction. The dimensionality of $\boldsymbol{\Phi}$ is $3Nn \times 3Nn$. For diamond and zincblende semiconductors $\kappa = 1$ or 2, so that $n = 2$). The matrix $\boldsymbol{M}$ is given by $M_{ij}(l\kappa, l'\kappa') = M(\kappa)\delta_{l\kappa i, l'\kappa' j}$, where $M(\kappa)$ is the mass of the κth atom in the unit cell. Equation (3.1) can be diagonalized numerically using standard packages for generalized eigenvalue problems. It is quite common to rewrite this equation as a more conventional eigenvalue problem of the form $H\Psi = E\Psi$ by defining a dynamical matrix $D_{ij}(l\kappa, l'\kappa') = \Phi_{ij}(l\kappa, l'\kappa')/(M_\kappa M_{\kappa'})^{-1/2}$, but this approach tends to obscure the physics of some common phenomena by combining in the dynamical matrix possible mass and force constant changes. This will become apparent below, as we discuss the effect of perturbations on the lattice dynamics of semiconductors.

The eigenvectors of (3.1) satisfy the completeness and orthonormality conditions

$$\sum_f^{3nN} \boldsymbol{M}\tilde{\varepsilon}^\dagger(f)\varepsilon(f) = I$$

$$\tilde{\varepsilon}^\dagger(f)\boldsymbol{M}\varepsilon(f') = \delta_{ff'} \tag{3.2}$$

so that any arbitrary collective displacement $\boldsymbol{u}$ can be written as a linear combination of these eigenvectors as

$$\boldsymbol{u} = \sum_f \varepsilon(f) Q_f. \tag{3.3}$$

The coefficients Q_f are known as the normal coordinates of the system. For a periodic solid Bloch's theorem requires that the eigenvectors be plane waves, which can be written as

$$\varepsilon_i(l\kappa \mid \boldsymbol{q}m) = \frac{e_i(\kappa \mid \boldsymbol{q}m)}{\sqrt{NM(\kappa)}} \exp\left[\mathrm{i}\boldsymbol{q} \cdot \boldsymbol{R}(l\kappa)\right]. \tag{3.4}$$

Here the e's are the so-called phonon polarization vectors. Substituting (3.4) into the orthonormality condition in (3.2), one finds that the phonon polarization vectors are normalized according to $\sum_{\kappa i} \mid e_i(\kappa \mid \boldsymbol{q}m) \mid^2 = 1$.

The ansatz represented by (3.4) decouples the eigenvalue problem in (3.1) into N sets of $3n \times 3n$ problems, one for each wave vector $\boldsymbol{q}$. The index f, which labels the $3Nn$ independent solutions to (3.1), is split into N wave vectors $\boldsymbol{q}$ and a branch index m that runs from 1 through $3n$. The $3n$ functions $\omega_m(\boldsymbol{q})$ represent the phonon dispersion curves. These curves can be measured using inelastic neutron scattering (INS), as shown in Fig. 3.1 for GaAs [19] and therefore represent the contact point between theory and experiment. Since Raman spectroscopy is kinematically restricted to $\boldsymbol{q} \cong 0$ phonons, its value as a probe of the lattice dynamics of semiconductors is much more limited. However, second order Raman scattering involves pairs of phonons with wave vectors $\boldsymbol{q}_1$ and $\boldsymbol{q}_2$ such that $\boldsymbol{q}_1 + \boldsymbol{q}_2 \cong 0$, and therefore it can be used to explore phonon frequencies away from $\boldsymbol{q} = 0$, as explained in the introductory chapter. In tetrahedral semiconductors the polarized (parallel incident and scattered light polarizations) second order Raman spectrum is essentially proportional to the phonon density of states,whose critical pointslead to sharp features in the Raman spectra [20]. The spectroscopy of these features can provide important lattice dynamical information in those cases for which it is not possible to obtain INS data. A good example is AlAs, which is only available as thin epitaxial layers due to the tendency of this material to oxidize rapidly in the presence of air. Since INS requires large crystalline samples, no INS spectra are available for AlAs. However, second order Raman scattering has provided strong evidence in support of the AlAs phonon dispersion curves calculated from density functional theory [21]. Second order Raman

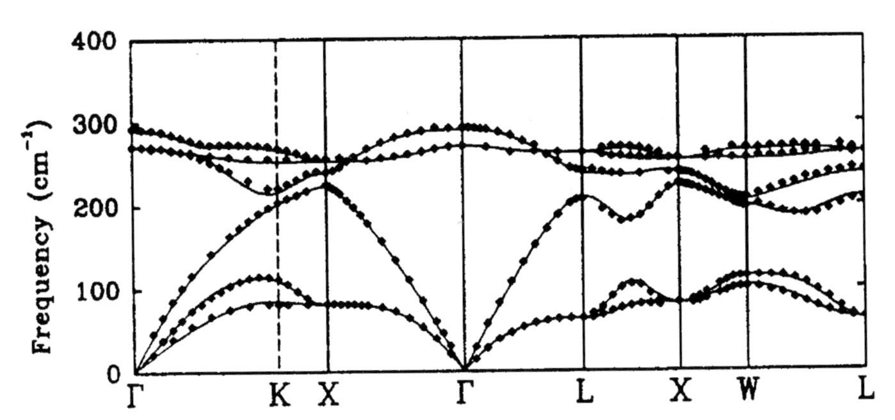

Fig. 3.1. *Diamonds:* experimental low temperature phonon dispersion relations for GaAs [19]. *Solid lines*: Calculated dispersion relations from density functional perturbation theory [30]

scattering has also played an important role in the study of the overbending of optical phonons in materials such as diamond [22] and AlSb [23].

The basic lattice dynamical problem is the determination of the force constants matrix $\boldsymbol{\Phi}$. If one considers the fact that in most crystalline solids – including tetrahedral semiconductors – the ionic displacements are extremely slow relative to typical times for electronic transitions, it can be assumed that the system remains in its ground electronic state as the ions move. Within this adiabatic approximation the instantaneous crystalline potential V is the total ground state energy of the crystal for a "frozen" displacement of the atoms from their equilibrium positions. Modern ab initio calculations of phonons – to be discussed below – are based on this idea. The traditional approach, however, has been to fit an empirical potential to the experimental phonon dispersion curves. The simplest such approach (sometimes called the Born–von-Karman model) takes the elements of the force constant matrix $\boldsymbol{\Phi}$ as adjustable parameters. The number of different parameters is expected to be manageable due to symmetry considerations and to the fact that distant atoms are not expected to interact strongly. However, the pioneering work of Frank Herman showed as early as 1959 that in the case of tetrahedral semiconductors interactions up to fifth neighbors had to be taken into account for a reasonable fit of the measured phonon dispersion relations [24]. These medium – range interactions are responsible-among other details of the phonon dispersion curves – for the dramatic flattening of the transverse acoustic branches as the wave vector approaches the Brillouin zone edge along same low-index directions [25]. This is clearly seen in Fig. 3.1 for GaAs, and it is observed in all tetrahedral semiconductors except diamond.

The physical interpretation of a Born–von-Karman fit to the phonon dispersion relations in tetrahedral semiconductors is complicated not only because of the large number of parameters needed but, more fundamentally, because the charge density in covalent materials is localized not only around the nuclei but also along the interatomic bonds. If interactions involving the bond charges play an important role in the lattice dynamics, they will impact the Born–von-Karman parameters in an indirect and in general very complicated way. Therefore, it appears that one can arrive at a much better understanding of phonons in tetrahedral semiconductors by explicitly constructing an interatomic potential that takes into account the important interactions. Once this potential is proposed, the Born–von-Karman parameters follow from simple differentiation and the phonon dispersion curves can be computed in terms of the adjustable parameters of the proposed interatomic potential. The most successful application of such an approach is Weber's Adiabatic Bond Charge model, which produces an excellent fit of the phonon dispersion relations in diamond, Si, and Ge with only 4 adjustable parameters [26]. The model can be extended to III-V and II-VI zincblende semiconductors by increasing the number of parameters to six [27,28]. The adiabatic bond charge model provides a straightforward interpretation for many of the peculiarities of the phonon dispersion relations in semiconductors. In particular, the flattening of the transverse acoustic branches can be traced back to the stiffness of the interaction between neighboring bond charges [25,26].

A completely new approach to the lattice dynamics of tetrahedral semiconductors became possible in the 1970's, as electronic total energy calculations based on density functional theory reached a high level of accuracy [29]. As mentioned above, for any crystalline distortion $\boldsymbol{u}$ the interatomic potential $V(\boldsymbol{u})$ can be equated (within the adiabatic approximation) to the system's energy eigenvalue $E(\boldsymbol{u})$ corresponding to the ground electronic state for frozen displacements $\boldsymbol{u}$. This makes it possible to obtain force constants from "first principles" without any fit to experiment. The first applications were restricted to phonons propagating along high symmetry directions, but the introduction of "Density Functional Perturbation Theory" (DFPT) made it possible to calculate phonon frequencies and eigenvectors for arbitrary wavevectors within the Brillouin zone [30]. The agreement with experiment is truly remarkable, as shown in Fig. 3.1. In addition to the excellent numerical agreement with experiment, the ab initio method offers new insights into the lattice dynamics of semiconductors. Figure 3.2 shows the phonon dispersion relations for GaAs calculated with GaAs and AlAs force constants. It is apparent that the resulting curves are virtually identical. A systematic study [31] finds that the force constants are a smooth function of the lattice parameter a, and that they change by no more than 20% from $a = 5.6$ Å through $a = 6.4$ Å. This transferability of force constants – which, as we will see below, is extremely useful for the modeling of semiconductor alloys – is not fully apparent when comparing empirical potentials fit to the experimental phonon dispersion relations. With hindsight, the reasons become quite obvious. The phonon dispersion relations represent the eigenvalues of the lat-

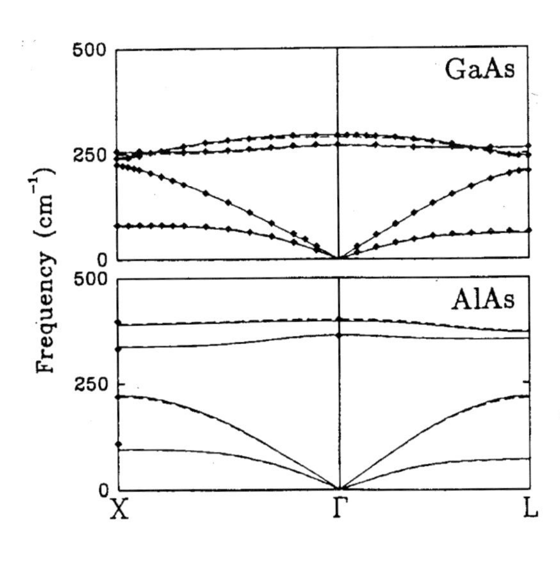

Fig. 3.2. *Top figure:* Phonon dispersion relations for GaAs using GaAs first-principles force constants (*solid line*) and AlAs first-principles force constants (*dashed line*). *Bottom figure:* Phonon dispersion relations for AlAs using AlAs first-principles force constants (*solid line*) and GaAs first principles force constants (*dashed line*) [30]

tice dynamical problem, but the interatomic potential is not fully determined by these eigenvalues: the vibrational eigenvectors are also needed. Unfortunately, phonon eigenvectors are difficult to measure (see for example [32] for Si), so that virtually all empirical determinations of interatomic potentials ignore the eigenvectors and fit the potential to the phonon frequencies. It is not surprising that this procedure leads to systematic errors and obscures the similarity of force constants for systems with similar lattice constants. Recently, the very successful adiabatic bond charge model has been refit using experimental phonon frequencies and the theoretical eigenvectors from ab initio calculations [33]. This procedure is justified because the excellent agreement between experiment and first-principles frequencies suggests that the first principles eigenvectors should be similarly accurate. This has in fact been verified in the few cases where experimental eigenvector determinations are available [32]. The newly fit adiabatic bond charge model has a slightly worse agreement with the experimental phonon dispersion relations but provides a much better account of the transferability of force constants among tetrahedral semiconductors.

Perturbations. The detailed knowledge of the lattice dynamics of semiconductors provides the appropriate starting point for the treatment of common perturbations, such as the addition of foreign atoms, the application of stress, etc. In general, these perturbations will change the force constant and mass matrices, so that (3.1) becomes

$$\left[(\boldsymbol{\Phi}+\Delta\boldsymbol{\Phi})-\omega_f^2(\boldsymbol{M}+\Delta\boldsymbol{M})\right]\boldsymbol{\varepsilon}(f)=0. \tag{3.5}$$

For small perturbations, as is often the case in practical applications, it is convenient to rewrite (3.5) in terms of the complete set of orthonormal unperturbed mode eigenvectors. We obtain

$$\sum_{f'}\left[\left(\omega_{0f}^2-\omega_f^2\right)\delta_{ff'}+\Delta\boldsymbol{\Phi}_{ff'}-\omega_f^2\Delta\boldsymbol{M}_{ff'}\right]c_{ff'}=0, \tag{3.6}$$

where we have introduced the matrix elements

$$\begin{aligned}\Delta\boldsymbol{M}_{ff'}&=\tilde{\varepsilon}_0^{\dagger}(f)\Delta\boldsymbol{M}\varepsilon_0(f')\\ \Delta\boldsymbol{\Phi}_{ff'}&=\tilde{\varepsilon}_0^{\dagger}(f)\Delta\boldsymbol{\Phi}\varepsilon_0(f')\end{aligned} \tag{3.7}$$

and the subscript "0" refers to unperturbed quantities. The coefficients $c_{ff'}$ are the expansion coefficients of the perturbed eigenvectors in terms of the unperturbed ones:

$$\varepsilon_i(l\kappa\mid f)=\sum_{f'=1}^{3Nn}c_{ff'}\varepsilon_{0i}(l\kappa\mid f'). \tag{3.8}$$

Equations (3.5) and (3.6) are very convenient as the starting point for a perturbation theory treatment. For cases in which non-degenerate perturbation theory can be used, expansions up to third order in the mass perturbation have been given [34]. It is important to stress that no additional approximations were made in going from (3.5) to (3.6). In other words, regardless of the size of the perturbation both equations can be solved exactly by numerical diagonalization – using the same generalized eigenvalue routines – and the solutions obtained are identical. Equations (3.5) and (3.6) can be used to treat isotopic disorder (in which case $\Delta\boldsymbol{\Phi} = 0$), stress (for which $\Delta\boldsymbol{M} = 0$), and alloying, for which in principle both the force constant and mass matrices change.

3.1.2 Anharmonic Effects

The theoretical description of vibrations based on (3.1) relies on the harmonic approximation, which neglects third- and higher-order derivatives of the crystalline potential. While this is a very good approximation for semiconductors, the neglected anharmonic terms are responsible for a number of important effects, some of them very useful for characterization purposes. This includes the temperature- and stress dependence of phonon frequencies. In the presence of small anharmonic perturbations the phonon picture remains approximately valid, but the phonon lifetime becomes finite (the lifetime is obviously infinite within the harmonic approximation, since phonons are the exact solution to the harmonic Hamiltonian). This lifetime manifests itself as a non-vanishing width of the phonon peaks measured with spectroscopic techniques such as Raman scattering. Particularly important for applications is the lifetime of long-wavelength longitudinal optical LO phonons in polar semiconductors, which is typically of the order of a few ps. These phonons couple very strongly to electrons, and the corresponding scattering rates depend directly on the steady-state non-equilibrium phonon population, which is largest (smallest) for the longest (shortest) anharmonic lifetimes [35].

A quantitative theory of anharmonic effects in semiconductors has not been available until very recently. Attempts to develop empirical potential models – in analogy with the empirical potential fits for the harmonic part of the potential – have not been very successful, in part because of the enormous proliferation of adjustable parameters as third and higher derivatives of the potential are included. Satisfactory agreement with experiment has only been obtained from ab initio calculations within density functional theory. The first successes of this theory were the calculation of third-order elastic constants [36], and the correct reproduction of the pressure and stress dependence of the frequency of high-symmetry phonons [37]. More recently, first principles calculations of the linewidth of Raman-active optical phonons in semiconductors have been published [38–40]. The agreement with experiment is truly remarkable if one considers that the accuracy of the results depends not only on a detailed knowledge of the anharmonic potential but also on a very accurate harmonic model for the same material. These conditions are so stringent that before the publication of the first principles results there was no other acceptable quantitative theory of Raman linewidths in semiconductors.

3.1.3 Raman Scattering by Phonons

The general theory of Raman scattering by phonons has been discussed in the introductory chapter; here we concentrate on aspects that are intrinsic to semiconductors.

The best way to visualize the ingredients needed to construct a theory of the Raman intensities is to start with the simpler case of elastically reflected light. When an incident field given by $\boldsymbol{E}(t) = \boldsymbol{E}_L \cos \omega_L t$ is present in a material, the system develops a polarization (dipole moment per unit volume) whose Cartesian components are given by

$$P_i(t) = \frac{1}{2} \sum_j \left(\chi_{ij} e^{-i\omega_L t} + \chi_{ij}^* e^{i\omega_L t} \right) E_{Lj} . \tag{3.9}$$

Here the tensor χ_{ij} is the system's electronic susceptibility, which is a function of the frequency ω_L. This time-dependent polarization produces electromagnetic radiation, and if we define the scattering cross section as the power radiated into a solid angle $\mathrm{d}\Omega$ divided by the incident intensity, we obtain the familiar expression [18]

$$\frac{\mathrm{d}\sigma}{\mathrm{d}\Omega} = \frac{\omega_L^4 V^2}{c^4} \left| \sum_{ij} e_{Si} \chi_{ij} e_{Lj} \right|^2 , \tag{3.10}$$

where V is the volume of the scattering medium and $\boldsymbol{e}_L$, $\boldsymbol{e}_S$ are unit polarization vectors for the incident and scattered light, respectively. The susceptibility χ depends on the electronic structure of the system, which changes as the atoms vibrate. The light scattering due to the phonon-induced change in susceptibility is what we know as Raman scattering. Since vibrational frequencies in semiconductors are typically much lower than the frequencies associated with electronic transitions, we can as a first approximation consider the electronic susceptibility for a frozen atomic configuration $\boldsymbol{u}$. For small displacements from equilibrium, as is usually the case for vibrations in crystals, one can expand the susceptibility as

$$\chi_{ij}(\boldsymbol{u}) = \chi_{ij}(\boldsymbol{0}) + \sum_f \chi_{ij,f} Q_f , \tag{3.11}$$

where

$$\chi_{ij,f} = \frac{\partial \chi_{ij}(\boldsymbol{u})}{\partial Q_f} = \sum_{l\kappa k} \frac{\partial \chi_{ij}(\boldsymbol{u})}{\partial u_k(l\kappa)} \epsilon_k(l\kappa \mid f) . \tag{3.12}$$

Substituting (3.11) into (3.10), one obtains a term that does not contain Q_f (which gives Rayleigh scattering) plus terms linear and quadratic in the normal coordinates. The linear terms vanish upon a thermal average, so that Raman scattering is produced by the quadratic terms only. Using standard results from the quantum theory of the harmonic oscillator for the thermal average of the quadratic terms, and introducing energy-conserving delta functions in an ad-hoc manner, one finally obtains a Raman differential cross section given by

$$\frac{\mathrm{d}\sigma}{\mathrm{d}\Omega\,\mathrm{d}\omega} = \frac{\hbar V^2}{2c^4}\sum_f \frac{(\omega_L-\omega_f)^4}{\omega_f}\{(n_f+1)\delta(\omega-\omega_f)+n_f\delta(\omega+\omega_f)\} \times \left|\sum_{ij} e_{Si}\chi_{ij,f}e_{Lj}\right|^2 . \tag{3.13}$$

Here ω_f is the Raman shift, and $n_f = [\exp(\hbar\omega_f/k_\mathrm{B}T)-1]^{-1}$ is the Bose–Einstein phonon occupation number for mode f. The first term in the curly bracket gives Stokes scattering (phonon creation) and the second term gives anti-Stokes scattering (phonon annihilation). A rigorous quantum mechanical derivation of the cross section reduces to (3.13) if the incident frequency satisfies $\omega_f << \omega_L << E_g(\mathrm{dir})/\hbar$ [41]. For visible light excitation the first condition is satisfied in semiconductors but the incident photon energy is *not* much smaller than the direct band gap $E_g(\mathrm{dir})$. This has little effect on the selection rules to be discussed below but can affect the comparison between predicted and observed intensities. Notice that a factor ω_S^4 – corresponding to the scattering frequency $\omega_S = \omega_L - \omega_f$ – appears in (3.13) instead of ω_L^4, as might be expected in view of (3.10). We have introduced this factor in an ad-hoc manner so that is consistent with the quantum theory of the Raman effect. The difference between the two factors is not important within the range of validity (3.13).

For the case of the three-fold degenerate zone-center optical phonons in diamond and zincblende semiconductors one can use (3.12) to find a simple expression for the susceptibility derivative tensor $\chi_{ij,f}$. From the orthonormality properties of the eigenvectors and the requirement that the center of mass of the unit cell should not move, one can write

$$\epsilon_i(l\kappa \mid \mathbf{0}k) = \frac{(-1)^\kappa}{M(\kappa)}\sqrt{\frac{\mu}{N}}\delta_{ik}, \tag{3.14}$$

where μ is the reduced mass of the unit cell, and we have used the composite symbol $f \equiv \mathbf{0}k$ with $k = x, y, z$ to label the three degenerate modes. Furthermore, the crystal periodicity requires that the spatial derivatives of the susceptibility be independent of the index l, and translational invariance requires that the sum of the susceptibility derivatives for all atoms within a unit cell be zero. Combining all these results we obtain

$$\chi_{ij,k} = \sqrt{\frac{N}{\mu}} \frac{\partial \chi_{ij}(\boldsymbol{u})}{\partial u_k(02)} = -\sqrt{\frac{N}{\mu}} \frac{\partial \chi_{ij}(\boldsymbol{u})}{\partial u_k(01)} \equiv \frac{1}{v_{\mathrm{c}}} \sqrt{\frac{1}{\mu N}} R_{ij,k}, \tag{3.15}$$

where the sign depends on our choice of eigenvectors in (3.14). We define here a "Raman tensor" $R_{ij,f}$, which, as will become more apparent later, is a volume-independent quantity. Here v_{c} is the volume of the unit cell (for diamond and zincblende semiconductors $v_{\mathrm{c}} = a^3/4$, where a is the lattice constant listed in Table 3.1). It is customary in the literature to express the strength of Raman scattering in tetrahedral semiconductors in terms of $R_{ij,f}$. Furthermore, since the cross section is proportional to the volume of the solid, it is useful to define a Raman cross section per unit volume $\mathrm{d}S/(\mathrm{d}\Omega\,\mathrm{d}\omega)$, called "scattering efficiency", so that we obtain the following expression for optic phonons in diamond and zincblende semiconductors:

$$\frac{\mathrm{d}S}{\mathrm{d}\Omega\,\mathrm{d}\omega} = \frac{\hbar}{2\mu v_{\mathrm{c}} c^4} \sum_{k=x,y,z} \frac{(\omega_L - \omega_k)^4}{\omega_k} \{(n_k+1)\delta(\omega-\omega_k) + n_k\delta(\omega+\omega_k)\} \times \left| \sum_{ij} e_{Si} R_{ij,k} e_{Lj} \right|^2 . \tag{3.16}$$

Symmetry considerations. Several selection rules follow from (3.13) or (3.16). The delta functions for the frequency insure energy conservation. The electronic susceptibility is a function $\chi_{ij}(\boldsymbol{\omega}, \boldsymbol{k})$ of the frequency and wave vector. The wave vector dependence was not indicated explicitly in (3.9) because we assumed the incident field to have an infinite wavelength, so that only the $\boldsymbol{k} = \boldsymbol{0}$ component plays a role in the subsequent derivations. The resulting susceptibility derivative $\chi_{ij,f}$ is symmetric in the indices i, j and will vanish if the phonon wave vector in $f \equiv (\boldsymbol{q}m)$ is nonzero. The latter represents the crystal momentum conservation rule. A more rigorous treatment would lead to the possibility of anti-symmetric terms and to expressions involving delta-functions of the form $\boldsymbol{\delta}(\boldsymbol{k}_S \pm \boldsymbol{q} - \boldsymbol{k}_L)$. However,the allowed phonon wave vectors resulting from these expressions are usually small compared to the size of the Brillouin zone of the crystal, so that the simple $\boldsymbol{q} = \boldsymbol{0}$ Raman selection rule is normally an excellent approximation (see the Chap. 1). Contributions beyond this "dipole approximation" can be expected to be proportional to qa (where a is the lattice constant) and therefore small. These considerations, however, may break down near resonances, as discussed in the introductory chapter. For example, the macroscopic electric field set up by longitudinal optic (LO) phonons in polar semiconductors makes a contribution to the Raman cross section of these modes that is proportional to $(qa)^2$ but so resonant that its strength can become comparable to the allowed $q = 0$ cross section. Several authors have reviewed this process [6–8].

A key feature of Raman scattering is its dependence on the polarization vectors of the incident and scattered light, and this dependence arises from the symmetry properties of the susceptibility derivative $\chi_{ij,f}$ or, equivalently, of the Raman tensor $R_{ij,f}$. These are second-rank tensors whose forms can be determined from group theory considerations. But the steep "learning curve" of group theory and its subtleties in the area of crystalline symmetries represent a serious barrier to the occasional user. Fortunately, tables with the form of the $R_{ij,f}$ tensor for crystals with different symmetries can be found in the literature, and this simplifies considerably the analysis of the experimental results. Less known is the fact that the form of the $R_{ij,f}$ tensor can be easily generated from the tables of characters for the corresponding point groups following an approach explained in detail by Callen [42].

The basic idea for any group theory analysis of the symmetry properties of the $R_{ij,f}$ tensor is the realization that the scattering cross section is a scalar that is invariant under any of the symmetry operations of the molecule or crystal. From (3.13), this cross-section could be written symbolically as $\mathrm{d}S/\mathrm{d}\Omega \propto \left|\sum_{ij} R_{ij,f} e_i e_j Q_f\right|^2$. Since the right-hand side of this expression must also be an invariant, we must find polynomials of the form $e_i e_j Q_f$ which are invariant under the symmetry properties of the crystal. The number of such polynomials gives the number of independent components of the tensor $R_{ij,f}$. In the jargon of group theory, this means that one must find polynomials of the form $e_i e_j Q_f$ which belong to the identical representation. Unsöld's theorem provides a systematic way of generating such polynomials. The theorem states that if the basis functions $\{\phi_1, \phi_2, \dots, \phi_n\}$ and $\{\psi_1, \psi_2, \dots, \psi_n\}$ generate exactly the same irreducible representation, the sum $\sum_{m=1}^{n} \phi_m \psi_m^*$ belongs to the identical representation. Since basis functions are listed in tables of characters, generating the Raman tensor can be as simple as inspecting the character table for the relevant group.

Let us apply the above ideas to the case of Raman scattering by optical phonons in zincblende and diamond structure semiconductors. For scattering by $\boldsymbol{q} = \mathbf{0}$ phonons, the point groups of each crystal structure are needed, and these are T_d and O_h for zincblende and diamond, respectively. Table 3.2 lists the irreducible representations and some of the corresponding basis functions for each of these groups. Once symmetrized, the products $e_i e_j$ transform like quadratic polynomials in x, y, z. Hence from Unsöld's theorem we can immediately see that all Raman-active phonons must belong to the irreducible representations for which these quadratic polynomials appear as basis functions. From Table 3.2, these are A_1, E, and F_2 in T_d and A_{1g}, E_g, and F_{2g} in O_h. Using the procedure outlined in Sect. 2.5 of the introductory chapter, we find that the $\boldsymbol{q} = \mathbf{0}$ optic phonons in zincblende and diamond structures belong to the three-dimensional F_2 and F_{2g} representations, respectively, and the corresponding eigenvectors are given in (3.14). Since $\{Q_x, Q_y, Q_z\}$ are basis functions for $F_2(T_d)$ and $F_{2g}(O_h)$, and the polynomials $\{yz, xz, xy\}$ are basis functions for the same representation, Unsöld's theorem implies

Table 3.2. Basis functions for the irreducible representations of the T_d (zincblende) and O_h (diamond) point groups. The functions are ordered and normalized according to Callen's prescription

T_d (zincblende)		O_h (diamond)	
$\Gamma_1(\mathrm{A}_1)$	$x^2+y^2+z^2$	$\Gamma_1(\mathrm{A}_{1g})$	$x^2+y^2+z^2$
$\Gamma_2(\mathrm{A}_2)$	xyz	$\Gamma_{1'}(\mathrm{A}_{1u})$	
$\Gamma_{12}(\mathrm{E})$	$[\sqrt{3}(x^2-y^2),$ $2z^2-x^2-y^2]$	$\Gamma_2(\mathrm{A}_{2g})$	
$\Gamma_{15}(\mathrm{F}_2)$	$(x,y,z);(yz,xz,xy);$ $(x^3,y^3,z^3);\ [x(y^2+z^2),$ $y(z^2+x^2),z(x^2+y^2)]$	$\Gamma_{2'}(\mathrm{A}_{2u})$	xyz
$\Gamma_{25}(\mathrm{F}_1)$	$[x(y^2-z^2),y(z^2-x^2),$ $z(x^2-y^2)]$	$\Gamma_{12}(\mathrm{E}_g)$	$[\sqrt{3}(x^2-y^2),2z^2-x^2-y^2]$
		$\Gamma_{12'}(\mathrm{E}_u)$	
		$\Gamma_{15'}(\mathrm{F}_{1u})$	
		$\Gamma_{15}(\mathrm{F}_{1u})$	$(x,y,z);(x^3,y^3,z^3);[x(z^2+y^2),$ $y(z^2+x^2),z(x^2+y^2)]$
		$\Gamma_{25'}(\mathrm{F}_{2g})$	$(yz,xz,xy),(yzx^2,zxy^2,xyz^2)$
		$\Gamma_{25}(\mathrm{F}_{2u})$	$[x(y^2-z^2),y(z^2-x^2),z(x^2-y^2)]$

$Q_x e_y e_z + Q_y e_z e_x + Q_z e_x e_y =$ invariant. There is only one such combination that can be formed. This means that $R_{ij,k}$ has the form

$$R_{ij,x}=\begin{pmatrix}0&0&0\\0&0&a\\0&a&0\end{pmatrix}\quad R_{ij,y}=\begin{pmatrix}0&0&a\\0&0&0\\a&0&0\end{pmatrix}\quad R_{ij,z}=\begin{pmatrix}0&a&0\\a&0&0\\0&0&0\end{pmatrix}. \tag{3.17}$$

Notice that these Raman tensors correspond to phonon eigenvectors along the x, y, and z axis, respectively. Since the three optical phonons are degenerate, any linear combination of these eigenvectors can also be chosen as eigenvectors. The corresponding Raman tensor can be easily obtained from (3.17) by expressing the new eigenvectors in terms of the old ones. For a phonon polarized along an arbitrary h-direction, one obtains

$$R_{ij,h}=\sum_{\lambda=x,y,z} e_{\lambda,h}R_{ij,\lambda}, \tag{3.18}$$

where $e_{\lambda,h}$ is the λth component of a unit vector in the h-direction.

For scattering by optic phonons in cubic diamond-structure systems, there is never a need to "rotate" the Raman tensor according to (3.18), since after the sum over the three degenerate optic modes in (3.16) one obtains exactly the same result for any set of three orthonormal polarization vectors. However, there are two important cases for which the rotation of the Raman tensor is convenient and even necessary. The first such case arises in the presence of external perturbations that lift the cubic degeneracy. The perturbation determines the form of the mode eigenvectors, which can no longer be chosen as arbitrary linear combinations. The Raman tensor corresponding to each of the perturbed modes can be approximately obtained from (3.18) once the perturbed mode eigenvectors are expressed in terms of the old ones. The application of these ideas to the case of stress perturbations is discussed below in Sect. 3.2.3.

The second important application of (3.18) is the case of polar semiconductors, for which the three zone center modes are split into two transverse optic modes (TO) and a LO mode. This splitting appears to contradict the group theory predictions of three-fold degeneracy for the optic modes of zincblende materials, but a closer examination reveals that the behavior of phonons in these materials is consistent with group theory. At exactly $q = 0$ the optic modes are indeed degenerate for an infinite crystal, and both transverse and longitudinal modes have the LO frequency. As q is increased, the photon frequency $\omega = cq$ approaches the frequency of optic vibrational modes. This leads to coupled photon-TO phonon modes, called polaritons [43]. The upper branch acquires an increasing photon character as q is increased further, whereas the lower branch becomes purely vibrational and approaches the frequency of the TO phonon. The phonon wave vector for Raman scattering is of the order of the wave vector of the incident photon, which for visible excitation is two orders of magnitude larger than the wave vector for which photon and phonon frequencies overlap and produce strong mixing. Hence the measured Raman frequencies correspond to the LO and TO frequencies. But since this "large" Raman wave vector is much smaller than the size of the Brillouin zone, the Raman tensors in (3.17) are still relevant for Raman scattering by LO and TO phonons. The tensor corresponding to the LO phonon is obtained from (3.18) by noting that this mode has an eigenvector parallel to the wave vector transfer (LO phonons). The component a is not necessarily the same for LO and TO phonons, since in the former case the polarizability is also a function of the macroscopic electric field. The relative magnitude of the two components a_{LO} and a_{TO} depends on the so-called Faust–Henry coefficient [44].

Our symmetry analysis based on Callen's approach can be easily generalized to treat more complicated cases. For example, we mentioned above that the macroscopic electric field associated with LO phonons contributes a term to the electron-phonon interaction – the so-called Fröhlich interaction – that makes strongly resonant $\boldsymbol{q}$-dependent contributions to the Raman tensor. In this case the quantity that must remain invariant is of the form $\mathrm{d}S/\mathrm{d}\Omega \propto \left|\sum_{ijk} R_{ijk,f} e_i e_j q_k Q_f\right|^2$. From inspection of Table 3.2 we can now form three invariants using Unsöld's theorem: $e_x e_x q_x Q_x + e_y e_y q_y Q_y + e_z e_z q_z Q_z$, $(e_y^2 + e_z^2)\, q_x Q_x + (e_z^2 + e_x^2)\, q_y Q_y + (e_x^2 + e_y^2)\, q_z Q_z$ and $e_y e_z q_y Q_z + e_z e_x q_z Q_x + e_x e_y q_x Q_y$. Thus the corresponding Raman tensors become

$$R_{ijx,x} = \begin{pmatrix} b & 0 & 0 \\ 0 & c & 0 \\ 0 & 0 & c \end{pmatrix} \quad R_{ijy,y} = \begin{pmatrix} c & 0 & 0 \\ 0 & b & 0 \\ 0 & 0 & c \end{pmatrix} \quad R_{ijz,z} = \begin{pmatrix} c & 0 & 0 \\ 0 & b & 0 \\ 0 & 0 & b \end{pmatrix} . \tag{3.19}$$

This means that the so-called "forbidden" Raman scattering by LO-phonons arises from a diagonal Raman tensor. Notice that the third invariant polynomial contributes off-diagonal components to Raman tensors of the form $R_{ijx,y}$, but these vanish in the specific case of the Fröhlich interaction, which is proportional to $\boldsymbol{q} \cdot \boldsymbol{\varepsilon}$.

Anastassakis and Burstein [45,46] have used Callen's approach for the determination of the form of the Raman tensor in a variety of situations, including the application of external stress and electric fields. The reader is referred to these very instructive papers for further illustrations of the power of this approach.

Table 3.2 cannot be used directly to predict what phonons will participate in second-order Raman scattering. This is because such phonons can have large wave vectors, for which the relevant symmetry group is not the point group of the crystal but the group of the wave vector $\boldsymbol{q}$. However, it still remains true that for any form of light scattering the allowed configurations correspond to the irreducible representations for which quadratic polynomials in x, y, z are basis functions. Thus the direct product of the irreducible representations to which the two phonons belong must contain A_1, E, and F_2 in T_d and A_{1g}, E_g, and F_{2g} in O_h. A remarkable feature of both zincblende and diamond semiconductors is that second-order Raman scattering is dominated by overtones belonging to the A_1 and A_{1g} irreducible representations, respectively. Thus second order Raman spectra resemble quite closely the phonon density of states (with a scale expanded by a factor of 2). This can be used to study the validity of lattice dynamical models, as indicated in Sect. 3.1.1.

Raman intensities. It is instructive to contrast our knowledge of the two sets of a priori unknowns in (3.13) and (3.16): the vibrational frequencies on the one hand and the components of the susceptibility derivative (or Raman) tensor on the other hand. While frequencies can be measured with extraordinary accuracy and calculated with very small errors, the measurement and the calculation of the Raman tensor components represent formidable problems. We now discuss the main experimental and theoretical issues surrounding the determination of absolute Raman cross sections.

The quantity measured directly by the experimentalist is the number of photons counted by a detector. Relating this experimental datum with the cross sections defined above is not trivial. Strictly speaking, (3.13) applies only to isolated molecules and is not valid for condensed matter, where the incident field "seen" by the scattering medium is not the laser field in air but the effective field in the medium. This requires complicated local-field corrections. In the quantum theory of Raman scattering in solids one can approximately correct for this effect by equating the (infinite) volume of the sample with the quantization volume for the electromagnetic field and replacing the vacuum permittivity by the medium's permittivity. This adds a factor of $(\eta_L \eta_S)^{-2}$ to the Raman cross section. Here $\eta_L(\eta_S)$ is the material's refractive index at the laser (scattered) photon frequency. In semiconductors this is a large correction of the order of 100, but it is almost entirely compensated by two other contributions. Since the cross section is defined *inside* the medium, we must use the speed of light in the medium. This gives additional dependencies on the index of refraction in such a way that the overall correcting factor becomes η_S/η_L, which is of the order of unity. However, large corrections do arise when we try to relate the cross section *inside* the sample with measurements that take place *outside* the sample. From a simple application of Snell's law it can be seen that the solid angle of scattering outside the sample $\mathrm{d}\Omega'$ is not equal to the solid angle inside the sample $\mathrm{d}\Omega$. For small numerical apertures, the two are related by $\mathrm{d}\Omega' = \eta_S^2\,\mathrm{d}\Omega$. In addition, a fraction of the incident power is reflected at the air/sample interface and a fraction of the scattered power is reflected at the sample/air interface. A third complication arises from the fact that semiconductors are usually strongly absorbing materials at visible wavelengths. For complete absorption in a backscattering geometry one can define an effective scattering volume given by $V_{\mathrm{eff}} = L^2/(\alpha_L + \alpha_S)$, where L^2 is the illuminated area and $\alpha_L(\alpha_S)$ is the absorption coefficient at the incident (scattered) frequencies. More complicated situations are discussed in [7] and [8]. Zeyher et al. have developed a rigorous quantum mechanical theory of Raman scattering in the absorption frequency range [47]. Combining all these corrections and defining $\mathrm{d}R'_S/\mathrm{d}\Omega'$ as the number of photons per unit time counted by the detector for unit solid angle of collection outside the sample (photon count rate), we obtain for backscattering experiments on a strongly absorbing material

$$\frac{\mathrm{d}R'_S}{\mathrm{d}\Omega'\,\mathrm{d}\omega} = \left[\frac{C(\omega_S)T_L T_S}{(\alpha_L + \alpha_S)\eta_S^2}\right]\left(\frac{P'}{\hbar\omega_S}\right)\frac{\mathrm{d}S}{\mathrm{d}\Omega\,\mathrm{d}\omega}, \tag{3.20}$$

where P' is the laser power incident on the sample, $T_L(T_S)$ is the power transmission coefficient of the sample at the incident (scattered) photon frequency; and $C(\omega_S)$ is a characteristic function of the experimental setup, which gives the ratio of photons counted by the detector divided by the number of photons scattered into $\mathrm{d}\Omega'$. Since $C(\omega_S)$ is difficult to measure, a popular approach for determining scattering efficiencies is the sample substitution method, whereby a sample with a known scattering efficiency is measured with the same setup in such a way that $C(\omega_S)$ can be eliminated after rationing the measured count rates. A detailed discussion of this and other methods for obtaining absolute Raman scattering efficiencies in semiconductors is given in [7] and [8]. Particularly noteworthy is the fact that these scattering efficiencies can be obtained from combined Raman and Brillouin measurements on the same sample (see box by M. Grimsditch). A truly absolute measurement of a Raman cross section (for the case of liquid benzene) is described in detail in a very instructive paper by Schomacker et al. [48].

A common source of confusion (although rarely a source of large errors) is the fact that many workers prefer to define the scattering cross section as the number of photons radiated into a solid angle $d\Omega$ divided by the number of incident photons per unit area. This "photon cross section" is equal to the "power cross section" used in (3.13) or (3.16) multiplied by ω_L/ω_S.

Given the experimental complications described above it is not surprising to find that absolute Raman efficiency measurements are rare. Moreover, large discrepancies between different groups are not uncommon. Table 3.3 shows selected experimental values for the single independent component of the first-order Raman tensor in diamond and zincblende semiconductors (called "Raman polarizability" in [7]). These values can be combined with (3.16) and (3.20) to estimate the strength of the measured Raman peaks. Notice that the scattering theory was developed by neglecting anharmonicity, so that the Raman peaks are predicted to be delta functions. Experimentally, the peaks are broadened by anharmonicity. The quantity to be compared with the theoretical count rate prediction is therefore the area under the peak (not the peak height).

Table 3.3. Selected values of the Raman polarizability a (in Å^2) for diamond and zincblende semiconductors

Material	ω_L(eV)	Experiment	Theory
Diamond	2.41	4.3 ± 0.6[a]	4.5[b]
Silicon	0.94/1.1	23 ± 4[c]	22[b]
Germanium	0.826	68 ± 14[d]	64[b]
GaAs (TO)	1	63 ± 10[e]	
GaAs (TO)	1.2	50[e]	
GaP (TO)	1	35 ± 6[e]	
GaP (TO)	1.92	23 ± 5[e]	

[a] M. Cardona: in *Light Scattering in Solids II*, ed. by M. Cardona, G. Güntherodt (Springer-Verlag, Berlin 1982) Vol. 50, p. 19–178
[b] W. Windl: *Ab-initio-Berechnung von Raman-Spektren in Halbleitern* (CH-Verlag, Regensburg 1995)
[c] J. Wagner, M. Cardona: Solid State Commun. **48**, 301 (1983)
[d] J. Wagner, M. Cardona: Solid State Commun. **53**, 845 (1985)
[e] Compiled by M. Cardona in (a)

Theoretical considerations. Since the intensity of Raman scattering is proportional to the change in the system's electronic susceptibility induced by the atomic displacements, the main ingredient of any theory of Raman intensities is a model for the system's electronic susceptibility. A simple approach that has been developed for molecular systems but is useful for insulators and semiconductors is the bond polarizability model [49,50]. This model assigns to each bond an axially symmetric polarizability given by

$$\Pi_{\alpha,\beta}(\boldsymbol{R}) = \frac{R_\alpha R_\beta}{R^2}\alpha_\parallel(R) + \left(\delta_{\alpha\beta} - \frac{R_\alpha R_\beta}{R^2}\right)\alpha_\perp(R), \quad (3.21)$$

and builds the system's polarizability as the sum of all bond polarizabilities. The coefficients $\alpha_\parallel$ and $\alpha_\perp$ are assumed to be functions of the bond length only. Since a general atomic displacement changes the bond orientation and length, the derivative of the polarizability relative to atomic displacements is not zero, and this explains the Raman activity of the system. Starting from (3.21), one finds that the susceptibility derivative defined in (3.12) is given by [34]

$$\begin{aligned}
\chi_{ij,f} = -\frac{1}{V}\sum_{nb}\Bigg\{ & \left[\left\{\frac{\alpha'_\parallel[R(nb)] + 2\alpha'_\perp[R(nb)]}{3}\right\}\hat{\boldsymbol{R}}\cdot\varepsilon(n|f)\right] \\
& + \left[\left\{\alpha'_\parallel[R(nb)] - \alpha'_\perp[R(nb)] - \frac{2(\alpha_\parallel[R(nb)] - \alpha_\perp[R(nb)])}{R(nb)}\right\}\right. \\
& \times \left.\left\{\hat{R}_i(nb)\hat{R}_j(nb) - \frac{\delta_{ij}}{3}\right\}\hat{\boldsymbol{R}}(nb)\cdot\varepsilon(n|f)\right] \\
& + \left[\left\{\frac{\alpha_\parallel[R(nb)] - \alpha_\perp[R(nb)]}{R(nb)}\right\}\right. \\
& \times \left.\left\{\hat{R}_i(nb)\varepsilon_j(n|f) + \hat{R}_j(nb)\varepsilon_i(nf) - \frac{2}{3}\hat{\boldsymbol{R}}(nb)\cdot\varepsilon(n|f)\delta_{ij}\right\}\right]\Bigg\}, \quad (3.22)
\end{aligned}$$

where the primed quantities are derivatives relative to the bond length and $\hat{\boldsymbol{R}}(nb)$ is a unit vector along the direction $\boldsymbol{R}(nb)$ of a bond b connecting atom n with one of its neighbors in the equilibrium configuration. The sum over n runs over all atomic sites. We use this index instead of the $(l\kappa)$ combination to emphasize that (3.22) is valid for any system, including molecules (in which case the expression is more conveniently written in terms of the polarizability derivative) and disordered semiconductors. When applied to crystalline solids, the right-hand side of (3.22) vanishes for modes with $\boldsymbol{q} \neq 0$, as expected. For the specific case of diamond and zincblende semiconductors, the sum of the unit vectors along the directions of the four bonds at each atom is zero. This immediately leads to the cancellation of the first and third bracket in (3.22) so that Raman scattering by the zone center phonons in these materials depends on a single parameter, as expected from group theory. Inserting (3.14) in (3.22) we find that the only independent component of the Raman tensor in (3.17) is given by

$$a = \frac{4}{3\sqrt{3}}\left[\left(\alpha'_{\parallel}(R) - \alpha'_{\perp}(R)\right) - \frac{2\left(\alpha_{\parallel}(R) - \alpha_{\perp}(R)\right)}{R}\right]. \tag{3.23}$$

The parameters of the bond polarizability model must be obtained from fits to Raman spectra and other dielectric properties. In the case of tetrahedral semiconductors, first-order Raman scattering depends on a single parameter, shown in (3.23), and the fits must also include second-order Raman scattering. This has been done by Go and coworkers [51]. It appears that the model is capable of providing a reasonably consistent explanation of Raman intensities, elasto-optic coefficients, and the dielectric function. Once the bond polarizability parameters are known for a few materials, one might be able to deduce their values in other materials by identifying chemical trends or using the so called "transferability" hypothesis [52], according to which similar bonds will have similar polarizabilities, even if they are placed in different environments. These expected chemical regularities represent the most appealing aspect of the bond polarizability model. For example, one could attempt to compute the Raman spectrum of semiconductor alloys using bond polarizability parameters for the parent compounds.

The susceptibility and its derivatives can also be calculated from the known electronic structure of the material, and this has been attempted using either empirical pseudopotentials [53] or tight-binding methods [54]. An even better approach to the susceptibility problem is to develop ab initio methods that do not rely on experimental input. If these are shown to reproduce the Raman intensities in known materials, there is no a priori reason why the predictions should be worse in new materials. An analogous situation arises in the context of the phonon structure. While good empirical models can reproduce the phonon structure of known materials with an accuracy comparable to that of ab initio methods, the extrapolation of the model parameters to new materials adds a significant level of uncertainty. By contrast, ab initio predictions for known and unknown materials should be of comparable quality.

The first important question that arises in the context of predicting Raman cross sections from first principles is whether once can use the local density approximation (LDA) to density functional theory, i.e., the same theoretical formalism that is known to yield extremely accurate phonon frequencies and eigenvectors. The reader familiar with the quantum mechanical expressions for the electronic susceptibility – which involve integrations over valence *and* conduction band states – will immediately recognize a serious problem that must be faced by the theorist: since the LDA underestimates the separation between conduction and valence bands, the predicted susceptibility (and its derivatives, which determine the Raman cross section) are likely to be in error. On the other hand, the static susceptibility for the equilibrium structure as well as for a frozen displacement pattern $\boldsymbol{u}, \chi_{ij}(\boldsymbol{u})$, must be a ground state property, and therefore, aside from possible numerical complications, it should be possible to compute it with the typical accuracy of other ground state properties calculated within the local density and adiabatic approximations. In other words, LDA ab initio calculations should be reliable within the range of validity of (3.13).

For the zone-center optic phonons in diamond and zincblende semiconductors the key quantity that must be calculated is the derivative of the susceptibility relative to the displacement of one atom in the crystal, as shown in (3.15). Since for computational purposes it is highly desirable to deal with a periodic system, one can displace all atoms within a sublattice by the same amount. If we refer to this collective displacement as $u_k(\kappa)$, it is apparent that

$$v_c \frac{\partial \chi_{xy}(\boldsymbol{u})}{\partial u_{rel,z}} = v_c \frac{\partial \chi_{xy}(\boldsymbol{u})}{\partial u_z(2)} = V \frac{\partial \chi_{xy}(\boldsymbol{u})}{\partial u_z(02)} = R_{xy,z} \equiv a \tag{3.24}$$

where $u_{rel,k} = u_k(2) - u_k(1)$, and the first identity follows from expressing the collective sublattice displacements in terms of their sum and differences and noting that the derivative of the susceptibility relative to the *sum* of the displacements must vanish because it corresponds to a uniform translation of the crystal. This expression is identical to the definition of the Raman polarizability in [7].

The basic procedure for the computation of the Raman polarizability from first principles is to calculate the crystal's susceptibility for different frozen displacements $u_{rel,k}$, from which a can be obtained by numerical differentiation. This method was pioneered by Baroni and Resta [55]. The results in Table 3.3, which are in impressive agreement with off-resonance Raman measurements in diamond, Si and Ge, are due to Windl, who has also extended the method to second-order Raman scattering, also with very satisfactory agreement with experiment [56]. It might be useful and instructive to fit a bond polarizability model to the more reliable first principles results. This should provide a deeper insight into the physical meaning of the bond polarizability parameters, in much the same way that fits of the adiabatic bond charge model to first-principles phonon calculations have contributed to the understanding of this model of the vibrational properties of semiconductors.

Unfortunately, neither the bond polarizability model predictions nor the first principles calculations can be expected to be very reliable for measurements on semiconductors under typical experimental conditions. Visible light photon energies are comparable to or higher than the band gaps of these materials, well beyond the limit of validity of (3.13) and (3.16). Resonance effects – not accounted for in the bond polarizability model or in the ab initio approach described above – make a significant contribution to Raman scattering by semiconductors under visible excitation. As shown by many authors [6–8,57], these resonance effects are a very useful spectroscopic tool by themselves, and therefore have received considerable theoretical and experimental attention. However, it is unlikely that a practical ab initio theory of Raman intensities will be developed in the near future for measurements in this regime. An accurate description of the electronic structure (including the conduction bands) becomes critically important, and this requires computer-intensive methods beyond the LDA. Even if such calculations can be performed, excitonic effects can play a significant role in determining the Raman cross section [7]. Only under extreme resonant conditions near a critical point in the joint valence-conduction density of states is it possible to limit the integrations to a few states near the critical points, which sometimes leads to simple analytical expressions for the Raman intensities [7,8].

3.2 Semiconductor Characterization

Our ability to predict the polarization properties, frequencies, and intensities of semiconductor Raman spectra provides a powerful tool for the unambiguous identification of the materials and their crystalline structure. Moreover, since many interesting phenomena can be described as perturbations to the well-understood vibrational properties of semiconductors, it is also possible to monitor these perturbations using Raman spectroscopy. In the remainder of this chapter we describe some of these applications: the use of Raman spectroscopy for the determination of crystalline orientations, the measurement of temperature and stress, the characterization of doping levels, and the study of alloy semiconductors.

3.2.1 Crystal Orientation

Although the traditional techniques for establishing the orientation of single crystals are X-ray diffraction, transmission electron microscopy, electron or ion channeling, and etch-pit measurements, Raman spectroscopy is an effective and sometimes superior alternative. The application of Raman spectroscopy to crystalline orientation is based on the fact that the three tensors in (3.17) refer to the crystalline cubic axes, whereas the light polarization vectors can be controlled in the laboratory system. Hence one can obtain crystalline orientations by studying the intensity of Raman scattering as the sample or the light polarization are rotated. A systematic methodology has been developed by Mizoguchi and Nakashima [58], who show that a precision of about 2° and a spatial resolution of the order of $1\,\mu m$ can be easily achieved. The method is also non-destructive and can be applied under ambient conditions, which represents an additional advantage over the traditional approaches. A good example of the use of Raman spectroscopy in crystal orientation measurements is the characterization of Si grown on insulators (SOI) [59,60]. Figure 3.3 compares the polarization-dependent Raman

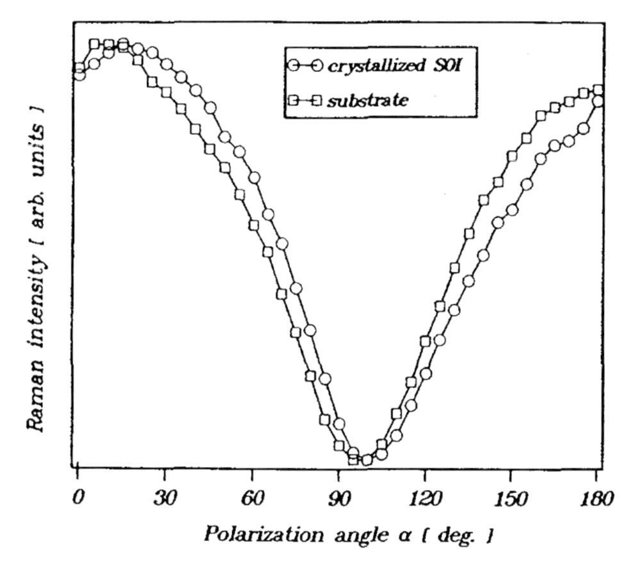

Fig. 3.3. Angular dependence of the Raman intensity for a silicon-on-insulator film grown on a Si substrate [60]

intensity between SOI and the (001)-oriented Si substrate. The relative displacement of the two curves reflects a backward-tilt in crystal orientation in the SOI of about 4° [60].

3.2.2 Temperature Monitoring

First-order Raman spectra change with temperature because the anharmonicity in the interatomic potential renders the phonon frequencies and lifetimes temperature-dependent [15]. In addition, the Stokes/anti-Stokes intensity ratios also depend on temperature, as is apparent from (3.15) and (3.16). Yet another temperature-induced effect can be observed under quasi-resonant excitation: the Raman intensities can change dramatically due to temperature-dependent changes in the band structure. In fact, temperature tuning of the band gaps was a way of measuring resonance excitation profiles before tunable lasers became available.

The Stokes/anti-Stokes method is very appealing because it does not rely on the details of the anharmonic interactions, which, as explained in the introductory chapter, can vary from material to material. It has been used, for example, to improve the design of semiconductor lasers [61]. It is important to emphasize, however, that temperature readings from Stokes/anti-Stokes ratios are very prone to systematic errors, since several corrections must be applied, and the ratios themselves are not very strong functions of the temperature. The corrections needed to extract temperatures from Stokes/anti-

Stokes in silicon have been considered in detail by Compaan and Trodahl [62]. This paper is an excellent illustration of the difficulties of such measurements.

The first important consideration is that for experiments in semiconductors under visible excitation resonance effects can be important, and the resonance enhancement is *not* the same for the Stokes and anti-Stokes signals. This is because the exact symmetry property of the Raman tensor is [63]

$$R_{ij,k}(\omega_L, \omega_S) = R_{ij,k}(\omega_S, \omega_L), \tag{3.25}$$

so that in order to obtain the same Raman tensor for Stokes and anti-Stokes scattering one needs two laser lines separated by the phonon frequency. Even if these lines were available, the experimental errors introduced by the change of excitation source would reduce the reliability of the measurements. The alternative to using two laser lines is to explicitly correct for the different Raman cross-sections for Stokes and anti-Stokes scattering. For the case of Si, this has been done by Compaan and Trodahl [62], taking advantage of the fact that in this material the Raman tensor is proportional to the square of the frequency derivative of the dielectric function, which can be easily determined from ellipsometry measurements.

In addition to the Raman tensor corrections, all factors in the square bracket in (3.20) depend on the light's frequency, so that their values can be different for Stokes and anti-Stokes scattering. Correcting for this dependence requires a detailed knowledge of the optical properties of the material, which is not always available.

Figures 3.4 and 3.5 show the temperature dependence of the Raman phonon frequency and linewidth for the elemental semiconductors Si, Ge, and α-Sn. It is quite apparent that the frequency is a sensitive function of temperature. A parameterization of this dependence can therefore be used to monitor temperature. This method has been used by Ostermeir et al. to study the temperature distribution in Si-MOSFETs [64]. Care must be exercised however, to verify that the temperature dependence of the Raman shift under local heating agrees with the shift measured on bulk samples without temperature gradients. For films grown on substrates, for example, the lattice mismatch is usually temperature-dependent, and this adds a stress contribution to the Raman shift. Even in a homogeneous sample, gradients of temperature produce different degrees of thermal expansion, which can also make stress contributions to the Raman phonon frequencies.

The Raman linewidth is also a strong function of temperature (particularly at high temperatures) and can be used for monitoring temperature. In principle, this approach has two distinct advantages over Raman shift measurements: stress effects represent a second-order perturbation (provided that there are no stress gradients leading to inhomogeneous broadening), and – unlike the case of the Raman shift – there are simple and physically meaningful expressions for the temperature dependence of the linewidth. For example, the full-width at half maximum (FWHM) of the Raman peak corresponding

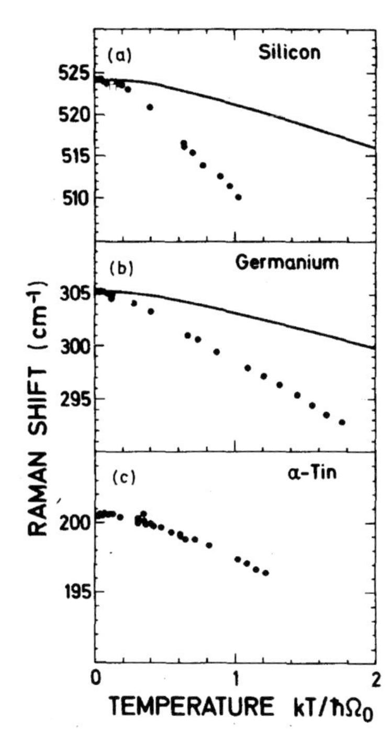

Fig. 3.4a–c. Temperature dependence of the Raman shift in Si, Ge, and α-Sn plotted as a function of a normalized temperature. The symbol $\boldsymbol{\Omega}_0$ denotes the low-temperature Raman mode frequency in each material. The *solid lines* represent the calculated thermal expansion contribution to the line shift [65]

to LO phonons in tetrahedral semiconductors is given by [65]

$$2\Gamma(T) = 2\Gamma^{in} + \sum_{\lambda} 2\Gamma_{\lambda}^{\text{hom}} \left[1 + n\left(\omega_{\lambda}\right) + n\left(\omega_{LO} - \omega_{\lambda}\right)\right], \tag{3.26}$$

where $n(\omega)$ is the Bose–Einstein occupation number for a phonon of frequency ω (for a standard Lorentzian with broadening parameter Γ, the FWHM is 2Γ). The sum runs in principle over all pairs of phonons into which the optic phonon can decay into, but in practice these frequencies tend to cluster around certain values, to the extent that for some semiconductors the temperature dependence can be fit with a single term. Good examples are Si, Ge, and α-Sn, for which the experimental temperature dependence of the Raman linewidth can be satisfactorily reproduced by assuming that the optic phonon with frequency ω_0 decays into pairs of phonons with frequencies $\omega_1 = 0.65\omega_0$ and $\omega_2 = 0.35\omega_0$ [65]. The first term in (3.26) is a temperature independent broadening that arises from defects, inhomogeneities, etc. The natural isotopic distribution in the material also makes a contribution, which is small in the case of the LO phonons but significant for TO phonons in polar semiconductors [66]. The recent availability of isotopically pure samples has made it possible to study the effect of isotopes on the Raman linewidth [67,68]. Table 3.4 shows low temperature Raman linewidths for some selected semiconductors. It is important to point out that (3.26) neglects

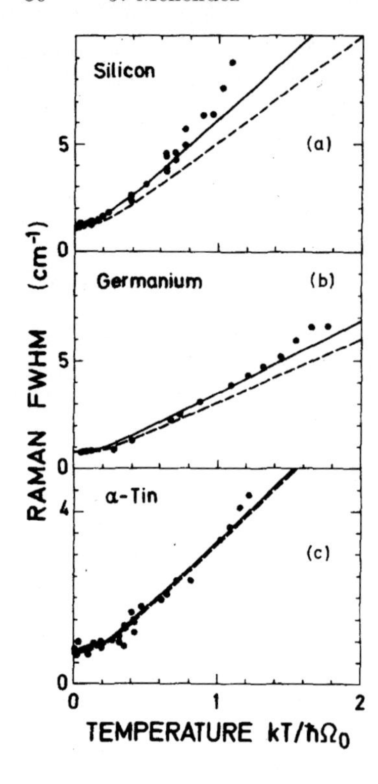

Fig. 3.5a–c. Measured full width at half maximum of the Raman peak as a function of the same normalized temperature used in Fig. 3.4. The *dashed line* is a prediction based on (3.26) and assuming that the optic phonon decays into two phonons of equal energy. The *solid lines*, in better agreement with experiment, assume a 65%/35% partition of the optic phonon energy [65]

fourth order anharmonicity (decay into three phonons) and therefore is not accurate at very high temperatures.

The possible presence of a temperature independent contribution to the Raman linewidth that varies from sample to sample limits the use of linewidths for temperature monitoring. By and large, however, the main limitation is the requirement that the instrumental broadening be significantly smaller than the natural width of the line. In view of the linewidth values in Table 3.4, this is a very stringent requirement, particularly for new generations of single-stage Raman spectrometers that are optimized for throughput.

3.2.3 Stress Measurements

The application of stress alters the phonon structure due to the anharmonic components of the interatomic potential. The resulting changes in Raman frequencies represent a powerful tool to monitor stress: micro-Raman spectroscopy is one of the techniques of choice for studying stress gradients in semiconductor devices.

The strain associated with the applied stress changes the equilibrium position of the atoms in the crystal. The phonon frequencies corresponding to the deformed crystal can be calculated within the quasi-harmonic approximation, whereby the force constants are evaluated at the new equilibrium

Table 3.4. Low temperature full width at half maximum (FWHM) of the first order Raman peak of selected semiconductors, in cm^{-1}. When known, the contributions $2\Gamma^{in}$ due to isotopic disorder and the FWHM (pure) for an isotopically pure sample are given

Material	Temperature	$2\Gamma^{in}$	FWHM (pure)	FWHM	Theory
Diamond				1.4[a]	1.01[b]
Si	10 K	0.024[c]	1.22 ± 0.02[c] (^{28}Si)	1.24 ± 0.07[d]	1.24[b]
Ge	10 K	0.03 ± 0.03[e]	0.62 ± 0.02[e] (^{73}Ge)	0.66 ± 0.02[e]	0.67[b]
α-Sn	10 K			0.81 ± 0.15[c]	
GaAs (LO)	12 K	0.12 ± 0.09[c]	0.55 ± 0.05[c] (^{69}GaAs)	0.67 ± 0.10[c]	0.66[f]
GaAs (TO)	4 K			0.83[g]	0.44[f]
AlAs (LO)	10 K	0	0.65[h]	0.65[h]	0.42[f]
GaP (LO)	5 K			0.20 ± 0.02[i]	0.18[f]
GaP (TO)				3.06[j]	
InP (LO)	6 K			0.026[k]	0.038[f]
InP (TO)	80 K			0.50[g]	0.49[f]

[a] M.A. Washington, H.Z. Cummins: Phys. Rev. B **15**, 5840 (1977); with instrument correction by the present author
[b] A. Debernardi, S. Baroni, E. Molinari: Phys. Rev. Lett. **75**, 1819 (1995)
[c] F. Widulle, T. Ruf, A. Göbel, I. Silier, E. Schönherr, M. Cardona, J. Camacho, A. Cantarero, W. Kriegseis, V.I. Ozhogin: Physica B **263-264**, 381 (1999)
[d] J. Menéndez, M. Cardona: Phys. Rev. B **29**, 2051 (1984)
[e] J.M. Zhang, M. Giehler, A. Göbel, T. Ruf, M. Cardona, E.E. Haller, K. Itoh: Phys. Rev. B **57**, 1348 (1998)
[f] A. Debernardi: Phys. Rev. B **57**, 12847 (1998)
[g] G. Irmer, M. Wenzel, J. Monecke: Phys. Stat. Sol. (b) **195**, 85 (1996)
[h] M. Canonico, J. Menéndez (unpublished)
[i] J. Kuhl, W.E. Bron: Solid State Commun. **49**, 935 (1984)
[j] B.K. Bairamov, Y.É. Kitaev, V.K. Negoduiko, Z.M. Khashkhozhev: Fiz. Tverd. Tela (Leningrad) **16**, 2036 (1974)
[k] F. Vallée: Phys. Rev. B **49**, 2460 (1994)

positions and are different from the force constants of the undeformed crystal due to the anharmonicity in the interatomic potential. For small values of the strain, it can be assumed that the difference $\Delta\boldsymbol{\Phi}$ between the two sets of force constants is linear in the strain, and a Taylor expansion makes it possible to evaluate $\Delta\boldsymbol{\Phi}$ in terms of the higher derivatives of the interatomic potential. The details of such derivation are given by Ganesan et al [69] and will be omitted here, since they require rather subtle arguments involving the different periodicities of the deformed and undeformed crystal and the fact that the macroscopic strain parameters do not fully determine the position of the atoms within a unit cell. However, it is easy to use Callen's method [42] to find the symmetry-allowed form of $\Delta\boldsymbol{\Phi}$, and this has been done by Anastassakis and Burstein [46]. For typical values of the stress, the induced frequency shifts are of the order of 0.1% of the mode frequency. This suggests that stress effects be treated perturbatively using (3.6) as the starting point. Since the zone-center optic phonons in cubic elemental semiconductors are triply degenerate, we must use degenerate perturbation theory, i.e. diagonalize the problem in the three-dimensional subspace of the zone center optic modes. If we use the index $k = x, y, z$ to label these modes, insert the symmetry-allowed form of $\Delta\boldsymbol{\Phi}$ from [46] and recall that in this case $\Delta\boldsymbol{M} = 0$, (3.6) becomes

$$\begin{pmatrix} pe_{11} + q\,(e_{22} + e_{33}) - \Delta\omega_g^2 & re_{12} & re_{13} \\ re_{12} & pe_{22} + q\,(e_{33} + e_{11}) - \Delta\omega_g^2 & re_{23} \\ re_{13} & re_{23} & pe_{33} + q\,(e_{11} + e_{22}) - \Delta\omega_g^2 \end{pmatrix} \begin{pmatrix} c_{gx} \\ c_{gy} \\ c_{gz} \end{pmatrix} = 0, \qquad (3.27)$$

where $\Delta\omega_g^2 = \omega_g^2 - \omega_0^2$ is the difference between the squares of the perturbed and unperturbed frequencies, e_{ij} the components of the strain tensor (defined by Kittel [70]), and p, q, r are the only symmetry-allowed coefficients of the perturbation. We use the index g to label the three solutions, which can be obtained analytically by setting the determinant of the coefficients in (3.27) equal to zero. The strain components are either known from the boundary conditions of the problem (for example, the in-plane strain components in epitaxial films) or can be derived from the applied or known stress using standard elasticity theory. The eigenvector components $\{c_{gx}, c_{gy}, c_{gz}\}$ can be inserted in (3.8) to obtain the phonon eigenvector corresponding to the perturbed mode g and also in (3.18) to obtain the corresponding Raman tensor. Note that a general stress will remove the three-fold degeneracy of the optic modes, so that arbitrary linear combinations of the displacement eigenvectors in (3.14) no longer yield acceptable phonon eigenvectors. It is therefore critical to properly rotate the Raman tensor to obtain the right selection rules for each individual Raman peak.

A common application of (3.27) is the case of a film epitaxially matched to a (001)-oriented cubic substrate. In this case the off-diagonal components of the strain are zero. The xx and yy diagonal components are given by $e_{xx} = e_{yy} = (a_{\mathrm{substrate}} - a_{\mathrm{film}})/a_{\mathrm{film}}$, and it is easy to conclude from the condition of vanishing stress at the growth surface that $e_{zz} = -(2C_{12}/C_{11})e_{xx}$ (here the C's are the elastic constants). This perturbation splits the three-fold degenerate modes into a singlet and a doublet whose eigenvalues are given by

$$\begin{aligned}\Delta\omega^2_{\mathrm{singlet}} &= \tfrac{2}{3}(p+2q)\left(1-\tfrac{C_{12}}{C_{11}}\right)e_{xx} + \tfrac{2}{3}(q-p)\left(1+\tfrac{2C_{12}}{C_{11}}\right)e_{xx}\\ \Delta\omega^2_{\mathrm{doublet}} &= \tfrac{2}{3}(p+2q)\left(1-\tfrac{C_{12}}{C_{11}}\right)e_{xx} - \tfrac{1}{3}(q-p)\left(1+\tfrac{2C_{12}}{C_{11}}\right)e_{xx}\,. \end{aligned} \qquad (3.28)$$

It is easy to show that the so-called Grüneisen parameter, defined as $\gamma = -\partial \ln\omega/\partial \ln V$, can be expressed in terms of the parameters of this theory as $\gamma = -(p+2q)/6\omega_0^2$. Hence, the first terms in (3.28) represent the hydrostatic contribution (which does not split the phonon frequencies) and the second term gives the frequency splitting caused by the biaxial component of the perturbation. Table 3.5 shows selected values of the p and q parameters that can be used to evaluate (3.28). A more complete list of these parameters was recently compiled in a review article by Anastassakis and Cardona [71]. This very instructive article also gives details on their measurement.

The phonon eigenvectors and Raman tensors for the singlet and doublet in (3.28) can be obtained using the general procedure outlined above, but it is intuitively clear in this case that the mode with eigenvector polarized along the (001) direction becomes the singlet and the modes polarized in the other two Cartesian directions become the doublet, so that the Raman tensor for the singlet is the third tensor in (3.17). This is the only mode that can be observed in the backscattering geometry for a (001)-oriented sample.

The foregoing discussion shows that it is relatively straightforward to evaluate the stress induced shifts and splittings in the Raman spectra for a *known* stress. On the other hand, the inverse problem (i.e., finding the stress tensor from measurements of the Raman spectra) is much more complicated, particularly in the presence of spatially inhomogeneous stresses with no obvious symmetry [72–76]. This is precisely the case in many important technological applications. An arbitrary stress has six independent components, so that at least six different measurements are needed to characterize the stress completely. In principle, this information is available: in silicon, for example, one can determine the shift of each the three split modes relative to the unperturbed optic phonon frequency, and one can measure the intensity of their corresponding Raman peaks relative to the intensity of the Raman line in unstressed Si. In practice, however, this is very difficult to do. Due to the high index of refraction of the semiconductor materials, the light propagates nearly parallel to the surface normal even for wide angles of incidence, and in this quasi-backscattering configuration some of the modes may not be observable. For example, for backscattering at the (001) surface of Si only

Table 3.5. Experimental anharmonicity parameters p and q for selected semiconductors. The parameters are normalized to the frequency ω^2 of the Raman mode in the unstressed material. When two sets of values are given, the first set is for the TO, the second for the LO phonon[a]

Material	p/ω^2	q/ω^2
Diamond	-2.81	-1.77
Si	-1.83	-2.33
Ge	-1.47	-1.93
GaAs	-2.40	-2.70
	-1.70	-2.40
InP	-2.5	-3.2
	-1.6	-2.8
AlSb	-2.1	-2.6
	-1.6	-2.6
GaSb	-1.9	-2.35
InAs	-0.95	-2.10
ZnSe	-2.75	-4.00
	-0.94	-2.28
InSb	-2.45	-3.04
	-1.72	-2.65
GaP	-1.35	-1.95
	-1.45	-2.5

[a] Adapted from Anastassakis and M. Cardona, in *High Pressure in Semiconductor Physics II, Semiconductors and Semimetals*, Vol. 55, ed. by T. Suski, W. Paul (Academic Press, 1998)

the mode polarized along the (001) direction is Raman-active, as indicated above. In spite of these difficulties it has been shown recently that even for (001)-oriented samples the other split phonons can be observed under wide angles of incidence if the measurements are performed very carefully. Using this "off-axis" technique the stress tensor can be determined directly from experiment [77,78], but the method is not directly applicable in the context of Raman microscopy, where there is a distribution of incident angles due to the high numerical aperture of the objectives.

The approach commonly followed to study stresses with Raman spectroscopy is to model the stress field using a numerical technique such as the finite element method, and to compare the spatially dependent Raman spectrum in the backscattering configuration with the Raman spectrum calculated from the assumed stress distribution. A case study is briefly described in the box by de Wolf.

3.2.4 Impurities

The key feature that explains the preeminent role of semiconductor materials in modern technology is the ease with which their electronic properties can be manipulated through doping. Quantifying doping levels in semiconductors is therefore a primary requirement in the analytical laboratory. The contactless nature of Raman spectroscopy and its high spatial resolution makes this technique an intriguing alternative for the determination of impurity concentrations. Impurities affect the Raman spectrum of semiconductors in different ways. On the one hand, the vibrational modes associated with motions of the impurity atoms can produce new Raman peaks. On the other hand, the presence of the impurities changes the Raman spectrum of the host material. This occurs through the change in mass and bond length (atomic effects) and, in the case of donors and acceptors, through the interaction of the carriers with the lattice (electronic effects). The latter are often much stronger.

Impurity vibrational modes. The vibrational modes of impurities in crystals and their light scattering properties have been discussed in Sect. 6 of the introductory chapter. A comprehensive review of this subject has been written by Barker and Sievers [79]. For concentrations of technological interest it is not easy to detect impurity modes using Raman scattering, since the signals are weak and often obscured by the second-order Raman spectrum of the host. Figure 3.6 [80] shows the local modes of boron in Si observed by Raman spectroscopy in a sample with a B concentration of $1.5 \times 10^{20}\,\mathrm{cm}^{-3}$. Considering the acceptable signal to noise ratios of the spectra and the progress in Raman instrumentation over the last 30 years, one can conclude that in spite of the technical difficulties even modest concentrations of impurities can be detected using Raman spectroscopy of their local modes. In the case of carbon in Si, for example, carbon impurities at concentration levels of $10^{17}\,/\mathrm{cm}^3$ can be observed [81].

The intensity of the impurity Raman peaks can be used to determine the impurity concentration. This has to be done on the basis of a previous calibration. Reliable calculations of Raman intensities for impurity modes are in principle possible (using the above described ab initio methods), but they are not available. Raman spectroscopy is particularly useful when the impurity concentration can be determined from ratios of intensities between the impurity and host Raman modes, thereby eliminating the need for absolute intensity measurements. A good example is C in Si [82]. The local mode of C is both Raman and infrared active, whereas Si itself is not infrared-active to first order. Thus the sample preparation requirements for measurements

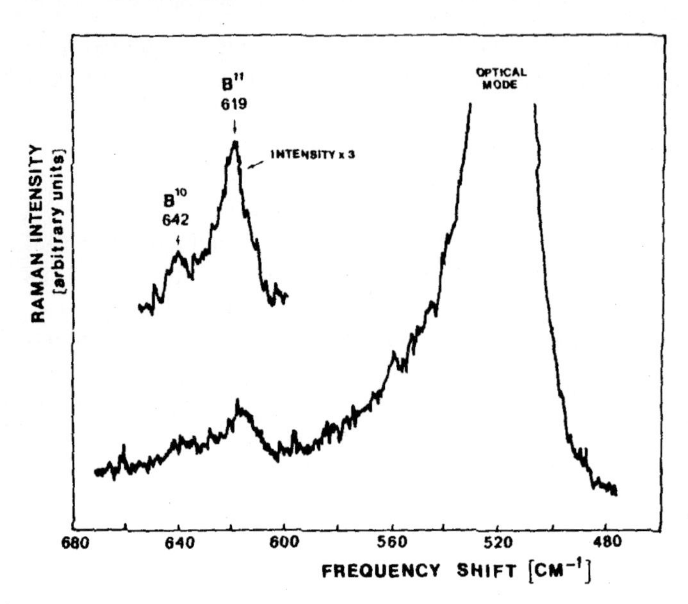

Fig. 3.6. Raman spectrum of B-doped Si at room temperature, showing the localized vibrational modes corresponding to the B impurities [80]

of C in Si are much less stringent for Raman than for infrared absorption spectroscopy.

Atomic effects. An order-of-magnitude estimate of atomic effects can be obtained by assuming that the mass defect is uniformly distributed over all atoms of the crystal. Substituting in (3.7) we obtain $\Delta M_{ff'} = x\,(\Delta M/M)\,\delta_{ff'}$. Here x is the fractional impurity concentration and $\Delta M = (M_{\mathrm{imp}} - M)$, where M is the mass of the Si atom and M_{imp} the mass of the impurity. Likewise, we can assume that the different atomic radii of the impurity and host atoms lead to a change in volume that is uniformly distributed over the crystal, so that we obtain a strain tensor whose components are given by $e_{ij} = x(\Delta R/R)\delta_{ij}$. Substituting in (3.27) , we find $\Delta\Phi_{ff'} = -6\gamma\omega_0^2\,(\Delta R/R)\,x\delta_{ff'}$, where γ is the Grüneisen parameter and ω_0 is the unperturbed Raman frequency. Since both the mass and strain perturbation are diagonal at this level of approximation, the mode eigenvectors remain unchanged and the only predicted effect is a frequency shift that is given by the diagonal elements of the perturbation:

$$\frac{\Delta\omega^2}{\omega_0^2} = -\left(\frac{\Delta M}{M} + 6\gamma\frac{\Delta R}{R}\right)x. \tag{3.29}$$

Figure 3.7 shows Raman spectra of p-type silicon for different B concentrations [83]. The main Si Raman peak is seen to broaden and shift to lower frequencies. But B is lighter and smaller than Si, so that, contrary to the

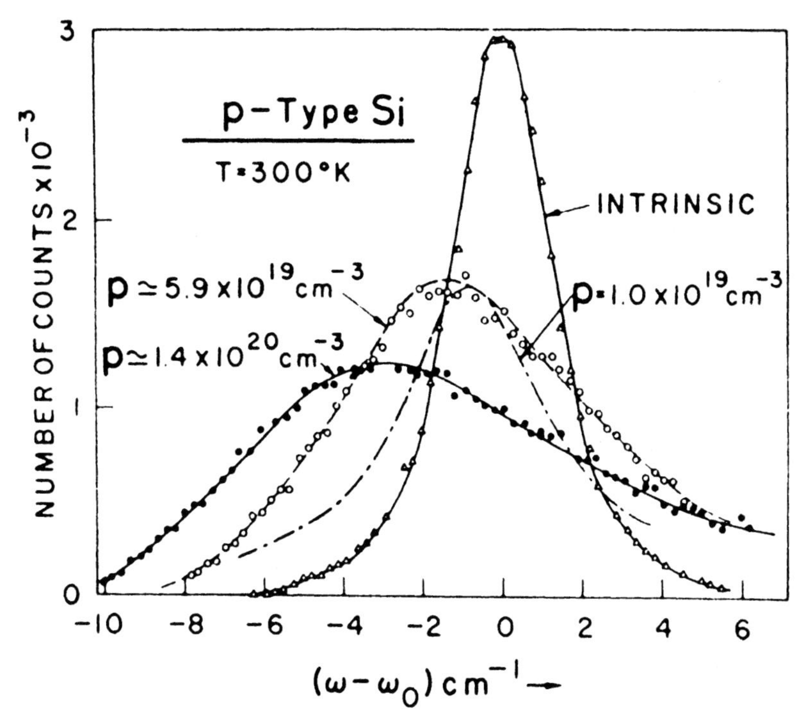

Fig. 3.7. Raman spectra of B-doped Si samples at room temperature, showing the effect of the impurities on the main Si Raman line [83]

experimental finding, (3.29) predicts an increase in the peak frequency. For our specific example of B in Si, this shift was estimated to be $+1\,\mathrm{cm}^{-1}$ for a concentration of 1.4×10^{20} [83]. Moreover, the observed shifts are strongly dependent on temperature, which can hardly be understood in terms of (3.29), and suggest that the main contribution to the boron effect on the Si Raman peak is electronic in character. Even for substitutional isovalent impurities, the agreement between (3.29) and experiment is poor. For example, for C in Si (3.29) predicts $\Delta\omega/\omega = 1.3x$, whereas the experimental value is $\Delta\omega/\omega = 0.4x$ [84].

One reason for the failure of (3.29) is that the assumption of a uniformly distributed mass excess or defect, used in the derivation, is not really justified. The real mass distribution leads to off-diagonal elements of the mass matrix $\Delta M_{ff'}$ which are not negligible relative to the diagonal terms. These off-diagonal elements $\Delta M_{ff'}$ (together with the off-diagonal elements $\Delta\Phi_{ff'}$ of the force constant matrix) mix the modes of the unperturbed crystal. One effect of this mixing is that all modes in the crystal acquire some Raman activity, so that one might expect to see peaks at the frequency of critical points in the phonon density of states of the unperturbed crystal [79].

The correlation between the local impurity mode and the atomic effects on the Raman spectrum of the host provides very useful information. This information has been used, for example to analyze the relative concentration of substitutional and interstitial C in $Si_{1-y}C_y$ [82]. The total C concentration in solid solutions with 1% or higher carbon content can be easily measured with techniques such as Rutherford Back-Scattering (RBS), but the fraction of these C atoms that occupy substitutional sites is very difficult to measure. In the Raman spectrum, a sharp local mode appears at $605\,\mathrm{cm}^{-1}$. This peak is related to substitutional C only. But substitutional as well as interstitial C can activate density of states features in the Raman spectrum of the host, so that the intensity of these features relative to the local mode provide semiquantitative information on the C distribution in the sample.

The theoretical treatment of the impurity problem is severely complicated by the lack of translational symmetry. Very sophisticated and ingenious methods have been developed to handle the atomic perturbation in a more rigorous way than in (3.29) [79], but given the computing power of today's personal computers, the simplest (and in principle exact) way to treat the problem is to diagonalize (3.5) or (3.6) for a large supercell containing as many atoms as possible. The size of these supercells used to be severely limited by the N^3 dependence of the time needed to diagonalize an $N \times N$ matrix, but today's home multimedia PCs have the memory and the processing power necessary to solve a 1000-atom problem in a few minutes. The results of such calculations, which will be discussed in more detail in the section on alloying, show that the "mass contribution" usually lowers the frequency of the main Raman line, contrary to the naive prediction in (3.29) for lighter impurity atoms.

The activation of otherwise silent modes by the impurity perturbation might be expected to be significant for the LO phonon branches in polar semiconductors. Since $\mathrm{d}\omega/\mathrm{d}q = 0$ near the Brillouin zone center, (see Fig. 3.1) no major distortion of the allowed Raman peak should be expected. However, the intraband matrix elements of the Fröhlich electron-phonon interaction are responsible for the so-called "forbidden" resonance Raman scattering by LO phonons [7,8]. As discussed above, the Raman tensor for this process is proportional to the magnitude of the phonon wave vector. Since the Fröhlich interaction is strongest for wave vectors much longer than $\boldsymbol{q} = \boldsymbol{k}_L - \boldsymbol{k}_S$ [35], violations of wave-vector conservation can lead to large Raman cross sections for the modes activated by the presence of impurities. This argument also applies to any other perturbation that breaks the translational symmetry. The effect of wave vector non-conservation manifests itself as a reduced interference between "forbidden" and allowed scattering as well as in an increase in the Raman cross section for the "forbidden" configuration [85–87]. In semiconductor superlattices the effect is even more dramatic. Since the optic mode frequencies in these non-cubic systems are angular dependent, wave vector non-conservation affects the Raman lineshape very strongly. In fact, the observation of the so-called "interface modes" has been ascribed almost entirely to this effect [88,89].

Electronic effects. The electronic effects that dominate the shift and broadening of the Raman spectrum in Fig. 3.7 are very complex and will not be discussed here in detail. The reader is referred to several excellent reviews [3,4,15]. The observed effects depend sensitively on the band structure of the host materials and are associated with the fact that the optic phonon can induce electronic transitions. The coupling with these transitions changes the phonon self-energy and affects its Raman lineshape. One of the most spectacular examples is the so-called Fano lineshape, the asymmetry in the Raman line that is caused by the interference between the Raman phonon and a continuum of intervalence electronic excitations in degenerate p-type semiconductors. A rigorous but readable account of this phenomenon is given by Wallis and Balkanski [15].

One particular situation that is worth discussing in more detail, due to its simplicity and its characterization potential, is the coupling of the optic modes in polar semiconductors with the carriers' collective modes of oscillations, called plasmons. For carriers in a single, parabolic, isotropic band, the plasmon frequency is given by

$$\omega_{\mathrm{p}}^2 = \frac{4\pi n e^2}{m^* \varepsilon_\infty}, \tag{3.30}$$

where n is the carrier concentration, ε_∞ the high-frequency dielectric function, and m^* the effective mass of the carriers. The macroscopic electric field associated with the plasmons interacts with the optic modes in polar semiconductors. This interaction leads to coupled modes of mixed phonon-plasmon character whenever the plasmon frequency approaches the optic phonon frequencies. Substituting typical values for semiconductors, one finds that this occurs for concentrations of technological interest in the $10^{17} - 10^{19}\mathrm{cm}^{-3}$ range. The mathematical description of coupled modes is similar to the description of the phonon-polaritons discussed above in regards to the LO-TO splitting at vanishing wave vector. The coupled system's eigenmodes are given by the zeros of the dielectric function, which can be written as the sum of a phonon and a plasmon part [3]:

$$\varepsilon(\omega) = \varepsilon_\infty \left[1 + \frac{\omega_{\mathrm{LO}}^2 - \omega_{\mathrm{TO}}^2}{\omega_{\mathrm{TO}}^2 - \omega^2 - \mathrm{i}\omega\Gamma} - \frac{\omega_{\mathrm{p}}^2}{\omega(\omega + \mathrm{i}\gamma)} \right], \tag{3.31}$$

where we have introduced, for completeness, broadening parameters for phonons and plasmons. The resulting modes are shown in Fig. 3.8 for the case of GaAs [90]. In the case of homopolar semiconductors there is no plasmon-phonon coupling, but the pure plasmon mode has been observed [4]. Yet another type of purely electronic Raman scattering is the observation of transitions between donor or acceptor energy levels [3].

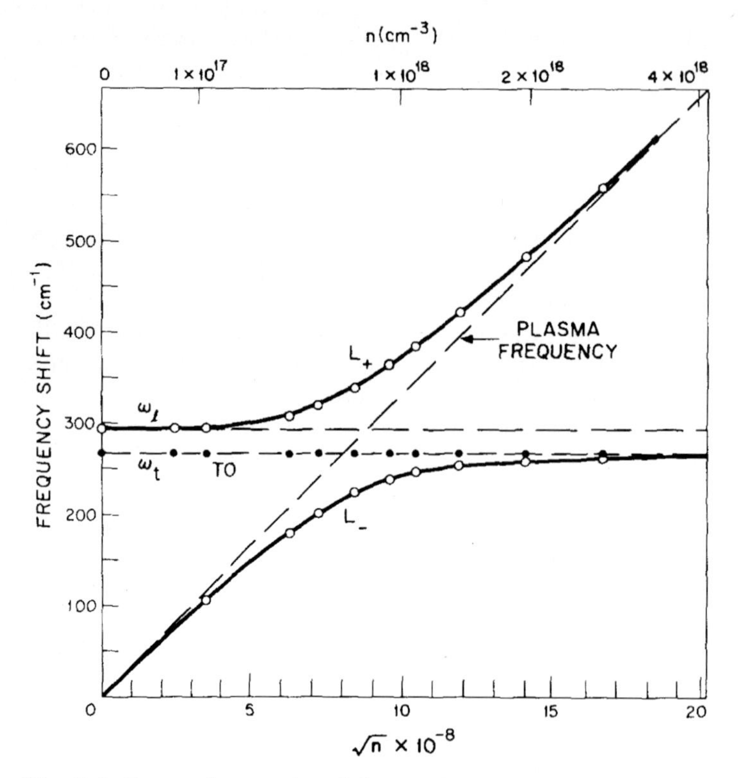

Fig. 3.8. Raman frequencies of the coupled plasmon-phonon modes in GaAs. labeled L^+ and L^-. The *solid lines* are calculated from the zeros of (3.31) [90]

The measurement of coupled or plasmon modes can be used to determine the carrier concentration, as is quite apparent from Fig. 3.8. It should be pointed out, however, that (3.30) may not be accurate enough if very precise measurements of concentrations are needed. Fortunately, empirical calibrations of plasmon frequencies are available for the most commonly used semiconductors [91].

3.2.5 Alloying

Most semiconductors compounds can be easily alloyed, and the spatial distribution of the intermixing atomic species can be made nearly random. This randomness removes the translational symmetry of the crystal, so that Raman spectra from semiconductor alloys might be expected to be similar to the Raman spectra from amorphous semiconductors [92], which, as explained in the introductory chapter, are roughly proportional to the density of phonon states. However, alloying is in general a much "gentler" perturbation than amorphization, as can be seen from diffraction studies, which clearly indicate the existence of a remaining order and a well-defined average lattice constant. The almost-crystalline structural properties of semiconductor alloys are reflected in their Raman spectra, which turn out to be dominated by relatively narrow peaks. Figure 3.9 illustrates the Si_xGe_{1-x} case, where one can see three strong peaks associated with Ge-Ge, Si-Ge, and Si-Si vibrations [93]. The compositional dependence of the frequency of the alloy Raman peaks, shown in Fig. 3.10 for the case of Si_xGe_{1-x} [94], is of great interest as a straightforward way to determine the alloy composition.

Two qualitatively different types of Raman spectra are observed from alloy semiconductors. In the first type one sees peaks that can be associated with the vibrations of an average crystal, whereas in the second type (see Fig. 3.9 as an example) the peaks observed are related to the vibrations of the individual materials that are being alloyed. In binary alloys with common anions or cations this is referred to as "one-mode" versus "two-mode" behavior. For individual alloy systems the compositional behavior can be quite complicated, and changes from one-mode to two-mode behavior as a function of the composition have been observed [79]. In Table 3.6 we summarize the experimental compositional dependence of Raman modes for selected semiconductor alloys. It is important to point out that an accurate determination of compositions presupposes an accurate measurement of phonon frequencies. Unfortunately, experimental errors of up to $2\,\mathrm{cm}^{-1}$ are frequent – particu-

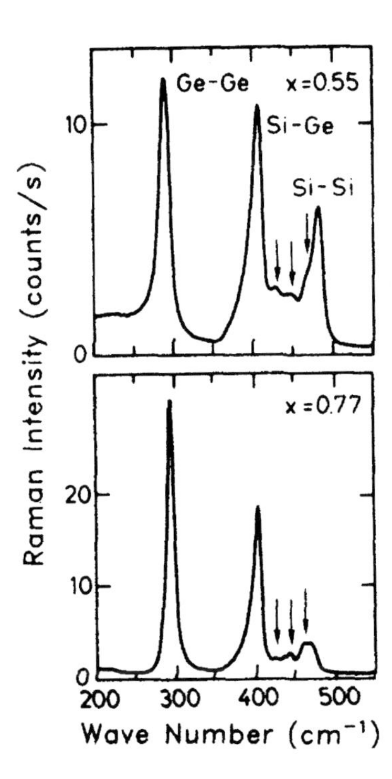

Fig. 3.9. Raman spectra from epitaxial $Si_{1-x}Ge_x$ alloys, showing the three main peaks and a few extra peaks indicated by *arrows* [93]

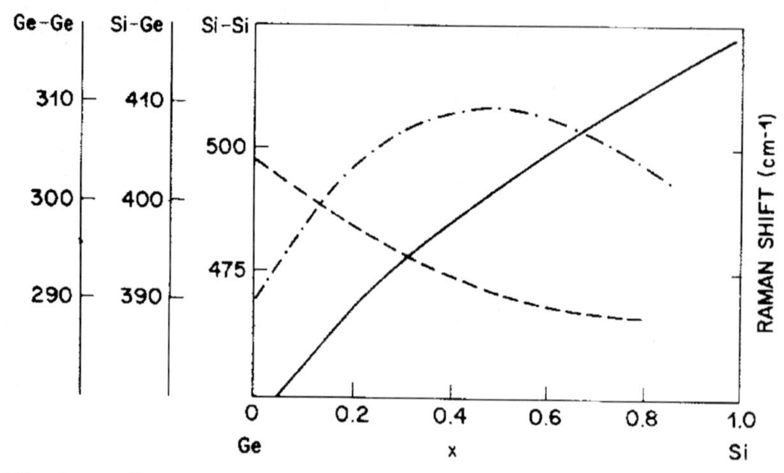

Fig. 3.10. Compositional dependence of the three main Raman peaks in bulk $Si_{1-x}Ge_x$ alloys. *Solid line*: Si-Si mode. *Dashed line*: Ge-Ge mode. *Dashed-dotted line*: Si-Ge mode [94]

larly when array detectors are used – , and this can lead to non-negligible compositional errors. In addition, sample heating effects can also shift the peak frequencies. A solution to these problems, proposed and implemented by Saint-Cricq et al. for the case of $Al_xGa_{1-x}As$ alloys [95], is to measure the *difference* in frequency between two Raman modes in the system. This eliminates most of the calibration problems, and in most cases it has the added benefit that the difference in Raman peak frequencies is a steeper function of composition. A more fundamental problem is that the degree of randomness in semiconductor alloys is seldom quantified, although it can be expected to affect the phonon frequencies. The seriousness of this effect can be inferred from the observation that many semiconductor alloy systems have a tendency to spontaneously form partially ordered structures [96]. Examples of the effect of ordering on the Raman spectrum of alloy semiconductors are the work of Tsang et al. [97] for the case of $Si_{1-x}Ge_x$ alloys and of Hassine et al. for $In_xGa_{1-x}P$ [98].

Given today's availability of powerful personal computers, as indicated above, the best theoretical approach to the study of phonons in semiconductor alloys is to perform supercell calculations. However, an accuracy comparable to Fig. 3.1 can only be expected if one can obtain interatomic force constants from first principles. This is still prohibitively expensive for large supercells, but one can take advantage of the transferability of force constants among tetrahedral semiconductors to perform calculations of phonons in alloy semiconductors with an accuracy comparable to full first-principles calculations. This approach has been used by Baroni et al. [99], who find excellent agreement with experiment for $Al_xGa_{1-x}As$. If one makes the additional

Table 3.6. Polynomial fits to the compositional dependence of the Raman frequencies in semiconductor alloys (in cm^{-1}). Data correspond to room temperature unless otherwise indicated. The functional dependencies are quoted from the original references, when given, or obtained from our own fits to published figures or tables. In the latter case we include the absolute value (in cm^{-1}) of the largest deviation between the fit and the data (number in parenthesis). Note that in several cases the authors quoted have themselves collected data from different sources, so that the original reference may not appear in our list

	$Si_{1-x}Ge_x$[a,b]		
Si-Si	Si-Ge ($x < 0.5$)		Ge-Ge
$521.2 - 67.91x$	$400.0 + 22.07x - 36.14x^2$ $+83.73x^3 - 88.54x^4$		$280.8 + 19.37x$
	$Al_xGa_{1-x}As$[c]		
AlAs-like (LO)	AlAs-like (TO)	GaAs-like (LO)	GaAs-like (TO)
$361.7 + 55.62x$ $-15.45x^2$	$361.07 - 9.78x$ $+9.81x^2(0.8)$	$292.1 - 39.96x$	$267.7 + 1.29x$ $-17.64x^2(0.9)$
	$Al_xGa_{1-x}Sb$[d]		
AlSb-like (LO)	AlSb-like (TO)	GaSb-like (LO)	GaSb-like (TO)
$312.2 + 8.3x$ $-2.1x^2$	$312.4 + 32.0x$ $-5.1x^2$	$234.9 - 22.2x$ $-6.9x^2$	$225.1 - 11.5x$ $-8.6x^2$
	$Al_xIn_{1-x}As$[e]		
AlAs-like (LO)	AlAs-like (TO)	InAs-like (LO)	InAs-like (TO)
$332.7 + 64.3x$ $+7.10x^2$	$334.2 + 41.37x$ $-14.62x^2$	$238.6 + 20.38x$ $-46.02x^2$	$217.3 + 27.85x$ $-28.27x^2$
	$In_xGa_{1-x}As$[f]		
InAs-like (LO)	InAs-like (TO)	GaAs-like (LO)	GaAs-like (TO)
$242.19 - 30.67x$ $+28.18x^2(1.8)$	$240.98 - 34.01x$ $+11.73x^2(1.0)$	$291.6 - 28.51x$ $-19.53x^2(1.0)$	$268.6 - 25.68x$ $-2.82x^2(1.5)$
	$In_xGa_{1-x}Sb$[g]		
InSb-like (LO) ($x > 0.3$)	InSb-like (TO) ($x > 0.3$)	GaSb-like (LO)	GaSb-like (TO)
$197.46 - 43.01x$ $+38.56x^2(1.4)$	$195.81 - 39.75x$ $+23.94x^2(1.3)$	$237 - 22.12x$ $-14.27x^2(1.1)$	$227 - 20.68x$ $-9.08x^2(1.9)$
	$In_xGa_{1-x}P$[h] (85 K)		
LO ($x < 0.17$)		TO ($x < 0.17$)	
$387.8 - 64.37x + 116.2x^2(0.5)$		$367.3 - 63.02x - 49.83x^2(0.7)$	

Table 3.6. *(Continued)*

	$AlAs_xSb_{1-x}$[i]		
LO		TO	
$342.97 + 58.70x + 2.83x^2(3.0)$		$321.91 + 46.19x - 6.00x^2(3.4)$	
	$GaAs_xP_{1-x}$[h,j]		
GaAs-like (TO) ($x > 0.5$)		GaP-like (LO) ($x < 0.17, 85\,K$)	GaP-like (TO) ($x < 0.17, 85\,K$)
$276.40 - 12.50x + 4.10x^2(0.2)$		$405.2 + 9.99x$ $-224.5x^2(0.3)$	$367 + 27.87x$ $-306.3x^2(0.3)$
		GaP-like (TO)	
		$367 - 43.74x + 16.71x^2(3.6)$	
	$GaAs_xSb_{1-x}$[k]		
GaAs-like (LO)	GaAs-like (TO)	GaSb-like (LO) ($x < 0.25$)	GaSb-like (TO) ($x < 0.25$)
$234.1 + 84.05x$ $-26.55x^2(3.8)$	$238.6 + 33.53x$ $-3.96x^2(2.4)$	$236.1 - 42.98x$ $+46.71x^2(0.4)$	$226.0 + 18.10x$ $-29.86x^2(1.0)$
		GaSb-like (LO and TO, $x > 0.25$)	
		$227.1 - 1.00x + 18.41x^2(3.5)$	
	$InAs_xP_{1-x}$[l] (77 K)		
InP-like (LO)	InP-like (TO)	InAs-like (LO)	InAs-like (TO)
$344 - 14.49x$ $-25.13x^2(6.0)$	$305.5 + 5.82x$ $-0.44x^2(5.2)$	$218.32 + 59.93x$ $-37.24x^2(4.2)$	$228.70 - 17.65x$ $+8.85x^2(1.0)$
	$Hg_xCd_{1-x}Te$[m]		
HgTe-like (LO)	HgTe-like (TO)	CdTe-like (LO)	CdTe-like (TO)
$127.38 + 15.68x$ $-5.66x^2(0.8)$	$129.75 - 13.74x$ $+1.69x^2(0.3)$	$166.9 - 11.10x$ $-4.13x^2(1.1)$	$138.8 + 10.42x$ $+3.20x^2(0.8)$
	$Zn_xCd_{1-x}Te$[n] (20–80 K)		
CdTe-like (LO, $x < 0.5$)	CdTe-like (TO, $x < 0.5$)	ZnTe-like (LO)	ZnTe-like (TO)
$167 - 39.41x$ $+41.69x^2(1.4)$	$147.3 + 18.11x$ $-5.01x^2(1.1)$	$173.10 + 56.90x$ $-20.00x^2(0.8)$	$172.95 + 16.28x$ $-7.23x^2(0.9)$

Table 3.6. *(Continued)*

ZnS_xSe_{1-x}[o] (12 K)			
ZnS-like (LO)	ZnS-like (TO)	ZnSe-like (LO, $x < 0.5$)	ZnSe-like (TO, $x < 0.5$)
$301.0 + 73.90x$ $-23.46x^2(2.9)$	$289.02 + 40.11x$ $-52.13x^2(0.9)$	$255.3 - 31.51x$ $+6.89x^2(0.9)$	$210.6 + 2.71x$ $+8.63x^2(2.4)$

[a] H.K. Shin, D.J. Lockwood, J.-M. Baribeau: Solid State Commun. **114**, 505 (2000)
[b] J.C. Tsang, P.M. Mooney, F. Dacol, J.O. Chu: J. Appl. Phys. **75**, 8098 (1994)
[c] Z.C. Feng, S. Perkowitz: Phys. Rev. B **47**, 13 466 (1993)
[d] G. Lucovsky, K.Y. Cheng, G.L. Pearson: Phys. Rev. B **12**, 4135 (1975)
[e] L. Pavesi, R. Houdré, P. Giannozzi: J. Appl. Phys. **78**, 470 (1995)
[f] J. Groenen, R. Carles, G. Landa, C. Guerret-Piécourt, C. Fontaine, M. Gendry: Phys. Rev. B **58**, 10 452 (1998)
[g] M.H. Brodsky, G. Lucovsky, M.F. Chen, T.S. Plaskett: Phys. Rev. B **2**, 3303 (1970)
[h] P. Galtier, J. Chevalier, M. Zigone, G. Martinez: Phys. Rev. B **30**, 726 (1984)
[i] H.C. Lin, J. Ou, C.H. Hsu, W.K. Chen, M.C. Lee: Solid State Commun. **107**, 547 (1998)
[j] Y.S. Chen, W. Shockley, G.L. Pearson: Phys. Rev. **151**, 648 (1966)
[k] T.C. McGlinn, T.N. Krabach, M.V. Klein, G. Bajor, J.E. Greene, B. Kramer, S.A. Barnett, A. Lastras, S. Gorbatkin: Phys. Rev. B **33**, 8396 (1986)
[l] R. Carles, N. Saint-Cricq, J.B. Renucci, R.J. Nicholas: J. Phys. C: Solid State Phys. **13**, 899 (1980)
[m] D.N. Talwar, M. Vandevyver: J. Appl. Phys. **56**, 1601 (1984)
[n] S. Perkowitz, L.S. Kim, Z.C. Feng, P. Becla: Phys. Rev. B **42**, 1455 (1990)
[o] D.J. Olego, K. Shahzad, D.A. Cammack, H. Cornelissen: Phys. Rev. B **38**, 5554 (1988)

assumption that the effective charges of the Al and Ga cations are the same (which is a very good approximation), then the potential term $\Delta\boldsymbol{\Phi}$ in (3.5) or (3.6) becomes zero and the $Al_xGa_{1-x}As$ alloy vibrational problem reduces to a mass perturbation problem, formally identical to the problem of treating different isotopes in an otherwise perfect crystal. The advantages of using (3.6) become now apparent: not only does this formulation of the problem conveniently single out the only significant component of the alloy perturbation, which is the mass disorder, but it is also numerically advantageous, since the long-range Coulomb interaction (which is difficult to treat in a disordered system) is fully included in the "unperturbed" problem. For a Raman calculation the relevant "unperturbed" phonons are the $\boldsymbol{q} = 0$ phonons in the Brillouin zone of al large supercell of pure GaAs (or pure AlAs). Of course, these phonons are easily obtained by folding the dispersion relations for pure GaAs (or pure AlAs) calculated for the true unit cell of these compounds.

A fast algorithm to perform this folding for arbitrary geometries has been developed by Kanellis [100].

The transferability of force constants works best for lattice-matched systems such as GaAs-AlAs. When two semiconductors with significantly different lattice constant form an alloy, a phonon calculation based on the transferability concept will not be as accurate as similar calculations for $Al_xGa_{1-x}As$ alloys. On the other hand, since most of these systems follow Vegard's law (linear interpolation of lattice constants) quite closely, one could use the lattice constant dependence of the force constants [31] to determine by interpolation the appropriate values for a fictitious material with the lattice constant of the alloy. This approach, however, would miss an important aspect of the physics of semiconductor alloys. Even though the average lattice constant of the alloy usually follows Vegard's Law, the individual bond lengths (which are much more important from the point of view of determining phonon frequencies) have a tendency to remain closer to their lengths in the parent compounds. The treatment of these local distortions from an ab initio perspective is difficult, but it has been carried out with considerable success by de Gironcoli and coworkers [101]. More recently, Rücker et al. have used an extended Keating valence force field model fit to first-principles calculations, and they are able to reproduce, with remarkable accuracy, the Raman spectrum of $Si_{1-x-y}Ge_xC_y$ alloys [102].

The relative simplicity of the Raman spectrum of semiconductor alloys has stimulated the quest for simple models that might account for the compositional dependence of the main peaks [14,16,79]. Even though these dependencies can be accurately reproduced using some of the computational approaches discussed above, there is always a need for simple, semi-quantitative arguments that capture the main physics of the problem. We believe that such a model can be easily built if we write the shift as the sum of a "mass perturbation" and "bond perturbation" term:

$$\Delta\omega^2(x) = \Delta\omega^2_{\mathrm{mass}}(x) + \Delta\omega^2_{\mathrm{bonds}}(x). \tag{3.32}$$

This is a generalization of (3.29) . The "mass perturbation" term has been referred to in the literature as "mass disorder" term [84,102,103], and also, borrowing concepts from superlattice physics [104], as the "confinement" contribution [105]. Indeed, both phonon confinement in superlattices and the mass perturbation in alloys tend to lower the vibrational frequency, and the physics behind this frequency lowering is the same in both cases. Let us consider, for example the case of $Al_xGa_{1-x}As$ alloys. For GaAs-like LO vibrations, the Al-atoms remain essentially at rest, so that the corresponding phonon eigenvector has zero amplitude at the sites occupied by Al-atoms. If one inserts such eigenvector in (3.5) (with $\boldsymbol{\Delta\Phi} = 0$) and uses its expansion in terms of pure GaAs eigenvectors (3.8), one finds that the frequencies of the GaAs-like optic modes are approximately given by

$$\omega_f^2 = \frac{\sum_{f'=1}^{3nN} |c_{ff'}|^2 \omega_{0f'}^2}{\sum_{f'=1}^{3nN} |c_{ff'}|^2}. \tag{3.33}$$

In other words, the GaAs-like mode frequencies are weighted averages of the pure GaAs mode frequencies. Since the Raman-active LO mode in GaAs (and in all other semiconductors with the possible exception of AlSb and diamond) is the highest-frequency phonon in the material, the GaAs-like modes in the alloys will necessarily have a lower frequency. If the coefficients $c_{ff'}$ are not too different from mode to mode, the alloy modes should approach the frequency of critical points in the phonon density of states of the unperturbed material. In the case of a superlattice, the mass perturbation is periodic, and the coefficients in (3.33) vanish except for the bulk GaAs modes that have the same wave vector when expressed in terms of the superlattice Brillouin zone. If one also takes advantage of the result that the frequency of phonons confined in a layer is insensitive to the thickness of the other material's layers, it is possible to find an equivalent problem in which the coefficient $c_{ff'}$ is different from zero only for a single bulk frequency $\omega_{0f'}$. Thus the superlattice frequency equals $\omega_{0f'}$. This property has been used to map phonon dispersion relations from measurements in superlattices [106].

We have used in the past the expression "microscopic strain" to refer to the second term in (3.32), but here we prefer to call it "bond perturbation" to avoid any confusion with the strain shifts given by (3.28), which would further modify the alloy mode frequencies if the system is grown epitaxially on a mismatched substrate. The bond perturbation term becomes important when two semiconductors with very different lattice constants are alloyed. This term can be represented by an expression similar to the second term in (3.29) [84,107], if one accounts correctly for the change in length of the relevant bond. This is done by writing $\Delta R/R = (1-a^{**})(\Delta a/a)$, where a^{**} is the so-called topological rigidity parameter [108,109]. The topological rigidity parameter measures the bond's tendency to conserve its length as a function of composition. For $a^{**} = 1$, the bond is perfectly rigid. For $a^{**} = 0$, the bonds follow the change in average lattice constant.

As explained in [105], the combination of the mass and bond perturbations explain all qualitative details of the compositional dependence of Raman modes in SiGe alloys, shown in Fig. 3.10. The mass perturbation lowers the frequency of the Ge-Ge mode as a function of the Si-concentration and also the frequency of the Si-Si mode as a function of the Ge-concentration. But the bond perturbation is equivalent to a compressive strain for the Ge-Ge bond and a tensile strain for the Si-Si bond. Thus the bond perturbation raises the frequency of the Ge-Ge mode and lowers the frequency of the Si-Si mode. In other words, the mass and bond contributions are additive for the Si-Si modes but tend to cancel each other for the Ge-Ge modes. This explains the much weaker compositional dependence of the Ge-Ge modes relative to the Si-Si modes. The remaining peak corresponds to Si-Ge vibrations. It has maximum frequency for a composition of 50%, for which the mass perturbation is smallest because the system is closest to a hypothetical SiGe zincblende compound, The noticeable asymmetry of the Si-Ge curve in Fig. 3.10 can be easily explained in terms of a reversal of the sign of the bond perturbation at $x = 0.5$ [105].

The model just described has been extended to polar alloy semiconductors by Groenen et al. [103]. Equation (3.32) is assumed to be valid for TO-like vibrations, and the LO mode frequencies can be obtained from the zeros of an average dielectric function. Groenen et al. have also calculated Raman intensities and explained a number of mysteries in the Raman spectra of $In_xGa_{1-x}As$ alloys. Hence it appears that it is indeed possible to develop a simple theory of alloy semiconductors that correctly accounts for the main features of the compositional dependence of the Raman peaks and even their experimental intensities.

Superlattices as a special case of alloying. The formalism described above to treat semiconductor alloys can be applied to the study of the vibrational properties of semiconductor superlattices, and in particular to the investigation of disorder at heterostructure interfaces, a topic that is also discussed in the chapter by D. Gammon. Fig. 3.11 shows calculated and experimental Raman spectra from a $(GaAs)_6(AlAs)_6$ superlattice in the frequency range of the AlAs-like optical modes. The superlattice phonons where

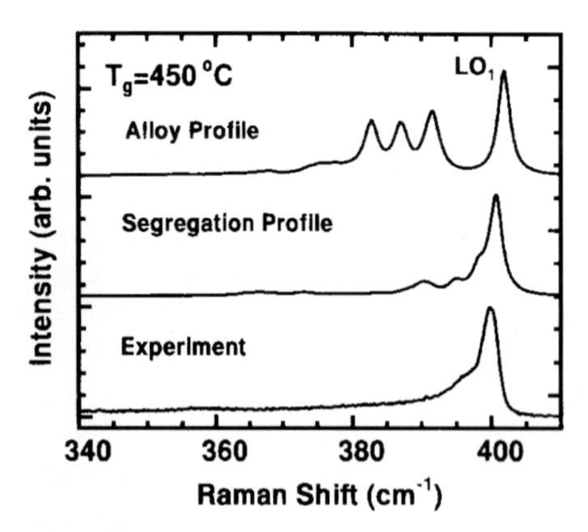

Fig. 3.11. Experimental Raman spectrum for AlAs-like modes in a $(GaAs)_6(AlAs)_6$ superlattice grown by molecular beam epitaxy at 450 °C (*bottom panel*). For a perfect superlattice one should see at narrow peak at 403 cm^{-1}. The *top panel* shows the prediction from a model that assumes alloying limited to the layers closest to the GaAs-AlAs interfaces. The *middle panel* shows the predicted Raman spectrum assuming that the main source of disorder is the tendency of Al to segregate to the surface during the growth of the superlattice [110]

obtained from (3.6) (with $\Delta\Phi = 0$) and the Raman spectra were calculated using (3.22). In the top spectrum in Fig. 3.11, the atoms close to the interfaces were randomly intermixed, whereas the atomic distribution for the middle spectrum was obtained from a model that assumes a tendency for Al to segregate to the growth surface [110]. Clearly, the Raman analysis provides strong support for the surface segregation hypothesis.

3.3 Conclusion

In this chapter we have sketched the applications of Raman scattering to the characterization of semiconductors. The emphasis has been on providing the most useful formulas and data for the analysis of Raman spectra, and on explaining the ideas underlying the applications of the technique and the possible pitfalls that the experimentalist must be aware off. Our coverage of the many applications of this technique has been necessarily short and quite arbitrary in our selection of examples, and the reader is encouraged to consult the recommended references for a more in-depth coverage.

Semiconductor physics is a mature field, and one could even claim that some of the problems that have intrigued physicists for decades, such as the vibrational properties, are on the verge of being definitively "solved". This, however, does not mean that Raman spectroscopy will fade in importance. Quite the contrary, the ability to predict Raman frequencies and intensities with excellent reliability eliminates the model-related ambiguities that have limited the applicability of the technique as a structural characterization tool. We hope that by providing a bird's eye view of these new capabilities, this article will stimulate the use of Raman spectroscopy for semiconductor characterization.

I would like to acknowledge the patience of the editors and the invaluable insight provided by Manuel Cardona and John B. Page.

References

1. K.W. Böer: *Survey of Semiconductor Physics: electrons and other particles in bulk semiconductors* (Van Nostrand Reinhold, New York 1990)
2. P.Y. Yu, M. Cardona: *Fundamentals of Semiconductors: Physics and Materials Properties* (Springer, Berlin 1996)
3. M.V. Klein: in *Light Scattering in Solids*, ed. by M. Cardona (Springer, Berlin 1975) Vol. 8, p. 147
4. G. Abstreiter, M. Cardona, A. Pinczuk: in *Light Scattering in Solids IV*, ed. by M. Cardona, G. Güntherodt (Springer, New York 1984) p. 5
5. A. Pinczuk, E. Burstein: in *Light Scattering in Solids*, ed. by M. Cardona (Springer, Berlin 1975) Vol. 8, p. 79
6. R.M. Martin, L.M. Falicov: in *Light Scattering in Solids*, ed. by M. Cardona (Springer, Berlin 1975) Vol. 8, p. 79

7. M. Cardona: in *Light Scattering in Solids II*, ed. by M. Cardona, G. Güntherodt (Springer, Berlin 1982) Vol. 50, p. 19
8. W. Richter: *Resonant Raman Scattering* (Springer, Berlin, Heidelberg, New York 1976)
9. J. Menéndez, A. Pinczuk: IEEE J. Quantum Electron **QE-24**, 1698 (1988)
10. O. Madelung: *Semiconductors: Physics of Group IV Elements and III-V compounds* (Springer, Berlin 1985) Vol. 17a
11. O. Madelung: *Semiconductors: Intrinsic Properties of Group IV Elements and III-V, II-VI, and I-VII compounds* (Springer, Berlin 1985) Vol. 22a
12. M.L. Cohen, J.R. Chelikowsky: *Electronic Structure and Optical Properties of Semiconductors* (Springer, Heidelberg, Berlin, New York 1989)
13. W.A. Harrison: *Electronic Structure and the Properties of Solids* (Dover, Toronto 1989)
14. P. Brüesch: *Phonons: Theory and Experiments II* (Springer, Berlin 1986)
15. R.F. Wallis, M. Balkanski: *Many Body Aspects of Solid State Spectroscopy* (North-Holland, Amsterdam 1986)
16. F.H. Pollak: in *Analytical Raman Spectroscopy*, ed. by J.G. Grasselli, B.J. Bulkin (Wiley, New York 1991) p. 137
17. G. Turrel, J. Corsent: *Raman Microscopy. Developments and Applications* (Academic Press, London 1996)
18. M. Born, K. Huang: *Dynamical Theory of Crystal Lattices* (Clarendon Press, Oxford, England 1954)
19. D. Strauch, B. Dorner: J. Phys: Condens. Matter **2**, 1457 (1990)
20. P.A. Temple, C.E. Hathaway: Phys. Rev. B **7**, 3685 (1973)
21. G.S. Spencer, J. Grant, R. Gray, J. Zolman, J. Menéndez, R. Droopad, G.N. Maracas: Phys. Rev. B **49**, 5761 (1994)
22. W. Windl, P. Pavone, K. Karch, O. Schütt, D. Strauch, P. Giannozzi, S. Baroni: Phys. Rev. B **48**, 3164 (1993)
23. W. Windl, P. Pavone, D. Strauch: Phys. Rev. B **54**, 8580 (1996)
24. F. Herman: J. Phys. Chem. Solids **8**, 405 (1959)
25. P. Brüesch: *Phonons: Theory and Experiments I* (Springer, Berlin 1986)
26. W. Weber: Phys. Rev. B **15**, 4789 (1977)
27. K.C. Rustagi, W. Weber: Solid State Commun. **18**, 673 (1976)
28. B.D. Rajput, D.A. Browne: Phys. Rev. B **53**, 9052 (1996)
29. L.J. Sham: in *Dynamical Properties of Solids*, ed. by G.K. Horton, A.A. Maradudin (North-Holland, Amsterdam 1974) p. 303.
30. P. Giannozzi, S. d. Gironcoli, P. Pavone, S. Baroni: Phys. Rev. B **43**, 7231 (1991)
31. T. Pletl, P. Pavone, U. Engel, D. Strauch: Physica B **263-264**, 392 (1999)
32. J. Kulda, D. Strauch, P. Pavone, Y. Ishii: Phys. Rev. B **50**, 13347 (1994)
33. L. Colombo, P. Giannozzi: Solid State Commun. **96**, 49 (1995)
34. J. Menéndez, J.B. Page: in *Light Scattering in Solids VIII: Fullerenes, Semiconductor Surfaces, Coherent Phonons*, ed. by M. Cardona, G. Güntherodt (Springer, Heidelberg 2000) p. 27
35. J. Kash, J. Tsang: in *Light Scattering in Solids VI*, ed. by M. Cardona, G. Güntherodt (Springer, Berlin 1995) Vol. 68, p. 423
36. O.H. Nielsen, R.M. Martin: Phys. Rev. B **32**, 3792 (1985)
37. B.A. Weinstein, G.J. Piermarini: Phys. Rev. B **12**, 1172 (1975)

38. A. Debernardi, S. Baroni, E. Molinari: in *22nd International Conference on the Physics of Semiconductors*, ed. by D.J. Lockwood (World Scientific, Vancouver 1994) Vol. 1, p. 373
39. A. Debernardi, S. Baroni, E. Molinari: Phys. Rev. Lett. **75**, 1819 (1995)
40. A. Debernardi: Phys. Rev. B **57**, 12847 (1998)
41. G. Placzek: in *Handbook der Radiologie* (Akademische Verlaggesellschaft, Leipzig 1934) p. 209.
42. H. Callen: Am. J. Phys. **36**, 735 (1968)
43. N.W. Ashcroft, N.D. Mermin: *Solid State Physics* (Holt, Rinehart and Winston, New York 1976)
44. W.L. Faust, C.H. Henry: Phys. Rev. Lett. **17**, 1265 (1966)
45. E. Anastassakis, E. Burstein: J. Phys. Chem. Solids **32**, 313 (1971)
46. E. Anastassakis, E. Burstein: J. Phys. Chem. Solids **32**, 563 (1971)
47. R. Zeyher, C.-S. Ting, J.L. Birman: Phys. Rev. B **10**, 1725 (1974)
48. K.T. Schomacker, J.K. Delaney, P.M. Champion: J. Appl. Phys. **85**, 4241 (1986)
49. M. Wolkenstein: Dokl. Akad. Nauk. SSSR **32**, 185 (1941)
50. A.A. Maradudin, E. Burstein: Phys. Rev. **164**, 1081 (1967)
51. S. Go, H. Bilz, M. Cardona: Phys. Rev. Lett. **34**, 580 (1975)
52. D. Bermejo, S. Montero, M. Cardona, A. Muramatsu: Solid State Commun. **42**, 153 (1982)
53. L.L. Swanson, A.A. Maradudin: Solid State Commun. **8**, 859 (1970)
54. L. Brey, C. Tejedor: Solid State Commun. **48**, 403 (1983)
55. S. Baroni, R. Resta: Phys. Rev. B **33**, 5969 (1986)
56. W. Windl: *Ab-initio-Berechnung von Raman-Spektren in Halbleitern* (CH-Verlag, Regensburg 1995)
57. A. Pinczuk, E. Burstein: in *Proceedings of the Tenth International Conference on the Physics of Semiconductors*, ed. by S.P. Keller, J.C. Hensel, F. Stern (United States Atomic Energy Commission, Cambridge, Massachusetts 1970) p. 727.
58. K. Mizoguchi, S. Nakashima: J. Appl. Phys. **65**, 2583 (1989)
59. J.B. Hopkins, L.A. Farrow: J. Appl. Phys. **59**, 1103 (1986)
60. G. Kolb, T. Salbert, G. Abstreiter: J. Appl. Phys. **69**, 3387 (1991)
61. F.U. Herman, S. Beeck, G. Abstreiter, C. Hanke, C. Hoyler, L. Korte: Appl. Phys. Lett. **58**, 1007 (1991)
62. A. Compaan, H.J. Trodahl: Phys. Rev. B **29**, 793 (1984)
63. R. Loudon: J. Raman Spectrosc. **7**, 10 (1978)
64. R. Ostermeir, K. Brunner, G. Abstreiter, W. Weber: IEEE Trans. Elect. Devices **39**, 858 (1992)
65. J. Menéndez, M. Cardona: Phys. Rev. B **29**, 2051 (1984)
66. F. Widulle, T. Ruf, A. Göbel, I. Silier, E. Schönherr, M. Cardona, J. Camacho, A. Cantarero, W. Kriegseis, V.I. Ozhogin: Physica B **263- 264**, 381 (1999)
67. H.D. Fuchs, C.H. Grein, R.I. Devlen, J. Kuhl, M. Cardona: Phys. Rev. B **44**, 8633 (1991)
68. H.D. Fuchs, C.H. Grein, C. Thomsen, M. Cardona, W.L. Hansen, E.E. Haller, K. Itoh: Phys. Rev. B **43**, 4835 (1991)
69. S. Ganesan, A.A. Maradudin, J. Oitmaa: Annals of Physics **56**, 556 (1970)
70. C. Kittel: *Introduction to Solid State Physics* (Wiley, New York 1996)
71. E. Anastassakis, M. Cardona: in *High Pressure in Semiconductor Physics II*, ed. by T. Suski, W. Paul (Academic Press, New York 1998) Vol. 55.

72. H. Brugger: in *NATO Advanced Research Workshop on Light Scattering in Semiconductor Structures and Superlattices*, ed. by D.J. Lockwood, J.F. Young (Plenum Press, Mont-Tremblant, Quebec, Canada 1991) Vol. B273, p. 259.
73. I.D. Wolf, J. Vanhellemont, A. Romano-Rodríguez, H. Norström, H.E. Maes: J. Appl. Phys. **71**, 898 (1992)
74. S. Narayanan, S.R. Kalidindi, L.S. Schadler: J. Appl. Phys. **82**, 2595 (1997)
75. I.I. Vlasov, V.G. Ralchenko, E.D. Obraztsova, A.A. Smolin, V.I. Konov: Appl. Phys. Lett. **71**, 1789 (1997)
76. M.C. Rossi: Appl. Phys. Lett. **73**, 1203 (1998)
77. G.H. Loechelt, N.G. Cave, J. Menéndez: Appl. Phys. Lett. **66**, 3639 (1995)
78. G.H. Loechelt, N.G. Cave, J. Menéndez: J. Appl. Phys. , (in print) (1999)
79. A.S. Barker, Jr., A.J. Sievers: Rev. Mod. Phys. 47, Suppl. No. 2, S1 (1975)
80. W. Nazarewicz, M. Balkanski, J.F. Morhange, C. Sébenne: Solid State Commun. **9**, 1719 (1971)
81. J.C. Tsang, K. Eberl, S. Zollner, S.S. Iyer: Appl. Phys. Lett. **61**, 961 (1992)
82. M. Meléndez-Lira, J. Menéndez, K.M. Kramer, M.O. Thompson, N. Cave, R. Liu, J.W. Christiansen, N.D. Theodore, J.J. Candelaria: J. Appl. Phys. **82**, 4246 (1997)
83. F. Cerdeira, M. Cardona: Phys. Rev. B **5**, 1440 (1972)
84. M. Meléndez-Lira, J. Menéndez, W. Windl, O.F. Sankey, G.S. Spencer, S. Sego, R.B. Culbertson, A.E. Bair, T.L. Alford: Phys. Rev. B **54**, 12866 (1997)
85. J. Menéndez, M. Cardona: Phys. Rev. Lett. **51**, 1297 (1983)
86. J. Menéndez, M. Cardona: Phys. Rev. B **31**, 3696 (1985)
87. C. Trallero-Giner, A. Cantarero, M. Cardona, M. Mora: Phys. Rev. B **45**, 6601 (1992)
88. R. Merlin, C. Colvard, M.V. Klein, H. Morkoç, A.Y. Cho, A.C. Gossard: Appl. Phys. Lett. **36**, 43 (1980)
89. A.K. Sood, J. Menéndez, M. Cardona, K. Ploog: Phys. Rev. Lett. **54**, 2115 (1985)
90. A. Mooradian, A.L. McWhorter: in *Light Scattering Spectra of Solids*, ed. by G.B. Wright (Springer, New York, Heidelberg, Berlin 1968) p. 273
91. P.A. Schumann, Jr.: Solid State Technology January **1970**, 50 (1970)
92. M.H. Brodsky: in *Light Scattering in Solids*, ed. by M. Cardona (Springer, Berlin, Heidelberg, New York 1975) Vol. 8, p. 208
93. M.I. Alonso, K. Winer: Phys. Rev. B **39**, 10056 (1989)
94. M. Renucci, J. Renucci, M. Cardona: in *Light Scattering in Solids*, ed. by M. Balkanski (Flammarion, Paris 1971) p. 326
95. N. Saint-Cricq, G. Landa, J.B. Renucci, I. Hardy, A. Muñoz-Yague: J. Appl. Phys. **61**, 1206 (1987)
96. A. Zunger, S. Mahajan: in *Handbook of Semiconductors*, ed. by S. Mahajan (North Holland, Amsterdam 1994) Vol. 3B, p. 1399
97. J.C. Tsang, V.P. Kesan, J.L. Freeouf, F.K. LeGoues, S.S. Iyer: Phys. Rev. B **46**, 6907 (1992)
98. A. Hassine, J. Sapriel, P.L. Berre, M.A.D. Forte-Poisson, F. Alexandre, M. Quillec: Phys. Rev. B **54**, 2728 (1996)
99. S. Baroni, S. d. Gironcoli, P. Giannozzi: Phys. Rev. Lett. **65**, 84 (1990)
100. G. Kanellis: Phys. Rev. B **35**, 746 (1987)
101. S. d. Gironcoli: Phys. Rev. B **46**, 2412 (1992)
102. H. Rücker, M. Methessel: Phys. Rev. B **52**, 11059 (1995)

103. J. Groenen, R. Carles, G. Landa, C. Guerret-Piécourt, C. Fontaine, M. Gendry: Phys. Rev. B **58**, 10452 (1998)
104. M.V. Klein: IEEE J. Quantum Electron **QE-22**, 1760 (1986)
105. J. Menéndez, A. Pinczuk, J. Bevk, J.P. Mannaerts: J. Vac. Sci. Technol. B **6**, 1306 (1988)
106. A.K. Sood, J. Menéndez, M. Cardona, K. Ploog: Phys. Rev. Lett. **54**, 2111 (1985)
107. R. Carles, G. Landa, J.B. Renucci: Solid State Commun. **53**, 179 (1985)
108. J.L. Martins, A. Zunger: Phys. Rev. B **30**, 6217 (1984)
109. N. Mousseau, M.F. Thorpe: Phys. Rev. B **46**, 15887 (1992)
110. G.S. Spencer, J. Menéndez, L.N. Pfeiffer, K.W. West: Phys. Rev. B **52**, 8205 (1995)

II Finding the Stress from the Raman Shifts: A Case Study

Ingrid de Wolf

Mechanical stress cannot be avoided during the processing of semiconductor devices. The growth of isolation oxides, the implantation of dopants, the deposition and growth of different films, and the fabrication and filling of trenches generate local mechanical stresses. The magnitude of these stresses depends on the geometry of the films, on their chemical properties and thermal expansion coefficient, on the deposition temperature, etc. Because stress may generate defects or indirectly affect device performance, considerable effort is spent in the microelectronics industry to find its magnitude and distribution in the device, to determine which processing steps are mainly responsible for its generation, and under what conditions the stress becomes critical. Most of these studies use finite element simulations of the stress generation during processing. Unfortunately, simulations may give a distorted image of the real situation if they are not properly validated. Raman spectroscopy is one of the few techniques that can be used for this validation. However, as explained in Sect 3.3 of Chap. 3, the complete spectroscopic information needed to compute an arbitrary stress is almost never available. A typical specimen to be analyzed is a Si-based device fabricated on a (001)-oriented Si crystal. In the backscattering configuration used for Raman microscopy one observes a single Si Raman peak, which is not sufficient to characterize the stress tensor. Broadening of the peak is sometimes observed, but this broadening is hard to interpret because it can be due to stress-induced splitting of peaks, as well as to large stress variations within the illuminated volume. This means that it is in general impossible to obtain quantitative or even qualitative information on the different strain tensor components from Raman data without assumptions or simulations. Therefore, since Raman data are needed to validate the stress simulations and the stress simulations are needed to interpret the Raman spectra, the goal of the industrial physicist is to achieve self-consistency between simulations and measurements.

The simplest assumption usually made is uniaxial (σ) or biaxial stress ($\sigma = \sigma_{xx} + \sigma_{yy}$). In this case there is a linear relation between the stress and Raman frequency shift $\Delta\omega$ and one obtains from (28) of Chap. 2 and the known elastic constants in Si $\sigma(\mathrm{MPa}) = -434\Delta\omega\,(\mathrm{cm}^{-1})$, for backscattering from a [001] surface with σ the in-plane stress [1]. Tensile stress ($\sigma > 0$) will result in a negative frequency shift, compressive stress ($\sigma < 0$) in a positive shift. However, the experimental situations are typically far more complicated. As a first case study, we look at mechanical stress induced by silicide lines in a Si substrate [2]. The stress induced by such a line in the substrate is not uniaxial or biaxial. At the edges of the line, shear stresses will also be present.

Figure II.1 shows the stress-induced Raman frequency shift measured on a Si wafer with 16 nm thick $CoSi_2$ lines. The lines are thin enough so that a Raman signal from underneath them can be obtained. The data clearly indicate tensile stress in the Si next to the lines ($\Delta\omega < 0$), and compressive stress underneath ($\Delta\omega > 0$). This is expected, because the silicide itself is under tensile stress. Notice also that when the lines are located closer to each other, the tensile stress between them increases.

The magnitude of the stress that produces the Raman shifts in Fig. II.1 can be determined from modeling. One can use finite element modeling to calculate the stress or strain induced by such lines in the substrate at each (x, z) position in the sampled volume. Next one solves (27) of Chap. 3 for all (x, z) to obtain the stress induced frequency shift and the Raman intensity of the three Si phonons. The last step is to integrate this locally produced signal over the probing volume of the focused laser beam. These calculations result in a 'model predicted Raman peak' as a function of the position, which can be compared with the experiment [3,4].

The full lines in Fig. II.1 result from fitting a simple analytical stress model to the Raman data, following the above procedure. The model was first fit to the two lines on the left side of the figure, with large spacing. A good fit was obtained, with some deviation from the data underneath the lines. Next, the same model parameters were used to see whether the stress model was still valid for lines with smaller spacing (right side). It turns out that this simple model does fit the Raman data rather well. The deviation under the lines is due to the fact that the analytical model assumed uniform stress in the lines, which is not entirely correct.

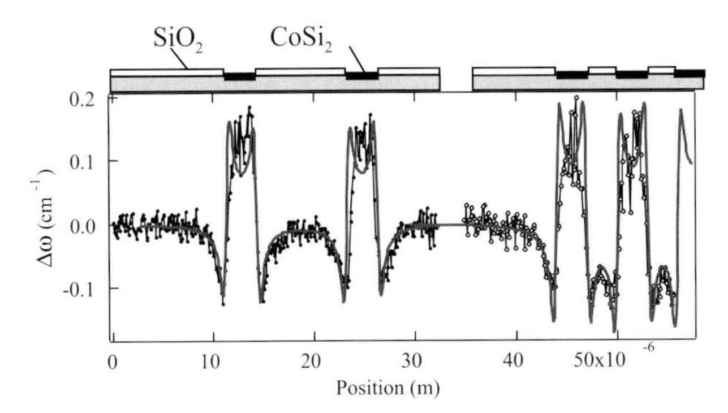

Fig. II.1. Raman spectroscopy experiment on 16-nm thick, 3-μm wide silicide lines ($CoSi_2$) with large spacing (*left*) and spacing equal to width (*right*) on Si. Laser light: 457.9 nm, 100× objective. The *full lines* show the result of a fit of an analytical stress model to the data

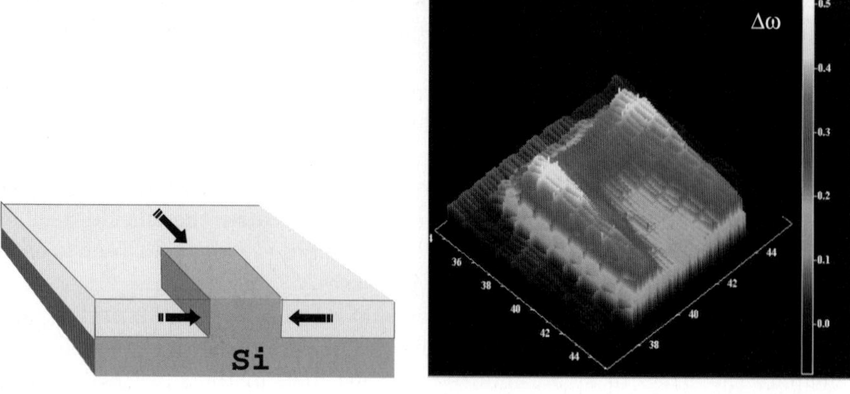

Fig. II.2. Raman shift $\Delta\omega$ measured during a 2D point-by-point scan across the edge of a shallow trench isolated Si line (width=5 µm)

Unfortunately, because it involves many calculations, modeling is very time consuming. Valuable information on stress can already be obtained by simple inspection of the Raman frequency shift. Figure II.2 illustrates the unique capability of Raman spectroscopy for the study of local stresses in Si devices [5]. It shows the result of a 2D measurement of stress in a 5 µm wide Si stripe surrounded by oxide. The distance between two measurement points was 0.2 µm along the width direction of the line and 1 µm along the length. A compressive stress ($\Delta\omega > 0$) is measured in the stripe, which is larger at the sides and decreases towards the center. At the corners it is clearly larger. This result is explained by the fact that the oxide, which surrounds the stripe, produces compressive stress on the Si as indicated by the arrows.

References

1. E. Anastassakis, A. Pinczuk, E. Burstein, F.H. Pollak, M. Cardona: Solid State Commun. **8**, 133 (1970)
2. I. De Wolf, D.J. Howard, K. Maex, H.E. Maes: Mat. Res. Soc. Symp. Proc. **427**, 47 (1996)
3. I. De Wolf, H.E. Maes, S.K. Jones: J. Appl. Phys. **79**, 7148 (1996); I. De Wolf, E. Anastassakis: [Erratum], J. Appl. Phys. **85**, 7484 (1999)
4. I. De Wolf, M. Ignat, G. Pozza, L. Maniguet, H.E. Maes: J. Appl. Phys. **85**, 6477 (1999)
5. I. De Wolf, G. Groeseneken, H.E. Maes, M. Bolt, K. Barla, A. Reader, P.J. McNally: Proc. 24th ISTFA, (1998) p.11

III Brillouin Scattering from Semiconductors

M. Grimsditch

Brillouin scattering, like Raman scattering, is the inelastic scattering of light by phonons. The difference between them is that Raman typically refers to optic phonons while Brillouin is reserved for interaction with acoustic modes. Because the frequency shifts produced by acoustic phonons are considerably smaller than for optic phonons, Brillouin scattering relies on the high resolution of a Fabry-Perot interferometer instead of a grating spectrometer typically used in Raman experiments. Until the early seventies Brillouin scattering was essentially a curiosity in solid-state physics, since it could only be performed on the highest quality, transparent materials. Two innovations, multipassing and tandem operation, introduced by Sandercock [1] in the early seventies transformed Fabry-Perot interferometry into its current state as a powerful tool in condensed matter physics.

Brillouin experiments typically give information on sound velocities, which can be analyzed to yield the elastic constants of the material; in this sense the technique provides the same information as that obtained from ultrasonic methods. The most significant advantages of Brillouin scattering over ultrasonic techniques are that it requires only very small samples and that it is non-contacting; its major disadvantage is the difficulty of dealing with opaque samples.

Brillouin scattering has contributed greatly to our understanding of wide-gap semiconducting materials, which are transparent to visible radiation. The elastic constants of materials that can only be grown as small crystals, notably the nitrides (BN, AlN, GaN, SiC and β-Si_3N_4, diamond, isotopically controlled diamond) have been determined using this technique. More recently, infrared Brillouin measurements have also become feasible [3], thereby making it possible to investigate many of the more conventional semiconductors; i.e., AlGaAs, $GaInP_2$, etc.

The interferometric technique, often identified as the Brillouin technique, has also recently been used to investigate, with very high resolution, low-lying Raman lines [4]. Figure III.1 shows the Raman active transition between the two lowest electronic ground states of boron impurities in diamond at 0 and 4 Tesla. The field dependence of the Raman lines has allowed both the degeneracy and the g-factors of the two lowest levels to be determined. Also observed in Fig. III.1b, as indicated with an asterisk, are two transitions between the Zeeman split levels of one of the ground states. These two Raman lines have frequencies below the Brillouin longitudinal acoustic (LA) and

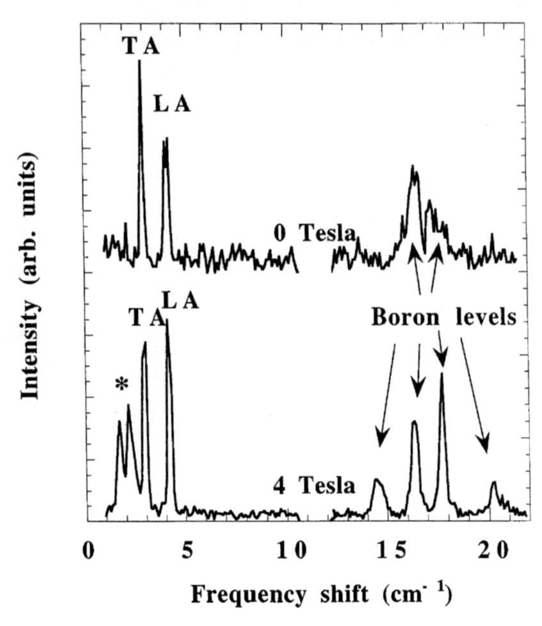

Fig. III.1. Spectra of electronic states of boron impurities in diamond at 6 K. LA and TA correspond to longitudinal and transverse sound waves, respectively (Brillouin peaks). The peaks near $16\,\mathrm{cm}^{-1}$ are transitions between two distinct electronic levels. Those identified with an asterisk are intra-level transitions between Zeeman sublevels

transverse acoustic (TA) peaks. These peaks correspond to the absorption conventionally observed in EPR experiments.

References

1. J.R. Sandercock: In *Light Scattering in Solids III, topics in Applied Phys.*, Vol. **51** ed. by M. Cardona, G. Güntherodt (Springer, New York 1978)
2. For a review of the results on ultrahard materials see M. Grimsditch: In *Proc. of the 16th Int. Conf. on Raman Spectroscopy*, ed. by A. Heyns (Wiley, New York 1998) p. 3, and references therein
3. S. Gehirsitz, H. Sigg, N. Herres, K. Bachem, K. Kohler, F. Reinhart: Phys. Rev. B **60**, 11601 (1999)
4. H. Kim, A.K. Ramdas, S. Rodriguez, M. Grimsditch, T.R. Anthony: Phys. Rev. Lett. **83**, 3254 and 4140 (1999)

4 Raman Scattering in Semiconductor Heterostructures

Daniel Gammon

Abstract. Applications of Raman scattering in the characterization and study of semiconductor heterostructures are reviewed with the focus on GaAs/AlAs quantum well structures. Vibrational and electronic Raman scattering are included. The use of resonant Raman scattering to gain sensitivity and selectivity is emphasized.

Semiconductor heterostructures such as GaAs/AlAs quantum wells, wires, and dots have attracted enormous interest starting in the early 1970s with the development of molecular beam epitaxy (MBE) [1–5]. Because these structures are made with one or more dimensions reduced to the size of micrometers or even nanometers they are often called semiconductor microstructures or nanostructures. The remarkable quality and control of structure in MBE-grown semiconductor heterostructures is made possible by the growth of one monolayer at a time and the capability to change from one semiconductor to another by simply closing and opening shutters in front of the source ovens [6]. This technology has created a vast array of new, artificial materials with engineerable properties and the opportunity to systematically explore physics in lower dimensions. Research in this material system has been driven in large part by the potential for producing useful devices such as light emitting diodes, lasers and high frequency transistors. However, the new and controllable properties obtained in these artificial semiconducting materials have led to a great deal of basic physics research in reduced dimensionality that has accompanied focused efforts to improve device capabilities.

Raman scattering spectroscopy is exceptionally useful in the study of the electronic and vibrational properties of heterostructures. Raman scattering is used to measure a large variety of low energy excitations of the semiconductor nanostructures, including confined and interface phonons, various elementary excitations of confined electron gases and liquids, and transitions of shallow impurities. Not surprisingly, Raman scattering has played leading roles in understanding the vibrational properties of quantum wells and the collective excitations of electron gases. Conversely, sufficient understanding of the elementary excitations has provided the opportunity to gain new information about the structure itself. For example, phonon scattering provides a probe of interface roughness while plasmon energies can characterize the density of charge carriers. These measurements can be made with high spatial and spectral resolution and selectivity.

Raman scattering is just one of many techniques used to study the electronic properties of nanostructures. Photoluminescence (PL), absorption and reflection techniques are often used to study electronic properties while far infrared absorption and various transport techniques often are also used to measure doped structures. The vast majority of phonon studies have used Raman scattering.

Although relatively weak when compared to photoluminescence, the Raman signal is enhanced greatly by resonance techniques. Many semiconductor nanostructures have strong optical resonances that can be used to increase the Raman signal enormously. By measuring the Raman intensity as a function of the laser frequency, excitonic properties and their interactions with vibrational or electronic elementary excitations are probed. Moreover, by resonating with specific confined electronic resonances of the nanostructure of interest, the Raman spectrum of a very small part of the sample is strongly enhanced over the background, thereby obtaining great selectivity. The high sensitivity and selectivity of resonant Raman scattering makes possible the measurement of the spectrum of electronic excitations or phonons in a single 10 nm layer or even in a single quantum dot. The scattering intensity, lineshape, and polarization dependence of the spectral lines often strongly depend on the resonance conditions. The excitons themselves are considerably modified by the nanostructure. Much of the recent progress has gone hand in hand with better understanding of the intermediate excitonic states involved in the resonance process. Raman spectroscopy is both complicated and enriched under strong resonance conditions.

The purpose of this chapter is to provide a short guide to Raman scattering in semiconductor heterostructures. A number of extensive reviews have been written in the last twenty years on specific aspects of Raman scattering in semiconductor heterostructures [7–12]. In this chapter there is no attempt to cover the literature comprehensively. In fact, only a handful of representative studies are considered. The goal is to introduce the capabilities of the technique to those new to the field. The discussion focuses on GaAs/$Al_xGa_{1-x}As$ quantum wells, thin flat layers of GaAs sandwiched between two layers of $Al_xGa_{1-x}As$. Because $Al_xGa_{1-x}As$ has a larger band gap than GaAs, the electrons and holes are confined in only one dimension and free in the other two, so that the electrons and holes behave in many ways as if they are two dimensional (2D). The discussion is extended at times to superlattices and 0D quantum dots.

In Sect. 4.1 a simple introduction to the physics of confined electrons and holes in semiconductor heterostructures is presented. This discussion is necessary in order to understand the resonance behavior of Raman scattering and also electronic scattering. Resonant Raman scattering and scattering geometry will be the focus of Sect. 4.2 and Sect. 4.3. In Sect. 4.4, phonon scattering is introduced and illustrated. Here, the effects of interface roughness will be emphasized. Electronic Raman scattering by impurities and the quasi-2D electron gas is discussed in Sect. 4.5.

4.1 Electrons in Semiconductor Heterostructures

With molecular beam epitaxy (MBE) or alternatively, organo-metallic chemical vapor epitaxy (OMCVD), a semiconductor crystal is grown one atomic layer at a time with low defect and impurity concentrations [6]. The chemical composition can be abruptly switched from one layer to the next to form heterojunctions. Layered crystals can be grown with layers ranging from less than a single monolayer to micrometers in thickness.

The most widely studied heterostructure is the GaAs/$Al_xGa_{1-x}As$ quantum well system (Fig. 4.1). When GaAs is sandwiched between $Al_xGa_{1-x}As$ layers the electrons and holes are both confined to the GaAs. Within the envelope approximation the electron wavefunctions are described as the product of a Bloch function, $u(\mathbf{r})$, and an envelope function, $\Phi_n(\mathbf{r})$ [1–5]:

$$\Psi_n(\mathbf{r}) = \Phi_n(\mathbf{r})u(\mathbf{r}) = \varphi_n(z)\mathrm{e}^{\mathrm{i}k_\varrho x}u(\mathbf{r}). \tag{4.1}$$

The quasi-2D envelope function consists of a localized quantum well function in the z-direction and a plane wave in the lateral direction (denoted as the ϱ-direction). For an infinitely-deep quantum well the envelope function must go to zero at the interfaces and $\varphi_n(z)$is a cosine or sine function. In the lateral directions the crystal remains periodic, wavevector remains a good quantum number, and the electron has energy dispersion described by an effective mass (Fig. 4.2). The bulk conduction band thus breaks up into a series of subbands. The conduction subband energies in an infinitely-deep quantum well are given by

$$E_{cn}(k_\varrho) = \frac{\hbar^2}{2m_e}\left(\left(\frac{(n+1)\pi}{d}\right)^2 + k_\varrho^2\right), \tag{4.2}$$

where d is the well width, $n = 0, 1...$, denoting the subband, and k_ϱ is the wavevector in the plane of the quantum well.

For holes in the valence band the band structure is more complicated. In bulk GaAs the valence bands consist of heavy and light-hole bands that are degenerate at $\boldsymbol{k} = 0$, and a split-off band that is shifted to higher energies by approximately 0.38 meV. The band gap between the lowest energy valence and conduction bands is called the E_0 gap while that associated with the split-off valence band is called the $E_0 + \Delta_0$ gap. In a quantum well, confinement leads to subband structure just as for the conduction bands (Fig. 4.2). However, the degeneracy at $\boldsymbol{k}_\varrho = 0$ is lifted because of the different masses of the bands. In addition, there is considerable band mixing between the heavy and light subbands, and strong nonparabolicity. Optical transitions between

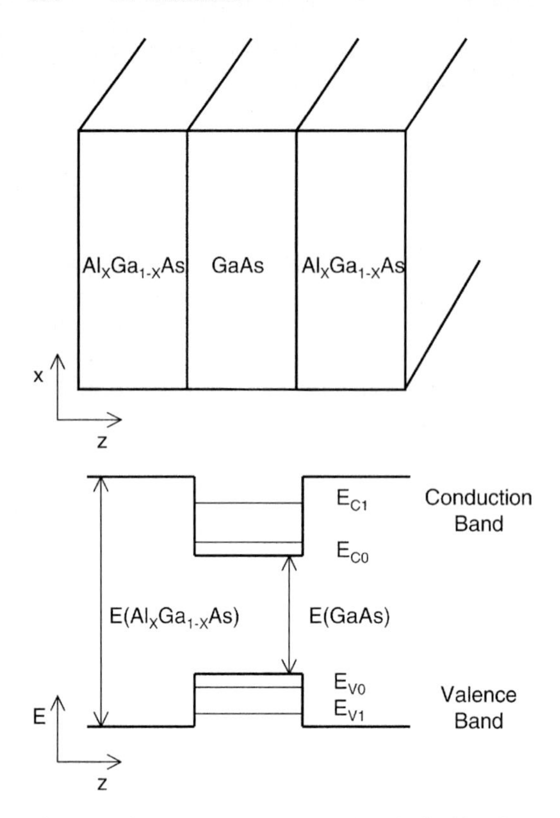

Fig. 4.1. Schematic diagrams of a GaAs/Al$_x$Ga$_{1-x}$As quantum well structure and its corresponding quantum well potential diagram. Confinement along the z-axis leads to splitting of the bands into subbands with minima at E_{cn} and E_{vn}

valence and conduction subbands are often called interband transitions. In GaAs/AlGaAs quantum wells and many other systems the subband structure has been extensively studied by optical techniques such as photoluminescence (PL) and absorption techniques [1,2].

A multiple quantum well structure is grown by repeating the quantum well period. If the barrier layers are sufficiently thin, the electrons will coherently tunnel between quantum wells. In this case the electronic states form minibands in the growth direction (along the k_z axis) and the electronic properties are described in terms of a superlattice model. In any case, the confinement energies and other properties are often described well in terms of a quantum well potential using the envelope function approach. It is worth noting that even if the quantum well structure does not form a superlattice structure as far as electronic states are concerned, it may still respond as a superlattice for acoustic phonons or for any collective excitations that involve the long-range electric field. The terms, multiple quantum well and super-

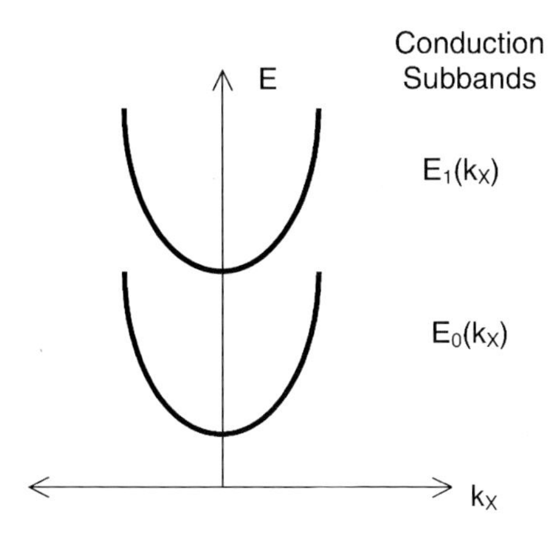

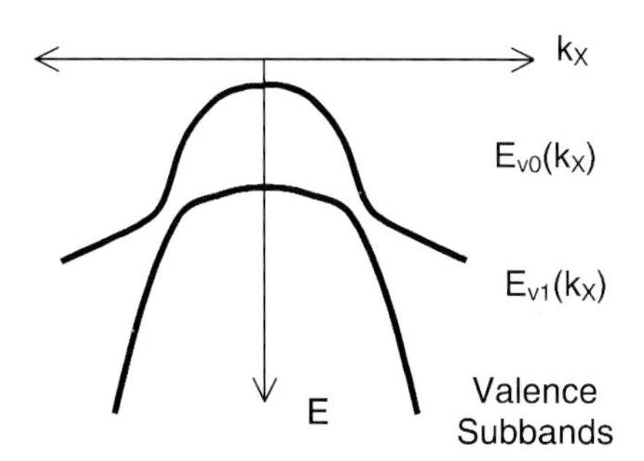

Fig. 4.2. Schematic diagram of the energy dispersion curves of the electron and valence subbands in a quantum well. The wavevector (k_x) is in the plane of the quantum well

lattice, are sometimes used interchangeably. Many other heterostructures are possible and have been studied in detail [1,2].

Additional confinement in the lateral dimensions in a quantum well leads to quantum wires and dots. For example, a 2D quantum well structure can be patterned and etched into 1D quantum wires or 0D quantum dots. In addition, it is possible to go the other way and build semiconductor quantum dots directly using colloidal chemistry [13]. This type of quantum dot is often called a nanocrystal. A great deal of research on nanocrystals has gone on in parallel with the study of MBE-based quantum wells, wires and dots. Although quantum wires again have subbands in their electronic structure, quantum dots have discrete electronic energy states as do atoms.

Quantum well samples can be doped during growth with shallow donors and/or acceptors with great control in position and density. One especially important type of structure is the modulation-doped quantum well in which

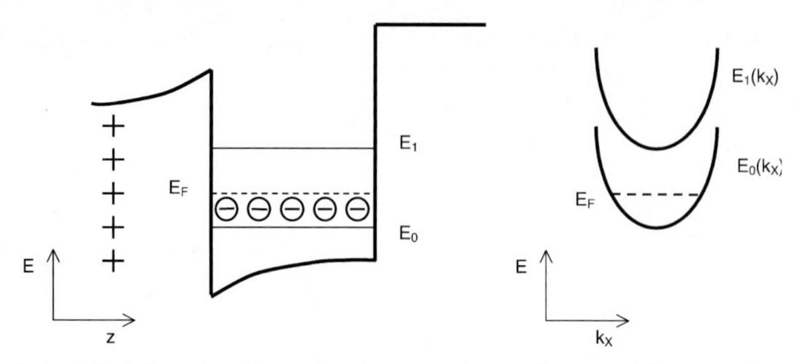

Fig. 4.3. Schematic diagram of the quantum well potential energy diagram with shallow donors modulation-doped into one of the barriers and the corresponding energies as a function of the wavevector in the plane of the quantum well. Electrons transfer from the donors into the quantum well up to the Fermi energy (E_F)

the impurities are put in the barriers only so that the electrons or holes leave the impurity and "fall" into the quantum well (Fig. 4.3). In this way a 2D electron or hole gas is formed separately from the ionized impurities (even at low temperatures and low concentration) resulting in very high electron mobility. Excitations of the 2D-electron gas have been extensively probed with many techniques, including Raman scattering. Transitions within or between conduction subbands are called intrasubband and intersubband transitions, respectively.

4.2 Resonant Raman Scattering

Semiconductor nanostructures often have very small volumes and are on top of much thicker buffer layers and substrates. Nevertheless, researchers have been able to study single quantum wells only a few monolayers thick. In many cases the sensitivity and multichannel capability of the charge-coupled device detector helps considerably [14]. But much of the Raman research on nanostructures has relied on resonant enhancements. By tuning the laser into near resonance with an optical band gap associated with the structure of interest, it is possible to enhance the Raman signal by many orders of magnitude. This is often necessary when working with single layers to obtain sufficient signal on top of a broad PL background, or to induce the Raman signal from one structure to dominate the Raman or PL signals from other layers. The selectivity obtained under resonance conditions allows the researcher to measure the response from a small subset of the entire collection of nanostructures that coexist in a sample. For example, in quantum dot samples, which in all cases have large random fluctuations in size and correspondingly large inhomogeneous broadening of their spectral lines, one can selectively measure only those dots that are within the bandwidth of the laser. Using excitonic resonances it is possible to measure the Raman spectrum of a single quantum well [15,16] or even a single quantum dot [17].

Raman scattering by phonons occurs through a three-step process involving the creation of an electron-hole pair (or exciton) via the electron-photon interaction, the scattering of the exciton through the creation or destruction of a phonon, and the reemission of the photon [18]. Higher order processes can also be important if elastic scattering is allowed in the intermediate state because of disorder [19,20]. The three-step resonant Raman scattering probability involves resonant denominators:

$$P \propto \left| \sum\nolimits_{\alpha\beta} \frac{\langle 0|H_{ex-l}|\beta\rangle\langle\beta|H_{ex-ph}|\alpha\rangle\langle\alpha|H_{ex-l}|0\rangle}{(E_s - E_\beta + \mathrm{i}\Gamma_\beta)(E_\mathrm{L} - E_\alpha + \mathrm{i}\Gamma_\alpha)} \right|^2 , \tag{4.3}$$

where E_α and E_β are the energies of the intermediate excitons and Γ_α and Γ_β are their homogeneous linewidths. H_{ex-l} and H_{ex-ph} are interaction Hamiltonian terms between exciton and light, and exciton and phonon, respectively. E_L and $E_S = E_\mathrm{L} \pm E_\mathrm{ph}$ correspond to the frequencies of the exciting and scattered light, respectively. The intermediate excitons (α and β) are associated with either the same subband or different subbands. Energy does not have to be conserved in the intermediate transitions, but wavevector does. Even wavevector conservation will break down if the intermediate exciton states are localized.

The resonant denominators in (4.3) lead to strong enhancements when either the frequency of the exciting or scattered light resonates with that of an exciton (see Fig. 4.4). These resonances are known as incoming and outgoing resonances, respectively. Experimentally, outgoing resonances are often much stronger. This result can be understood from breakdown in wavevector conservation induced by interface disorder [19,20]. However, the contribution of other exciton states associated with higher subbands can also contribute to the asymmetry, and can also lead to the opposite case where the ingoing resonance is stronger [21]. In samples with subband splittings equal to an optical phonon energy, (4.3) leads to double resonance [22,23].

The resonant Raman profile (Fig. 4.4) is obtained by measuring the intensity of a specific phonon as a function of the photon energy (either the exciting or scattered photon). In this way the excitonic states and their interactions with the phonons, or alternately, the excitations of the electron gas are probed. In semiconductor nanostructures the excitons, like the phonons and plasmons, are strongly dependent on the size, shape and other characteristics of the nanostructure. Conclusions obtained from the optical spectra complement those obtained from the Raman spectrum. In these cases Raman studies are greatly enriched [20,24].

In the case of optical phonons in a polar semiconductor, the interaction between exciton and phonon in the resonant scattering process arises through both the deformation potential and the depolarization field of the phonon (Fröhlich interaction) [25]. The deformation potential scattering is local in

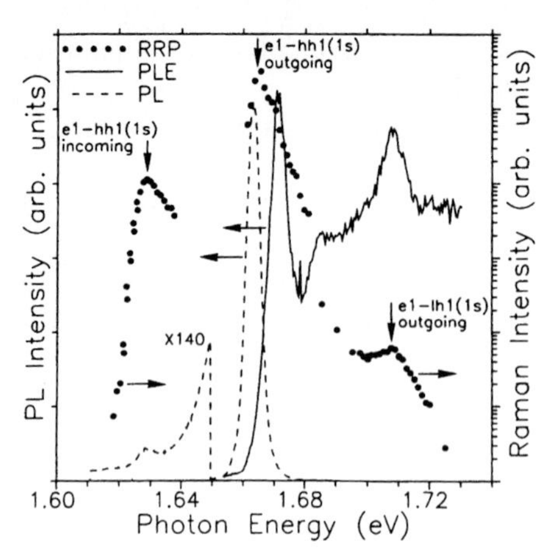

Fig. 4.4. Resonant Raman intensity of the LO_2 phonon from a 4.6 nm GaAs/ 4.1 nm AlAs multiple quantum well sample at 9 K as a function of scattered photon energy. Ingoing and outgoing resonances are denoted. Photoluminescence (PL) and PL excitation spectra (PLE) are also shown. From [20]

real space, and correspondingly k-independent. On the other hand, the Fröhlich interaction involves the interaction of the exciton with the long-range dipolar field of the LO phonon and is wavevector dependent and goes as k^2. Because wavevector conservation leads to scattering only at $k \sim 0$, the Fröhlich interaction, though not normally important, is strongly dependent on breakdown of wavevector selection rules and is strongly enhanced, for example, by impurities [26] or interface roughness [19,20].

Resonant Raman scattering from excitations of the electron gas also can arise from the three-step process discussed above [27]. For example, scattering of intersubband plasmons occurs through the scattering of the virtual photoexcited electron or hole that is accompanied by the creation or destruction of a plasmon. Likewise, a collective spin density excitation also can be scattered in this way (Fig. 4.5). These scattering process are mediated by the direct and exchange Coulomb interactions [27], although they have been studied much less than in the case of phonons.

Scattering of electrons can occur also in a two-step process [28]. This process, which has been discussed in detail for the case of 3D semiconductor plasmas, can lead to both charge and spin density excitations. In this case, scattering of light occurs directly from fluctuations of the electron gas – analogous to Rayleigh scattering from density fluctuations in a classical gas. In this case, light can couple to the spin density fluctuations of the electron gas because the spin and orbital degrees of freedom in the wavefunction of the hole in the intermediate state of the scattering process are mixed by the spin-orbit interaction.

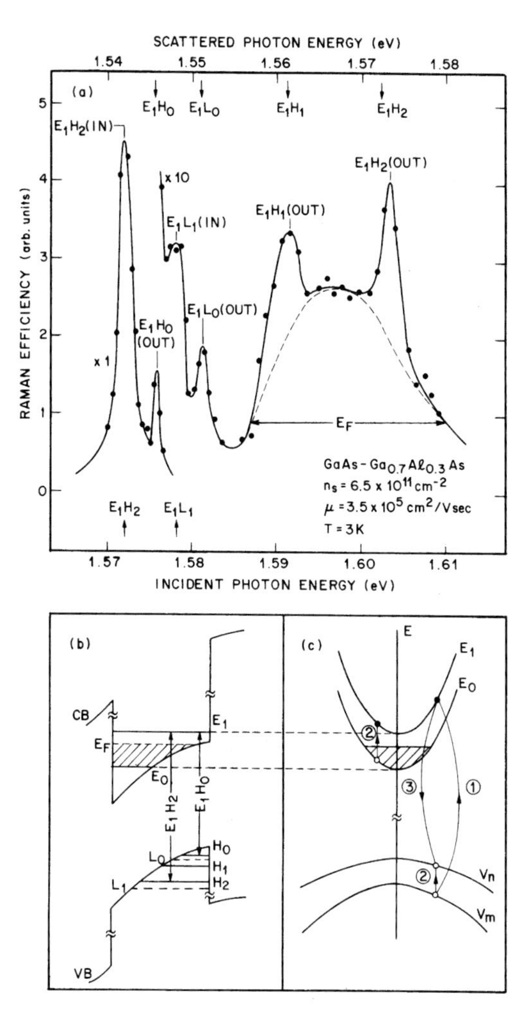

Fig. 4.5a–c. Profile of resonant Raman enhancement of the spin-density intersubband excitation as a function of the incident photon energy. Schematic of the quantum well potential. Schematic of a 3-step resonant Raman process. From [27]

4.3 Kinematics

The majority of Raman experiments on semiconductor heterojunctions have been performed in the backscattering geometry along the normal to the quantum well plane (z-direction). To probe the dependence of the scattering on the wavevector in the plane of a quantum well the sample can be rotated with respect to the wavevector of the incident and scattered light (Fig. 4.6). The components of the scattering wavevector within the crystal are

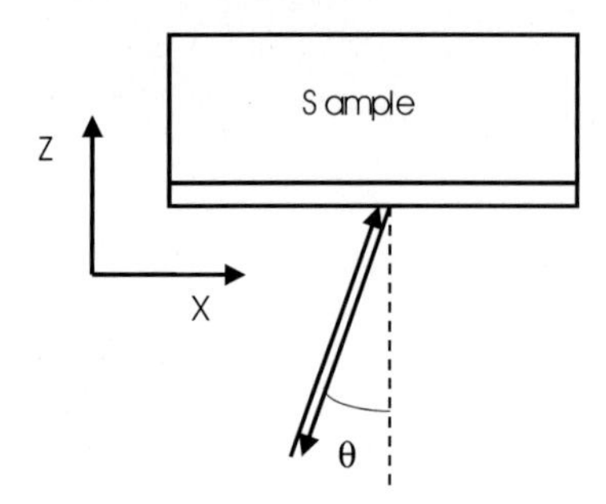

Fig. 4.6. Raman scattering geometry to increase the wavevector transfer into the plane of the quantum well

$$k_\varrho = \frac{4\pi}{\lambda_\mathrm{L}} \sin\theta \,, \tag{4.4}$$

$$k_z = \frac{4\pi}{\lambda_\mathrm{L}} n(\lambda_\mathrm{L}) \left[1 - \left[\frac{\sin\theta}{n(\lambda_\mathrm{L})} \right]^2 \right]^{1/2} , \tag{4.5}$$

where λ_L is the laser wavelength and $n(\lambda_\mathrm{L})$ is the refractive index. Because the index of refraction is large ($n_\mathrm{GaAs} \approx 3.6$), the angle of the laser beam within the crystal remains relatively small even for large sample rotations. The maximum k_z is 5.5 x $10^5\,\mathrm{cm}^{-1}$ while the maximum $k_\varrho < 1.510^5\,\mathrm{cm}^{-1}$ for GaAs at the E_0 band gap ($\lambda_\mathrm{L} \approx 820\,\mathrm{nm}$). A transmission geometry can be used to increase further the angle of the scattering wavevector, although the substrate must be removed if it is opaque [29]. Alternatively the laser can be focused on the side of a superlattice with a microscope to measure excitations with wavevector completely in the x-direction [30,31].

4.4 Vibrational Raman Scattering in Semiconductor Heterostructures

Vibrational excitations of semiconductors are modified in heterostructures, and new modes associated with the interfaces arise. In this section the nature of phonons in a quantum well is discussed. Raman scattering is the most powerful way to study phonons in nanostructures and there has been extensive work, especially in layered semiconductors. Because of the small wavevector of light, only phonons near $\boldsymbol{k} = 0$ can be studied. Nevertheless, this limited view has led to a good understanding of phonons in superlattices as probed by Raman spectroscopy. There are several extensive reviews focused on phonon scattering [7,12,25], and an introductory textbook discussion has recently been published [5].

4.4.1 Phonons in Semiconductor Quantum Wells

In zincblende semiconductors, with two atoms per unit cell, (for example, GaAs and AlAs) there are both acoustic and optic phonons. If a superlattice crystal is formed out of these two materials with a new unit cell, $GaAs_n/AlAs_m$ (where n and m are the number of monolayers in each layer), both the acoustic and optical phonons are modified. The acoustic phonons become folded acoustic phonons and the optical phonons become confined and interface phonons. Most Raman experiments in this system have been performed in the backscattering geometry on samples grown along the [001] direction. In this case folded longitudinal acoustic (LA) modes and confined longitudinal-optical (LO) modes are allowed. Disorder-induced scattering due to interface phonons, and to a lesser extent, TA and TO phonons is also observed, however. Interface phonons and also transverse modes are allowed in other scattering geometries (for example, focusing on the edge of a superlattice with a microscope [30,31]), or on quantum well samples grown along different crystal directions [12].

There are simple models to describe each of these modes, as summarized below. The simplest models are often very useful, although, in the case of optical phonons, the identities of the confined and interface phonons become mixed as the wavevector is rotated to lie in the plane of the quantum wells.

Folded Acoustic Phonons. In the case of acoustic phonons, because the bulk dispersion curves of the two semiconductors are similar, there is not much reflection at the interfaces, and the resulting superlattice acoustic phonons are propagating modes with a sound velocity that is approximately the weighted average of the constituents. However, because of the new periodicity, the bulk Brillouin zone is folded into a new superlattice Brillouin zonee with zone edge at$k_z = \pi/d$ (where $d = d_1 + d_2$ is the superlattice period) [32]. The folding of the phonon dispersion curves is shown in Fig. 4.7. In addition, a small band gap opens up at the zone center and edge due to the normally small acoustic mismatch at the interfaces. Scattering is strong when incident and scattered light is parallel polarized. For wavevectors much less than the bulk reciprocal lattice vectors the acoustic dispersion curves for the superlattice can be obtained from a continuum model using a linear approximation. The dispersion is given by

$$\cos k_z d = \cos\left[\omega\left(\frac{d_1}{\nu_1} + \frac{d_2}{\nu_2}\right)\right] - \frac{\alpha^2}{2}\sin\left(\omega\frac{d_1}{\nu_1}\right)\sin\left(\omega\frac{d_2}{\nu_2}\right) ; \tag{4.6}$$

with

$$\alpha = \frac{\varrho_2\nu_2 - \varrho_1\nu_1}{\sqrt{\varrho_2\nu_2\varrho_1\nu_1}} , \tag{4.7}$$

where ϱ_i and ν_i are the bulk densities and sound velocities, respectively. The parameter α describes the amplitude of the acoustic modulation through the acoustic impedances, $\varrho_i\nu_i$. These equations are valid for both LA and TA modes provided the appropriate constants are used.

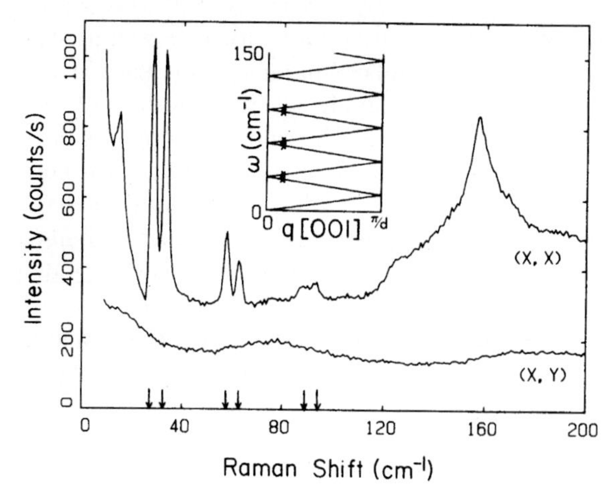

Fig. 4.7. Folded acoustic phonon spectrum of a $(GaAs)_{15}/(Al_{0.3}Ga_{0.7}As)_3$ superlattice. A series of doublets are observed with energies determined by the folded dispersion curves and the wavevector of the light. The *inset* shows the calculated dispersion curve of the LA modes, which are folded over by the superlattice periodicity. From [32]

The acoustic phonon dispersion curves are folded into a number of optic-like phonon branches that can be measured with Raman scattering near the zone center (see Fig. 4.7). Doublets are observed in the parallel polarization geometries with frequencies determined by the wavevector of light and the scattering geometry. In backscattering geometry

$$\omega_m^{\pm} = \left(\frac{m\pi}{d} \pm \frac{4\pi n(\lambda_\mathrm{L})}{\lambda_\mathrm{L}}\right) \nu_\mathrm{SL} \,, \tag{4.8}$$

in which the velocity, v_{SL}, is an average over the two superlattice components

$$\nu_{SL} = d\left(\frac{d_1}{\nu_1} + \frac{d_2}{\nu_2}\right)^{-1} . \tag{4.9}$$

This is an approximation that is found to work in normal backscattering geometries. The measurement of the frequencies of the folded acoustic phonons provides a characterization of the period.

The relative intensities provide a measure of the relative thickness of the layers much like in x-ray diffraction experiments. In a similar way the intensities of the folded acoustic phonons can also provide a measure of the thickness of the interfaces. However, the accuracy of these parameters obtained in a Raman experiment is much less than that obtained by x-ray diffraction. In single quantum wells, wires and dots there is no periodicity and folding does not occur.

Confined Optical Phonons. The optical phonons are strongly modified by structuring of the semiconductor. We consider first the case for phonons with $k_\varrho = 0$. These are the phonons probed in a backscattering experiment if wavevector conservation does not break down. If the dispersion curves of the optical phonons of the two constituent semiconductors do not overlap, the vibrations of one layer cannot propagate into the other and the phonons will form standing waves (confined modes). The quantum frequencies of the confined optical phonons can be understood in the same way as a vibrating string with clamped ends, or as an electron confined to a infinitely deep quantum well. The phonon amplitude involves an envelope function that goes to zero near the interface of the quantum well (see Fig. 4.8). Therefore, an integer number of half wavelengths of the phonon must fit in the well (n monolayers), and the allowed wavevectors are found from

$$q_m = \frac{m\pi}{(n+\delta)a_0}, \quad m = 1, 2, ..., n \,, \tag{4.10}$$

where a_0 is the monolayer thickness (2.83 Å in [001] GaAs) and δ accounts for the penetration of the vibration into the neighboring layer. In the case of GaAs/AlAs: $\delta = 1$ monolayer (only the common As layer at each interface contributes). From these wavevectors the energies of the confined modes can be found from the dispersion relations of the bulk phonons of the constituent

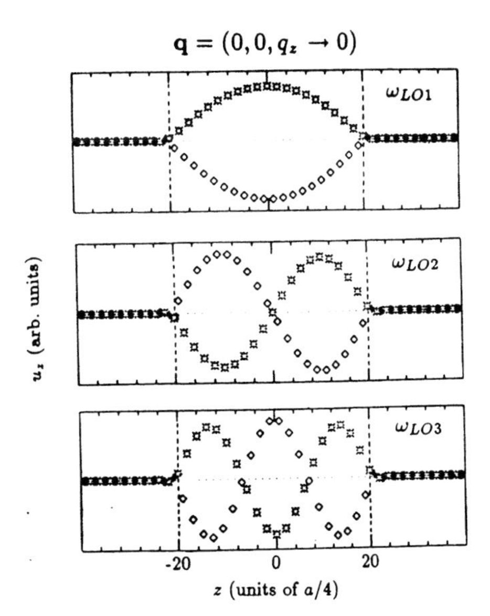

Fig. 4.8. Atomic displacements of the GaAs optic phonons for a $(GaAs)_{20}/(AlAs)_{20}$ superlattice with wavevector, **q**. The *dashed vertical lines* mark the As interface planes. *Diamonds*, *stars* and *asterisks* indicate the positions of Ga, Al and As planes, respectively. From [89]

semiconductors, $E_m = E(q_m)$. This is exactly the way that one finds the energy of a quantum particle in an infinitely deep, square quantum well at $k_\varrho = 0$. The energy shifts from the value of the energy in the bulk semiconductor at $k \approx 0$ are confinement energies. The confinement energies are negative in the case of GaAs and AlAs because the dispersion for optical phonons in these materials is negative.

An example of the optical phonon spectra from a multiple quantum well sample is shown in Fig. 4.9 for several polarization configurations. The phonons in the region around $E = 280\,\mathrm{cm}^{-1}$ arise from the GaAs layers and those near $E = 400\,\mathrm{cm}^{-1}$ arise from the AlAs layers. The TO modes are not allowed in backscattering from a (001) surface. Moreover, when out of resonance the odd confined modes are measured in the (x, y) or (y, x) geometries while the even modes are observed in the (x, x) or (y, y) geometries, where x and y correspond to polarization of the incident and scattered light polarization along the [110] crystal axes. In resonance the even modes and strong interface modes are observed to dominate in all scattering geometries [25].

This simple approach for calculating the energies breaks down if there is sufficient interface roughness [12,33]. The interfaces become blurred and the

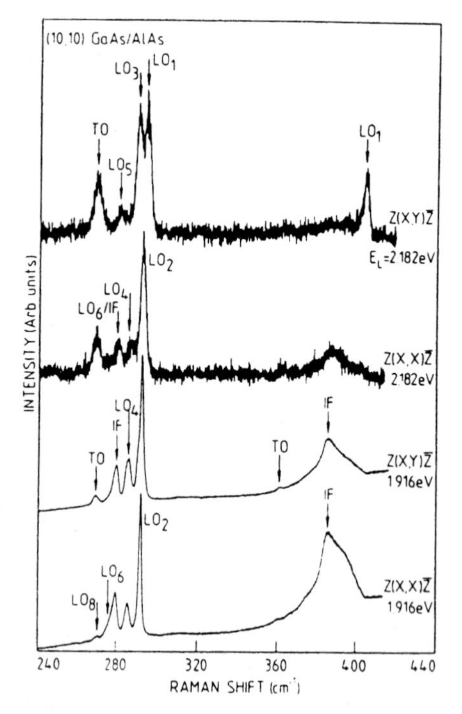

Fig. 4.9. Confined optical phonon spectrum of a $(GaAs)_{10}/(AlAs)_{10}$ superlattice for different polarization configurations for resonant ($E_L = 1.916\,\mathrm{eV}$) and nonresonant ($E_L = 2.182\,\mathrm{eV}$) excitation. From [90]

quantum well is no longer square. To test the model for a particular sample or class of samples, the energies of the confined modes can be plotted as a function of q_m, as determined by (4.10), and compared to the bulk dispersion curves. In the case of most narrow layers there is relatively poor agreement. Moreover, it is possible that disorder due to interface roughness may lead to partial breakdown in wavevector conservation and allow scattering from excitations with $k_\varrho \neq 0$. These issues will be discussed in more detail in later sections.

The phonons can also be calculated within a linear chain model. This is relatively simple in one dimension and can be extended to 2 or 3 dimensions if desired. It is also possible to model the interface to some extent by including an extra alloy layer at the interface with appropriate masses and spring constants.

For phonons with $k_\varrho \neq 0$ it is necessary to consider dispersion. Close to the $k_\varrho = 0$ confined optical modes have little mechanical dispersion, however, the long range electric field is sensitive to the interfaces and new modes arise that are known as interface modes.

Interface Optical Phonons. In compound semiconductors like GaAs and AlAs the LO phonon has an electric field associated with it. In the presence of interfaces, the electric field gives rise to interface modes in addition to the confined optical phonons (Fig. 4.10). These vibrations, which arise when $k_\varrho \neq 0$, have maximal values at the interfaces and slowly decay perpendicular to the interface. The energies of these modes can be found from Maxwell's equations, using for each material an appropriate dielectric function,

$$\varepsilon(\omega) = \varepsilon(\infty)\frac{\omega^2 - \omega_{LO}^2}{\omega^2 - \omega_{TO}^2}. \tag{4.11}$$

For an infinite superlattice the energies of the interface modes are found as a function of (k_ϱ, k_z) from

$$\begin{aligned}\cos(k_z d) &= \cosh(k_\varrho d_1)\cosh(k_\varrho d_2) \\ &\quad + \frac{1}{2}\left(\frac{\varepsilon_1}{\varepsilon_2} + \frac{\varepsilon_2}{\varepsilon_1}\right)\sinh(k_\varrho d_1)\sinh(k_\varrho d_2).\end{aligned} \tag{4.12}$$

In the limit as $k_z \to 0$, (4.12) reduces to equations for two bands,

$$-\frac{\varepsilon_1(\omega)}{\varepsilon_2(\omega)} = \begin{cases}\tanh(k_\varrho d_1/2)\coth(k_\varrho d_2/2) \\ \tanh(k_\varrho d_2/2)\coth(k_\varrho d_1/2)\end{cases}. \tag{4.13}$$

For $d_1 = d_2$, (4.13) reduces

$$\varepsilon_1(\omega) = -\varepsilon_2(\omega), \tag{4.14}$$

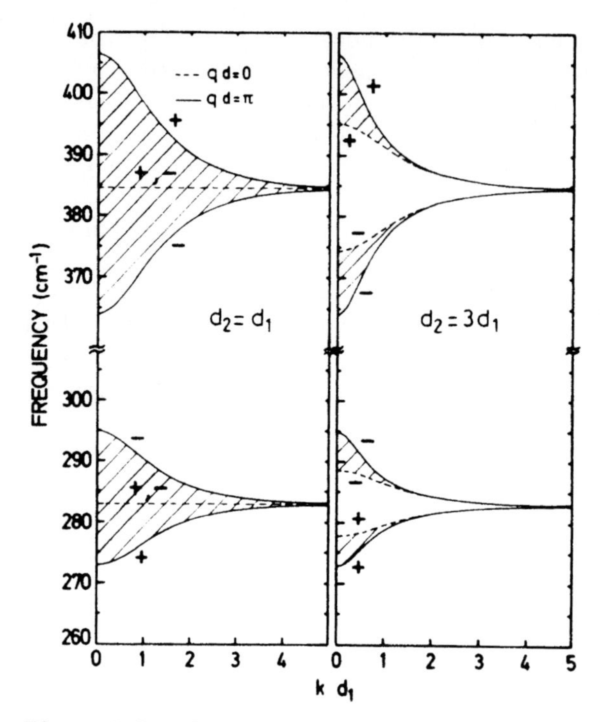

Fig. 4.10. Interface phonon frequencies for GaAs/AlAs superlattices as a function of wavevector in the plane of the layers ($k_\varrho = k$) for different values of wavevector normal to the layers ($k_z = q$) and superlattice period (d). From [49]

which is the equation for an interface mode localized at the interface between two semi-infinite dielectric media.

The energies of the interface phonons change strongly as the angle of the phonon wavevector rotates to lie along k_ϱ. This behavior was measured directly in a microscopic Raman scattering study of multiple GaAs/AlAs quantum wells. In this study the dispersion of the interface modes was measured by focusing a microscopic laser spot onto the edge of a superlattice that was polished at various angles with respect to the superlattice direction (Fig. 4.11) [30,31]. Because the dispersion of the confined modes is relatively flat, the interface and confined modes cross. Mixing occurs between the interface modes and the odd confined modes and leads to anti-crossing of the frequencies, as described in more complete models [34]. The even confined modes have the wrong symmetry and do not mix with the interface modes.

4.4.2 Phonons as a Probe of Interface Roughness in a Quantum Well

One of the most critical issues in semiconductor nanostructures is the nature of the interfaces. Raman scattering has the potential to characterize the

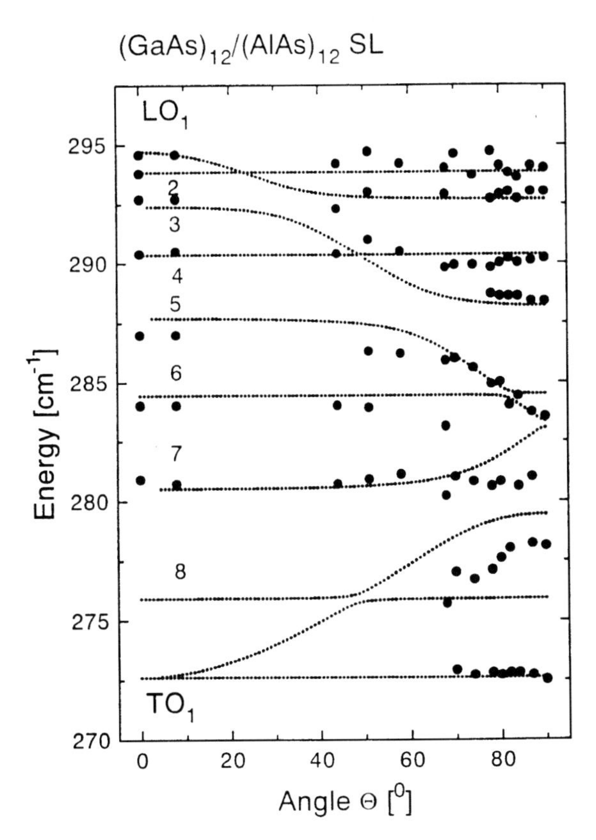

Fig. 4.11. Microscopic Raman results (*circles*) obtained in backscattering geometry from the edge of a $(GaAs)_{12}/(AlAs)_{12}$ superlattice polished at different angles with respect to the superlattice direction are compared to calculations (*lines*). Dispersion due to the interface modes and also mixing between the interface modes and the even confined modes is observed. From [31]

interfaces, and a number of experimental studies have been reported on quantum well structures [12,25,35]. In particular, the nature of the interfaces in GaAs/AlAs quantum wells has been studied in great detail by Raman scattering and many other techniques. In this system, with a common anion and little mismatch in lattice parameters, the interface is characterized by the position of the Ga and Al cations. In a perfect structure, the interface would abruptly change from a complete Al monolayer to a complete Ga monolayer. Real structures have mixing of the two cations over several monolayers. In the simplest picture this blurred interface layer is modeled as a thin alloy layer.

One very simple measure of the thickness of the interface layer is the magnitude of the energy splittings in the spectrum of the confined optical phonons. If the interfaces are perfect, the energies of the confined optical modes are obtained from an infinite *square* potential well model and the confinement shifts of the modes goes as m^2 where m is mode index. If there is extreme mixing of the Ga and Al, the quantum well becomes more parabolic than square and the mode spacing becomes constant. A figure of merit was suggested to quantify this measure in order to characterize the quality of quantum well samples grown under different conditions [36]:

$$\varrho = \frac{(\omega_5 - \omega_3)}{(\omega_3 - \omega_1)} . \tag{4.15}$$

A square well will lead to $\varrho = 2$ whereas a parabolic well leads to $\varrho = 1$. A very clear example of this effect was demonstrated by measuring the phonon spectrum of a quantum well sample after heating the sample to high temperatures for varying periods of time [37]. At high annealing temperatures the Al and Ga anions can exchange positions and the interface becomes thicker. The Raman spectrum gave $\varrho = 2.0$ for the unannealed sample and $\varrho = 1.35$ after annealing for 3 hours at 850 °C.

The spectrum of acoustic phonons in a multiple quantum well is also sensitive to the nature of the interface. The energies of the folded acoustic modes depend only on the quantum well period. However, the shape of the anion modulation, in other words, how blurred are the interfaces, leads to a significant change in intensities that can be quantitatively modeled using the photoelastic model. This effect is very similar to x-ray diffraction and the analysis is similar. The spectral intensities of the folded acoustic modes are related to the Fourier transform of the real-space cation composition modulation.

When x-ray diffraction characterization is possible it generally provides much higher precision. However, x-ray diffraction becomes more difficult when working with small structures, such as a single quantum well, or non-periodic complex structures, such as a quantum well sample with many different single quantum wells with different widths. With Raman scattering it is possible to resonate with a particular layer by tuning the laser frequency to the layer's energy band gap, thereby increasing strongly the scattering efficiency to the point where it is measurable and also dominates the Raman spectrum. In this way the necessary sensitivity and selectivity can be achieved to probe small individual volumes within large or complicated crystals. As an example, a Raman study of the lateral nature of the interface using Raman scattering will be discussed next.

Lateral Interface Structure. In quantum well samples of the highest quality, the thickness of the interface is on the order of a single monolayer [38]. In such samples the lateral structure of the interface within the monolayer varies dramatically, leading to significant changes in the optical properties of the quantum well. For example, if the growth of the sample is interrupted for approximately a minute before the interface is formed, the surface forms monolayer-high islands as the atoms migrate on the growing surface to lower energy positions at the edges of islands (Fig. 4.12). These "monolayer-high islands", which can reach diameters of 100's of nanometers, can be measured directly with STM if the growth is terminated [39].

Although buried interfaces are difficult to image directly with nanometer resolution, the interface can be probed through optical spectroscopies such as photoluminescence (PL) [38]. In quantum wells the PL is dominated by recombination of excitons (bound electron-hole pairs). The lineshape of the spectral lines varies dramatically depending on the length scale of these large monolayer high islands relative to the exciton's Bohr diameter ($\approx 20\,\mathrm{nm}$ in GaAs). When the islands are less than the Bohr diameter of the exciton the interface is averaged over and the spectral lines consist of a single broad peak. However, if the island structure becomes larger than the Bohr diameter in a narrow quantum well, the PL lines break up into doublets or triplets with splittings corresponding to differences in well width of one monolayer. This splitting arises from regions of the quantum well that differ by one monolayer due to the monolayer-high islands at the interfaces.

From PL and other optical spectroscopies it is obvious when a sample has lateral interface structure large compared to the exciton's Bohr diame-

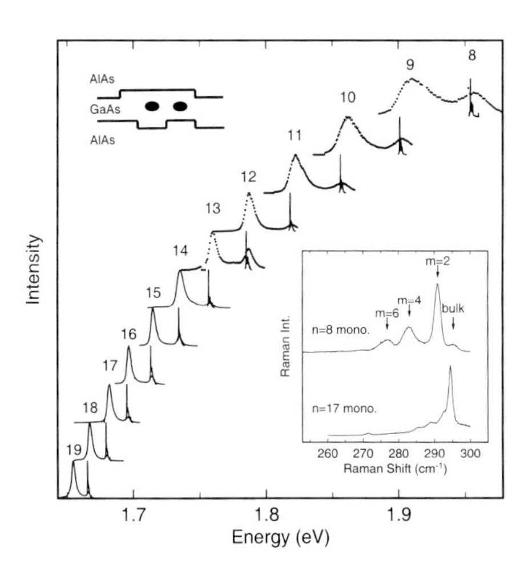

Fig. 4.12. Optical spectra of lowest energy exciton from a $(\mathrm{GaAs})_n/(\mathrm{AlAs})_{18}$ single quantum well as a function of well width (n monolayers). Large monolayer-high islands at the interfaces lead to a splitting of the exciton line into a doublet with an energy difference corresponding approximately to the difference in well width of one monolayer. Also shown (sharp spectral structure) are the LO phonon spectra obtained when the laser is tuned in frequency one phonon above the exciton. *Lower inset*: The resulting resonant Raman spectra with $n = 17$ and 8 monolayers for well widths. *Upper inset*: Highly simplified schematic of the quantum well illustrating excitons existing within regions differing in width by one monolayer

ter. However, because the exciton averages over smaller structure, it is not clear whether the interfaces also contain a significant amount of small-scale structure within these large islands [40,41]. One way to address this question is to compare the energies of the optical phonons in a Raman spectrum with those expected with and without alloy layers at the interfaces. To avoid complicating the spectrum with contributions from regions of the quantum well that have different well widths (arising from large monolayer islands at the interfaces) resonance Raman scattering can be used. By tuning the laser into resonance with a specific exciton energy, phonons associated with a particular well width can be measured (to within a monolayer) [15,16].

In one experiment (Fig. 4.12) the optical phonon spectrum was measured monolayer by monolayer from a single quantum well in the range from 8 to 19 monolayers (2.3 to 5.4 nm) [16]. This was done on a sample that was grown like a wedge so that the average well width varied from one side of the sample to the other, in addition to fluctuating by one or two monolayers at each position. Thus, by scanning the laser across the sample the exciton spectrum could be measured monolayer by monolayer. The energy of the exciton provided a characterization of the well width. Moreover, by tuning the laser frequency into resonance with specific monolayer resonances as the well width changed, the phonon spectrum could be recorded monolayer by monolayer. In Fig. 4.12 both the exciton spectra and the resonant Raman spectra are shown as a function of well width in monolayer steps. In the lower inset to Fig. 4.12 are shown the resonant Raman spectra from regions with $n = 17$ and 8 monolayers.

The resulting energies for the LO phonon modes are plotted as a function of well width in Fig. 4.13. The data was compared with calculations based on a one dimensional linear chain model with thin alloy layers at the in-

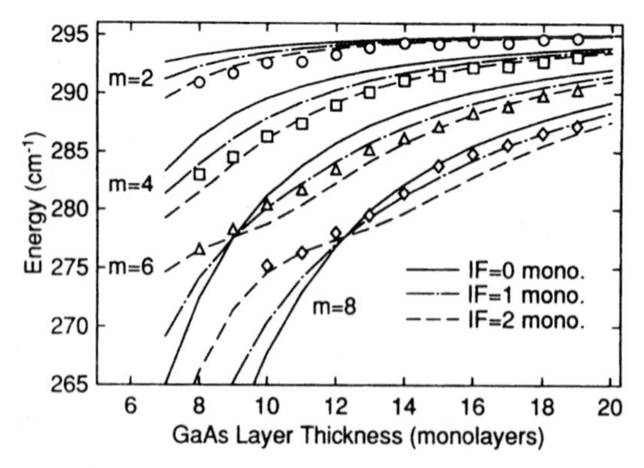

Fig. 4.13. The measured (*symbols*) and calculated (*lines*) energies of several confined LO modes as a function of well width. The calculations were performed with interfaces modeled as uniform alloy layers with concentrations $x = 0.5$, and widths 0, 1 or 2 monolayers. From [24]

terfaces. The experimental results were in good agreement with calculations for a structure with an alloy-like layer about one monolayer thick at the two interfaces. From this resonant Raman study it is clear that the interface structure in the highest quality GaAs/AlAs quantum wells consists of both large-scale island structure and small-scale alloy-like structure, both about a monolayer thick. The large-scale structure arises from the migration of the anions on the growing surface to the edges of large monolayer-high islands that grow to diameters on the order of 10–100 nm. The small-scale structure is most likely atomic-scale roughness that develops due to the exchange of Al and Ga on the growing surface as the interface is being formed [42]. As one might expect, this atomic scale roughness depends on growth temperature among other factors [16,43,44].

Phonon Spectrum Within a Single Interface Island. By resonating with the exciton, large increases in Raman intensities are achieved. In this way a small fraction of a single quantum well can be studied. It is interesting to estimate how small this fraction is in the example given in the last section. The total number of islands within the laser spot is determined roughly by the ratio of the area of the laser spot ($1000\,\mu m^2$) to the area of a typical island within this single quantum well ($10^{-3}\,\mu m^2$); about 10^7 islands. However, the fraction of these islands whose excitons are in resonance is determined by the ratio of the exciton's homogeneous ($30\,\mu eV$) to inhomogeneous linewidth (3 meV), or about 0.01. This leads to an estimate of about 10^5 islands that are excited in a macroscopic resonant Raman experiment on this type of quantum well structure.

Remarkably, this measurement can be pushed to the limit of a single island [17]. If the laser spot is reduced to a sufficiently small area it is possible to resolve the PL and absorption lines arising from individual excitons localized laterally within single islands (Fig. 4.14) [45–47]. The PL spectra shown in this figure were obtained through small apertures in a metal shadow mask. These masks were patterned directly on the sample using evaporation and electron beam lithography and range in diameter from $25\,\mu m$ down to 200 nm. PL was excited and detected through the same aperture in a backscattering geometry. As the laser spot was reduced from a macroscopic $25\,\mu m$ aperture into the optical near field regime of 200 nm, the inhomogeneous spectrum broke up into a decreasing number of extraordinarily sharp PL spikes. These PL lines arose from single exciton states localized laterally within individual islands.

In many ways the properties of these spatially localized excitons can be described in terms of quantum dot potentials [47,48]. The lateral confinement leads to additional confinement energy. It is fluctuations in these lateral confinement energies that lead, in part, to the inhomogeneous linewidth observed in the optical spectra of narrow quantum wells. By reducing the focused la-

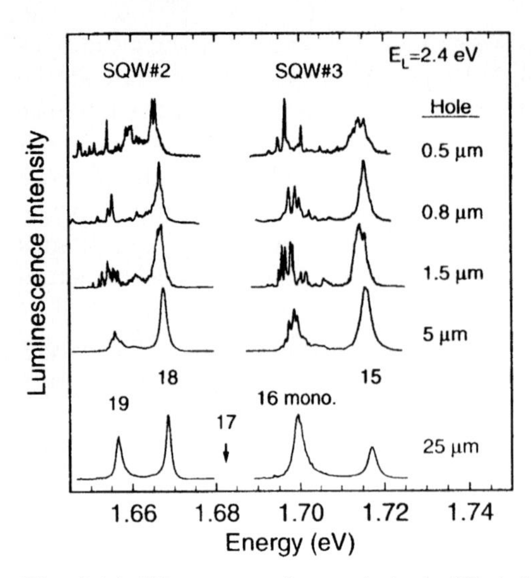

Fig. 4.14. PL spectra of several single $(GaAs)_n(AlAs)_{18}$ quantum wells as a function of aperture size on the sample. The broad inhomogeneous exciton spectrum breaks up into a single exciton lines when a sufficiently small area is probed. Both exciting and detected light passed through the same small hole in a metal shadow mask processed directly on the sample. From [91]

ser's spot size to the microscopic or to the submicroscopic regime individual excitons can be excited and detected.

By tuning the laser into strong resonance with an individual exciton localized within a single island, the associated phonon spectrum was measured (Fig. 4.15) [17]. This was accomplished in the outgoing resonance. A spectrometer was tuned to the exciton resonance while the laser was scanned through the optical phonon regime at higher energy. The intensity of the emitted light as a function of the difference between the laser and the spectrometer energies was measured, providing the phonon spectrum associated with this single localized exciton. In Fig. 4.15 the macroscopic Raman spectrum measured through a 25 μm aperture is also plotted. This spectrum was obtained in the conventional way; fixing the laser and scanning the spectrometer. The energies and linewidths of the optical modes are in good agreement between the two results. This implies that the energies of the phonons, unlike those of the excitons, are insensitive to the islands at the interfaces. This is not surprising because a change in well width of one monolayer leads to a change in phonon confinement energy that is less than the homogeneous linewidth. Nevertheless, because the scattering is directly mediated by the localized exciton, the spatial envelope of phonons excited in the scattering process with the localized exciton must also be localized, and the measurement serves as a local probe of phonons.

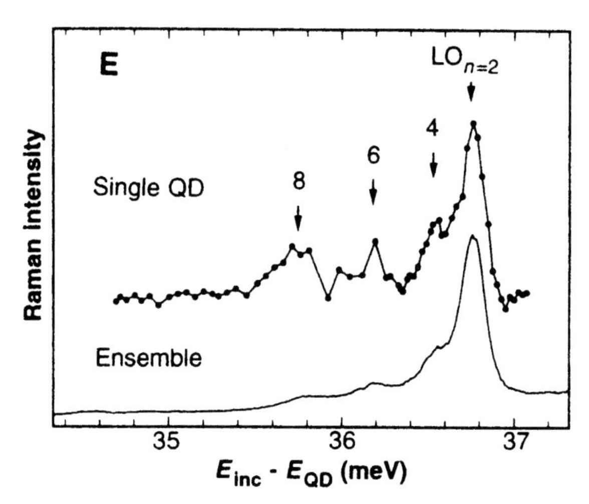

Fig. 4.15. Resonant Raman spectra of LO phonons obtained from the 19 monolayer quantum well shown in Fig. 4.14. The bottom trace shows the macroscopic spectrum obtained in a conventional resonant Raman scattering experiment (as in Fig. 4.12). The bottom trace shows the phonon spectrum obtained through a 500 nm aperture. In this case the spectrometer was tuned to the frequency of a single localized exciton while the laser was scanned in frequency. From [17]

Although there was no measurable change in the energies of the phonons, there were fluctuations in the relative intensities from dot to dot. This was particularly noticeable in the intensity of the TO modes. This result was not well understood, however, because the TO phonon is normally forbidden in this geometry, the origin of this observation may lie in the breakdown of the selection rules arising from lateral localization.

This experiment serves to illustrate one potential direction of interest in the Raman characterization of semiconductor nanostructures. With resonant Raman spectroscopy, the phonon spectra of individual nanostructures with dimensions down to a few tens nm's can be selectively measured. In some cases it should be possible to obtain not only the spectra but also the spatial images of the phonon intensity with resolution in the optical near field regime.

Raman Scattering Induced by Interface Roughness. The roughness of the interfaces leads to partial breakdown of selection rules. Interface roughness provides a mechanism by which wavevector conservation can be violated. An especially clear example of this is the common observation of interface (IF) phonons observed in backscattering geometry from multiple quantum well samples [29,49–51]. Interface phonon scattering requires a component of the scattered wavevector in the plane of the quantum well. Transfer of wavevector into the plane of the quantum wells through localized excitons in the intermediate state of the Raman scattering process can explain the observation of IF phonon scattering in the backscattering geometry. The scattering arises from coupling of the light into a distribution in phonon wavevectors. Roughly, the spectrum will reflect the density of states in the phonon's dispersion curve out to a wavevector given by the inverse of the exciton's localization size.

Early on, the strong interface phonon scattering in pure backscattering geometry was recognized as arising from a breakdown in wavevector conservation, probably due to interface roughness [25,29,49]. This effect was especially dramatic under outgoing resonance conditions. The Raman spectrum was interpreted as a combination of peaks due to confined phonons along with a peak due to forbidden interface scattering in many cases. Recently, however, Shields et al. [50,51] pointed out that the scattering intensities could not be understood under close examination with this simple interpretation, and that interface roughness plays an even stronger role than previously appreciated. They were able to show that under resonant conditions the entire spectrum, except for the $m = 2$ confined mode, could be understood as scattering from forbidden interface phonons. They demonstrated that the optical phonon spectrum should be viewed as a broad feature due to forbidden interface phonon scattering, originating from the entire dispersion curve of the interface phonon (Fig. 4.16). *Dips* in the spectral structure could be associated with anti-crossing between the interface modes and the odd confined modes. At these energies there is a lower density of states. In complete contrast to previous studies, *peaks* in the structure were reinterpreted as regions between the odd confined modes. It is coincidental that these peaks in the spectra occur roughly at the energies of the even confined modes. It is because of this coincidence that the simple models previously used provided satisfactory agreement with the measured energies. According to their calculations, the contribution to the resonant spectra due to the even confined modes, with the exception of the $m = 2$ mode, is much smaller than that due to the forbidden interface modes. This reinterpretation was supported with a detailed calculation of the spectrum.

Recently, Ruf et al. [12,52,53] studied the structure in the folded acoustic phonons under resonance conditions. Disorder-induced scattering led to a broad continuum in the spectra in addition to the sharp allowed modes. Just as in the case of the optical phonons [19–21], considerable information can be obtained on the intermediate excitonic states involved in the resonant Raman process from the intensity. The phonons provide a probe of disorder.

4.5 Electronic Raman Scattering in Semiconductor Heterostructures

If a quantum well is doped with shallow donors or acceptors, a variety of low energy excitations can be excited and studied in detail with Raman scattering [10,54]. If impurities are doped directly into the middle of the quantum

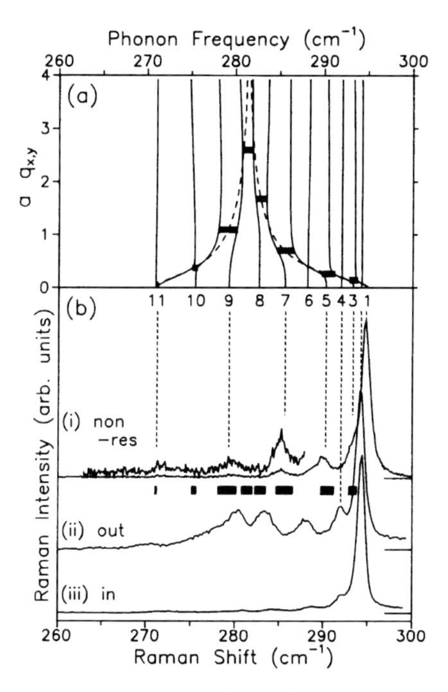

Fig. 4.16a,b. Calculated dispersion of optical phonons with in-plane wavevector ($q_{x,y}$) in a 46 Å/46Å GaAs/AlAs MQW. Raman spectra taken under conditions with (i) nonresonant and z(x,y)z (ii) outgoing resonance and $z(x,x)z$, or (iii) ingoing resonance with $z(x,x)z$. The horizontal bars indicate gaps opened in the IF phonon dispersion by anticrossing with odd-order confined modes, which produce minima in the resonant Raman spectra. From [50]

wells the electrons or holes remain bound but the impurity spectra are modified due to the additional confinement of the quantum well. Alternatively, if the shallow impurities are doped into the barriers a few tens of nanometers away from the quantum well interface (modulation doping), the electrons or holes fall into the lower energy quantum well and form a very high quality quasi-2D electron or hole gas. Excitations of these 2D electron and hole gases have been studied in great detail [10,55].

4.5.1 Shallow Impurities

Dopants such as Si and Be can be selectively doped into the structure using MBE with remarkable precision. Transitions of both neutral donors and acceptors in quantum wells have been measured spectroscopically with photoluminescence [56,57], far infrared absorption [58], and Raman scattering spectroscopy [59–61].

Shallow Donors. Silicon behaves primarily as a shallow donor in bulk GaAs and can be treated as a hydrogen-like atom with one electron orbiting the ion with a binding energy of approximately 6 meV and a Bohr radius of 10 nm. If the Si donor is doped into a GaAs/AlGaAs quantum well the electron wavefunction is forced to overlap the positive ion even more. As a result, the binding energy increases with decreasing well width as the width becomes less than the Bohr radius. In the 2D limit the binding energy goes to 4 times the 3D binding energy [62]. Transitions between 1s- and 2s-like bound states have been measured with Raman scattering [63,64], while transitions between $1s-$ and $2p-$ like states have been measured with far-infrared absorption [65]. An example of measured transition energies for GaAs/AlGaAs quantum wells doped in the center portion with Silicon donors is shown in [54]. Measured values are in good agreement with effective mass calculations [66,67].

The binding energy and symmetry of its wave function depend on the position of the donor ion within the quantum well [54,62]. If the donor ion is at the interface of a quantum well, the electronic wavefunction must go to zero at the donor ion, assuming infinite barrier height. If the well width is much larger than the Bohr radius, the envelope function of the lowest energy state is identical with the $2p_z$-like state of the bulk, and the binding energy is reduced to 1/4 of the value it would have at the center. The transition energy of the impurity thus provides a measure of its position relative to an interface, which could be used to measure impurity segregation during growth.

Shallow Acceptors. Shallow acceptors in GaAs quantum wells have also been measured with Raman scattering [61,68,69,73,74]. Because acceptor wavefunctions are derived from the valence bands, the energy spectrum is more complex than donors (Fig. 4.17) [70,71]. Just as the heavy and light hole bands are split in the quantum well, the bulk four-fold degenerate $1s$ and $2s$ levels [72] each split into doublets in a quantum well. However, because of the relatively strong acceptor binding energy, light and heavy hole mixing within the acceptor $1s$ states is very high and the splitting is much less than that of the heavy and light-hole bands. Transitions between the split 1s states and the $1s - 2s$ states have been measured (Fig. 4.18). Each of these states maintains a two-fold degeneracy that can be split with a magnetic field. In this way spin flip Raman scattering can than be measured (Fig. 4.19) [68,69].

Most electronic Raman spectroscopy studies of quantum wells have relied on resonant enhancement to obtain sufficient signal. With the laser at or close to the band edge, photoluminescence can confuse or obscure the Raman peak. In the case of donors, photoluminescence occurs when a photo-excited hole recombines with the donor electron (Fig. 4.20). The energy of this PL transition is independent of the laser frequency. In contrast, a Raman transition should track the laser frequency (as long as there is a distribution of resonant energies). This sort of behavior, seen in Fig. 4.21, allowed the authors to differentiate between PL and Raman, and obtain a measure of the

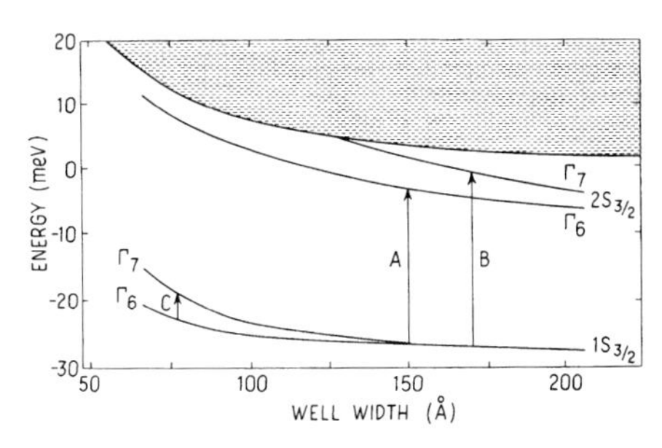

Fig. 4.17. The calculated well-width dependence of the energies of the 1s and 2s-like states of an acceptor at the center of a GaAs/$Al_{0.3}Ga_{0.7}$As quantum well. The *dashed region* denotes the continuum of free-hole energies. From [61]

$1s - 2s$ transition energy of the silicon donor as a function of well width [63]. Similar approaches were used in the case of acceptor studies [73,74].

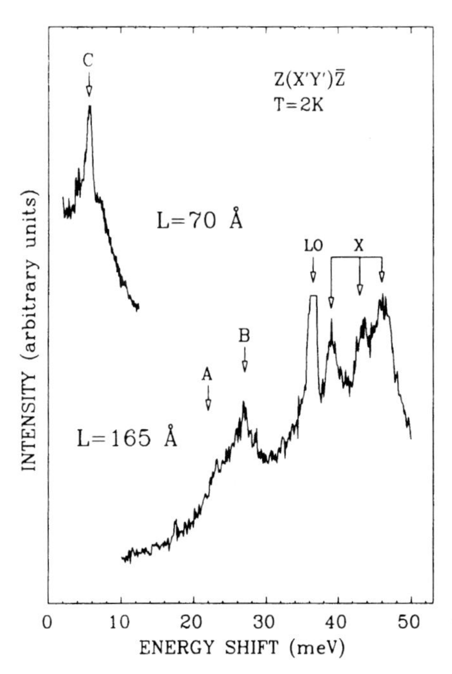

Fig. 4.18. Raman Spectra from two center-doped GaAs/$Al_{0.3}Ga_{0.7}$As quantum well samples with well widths L showing acceptor transitions A, B and C from Fig. 4.17. The features labeled X were not understood. From [92]

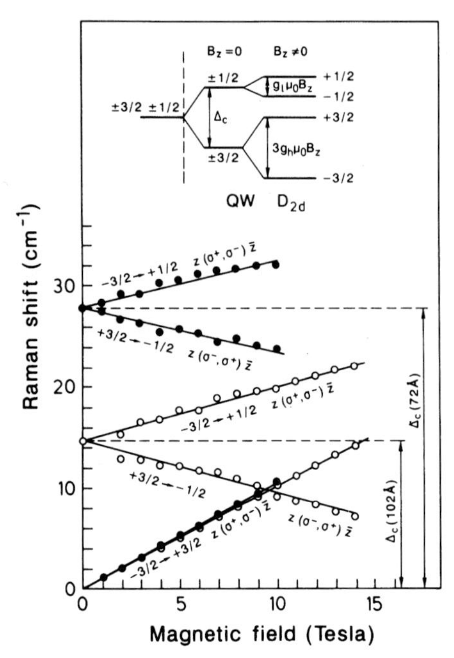

Fig. 4.19. Raman transition energies as a function of magnetic field along the z-axis. Transitions labeled Δc correspond to the C transition of Figs. 4.18 and 4.19 and are between the $\Gamma_7(m = \pm 3/2)$ and $\Gamma_6(m = \pm 1/2)$ acceptor states. Results from two center-doped GaAs/$Al_{0.3}Ga_{0.7}$As quantum well samples with well widths of 10.2 nm and 7.2 nm are shown. The *inset* shows the acceptor energy levels going from bulk to quantum well with and without a magnetic field. From [69]

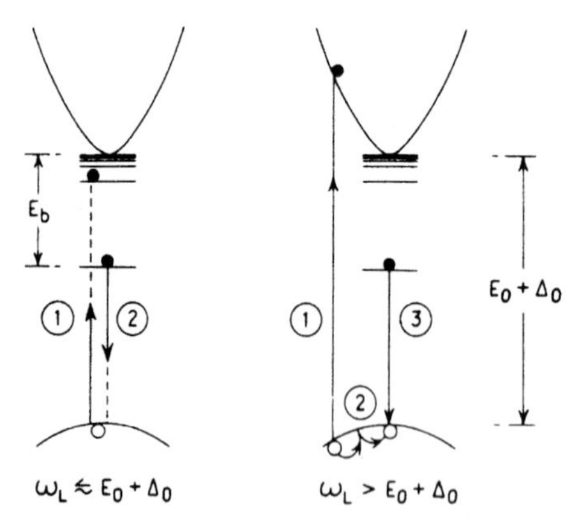

Fig. 4.20. Schematic single particle diagram of conduction and valence subbands and donors levels, showing transitions that lead to Raman scattering (*left*) and photoluminescence (*right*). From [59]

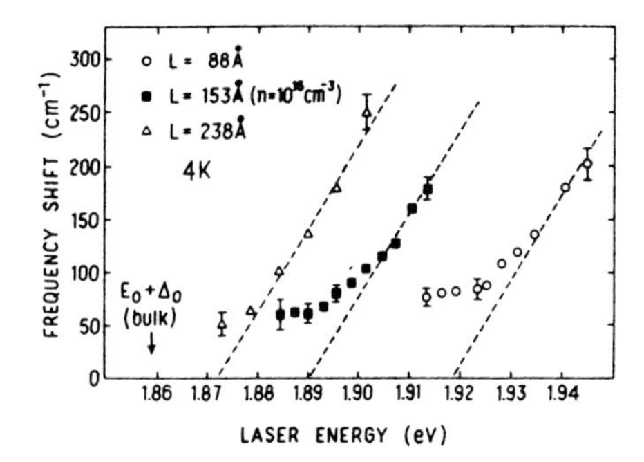

Fig. 4.21. Raman frequency shift of the donor transition as a function of the laser energy for three different samples. The dashed lines with slopes equal to unity give the expected shifts of the photoluminescence. As the laser energy approaches the band gap and the condition of strong resonance the transition becomes Raman-like with a constant energy relative to the laser energy (see Fig. 4.20). From [63]

4.5.2 Quasi-Two-Dimensional Electron Gas

A 20 nm quantum well has several conduction subbands that are separated by several 10's of meV. If the quantum well is modulation-doped with electrons, the first conduction subband is filled up to a 2D Fermi energy of $E_F = 2\pi\hbar^2 n_S/m$, where $m = 0.068m_0$ is the electron effective mass in GaAs. With $n_S = 10^{11}\,\mathrm{cm}^{-2}$, $E_F = 3.6\,\mathrm{meV}$. This degenerate quasi-2D electron gas has a number of excitations whose energy dispersions are shown in Fig. 4.22. Transitions within the first subband from below the Fermi level to empty states above are called intra-subband transitions. Transitions between subbands are called inter-subband transitions. Both of these single-particle transitions have a continuum of transition energies as shown in Fig. 4.22. The spread in single particle energies as a function of wavevector is given by [75,76]

$$0 \le E(k_\varrho) \le \nu_f \hbar k_\varrho + \frac{\hbar^2 k_\varrho^2}{2m} \quad \text{intrasubband}\,, \quad (4.16)$$

$$E_{01} - \nu_f \hbar k_\varrho + \frac{\hbar^2 k_\varrho^2}{2m} \le E(k_\varrho) \le E_{01} + \nu_f \hbar k_\varrho + \frac{\hbar^2 k_\varrho^2}{2m} \quad \text{intersubband}\,, \quad (4.17)$$

where the Fermi velocity is $\nu_f = (\hbar/m)(2\pi n_S)^{1/2}$. The wavevector of the excitation, k_ϱ, is in the plane of the quantum well. If superlattice effects are considered there will also be a z-component of the wavevector (perpendicular to the plane of the quantum well) [55,75].

In Raman scattering the light can couple to the electrons through the electric field associated with their charge density. In this way the spectrum of charge density excitations is measured. The charge density excitation spectrum is also measured in far-infrared absorption. With Raman scattering, however, it also is possible to couple to the excitations of the spin density of

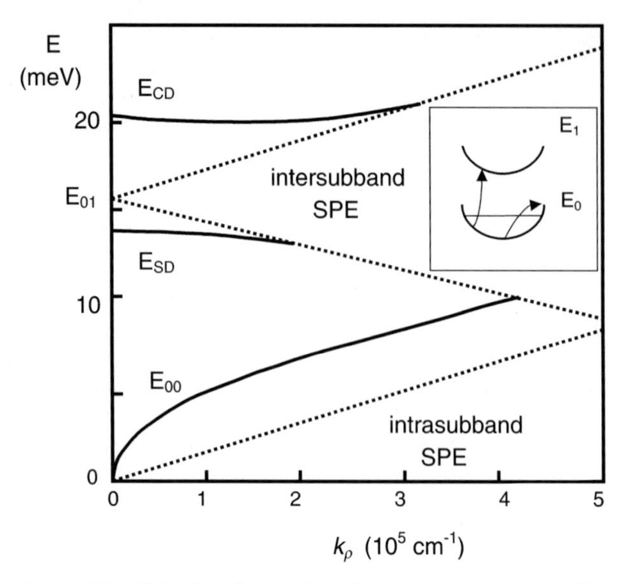

Fig. 4.22. Calculated energies of excitations of quasi-2D electron gas as a function of wavevector (k_ϱ) in the plane of the quantum well for well width of 26 nm and electron density $n_s = 3.6 \times 10^{11}\,\mathrm{cm}^{-2}$. The collective intrasubband charge density excitation (E_{00}), and collective intersubband charge density (E_{CD}) and spin density (E_{SD}) excitations are shown. Dotted lines delimit the continua of single particle excitations (SPE). Adapted from [85]

the electrons in addition to charge density [77]. The spin density excitation involves an excitation in which the two electronic spin components of the electron gas are out of phase so that the total electron density, and therefore the charge density, is constant in time - only the spin density oscillates.

The scattering intensity of the excitations is dependent on the relative polarization of the incident and scattered light. For example, charge density excitations of the electron gas are allowed only when the polarization of the incident and scattered light are parallel to each other. In contrast spin density excitations are allowed only with polarizations perpendicular to each other [28,76].

The number of conduction band electrons is very small, and the vast majority of Raman measurements of electronic excitations in quantum wells have been performed with the laser in resonance with a band gap to obtain sufficient signal. Usually this is done with excitonic states associated with the E_0 or the $E_0 + \Delta_0$ gap.

Inter-Subband Excitations. The excitation spectrum of the electron gas can be calculated from the imaginary part of its response function [28,78]:

$$I_j \propto \mathrm{Im}\chi_j(k, \omega). \tag{4.18}$$

The index j refers to charge or spin density response. Consider only the case of the intersubband transition between the first and second subband, and ignore other subbands. If electron-electron interactions are neglected the response is given by [79,80]

$$\chi_0(k_\varrho,\omega) = D_{10}(k_\varrho,\omega) + D_{01}(k_\varrho,\omega) \tag{4.19}$$

$$D_{\mathrm{i}j} = 2\sum_{k=0}^{\infty} \frac{f_j(E_j) - f_i(E_i)}{E_j(k+k_\varrho) - E_{\mathrm{i}}(k_\varrho) - \hbar(\omega + \mathrm{i}\Gamma)}. \tag{4.20}$$

Here $f(E)$ is the Fermi-Dirac distribution function and Γ is a phenomenological broadening parameter. This produces a single band of single particle excitations (SPE) peaked and centered at $E_{01} + \hbar^2 k_\varrho^2/2m$. At $k_\varrho = 0$ the intersubband SPE for all electrons in the first subband is at E_{01}. In this case the response function can be written (for $\Gamma = 0$),

$$\chi_0(k_\varrho = 0,\omega) = \frac{2n_S E_{01}}{E^2 - E_{01}^2}. \tag{4.21}$$

Coulomb interactions between electrons lead to collective modes of excitation in addition to single particle transitions. The direct Coulomb interaction leads to oscillations of the charge density (called charge-density waves or plasmons), and the exchange Coulomb interaction leads to oscillations of the spin density (called spin-density waves or spin waves) [81]. The response given by (4.18) is modified when electron interactions are considered. A simple method for incorporating interactions is through the generalized random phase approximation, which includes the exchange Coulomb interaction (β_{01}) in addition to the direct Coulomb interaction (α_{01}) [82]. The response function becomes [81]

$$\chi_j(k_\varrho,\omega) = \frac{\chi_0(k_\varrho,\omega)}{1 - \gamma_j(k_\varrho)\chi_0(k_\varrho,\omega)}. \tag{4.22}$$

The $\gamma_j(q)$ describe the Coulomb interactions:

$$\begin{aligned}\gamma_{CD}(0) &= \frac{\alpha_{01}}{\varepsilon(\omega)} - \beta_{01},\\ \gamma_{SD}(0) &= -\beta_{01}.\end{aligned} \tag{4.23}$$

The charge density excitations interact through the direct term, α_{01}, although it is reduced by the exchange term, β_{01}. The interaction potentials are calculated using the electron envelope functions in the quantum well. The envelope functions are found by solving Schroedinger's equation self-consistently with Poisson's equation to account for the Coulomb potential. Exchange is included through the local density approximation. This calculation for the envelope functions accounts not only for the square well potential but also for the static Coulomb interaction between the electrons and between the electrons and the ionized donors in the barrier. Using these envelope functions, $\varphi_i(z)$, the interaction parameters are given by

$$\alpha_{01} = \frac{2\pi e^2}{\varepsilon_\infty k_z} \int \mathrm{d}z \int \mathrm{d}z' \varphi_0(z)\varphi_1(z) e^{-k_z|z-z'|} \varphi_0(z')\varphi_1(z') \tag{4.24}$$

and

$$\beta_{01} = \int \mathrm{d}z U(z) \phi_0^2(z) \phi_1^2(z), \tag{4.25}$$

where $U(z) \approx 1.7 a_B^3 r_S^2$ is the derivative of the exchange potential given in units of the effective Rydberg (6 meV in GaAs), $r_s(z) = [4\pi a_B^3 n_S \phi_0^2(z)/3]^{-1/3}$, and a_B is the effective Bohr radius (10 nm in GaAs) [83].

The direct Coulomb interaction is screened by LO phonons through the phonon's contribution to the dielectric function,

$$\varepsilon(\omega) = \frac{E^2 - E_{LO}^2}{E^2 - E_{TO}^2}. \tag{4.26}$$

This phonon screening becomes resonant if the energy of the plasmon is close to that of the LO phonon, and leads to plasmon-phonon coupled modes. The zeros of the denominator of (4.22) give the energies of the collective excitations of the electron gas associated with the transition between the first two subbands. In addition there are still single particle excitations but they are now strongly "screened" by the denominator and much weaker (Fig. 4.23). At $k_\varrho = 0$, the energy of the collective intersubband charge density excitation (CDE) is given by

$$E_{CD}^2 = E_{01}^2 + 2n_s E_{01} \left(\frac{\alpha_{01}}{\varepsilon(\omega) - \beta_{01}} \right), \tag{4.27}$$

and the energy of the collective intersubband spin density excitation (SDE) is given by

$$E_{SD}^2 = E_{01}^2 - 2n_s E_{01} \beta_{01}. \tag{4.28}$$

The CDE (inter-subband plasmon) is shifted up by the direct Coulomb interaction, α_{01}, which is screened by LO phonons through the phonon's dielectric function. It is reduced somewhat by the exchange interaction, β_{01}. The SDE is shifted down from the bare intersubband excitation energy by the exchange interaction.

The inter-subband excitations correspond to a charge- or spin-density that is oscillating back and forth in the z-direction. At $k_\varrho = 0$ all the electrons have single particle energies equal to E_{01}. The collective charge-density excitations (inter-subband plasmons) are shifted up in energy. This increase in

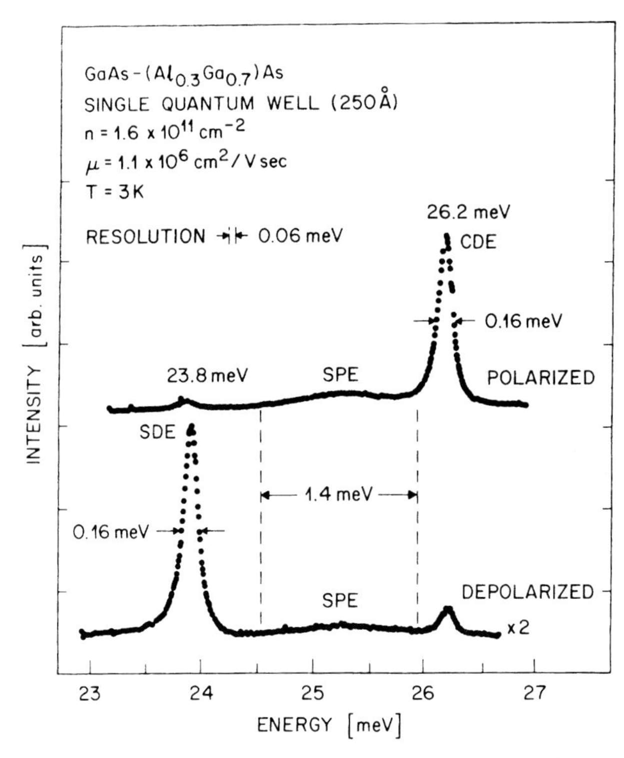

Fig. 4.23. Raman spectra of the charge density and spin density excitations. Collective charge density (CDE), spin density (SDE) and single particle excitations (SPE) are observed. From [81]

frequency can be understood as the additional restoring force arising from the electric field (depolarization field) associated with the oscillating charge density. This depolarization field can be understood intuitively as arising from depolarization surface charges at the interfaces of the quantum well.

The Coulomb interactions can be measured directly from the spectra,

$$2n\alpha_{01} = E_{CD}^2 - \frac{E_{SD}^2}{E_{01}} \,, \tag{4.29}$$

$$2n\beta_{01} = E_{01}^2 - \frac{E_{SD}^2}{E_{01}} \,. \tag{4.30}$$

The magnitudes of the Coulomb interactions have been measured for the electron intersubband transitions in GaAs/AlGaAs quantum wells modulation-doped asymmetrically on one side. These experimental results were compared with detailed calculations using the random phase approximation combined with the local density approximation for the exchange interactions, and also a nonlocal theory involving a variational solution of the Bethe-Salpeter equation for the density-density correlation function [84,85]. The results are reproduced in Fig. 4.24.

By rotating the sample with respect to the incident and detected wavevector of the light, the dispersion of the excitations can be measured up to a maximum wavevector transfer of $k \approx 2 \times 10^5\,\mathrm{cm}^{-1}$. In this way it is possible to observe broadening of the SPE and shifts of the collective excitations as expected from the dispersion diagram shown in Fig. 4.23. When the collective modes shift into resonance with the continuum of SPE, the collective modes become Landau-damped and broaden out.

The above model can be extended to include nonparabolicity of the electron subbands [86] and applied magnetic fields [87]. Raman scattering has been used with great success to study the excited states of the electron gas in the quantum Hall regime [88].

The discussion given above for the scattering of light by the electron gas was based on the random phase approximation, which has been able to explain most aspects of the data. There are some discrepancies though. One long-standing problem between this theory and measurement concerns the relative intensities of the single particle excitations compared to the collective excitations in the spectra [81]. Measurements have demonstrated that the SPE are orders of magnitude stronger than expected from (4.18) and (4.22). One possible explanation given for this problem was that disorder in the system leads to mixing of high k-vector excitations into the low k-vector excitations probed by Raman scattering [81]. However this theory is clearly too limited because it ignores the presence of the valence band even in the

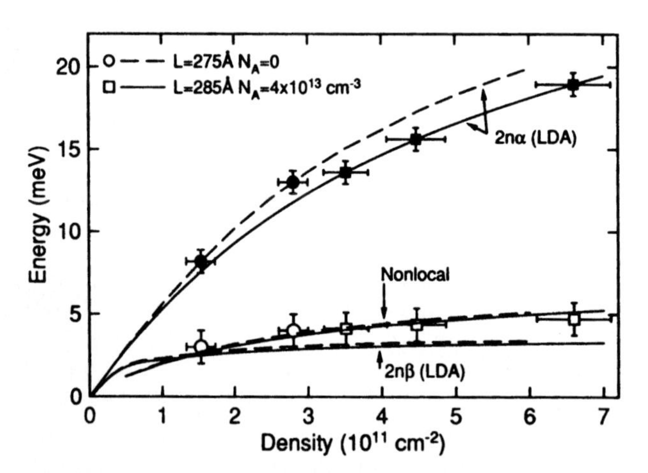

Fig. 4.24. Comparison between measured (*symbols*) and calculated (*lines*) values of the direct (α) and exchange parameters (β). For this comparison wavefunctions were calculated with well widths (L) and background acceptor concentrations (N_A) obtained by fitting $2n\alpha$. In this way a good test of the models for calculating the exchange interaction were made. From [84]

case of strong resonance. Recently an improved theory has been developed which accounts for the valence band in the resonant Raman scattering process and can quantitatively explain the strong SPE scattering [93].

Intra-Subband Excitations. For a single quantum well, intra-subband excitations involve electron density oscillating within the plane of the quantum well. As $k_\varrho \to 0$, the single-particle transition energies go to zero (see Fig. 4.22). In addition, the energy of the collective intra-subband excitation also goes to zero because the depolarization field is zero for intra-subband excitations at $k_\varrho = 0$. This can be understood intuitively from the surface depolarization charges at the edges of the quantum well arising from the $k_\varrho = 0$ charge oscillation. The electric field between two thin strips of charge is zero if they are far apart. Away from $k_\varrho = 0$ an intra-subband plasmon is set up, which for a single quantum well is 2D-like with an energy given by

$$E_{00}^{2D}(k_\varrho) = \hbar \left(\frac{2\pi n_s e^2}{\varepsilon_0 m} k_\varrho \right)^{1/2} . \tag{4.31}$$

If there is more than one quantum well, the intra-subband response is modified. For an infinite stack of quantum wells the intra-subband plasmon energy becomes

$$E_{00}(k_\varrho, k_z) = E_{00}^{2D}(k_\varrho) S(k_\varrho, k_z)^{1/2} . \tag{4.32}$$

The superlattice structure factor, $S(k_\varrho, k_z)$, is given by

$$S(k_\varrho, k_z) = \frac{\sinh(k_\varrho d)}{\cosh(k_\varrho d) - \cos(k_z d)} . \tag{4.33}$$

When $k_\varrho d \gg 1$ the distance between planes is large relative to the wavelength and the plasmons are not coupled. In this case $S(k_\varrho, k_z) \approx 1$ and the plasmon reduces to the 2D plasmon. In the limit $k_\varrho d \ll 1$ the intra-subband plasmons of different quantum wells are coupled. At $k_z d = 0$ the plasmons within each quantum well are in phase and the intra-subband plasmon becomes 3D-like with a nonzero energy at $k_\varrho d = 0$.

$$E_{00}(0,0) = \hbar \left(\frac{4\pi n_S e^2}{d \varepsilon_0 m} \right)^{1/2} \tag{4.34}$$

In the limit $k_\varrho d \ll 1$ with $k_z d \neq 0$ the intra-subband plasmon is neither 2D-like nor 3D-like. It becomes acoustic-like with linear dispersion given by

$$E_{00}(k_\varrho, k_z) = \hbar \nu(k_z) k_\varrho , \tag{4.35}$$

with

$$\nu(k_z) = \left(\frac{2\pi n_S e^2}{\varepsilon_0 m} \frac{d}{1 - \cos(k_z d)} \right)^{1/2} . \tag{4.36}$$

4.6 Conclusion

There has been enormous effort to understand the physics of semiconductor heterostructures over the last three decades. In this research Raman scattering has played an important role in the study of the vibrational and low energy electronic excitations of these lower dimensional structures. It has been the primary technique for studying phonons and has been very important in the study of the excitations of the 2D electron gas in semiconductor quantum wells.

The vast majority of this research has been carried out on planar quantum well structures. In this chapter we have focused exclusively on [001] $GaAs/Al_xGa_{1-x}As$ quantum wells for the sake of clarity and because it has been in many ways the prototypical quantum well system. But the range of material systems studied is very broad. One important class of materials not considered here is the strained layer heterostructures. Because strain will shift the optical phonon energies, Raman scattering can provide a characterization of strain that may be useful in structures too small to measure with other techniques. This capability is reviewed in [25].

Many Raman scattering studies have been carried out with laser frequencies in near resonance with the electronic band gap of the semiconductor to take advantage of the large resonant enhancements in scattering intensities. This trend will certainly continue in the future as researchers focus more on quantum wires and dots. Often resonant Raman scattering has been necessary to obtain sufficient signal-to-noise ratios, or to selectively measure a subset of the overall sample. However, resonant Raman scattering also probes the electronic behavior of the semiconductor at the frequency of the laser in addition to providing the spectra of the low energy excitations themselves.

References

1. G. Bastard: *Wave Mechanics Applied to Semiconductor Heterojunctions*, (Les Editions de Physique, Les Ulis 1988)
2. C. Weisbuch, B. Vinter: *Quantum Semiconductor Structures, Fundamentals and Applications* (Academic, San Diego, CA 1991)
3. J. Singh: *Physics of Semiconductors and their Heterostructures*, (McGraw-Hill, New York, 1993)
4. E.L. Ivchenko, G. Pikus: *Superlattices and Other Heterojunctions: Symmetry and Other Optical Properties*, Springer Ser. Solid State Sci., Vol 110 (Springer, Berlin, Heidelberg 1995).
5. P.Y. Yu, M. Cardona: *Fundamentals of Semiconductors: Physics and Materials Properties* (Springer, Berlin, Heidelberg 1996) p. 480
6. M.A. Herman, H. Sitter: *Molecular Beam Epitaxy*, 2nd ed., Springer Ser. Mater. Sci., Vol 7 (Springer, Berlin, Heidelberg 1996)
7. M.V. Klein: IEEE J Quantum Electron. **22**, 1760 (1986)
8. J. Menéndez: J. Lumin. (Netherlands) **44**, 285 (1989)

9. D.J. Lockwood, J.F. Young (eds.): NATA ASI Ser. B , Phys. (USA Vol. 273 (1990)
10. G. Abstreiter, R. Merlin, A. Pinczuk: IEEE J. Quantum Electron. **22**, 1771 (1986)
11. *Light scattering in Solids V: Superlattices and other Microstructures*, ed. by M. Cardona, G. Güntherodt, Topics Appl. Phys. Vol. 66 (Springer, Berlin, Heidelberg 1989)
12. T. Ruf: *Phonon Scattering in Semiconductors, Quantum Wells and Superlattices: Basic Results and Applications* (Springer, Berlin Heidelberg 1998)
13. L.E. Brus, A.L. Efros, T. Itoh: *Special Issue on Spectroscopy of Isolated and Assembled Semiconductor Nanocrystals*, J. Lumin. **70** (V70) (1996)
14. J.C. Tsang: in [11], p. 233.
15. G. Fasol, M. Tanaka, H. Sakaki, Y. Horikoshi: Phys. Rev. B **38**, 6056 (1988)
16. D. Gammon, B.V. Shanabrook, D.S. Katzer: Phys. Rev. Lett. **67**, 1547 (1991)
17. D. Gammon, S.W. Brown, E.S. Snow, T.A. Kennedy, D.S. Katzer, D. Park: Science **277**, 85 (1997)
18. A. Pinczuk, E. Burstein: *Fundamentals of Inelastic Light Scattering in Semiconductors and Insulators*, in *Light Scattering in Solids*, ed. by M. Cardona, Topics Appl. Phys. Vol.8 2nd ed. (Springer, Berlin, Heidelberg 1983) p. 23
19. A.J. Shields, M. Cardona, R. Notzel, K. Ploog: Phys. Rev. B **46**, 6990 (1992)
20. A.J. Shields, M. Cardona, R. Notzel, K. Ploog: Phys. Rev. B **46**, 10490 (1992)
21. J.E. Zucker, A. Pinczuk, D.S. Chemla, A. Gossard, W. Wiegmann: Phys. Rev. Lett. **51**, 1293 (1983) ; Phys. Rev. B **29**, 7065 (1984)
22. D.A. Kleinman, R.C. Miller, A.C. Gossard: Phys. Rev. B **35**, 664 (1987)
23. A. Alexandrau, M. Cardona, K. Ploog: Phys. Rev. B **38**, 2196 (1988)
24. D. Gammon, B.V. Shanabrook, D.S. Katzer: Phys. Rev. Lett. **67**, 1547 (1991)
25. B. Jusserand, M. Cardona: in *Light Scattering in Solids V*, ed. by M. Cardona, G. Güntherodt, Topics Appl. Phys. Vol.66 (Springer, Berlin, Heidelberg 1989) p. 49
26. W. Kaushke, A.K. Sood, M. Cardona, K. Ploog: Phys. Rev. B **36**, 1612 (1987)
27. G. Danan, A. Pinczuk, J.P. Valladares, L.N. Pfeiffer, K.W. West: Phys. Rev. B **39**, 5512 (1989)
28. M.V. Klein: in *Light Scattering in Solids*, ed. by M. Cardona, Topics Appl. Phys. Vol.8 2nd ed. (Springer, Berlin, Heidelberg 1983) p. 147
29. R. Merlin, C. Colvard, M.V. Klein, H. Morkoc, A.Y. Cho, A.C. Gossard: Appl. Phys. Lett. **36**, 43 (1980)
30. G. Scamarcio, M. Haines, G. Abstreiter, E. Molinari, S. Baroni, A. Fischer, K. Ploog: Phys. Rev. B **47**, 1483 (1993)
31. M. Zunke, R. Schorer, G. Abstreiter, W. Klein, G. Weimann, M.P. Chamberlain: Solid State Commun. **93**, 847 (1995)
32. C. Colvard, T.A. Gand, M.V. Klein, R. Merlin, R. Fischer, H. Morkoc, A.C. Gossard: Phys. Rev. B **31**, 2080 (1985)
33. E. Molinari, S. Baroni, P. Giannozzi, S. De Gironcoli: Phys. Rev. B **45**, 4280 (1992)
34. M.P. Chamberlain, M. Cardona, B.K. Ridley: Phys. Rev. B **48**, 14356 (1993)
35. B. Jusserand, F. Mollot, R. Planel, E. Molinari, S. Baroni: Surf. Science **267**, 171 (1992)
36. B. Jusserand, F. Alexandre, D. Paquet, G. LeRoux: Appl. Phys. Lett. **47**, 301 (1985)

37. D. Levi, Shu-Lin Zhang, M.V. Klein, J. Klem, H. Morkoc: Phys. Rev. B **36**, 8032 (1987)
38. D. Gammon, B.V. Shanabrook, D.S. Katzer: Appl. Phys. Lett. **57**, 2710 (1990)
39. A. Zrenner et al.: Phys. Rev. Lett. **72**, 3382 (1994)
40. A. Ourmazd, D.W. Taylor, J. Cunningham, C.W. Tu: Phys. Rev. Lett. **62**, 933 (1989)
41. C.A. Warwick, W.Y. Jan, A. Ourmazd, T.D. Harris: Appl. Phys. Lett. **56**, 2666 (1990)
42. J.M. Moison, C. Guille, F. Houzay, F. Barthe, M. Van Rompay: Phys. Rev. B **40**, 6149 (1991)
43. D.S. Katzer, D. Gammon, B.V. Shanabrook: J. Vac. Sci. Technol. B **12**, 1056 (1994)
44. D. Gammon, B.V. Shanabrook, D.S. Katzer: in *Semiconductor Interfaces and Microstructures*, ed. by Z.C. Feng (World Scientific, Singapore 1992) p.149.
45. H.F. Hess, E. Betzig, T.D. Harris, L.N. Pfeiffer, K.W. West: Science **264**, 1740 (1994)
46. D. Gammon, E.S. Snow, B.V. Shanabrook, D.S. Katzer, D. Park: Phys. Rev. Lett. **76**, 3005 (1996)
47. D. Gammon, E.S. Snow, B.V. Katzer, D. Park: Science **273**, 87 (1996)
48. D. Gammon: MRS Bulletin **23**, 44 (1998)
49. A.K. Sood, M. Menéndez, M. Cardona, K. Ploog: Phys. Rev. Lett. **54**, 2115 (1985)
50. A.J. Shields, M. Cardona, K. Eberl: Phys. Rev. Lett. **72**, 412 (1994)
51. A.J. Shields, M.P. Chamberlain, M. Cardona, K. Eberl: Phys. Rev. B **51**, 17728 (1995)
52. T. Ruf, V.I. Belitsky, J. Spitzer, V.F. Sapega,, M. Cardona, K. Ploog: Phys. Rev. Lett. **71**, 3035 (1993)
53. T. Ruf, J. Spitzer, V.F. Sapega,, V.I. Belitsky, M. Cardona, K. Ploog: Phys. Rev. B **50**, 1792 (1994)
54. B.V. Shanabrook: Physica **146B**, 121 (1987).
55. A. Pinczuk, G. Abstreiter: in [11], p.153.
56. R.C. Miller, A.C. Gossard, W.T. Tsang, O Munteanu: Phys. Rev. B **25**, 3871 (1982)
57. B.V. Shanabrook, J. Comas: Surf. Sci. **142**, 504 (1984)
58. N.C. Jarosik, B.D. McCombe, B.V. Shanabrook, J. Comas: Phys. Rev. Lett. **54**, 1283 (1985)
59. B.V. Shanabrook, T. Comas, T.A. Perry, R. Merlin: Phys. Rev. B **29**, 7096 (1984)
60. T.A. Perry, R. Merlin, B.V. Shanabrook, T. Comas: Phys. Rev Lett. **54**, 2623 (1985)
61. D. Gammon, R. Merlin, W.T. Masselink, H. Morkoc: Phys. Rev. B **33**, 2919 (1986)
62. G. Bastard: Phys. Rev. B **24**, 4714 (1981)
63. B.V. Shanabrook, J. Comas, T.A. Perry, R. Merlin: Phys. Rev. B **29**, 7096 (1984)
64. T.A. Perry, R. Merlin, B.V. Shanabrook, J. Comas: Phys. Rev. Lett. **54**, 2623 (1985)
65. N.C. Jarosik, B.D. McCombe, B.V. Shanabrook, J. Comas, J. Ralston, G. Wicks: Phys. Rev. Lett. **54**, 1283 (1985)

66. C. Mailhiot, Y.C. Chang, T.C. McGill: Phys. Rev. B **26**, 4449 (1982)
67. R.L. Greene, K.K. Bajaj: Solid State Commun. **45**, 825 (1983)
68. V.F. Sapega, M. Cardona, K. Ploog, E.L. Ivchenko, D.N. Merlin: Phys. Rev. B **45**, 4320 (1992)
69. V.F. Sapega, T. Ruf, M. Cardona, K. Ploog, E. L Ivchenko, D.N. Merlin: Phys. Rev. B **50**, 2510 (1994)
70. W.T. Masselink, Y.C. Chang, H. Morkoc: Phys. Rev. B **32**, 5190 (1985)
71. A. Pasquarello, L.C. Andreani, R. Buczko: Phys. Rev. B **40**, 5602 (1989)
72. K. Wan, R. Bray: Phys. Rev. B **32**, 5265 (1985)
73. P.O. Holtz, M. Sundaram, R. Simes, J.L. Merz, A.C. Gossard, J.H. English: Phys. Rev. B **39**, 13293 (1989)
74. G.C. Rune, P.O. Holtz, M. Sundaram, J.L. Merz, A.C. Gossard, B. Monemar: Phys. Rev. B **44**, 4010 (1991)
75. A. Pinczuk, G. Abstreiter: in [11], p.153.
76. G. Abstreiter, M. Cardona, A. Pinczuk: in *Light scattering in Solids IV: Introductory Concepts*, ed. by M. Cardona, Topics Appl. Phys. Vol.54 (Springer, Berlin, Heidelberg 1984) p. 5.
77. M.V. Klein: in *Light scattering in Solids I: Introductory Concepts*, ed. by M. Cardona, Topics Appl. Phys. Vol.8 2nd ed. (Springer, Berlin, Heidelberg 1983) p. 147.
78. P.M. Platzman, P.A. Wolff: *Waves and Interactions in Solid State Plasmas*, (Academic Press, New York 1973).
79. J.K. Jain, S. Das Sarma: Phys. Rev. B **36**, 5949 (1987)
80. A.C. Tselis, J.J. Quinn: Phys. Rev. B **29**, 3318 (1984)
81. A. Pinczuk, S. Schmitt-Rink, G. Danan, J.P. Valladares, L.N. Pfeiffer, K.W. West: Phys. Rev. Lett. **63**, 1633 (1989)
82. T. Ando: Solid State Commun. **21**, 133 (1977)
83. D. Gammon, B.V. Shanabrook, J.C. Ryan, D.S. Katzer: Phys. Rev. B **41**, 12311 (1990)
84. D. Gammon, B.V. Shanabrook, J.C. Ryan, D.S. Katzer, M.J. Yang: Phys. Rev. Lett. **68**, 1884 (1992)
85. J.C. Ryan: Phys. Rev. B. **43**, 4499 (1991)
86. G. Brozak, B.V. Shanabrook, D. Gammon, D.A. Broido, R. Beresford, W.I. Wang: Phys. Rev. B **45**, 11399 (1992)
87. G. Brozak, B.V. Shanabrook, D. Gammon, D.S. Katzer: Phys. Rev. B **47**, 9981 (1993)
88. See, for example, M.A. Eriksson, A. Pinczuk, B.S. Dennis, S.H. Simon, L.N. Pfeiffer, K.W. West: Phys. Rev. Lett. **82**, 2163 (1999)
89. H. Rücker, E. Molinari, P. Lugli: Phys. Rev. B **45**, 6747 (1992)
90. D.J. Mowbrey, M. Cardona, K. Ploog: Phys. Rev. B **43**, 1598 (1991)
91. D. Gammon, E.S. Snow, D.S. Katzer: Appl. Phys. Lett. **67**, 2391 (1995)
92. D. Gammon, R. Merlin, D. Huang, H. Morkoc: J. Crystal Growth **81**, 149 (1987)
93. S. Das Sarma, Daw-Wei Wang: Phys. Rev. Lett. **83**, 816 (1999)

IV Raman Scattering Enhancement by Optical Confinement Semiconductor Planar Microcavities

B. Jusserand and A. Fainstein

After some years of studies in the domain of atomic physics, optical microcavities have received much attention recently in semiconductor science [1]. An example of such high finesse structures consists of quantum wells embedded in a λ cavity spacer between two Bragg mirrors made of alternating quarter-wavelength GaAlAs and AlAs layers. The strong confinement of the optical field near such light emitting dipole significantly modifies the intrinsic emission rate, opening new possibilities in terms of optical devices. Nonlinear interactions between light and matter can also be significantly enhanced due to the light confinement. In fact, a double *optical* Raman scattering resonance configuration can be attained in a semiconductor microcavity [2]. Taking advantage of the angular dependence of the frequency of the cavity mode in a planar microcavity, the incident and scattered frequencies can be tuned to this resonance at two different angles and their frequency difference to a Raman active vibration in the material.

In Fig. IV.1 we show the Raman spectra obtained when this optical double resonance is tuned through the LO-phonon spectral region corresponding to three InAs QWs embedded in a microcavity. Thanks to the high spectral selectivity and strong enhancement (up to 10^4), several narrow lines are discerned. These lines correspond to interface optical phonons of the whole stratified structure, including both QWs and mirror layers. Because of the finite size and complex arrangement of the stack, the interface modes are discretized and surface modes are observed. The modifications of the Raman efficiency can be quantitatively described based on Maxwell equations in stratified media and they directly reflect the finesse of the cavity mode [3]. In turn, the phonon spectra can be reproduced using a transfer matrix treatment of the interface phonon dielectric model [4].

In a microcavity with a multiple quantum well fully filling the $3\lambda/2$ spacer we have also observed folded acoustic phonons in experimental conditions where they cannot be usually observed (80 K, well below the band gap). In addition, due to internal reflections, contributions of both backward and forward scattering are simultaneously recorded, and the in-plane dispersion can be more easily derived. These results illustrate the interest of cavity-enhanced Raman scattering to selectively extract contributions of weak scatterers in semiconductor heterostructures.

Besides the above application for Raman efficiency enhancement, microcavities have enabled a renewed perspective into fundamental aspects of

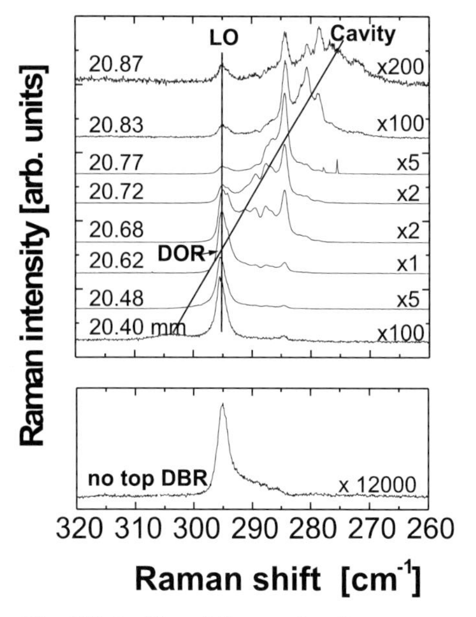

Fig. IV.1. *Top:* Microcavity Raman spectra as a function of spot position (i.e., cavity mode energy). It shows the selective enhancement of particular features of the phonon Raman spectra that can be obtained by detailed tuning of the cavity mode. The crossing of the solid lines denotes the exact double optical resonance (DOR) for the chosen incidence angle ($\sim 50°$). *Bottom:* Basically non-amplified spectrum obtained in a half-cavity (without top mirror). It highlights the enhancement of over four orders of magnitude, and the observation of fine spectral details, in the full cavity case

the light-matter interaction in semiconductors. When the cavity mode approaches the exciton transition in the cavity medium, strong coupling can appear and cavity polaritons become the correct description of the coupled photon–exciton system, with two branches, separated by the so-called Rabi splitting [1]. When Raman scattering is performed close to these energies, modifications of the scattering cross sections are expected, which can be described by first order perturbation in the polariton–phonon interaction. In bulk semiconductors, there has been no conclusive evidence of the bulk polariton mediation of first-order Raman scattering. In microcavities, with Rabi splittings of 5 to 10 meV, we have observed specific signatures of the polariton mediation (see Fig. IV.2). The resonant Raman cross section can be simply expressed within the perturbation theory as $S_p^i S_x^i S_p^s S_x^s$, where $S_{p,x}^{i,s}$ represent the photonic or excitonic strength of the incident or scattered polaritons, respectively [5,6].

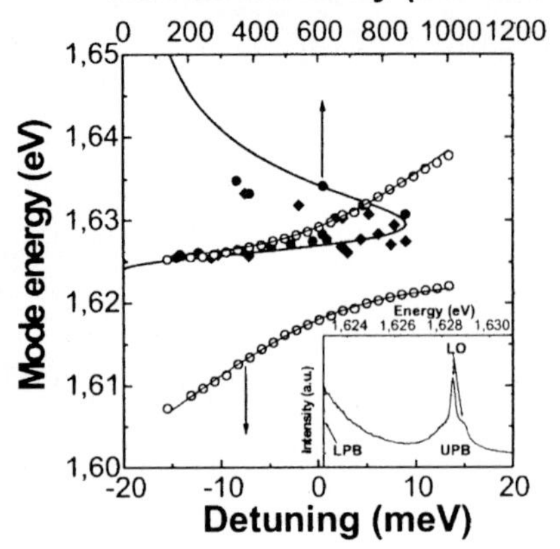

Fig. IV.2. *Bottom axis:* Microcavity coupled exciton-cavity-mode energies (*open circles*). The *thin solid curve* is a fit with a two-coupled-modes Hamiltonian. *Top axis:* Resonant Raman intensity as a function of detuning for two LO-phonon lines (*solid circles* and *squares*). The *thick solid curve* is a calculation based on the cavity-polariton mediated scattering model [6]. *Inset:* LO phonon RRS spectrum tuned for outgoing resonance with the upper polariton branch (UPB). The LO phonon is the narrow line on top of the broader UPB luminescence. The second LO line can also be seen slightly out of resonance. The high-energy tail of the lower polariton band (LPB) is also visible

References

1. Confined Electrons and Photons: New Physics and Applications, ed. by E. Burstein, C. Weisbuch (Plenum, New York 1995)
2. A. Fainstein, B. Jusserand, V. Thierry Mieg: Phys. Rev. Lett. **75**, 3764 (1996)
3. A. Fainstein, B. Jusserand: Phys. Rev. B **57**, 2402 (1998)
4. A. Fainstein, B. Jusserand: Phys. Rev. B **54**, 11505 (1996)
5. A. Fainstein, B. Jusserand, V. Thierry Mieg: Phys. Rev. Lett. **78**, 1576 (1997)
6. A. Fainstein, B. Jusserand, R. André: Phys. Rev. B, Rapid Comm. **57**, R9439 (1998)

5 Raman Scattering in High-T_c Superconductors: Phonons, Electrons, and Magnons

M. Cardona

It was realized very soon after the discovery of high-T_c superconductors that Raman spectroscopy is an excellent technique for the investigation of low energy elementary excitations in these materials and also for the characterization of their structural, electronic and vibronic properties. The primitive cell of high-T_c superconductors is usually centrosymmetric and contains a large number of atoms. While this large number of atoms makes phonon spectra rather complex, the existence of an inversion center allows us to classify the phonons into odd and even with respect to the inversion. Odd phonons are Raman forbidden while even phonons are infrared (IR) forbidden. This feature is of considerable help for the assignment of the various phonon related peaks observed in the Raman and also in the IR spectra (because of the nature of this treatise we shall only discuss IR spectra when needed for the interpretation of phenomena related to Raman spectra). Forbidden phonons may become allowed in a spectrum due to the presence of lattice defects that lower the symmetry around the atoms participating in the vibrations of the phonon under consideration. This often happens in high-T_c superconductors as a result of the nonstoichiometric nature of the oxygen composition, which is related to the concentration of superconducting carriers (holes). Although the even phonons (requiring the involvement of at least two equivalent atoms in the mode being considered) can be Raman active, not all of them are. Additional selection rules specific to the crystal point group can also reduce the number of Raman active even phonons (this reduction is more drastic in the case of IR spectra, since only those symmetries that correspond to the three components of the electric field can lead to dipole allowed transitions).

Besides the selection rules just mentioned, related to the crystallographic point group, the translational symmetry of the crystals requires conservation of the wavevector $\boldsymbol{k}$ in an absorption or scattering process. In the optical region, the wavelength of the photons is very small compared with the size of the primitive cell and, consequently, only phonons with $\boldsymbol{k} \simeq 0$ (i.e. near the center of the Brillouin zone) can participate in first-order (i.e. involving only one phonon) optical processes. Optical spectroscopies are therefore not suitable for the determination of the full phonon dispersion relations (i.e. the frequency ω versus $\boldsymbol{k}$). This is the typical domain of inelastic neutron scattering which, however, requires large single crystals often not available for high-T_c superconductors. Although inelastic neutron scattering experiments have been performed for high-T_c superconductors, most of the experimental knowledge available for these materials originates from their Raman and IR spectra. The recently developed technique of Raman scattering of x-rays obtained by strong monochromatization of synchrotron radiation also allows the scanning of the full Brillouin zone [1]. However, it has not yet been applied to high-T_c superconductors.

Raman scattering can also be used to investigate low frequency electronic excitations. It was one of the first spectroscopic techniques that revealed the existence of a gap for electronic excitations in the superconducting phase of high-T_c superconductors [2]. Soon after, it was realized that gaps observed in the electronic Raman spectra exhibit different features depending on the symmetry of the polarization configuration used for the scattering measurements when the Raman shift tends to zero [3,4]. The corresponding Raman spectra provided some of the early hints as to the anisotropic nature of the gap and its $(k_x^2 - k_y^2)$-type of symmetry.

The electronic excitations responsible for the observed Raman gap couple noticeably to some of the Raman active phonons via electron–phonon interaction, a fact that results in changes in phonon self-energies and spectral strengths when crossing T_c. Particularly strong effects that reveal a remarkably large electron–phonon interaction have been recently observed for Hg1234 and the isomorphic material (CuC)-1234 [5,6]. Raman spectroscopy thus provides rather direct evidence of strong electron–phonon interaction in high-T_c superconductors. We should keep in mind, however, that only $\boldsymbol{k} = 0$ phonons are accessible to Raman spectroscopy and, therefore, the fact that a few of them show evidence of strong electron–phonon coupling cannot be construed to imply that the average interaction of electrons with *all* phonons, responsible for the superconductivity in the conventional BCS theory, is strong and a possible candidate for the pairing mechanism in high-T_c superconductors.

Another type of electronic scattering has been observed and extensively investigated in the high-T_c superconductors that contain rare earth elements, in particular those of the Nd_2CuO_4 family which are believed to be n-type conductors in their normal state. These rare earth ions possess an unfilled f-shell whose electrons occupy strongly correlated many-body states split, even in the free ion, by Coulomb and spin–orbit interaction. The crystalline electric field lowers the symmetry of the free ion levels, producing what are known as crystal field states. Transitions from the ground state of the $4f$ ions to excited states have even parity and are, in principle, Raman active at low temperature. At higher temperatures, electrons are transferred from the ground state to excited states and the intensity of the ground state transitions decreases while transitions with higher lying states as initial states begin to appear in the Raman spectra. Two types of such *crystal field* (CF) transitions are observed: those between initial and final states belonging to the lowest (ground) spin multiplet and those with the final states in excited multiplets. The former have typical energies of less than $100\,\mathrm{meV}(\simeq 800\,\mathrm{cm}^{-}1)$ while the latter have energies of the order of several hundred meV. The former are also observed by means of inelastic neutron scattering while the energy of the latter is too large to contribute to scattering by thermal neutrons from conventional reactors. They can, however, be observed by using neutron spallation sources [7]. The first multiplet transitions ($\lesssim 100\,\mathrm{meV}$) often take place in the neighborhood of Raman active phonons. The mutual interaction of transitions with phonons can lead to strong frequency and intensity renormalizations.

Most high-T_c superconductors can be prepared as metallic or semiconducting (nonsuperconducting) crystals by varying the doping, e.g., the oxygen concentration. In the latter modification, the copper ions have 2+ valence (Cu^{2+}), i.e., a $3d^9$ configuration which can be regarded as a 3*d hole* in an otherwise filled $3d^{10}$ shell leading to a localized magnetic moment ($S = 1/2$). The magnetic moments of the Cu^{2+} ions are coupled antiferromagnetically in each CuO_2 plane. Fluctuations of the magnetic moments around the equilibrium positions correspond to excitations called magnons. Cuprates with only one CuO_2 plane per unit cell (e.g. $Bi_2Sr_2CuO_6$) have only one magnon branch, with acoustic character, in the magnetic Brillouin zone while those with n CuO_2 planes have n magnon branches. Scattering by one magnon with $\boldsymbol{k} = 0$, while in principle allowed for $n \geq 2$, is very weak since it arises only from the weak spin–orbit interaction. Scattering by two magnons involves the exchange interaction [8] and therefore can be much stronger. It also does not require magnons with zero wavevector, the corresponding selection rule for two-magnon processes being $\boldsymbol{k}_1 + \boldsymbol{k}_2 = 0$. A broad peak thus appears on the Raman spectrum corresponding to two magnons near the edge of the Brillouin zone [8–10]. The Raman shift is $\approx 3\,J$, where J is the in-plane exchange constant. The frequency of the two magnon Raman peak, $\simeq 3\,J$, led to one of the earliest determinations of the exchange constant $J \simeq 80\,\mathrm{meV}$ for high-T_c superconductors.

Remnants of the two magnon peak persist even in superconducting crystals of the cuprates under consideration [10]. They represent antiferromagnetic fluctuations with a small coherence length, to which some authors attribute the origin of the pairing mechanism that leads to superconductivity [11].

The high-T_c superconducting cuprates constitute a class of materials which display the broad range of light scattering phenomena mentioned above. As such, they can be used to illustrate many of the techniques and applications of Raman spectroscopy. This is the aim of the present chapter. In its various sections we shall review the theory and the main experimental results concerning light scattering by phonons, electron, and magnons and their mutual interactions, from the vantage point of Raman spectroscopy.

A few reviews on the applications of Raman scattering to the study and characterization of high-T_c superconductors have appeared. Among the most comprehensive ones are [12–17]. For a more general survey of the properties of high-T_c superconductors we suggest the series "Physical Properties of High-T_c Superconductors" [18] and also [19].

5.1 High-T_c Superconductors: Chemical Composition and Crystal Structure

Most readers of this treatise will be interested in the general aspects and applications of Raman spectroscopy. This chapter, however, is devoted to applications of Raman spectroscopy to a very topical but special kind of materials: the high-T_c superconductors. We therefore include a brief description of the chemical and crystallographic structures of these materials and their properties.

Superconductors with $T_c \gtrsim 40\,K$ are called high-T_c superconductors (HTSC). In spite of some so far unconfirmed reports to the contrary, all known HTSC belong to a family of materials generically called "cuprates". They all contain CuO_2 planes as illustrated in Fig. 5.1 for the case of $YBa_2Cu_3O_7$, a superconductor with $T_c \simeq 90\,K$ (note that these planes can be slightly warped, as is the case in Fig. 5.1). In this figure yttrium has been replaced by a generic rare earth, RE. The crystal structure is either tetragonal (point group D_{4h}, e.g., $Tl_2Ba_2CuO_6$) or orthorhombic (see Fig. 5.1, point group D_{2h}). All HTSC, in their stoichiometric form, are centrosymmetric (inversion centers in Fig. 5.1 are the RE, Cu1, and O1 sites). Their vibrational excitations (phonons) for $\boldsymbol{k} = 0$ are therefore either Raman active, IR active or, in a few cases, optically silent.

The number of CuO_2 planes in Fig. 5.1 is two per unit cell (in this case the unit cell is the same as the primitive cell, PC). This number is fixed

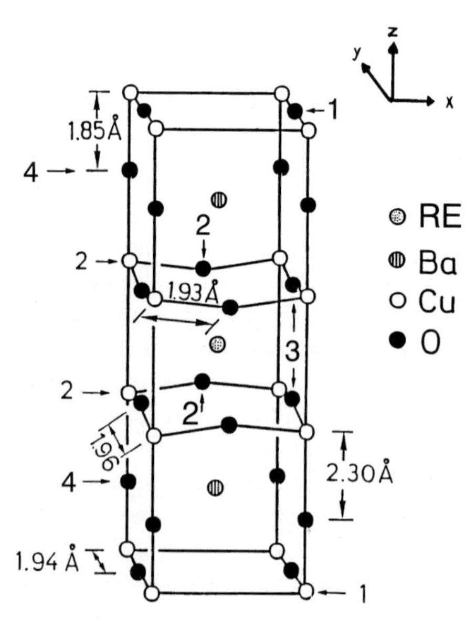

Fig. 5.1. Primitive unit cell (PC) of the RE-123 superconductors $REBa_2Cu_3O_7$, indicating some of the important bond lengths

for the $REBa_2Cu_3O_7$ family. Other families of HTSC, such as the one that holds the record of the highest T_c (134 K at atmospheric pressure, $\simeq$ 164 K under a hydrostatic pressure of 30 GPa [20]): $HgBa_2Ca_{n-1}Cu_nO_{2n+2+\delta}$, have members with several possible numbers n of CuO_2 planes. These planes are separated by spacer layers which, in the case of $REBa_2Cu_3O_7$ (RE-123 for short), [21] consist of one rare earth atom per unit cell while for the mercury family mentioned above ($HgBa_2Ca_{n-1}Cu_nO_{2n+2+\delta}$, Hg-12($n-1$)$n$ for short) as many as $n = 6$ CuO_2 planes may be present in each unit cell. When several n's are possible, T_c increases with n, reaching a maximum for $n \simeq 3$ [22]. In addition to the CuO_2 planes and the Ca or RE spacers just mentioned, a few other planes of atoms are required to stabilize the structure of a HTSC. Figure 5.1 shows two BaO planes, with a structure similar to the Na–Cl planes in a rock salt crystal, placed symmetrically on both sides of the CuO_2–Y–CuO_2 moiety. Some materials, such as the Bi-compounds Bi-22($n-1$)n, have SrO- instead of BaO-planes. Finally, at the top and the bottom of the PC, always similar to that in Fig. 5.1, one or two additional planes are found. They can contain simply metal ions (RE, Cu) or metal oxides (BiO, TaO, PbO).

The critical temperature T_c is determined not only by n but also by the oxygen content represented by δ in $HgBa_2Ca_{n-1}Cu_nO_{2n+2+\delta}$, i.e., by the electronic balance between the cations Hg^{2+}, Ba^{2+}, Ca^{2+}, and Cu^{2+} (a total charge of $+4(n+1)$) and the oxygen anions ($O^{2-}_{2n+2+\delta}$, a total charge of $-[4(n+1)+2\delta]$) which amounts to a net charge balance of -2δ per PC in the mercury family, the minus sign implying that electrons are missing, i.e., that the materials are *hole* conductors. This is the case for most HTSC, with the possible exception of the Nd_2CuO_4 (Nd-214) family, which are believed to be n-type [23,24].

In the Hg-12($n-1$)n HTSC the hole doping is proportional to δ, i.e., to the deviation from stoichiometry. The RE-123 HTSC, however, have a charge balance of -1.0 in the stoichiometric form of Fig. 5.1, i.e., they are hole conductors and superconductors even when stoichiometric. The RE can be either yttrium or almost any rare earth ion, with the exception of terbium (the structure of Fig. 5.1 is not stable when the RE ion is Tb) and probably praseodymium, although several reports have appeared recently claiming that Pr-123 is also a HTSC when prepared in a specific way [25]. Although stoichiometric $REBa_2Cu_3O_7$ is a HTSC (with the exceptions just mentioned), it can be prepared with a nonstoichiometric composition through annealing at $T \simeq 900\,°C$, at which temperature less oxygen is incorporated, followed by more or less rapid quenching. It then becomes $REBa_2Cu_3O_{7-\delta}$. The parameter δ can be as large as $\delta = 1$, in which case (as well as for $\delta \gtrsim 0.6$) RE-123 is a nonsuperconducting semiconductor. For $\delta \simeq 1$ all O1 ions (the so-called chain oxygens) are missing and the crystal structure becomes exactly tetragonal (D_{4h} point group). Throughout most of the nonsuperconducting range of δ, and also at high temperatures for lower values of δ, the remaining

O1 ions are distributed at random among the four equivalent positions of the D_{4h} group. On average the crystal retains D_{4h} symmetry but the O1 disorder does have implications for optical spectroscopies. The charge balance for $REBa_2Cu_3O_{7-\delta}$ is $-1+2\delta$; it also implies that these materials are hole-conducting, at least for $\delta < 0.5$.

We have mentioned that in RE-123 superconductivity disappears for large values of δ, which implies that T_c must decrease for intermediate values of δ. Actually, there is in almost any HTSC with variable stoichiometry an *optimum* value of δ, δ_m for which T_c has a maximum. The materials with $\delta = \delta_m$ are called *optimally doped*. For larger values of δ (*overdoped* HTSC) T_c decreases even though the hole concentration nominally increases. For $\delta < \delta_m$ (*underdoped* HTSC) both T_c and the hole concentration decrease. It is this region of δ's that poses the greatest difficulties for the theoretical understanding of HTSC: even the normal state seems to be highly anomalous, differing strongly from the Fermi liquid behavior of conventional metals. The *overdoped* region seems to obey Fermi liquid behavior.

A modification of the RE-123 structure leads to the so-called RE-124 HTSCs. These materials possess two CuO chains per PC instead of the single chain of the RE-123 structure of Fig. 5.1 [26]. Two chains are strongly bonded to each other, being placed on top of each other along the c-axis and shifted by one-half of the b lattice constant with respect to each other. Thus each chain copper atom is bonded to an oxygen of the nearest neighboring chain. These strong Cu–O bonds stabilize the stoichiometry of the structure: The concentration of chain oxygens is the stoichiometric one and cannot be changed through annealing. Although the nominal charge balance is -1, these materials are *underdoped*: $T_c \simeq 80\,\mathrm{K}$ at atmosphere pressure. With increasing pressure, T_c increases as the material approaches the optimally doped state at a pressure of 8 GPa (1 GPa $\simeq$ 10 kbar) [27]. The yttrium spacer of either Y-123 or Y-124 can be partially replaced by calcium. The hole density is enhanced as Y^{3+} is replaced by Ca^{2+} for a given oxygen concentration. In this manner, a given material, including $(Y_{1-x}Ca_x)$-124, can be swept from underdoped through optimally doped to overdoped with increasing x.

Of interest for optical spectroscopy is the modification of the center of inversion sites that takes place between RE-123 and RE-124. In the latter, the RE ions remain centers of inversion, but the chain oxygen and copper ions do not. Because of the existence of bonded double chains, the centers of inversion that were at the Cu1 (O1) sites in RE-123 shift to positions midway between the two Cu (O) of the double chains. The phonons involving O1 and Cu1 chain atoms, Raman forbidden in RE-123, become allowed in RE-124 [28].

Most HTSCs exhibit metallic conductivity *along the* CuO_2 *planes* above T_c. Perpendicular to the planes, however, they behave like semiconductors (except, possibly, in the overdoped regime), the mean free path being smaller than the lattice constant c. This fact suggests that they can be regarded as two-dimensional conductors in their normal state. With increasing doping, the 2D conduction tends to become 3D, especially in the highly overdoped regime [29].

The first discovered HTSC is $La_{2-x}Ba_xCuO_4$ (La-214) [30]. It reaches critical temperatures $T_c \simeq 38$ K for $x = 0.15$. The stoichiometric structure possesses a CuO_2 layer per PC with two LaO layers, one on each side of the CuO_2 layer. The stoichiometric compound has exact charge balance between the cations (+8) and the four oxygens (−8): therefore it is not conducting. Conduction, and superconductivity, are achieved by replacing x trivalent La^{3+} ions by the same number of divalent Ba^{2+} (or Sr^{2+}) ions: the charge balance becomes $-x$ and the material is a hole conductor ($x \simeq 0.15$ electrons are missing per PC in the optimally doped case).

The La-214 structure is tetragonal at high temperatures (point group D_{4h}, see Fig. 5.2). At low temperatures, and depending on x, the PC becomes orthorhombically distorted (point group D_{2h}). Figure 5.2 also shows the PC of a material chemically but not crystallographically isomorphic to La-214: Nd_2CuO_4 (Nd-214, Nd can be replaced by either Pr, Gd, or Sm). Its point group is also D_{4h} but the arrangement of the atoms in the PC is rather different in Nd-214 than in La-124: a CuO_2 plane is flanked by two Nd planes, the PC being completed by oxygen planes (one plane with 2 oxygens per PC) (see Fig. 5.2). This material, in its stoichiometric form, has zero charge balance and thus is a semiconductor. When doped with Ce ($Nd_{2-x}Ce_xCuO_4$, one can also use Th instead of Ce) it is believed to become an electron conductor because the trivalent Nd^{3+} is replaced by a tetravalent Ce^{4+} [24].

To conclude this survey, we mention the $TlBa_2Ca_{n-1}Cu_nO_{2n+3-\delta}$ (Tl-12$(n-1)n$) [31] and $Tl_2Ba_2Ca_{n-1}Cu_nO_{2n+4-\delta}$ (Tl-22$(n-1)n$) [32] families, with T_c up to ≈ 125 K and the tetragonal point group D_{4h}. The corresponding Bi-based materials $Bi_2Sr_2Ca_{n-1}Cu_nO_{2n+4+\delta}$ [33] have a distorted D_{4h}-like PC. These materials are actually orthorhombic (D_{2h}) because of a distortion along the $(x+y)$ direction [34] (note that in RE-123 and RE-124 the distortion is along the chain direction [4,35], i.e., along y. This difference has important consequences in Raman spectroscopy [4].)

All materials just mentioned are opaque to the near-IR, visible and UV lasers usually employed for Raman scattering experiments. Therefore Raman experiments must be performed in a backscattering configuration. In this configuration the optical and crystalline quality of the scattering surface are of paramount importance. The penetration depth of these lasers into HTSC crystals is about 20 nm, which means that, since the incident light has to get into the HTSC and the scattered light out, the sampling depth is about 10 nm. This depth amounts to about 10 PCs along the c-axis and 30 PCs along either y or x. Some care is therefore required in preparing clean surfaces that are representative of the bulk. This problem, however, is by far not as acute as for techniques such as photoelectron [36] or tunneling spectroscopies [26] where the sampling depth is less than 1 nm, i.e., a primitive

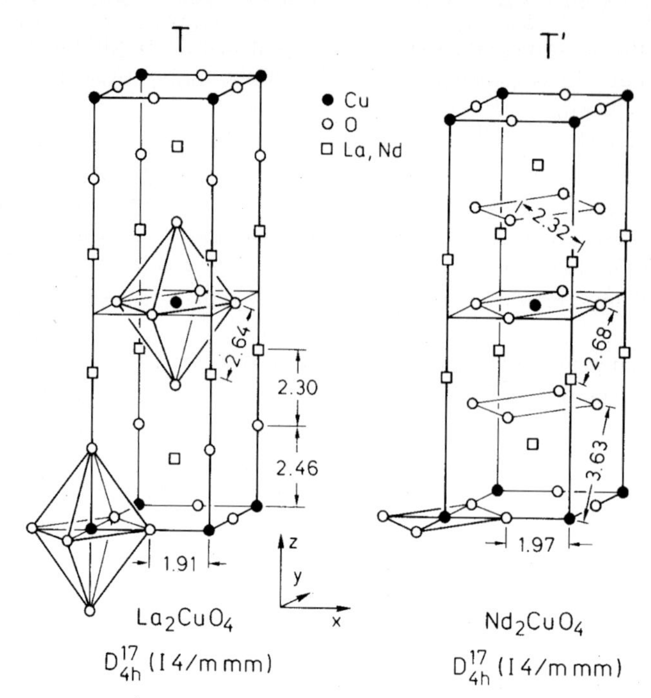

Fig. 5.2. Structure of La_2CuO_4 (T) and Nd_2CuO_4 (T'). The two structures differ in the atomic site of one oxygen atom which is part of CuO_6 octahedra (*left*) or simple CuO_4 square planes (*right*). Distances given in Å (Ref. [13], p. 300). The respective point groups are given under each of the structures. Through partial replacement of La by Ba or Sr La-214 becomes a p-type conductor and a HTSC while Nd:214 doped with either Ce or Th becomes an n-type conductor and also, at $T_c \lesssim 30\,K$, a HTSC

cell. In the latter cases one is always exposed to criticism that surface and not bulk properties are being measured. For single crystal HTSC, a sample depth of 10 nm should be sufficient to guarantee that bulk properties are being measured. For polycrystalline (ceramic) samples this is not always the case although, with adequate care, Raman spectra representative of the bulk can be obtained. It is even possible, using a micro-Raman setup (a spectrometer with a confocal microscope arrangement) to select scattering surfaces of microcrystallites with a specific orientation [6].

A few comments concerning the growth of HTSC are also in order. The original HTSC, the La-214 materials (the so-called Zurich salt [30]), were first prepared by solid state reaction in a furnace using La_2O_3, CuO, and $BaCO_3$ as precursors. They had the morphology of a ceramic material composed of sintered microcrystallites. The first Raman measurements on these materials were rather irreproducible and unreliable, although, with increasing experience, Raman spectroscopists were able to discriminate the bulk peaks of

La-214 from those related to surface effects and foreign phases. Fortunately, few problems of this type have occurred in the HTSC discovered later, in particular the RE-123 family, for which Raman spectra full of information have been obtained even with the originally synthesized ceramic samples. Measurements on such samples, however, do not allow the Raman spectroscopist to make full use of the polarization selection rules, applicable to single crystals, and thus to separate the various independent symmetry components of the Raman spectra.

Single crystals of HTSC became available shortly after the first ceramic samples had been prepared. Such single crystals fall into two categories: bulk samples and thin films. The bulk single crystals can be prepared by melting in a flux that contains a surplus of one of the precursors. Slow cooling yields a slag whose interior or surface contains rather beautiful small crystallites of the material under consideration (see Fig. 5.3) often with the shape of a parallelepiped whose faces are perpendicular to the orthorhombic axes. Another powerful technique for growing large single crystals, and metallic oxides in general, in particular high-T_c superconductors, is the travelling-zone method which is implemented by zone-melting of a precursor bar in a mirror furnace (quartz-iodine incandescent lamps focused on the zone to be melted). The molten zone is then made to travel vertically from one end of the bar to the other.

Thin films of HTSC are prepared by any of several techniques for growing thin films epitaxially or in a textured way (i.e., with a preferred orientation) onto a substrate. As a substrate, a perovskite-type material is chosen (e.g., $SrTiO_3$, $NdGaO_3$). In this manner, highly oriented, high quality films have been grown with several orientations [38–40]. Raman spectroscopy, as a non-destructive technique requiring small sampling volumes, has been very helpful

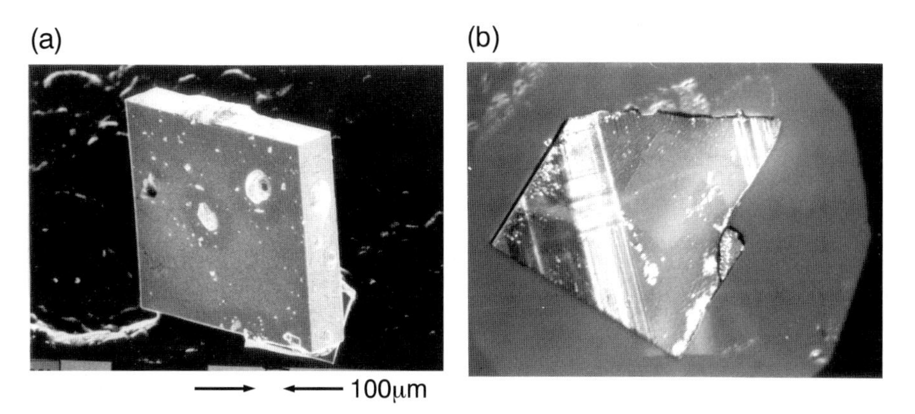

Fig. 5.3. Typical examples of two $YBa_2Cu_3O_{7-\delta}$ crystals grown from flux. (**a**) Optical micrograph of a twinned crystal obtained with unpolarized light. (**b**) Micrograph with polarized light of a largely untwinned crystal obtained from flux. The two orthogonal domains are displayed as light and dark areas [37]

for characterizing such films, in particular their degree of orientation [40]. The most common techniques for the preparation of HTSC are laser ablation from a ceramic target (using an excimer laser), sputtering, and molecular beam epitaxy (MBE).

We would like to mention an important feature of *the orthorhombic* HTSCs such as those of the RE-123 family. As shown in Fig. 5.1, the orthorhombicity is determined by CuO chains which run (by definition) along the y axis. Since the a and b lattice constants are nearly the same, the chains can be oriented along either one of the horizontal axes of Fig. 5.1. This results in *twinning*: a crystal with the morphology of Fig. 5.3 can consist of an ensemble of twinned crystallites, some with the chains along y and others with the chains along x. Most *as-grown* crystals are twinned, with equal amounts of both orientations. In bulk crystals the size of those crystallites is a few microns and therefore they can be seen with an optical polarizing microscope. In thin films, the twin size is usually smaller; they cannot be observed with an optical microscope. In this case Raman spectroscopy, which measures an average of the spectra of both types of twins, comes in handy. For 50% twinned crystals, the spectra in xx configuration (incident light polarized with the E-field along x, scattered field also along x) and yy polarizations are nearly identical. In the case of untwinned crystals, which may be found accidentally among the many retrieved from a slug, considerable and quantifiable differences appear between the two polarization configurations. With this technique the first untwinned crystals were identified [41].

Finally we note that the usefulness of a HTSC depends not only on its high T_c but also on its critical current density, i.e., the maximum current density a superconductor can carry at a temperature $T < T_c$ without switching to normal. HTSC are type-II superconductors: they have a very low threshold magnetic field H_{c1} at which flux penetration takes place in the form of *flux tubes* with *normal cores*. For a perfect crystal structure these tubes move upon application of a current, thus producing ohmic-like losses because of their normal cores. In the absence of an external magnetic field, the self-field of the circulating current may also create flux tubes (vortices); their motion determines the critical current density j_c, which for a perfect crystal, would be rather low (because H_{c1} is only a few Gauss). Crystal defects, however, can pin the flux tubes and thus enhance j_c. The highest critical current ($j_c \simeq 10^8$ A/cm^2 at 4 K) has been observed for thin films [42]. Ceramic materials have j_c's as high as 10^6 A/cm^2 [43]. Several articles on the subject of vortex pinning and critical currents can be found in [44].

5.2 Raman Scattering by Phonons in High-T_c Superconductors

5.2.1 Vibrational Frequencies and Eigenvectors

As an illustration we will consider primarily the canonical Y-123 superconductor ($YBa_2Cu_3O_{7-\delta}$, $0 \lesssim \delta \lesssim 0.6$). Its unit cell has been shown in Fig. 5.1. It is composed of two CuO_2 planes, perpendicular to the c-axis, which are generally believed to be responsible for the superconductivity, a linear CuO chain (for $\delta = 0$), which determines the hole doping of the planes, and three spacer layers (one Y and two BaO planes). As already discussed at the beginning, the sites of the Y, the Cu and the O of the chains (Cu1, O1) are inversion centers. The vibrations of these atoms for $\boldsymbol{k} = 0$ are thus odd upon inversion and therefore Raman inactive (but IR-active).[1] The remaining atoms are not at centers of inversion. They thus appear as pairs of equivalent atoms connected by the inversion. They give rise to odd (both atoms moving in the same direction) and even (motion in opposite directions) vibrational patterns. The corresponding pairs of IR- and Raman-active vibrations are sometimes called *Davidov doublets*.

The atoms of Fig. 5.1 can vibrate along the z as well as along the x and y directions. The even vibrations along z appear to be much stronger in the Raman spectra than their x-y counterparts [45]. Similarly, in the superconducting crystals ($\delta \gtrsim 0.6$) the IR spectra polarized along z reveal phonon structure while along x and y the coupling to the phonon is too weak to allow observation in the IR spectra. The latter happens because of the strong electronic conductivity which manifests itself in a very small penetration depth. The weak Raman activity of the (x, y)-polarized in-plane vibrational modes is likely to result from the forbidden nature of the phonon scattering for an isolated CuO_2 plane.

If one removes the O1 atoms of the chains, the unit cell of Fig. 5.1 becomes tetragonal (point group D_{4h}) and the x, y vibrations become degenerate by symmetry. The existence of intact Cu chains, however, lowers the point group symmetry to orthorhombic (D_{2h}). Under these conditions the x, y degeneracy splits and all phonons at the zone center (Γ) become nondegenerate [13,45,46]. In crystals with large δ ($\delta \gtrsim 0.6$) the chain oxygens can occupy all equivalent positions in the planes of the chains and, on the average, the crystals have tetragonal symmetry. The concomitant disorder can, however, make certain Raman forbidden modes allowed.

For the structure of Fig. 5.1 five z-polarized Raman phonons exist. They correspond to the irreducible representation A_g of the D_{2h} group (or A_{1g} in the case of D_{4h} symmetry, see Table 5.1). They can be observed for (z, z) (incident, scattered) or for (x, x) and (y, y) polarizations. The lowest frequency z-polarized Raman phonon involves mainly the vibrations of the hea-

[1] As mentioned in Sect. 5.1, Y-124 has double chains which lead to both IR- and Raman-active phonons [28].

Table 5.1. Character table for the D_{2h} point group which corresponds to the PC of Fig. 5.1. Also given are the symmetry operations, selection rules for Raman- and IR-active modes and compatibility relations (comp.) for the corresponding tetragonal (D_{4h}) group and from this group to an orthorhombic one rotated by 45° [$D_{2h}(45°)$], that of Bi-22$(n-1)n$. Note that the selection rules refer only to the D_{2h} point group. Note also that for RE-123 the x and y axes are along the bond directions while in Bi-22$(n-1)n$ they are rotated by 45°. The symbols have their usual meaning. From [13]

D_{2h}	E	C_2^z	C_2^y	C_2^x	i	σ^{xy}	σ^{xz}	σ^{yz}	Selection rules	Comp. D_{4h}	Comp. $D_{2h}(45°)$
A_g	1	1	1	1	1	1	1	1	xx, yy, zz	$A_{1g} B_{1g}$	$A_g B_{1g}$
A_u	1	1	1	1	−1	−1	−1	−1	silent	$A_{1u} B_{1u}$	$A_{1u} B_{1u}$
B_{1g}	1	1	−1	−1	1	1	−1	−1	xy	$A_{2g} B_{2g}$	$B_{1u} A_u$
B_{1u}	1	1	−1	−1	−1	−1	1	1	ir(z)	$A_{2u} B_{2u}$	$B_{1u} A_u$
B_{2g}	1	−1	1	−1	1	−1	1	−1	xz	E_g	B_{2g}
B_{2u}	1	−1	1	−1	−1	1	−1	1	ir(y)	E_u	B_{2u}
B_{3g}	1	−1	−1	1	1	−1	−1	1	yz	E_g	B_{3g}
B_{3u}	1	−1	−1	1	−1	1	1	−1	ir(x)	E_u	B_{3u}

viest atom (Ba),[2] the next lowest involving those of the Cu2 atoms of the CuO_2 planes (150 cm^{-1}). The remaining three modes are dominated by vibrations of the O2-O3 plane oxygens (340, 440 cm^{-1}) and the apical oxygens (500 cm^{-1}).

In the tetragonal RE-123 structure (i.e., $REBa_2Cu_3O_6$, without chains) the O2 and O3 atoms are symmetry equivalent: a fourfold rotation transforms one into the other. The corresponding z-polarized Raman phonons are thus expected to be mixed (O2 and O3). They split into two linear combinations, one in which O2 and O3 of a given plane vibrate in the same direction (fully symmetric, A_{1g} rep of D_{4h}) and the other in which they vibrate in opposite directions. The latter change sign upon a 90° rotation and thus behave like (x^2-y^2) under the symmetry operations of D_{4h} (they are said to belong to the B_{1g} rep (rep = irreducible representation) of the D_{4h} group, see Fig. 5.4).

In the presence of the chains, the tetragonal symmetry is lowered to orthorhombic (D_{2h}). The A_{1g} and B_{1g} modes both become A_g (in the D_{2h} notation) and mix. The amplitudes of the O2 and O3 displacements are then no longer equal. We recall, however, that the symmetry can be lowered to D_{2h} in a different way, i.e., by applying a distortion along the $(x+y)$ diagonal of the unit cell. In this case, the B_{1g} mode of Fig. 5.4 remains odd upon reflection on the $(x+y)$ plane which still is a symmetry element: The amplitudes of the O2-O3 vibrations remain equal in the mixed modes. This case obtains

[2] Note that modes of the same symmetry mix and, strictly speaking, cannot be assigned to the vibrations of a single type of ion.

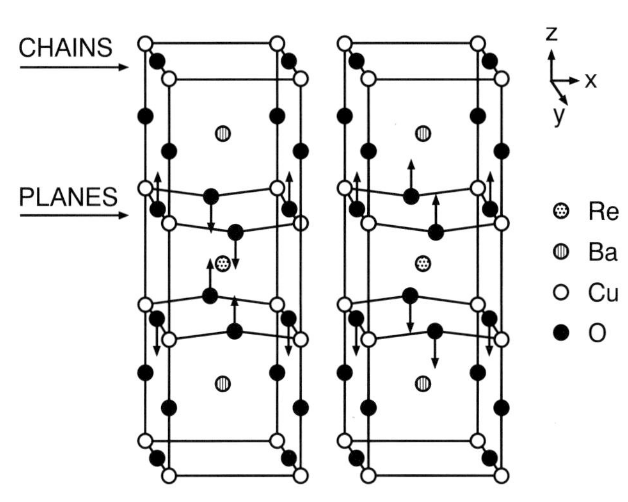

Fig. 5.4. Schematic diagram of the eigenvectors of the A_{1g} (*right PC*) and B_{1g} (*left PC*) Raman-active phonons (D_{4h} notation) of Y-123. These eigenvectors are only approximate in the D_{2h} modification of Y-123 but exact in $D_{2h}(45°)$ Bi-2212

for the bismuth superconductors[3] (e.g. Bi-22$(n-1)n$). A more detailed discussion of phonons in high-T_c superconductors and their symmetries can be found in [12,13,46].

Table 5.1 displays the characters of the irreducible representations (reps) of the orthorhombic D_{2h} point group which corresponds to the symmetry of $\boldsymbol{k}=0$ excitations in RE-123, RE-124, and Bi-22$(n-1)n$ (keep in mind, however, the 45° axes rotation in the latter HTSC, as mentioned in the caption of Table 5.1). This table also shows the polarization configurations allowed in the first order Raman and IR spectra and the compatibility relations for reps when the symmetry is raised from D_{2h} to D_{4h}. These compatibility relations in the D_{2h} case are different from those in the $D_{2h}(45°)$ case.

Because of their dominant strength, we discuss the Raman tensors corresponding to the A_g modes of Table 5.1 (D_{2h} symmetry). They have the general form:

$$\boldsymbol{R}(A_g) = \begin{pmatrix} \alpha & 0 & 0 \\ 0 & \beta & 0 \\ 0 & 0 & \gamma \end{pmatrix} . \qquad (5.1)$$

Let us recall that the Stokes–Raman efficiencies S (in appropriate units) are obtained by contracting the corresponding tensor $\boldsymbol{R}$ with the unit vectors

[3] The Raman-active phonons of Bi-2212 have been recently observed also in the inelastic tunneling spectrum [47].

of the incident and scattered electric fields, $\hat{\mathbf{e}}_\mathrm{L}$, and $\hat{\mathbf{e}}_\mathrm{S}$ and squaring the magnitude of the contracted product:

$$S_\mathrm{S} \propto \left|\hat{\mathbf{e}}_\mathrm{S} \cdot \boldsymbol{R} \cdot \hat{\mathbf{e}}_\mathrm{L}\right|^2 \cdot (n_\mathrm{B}+1) , \quad (5.2)$$

where n_B is the Bose–Einstein statistical factor (for anti-Stokes scattering $n_\mathrm{B}+1$ must be replaced by n_B, see Chap. 1).

We have mentioned that for D_{2h} symmetry there are five A_g phonon modes for the structure of Fig. 5.1. In the D_{4h} case (and also for $D_{2h}(45°)$) only four A_{1g} (A_g for $D_{2h}(45°)$) obtain. The fifth mode (B_{1g} symmetry) has the pattern shown on the left in Fig. 5.4, which corresponds to x^2-y^2 symmetry (this also seems to be the symmetry of the superconducting gap, see Sect. 5.3.2, also that in which the two-magnon spectrum is strongest, see Sect. 5.6). The Raman tensor of B_{1g} phonons has the form

$$\begin{pmatrix} \alpha & 0 & 0 \\ 0 & -\alpha & 0 \\ 0 & 0 & 0 \end{pmatrix} . \quad (5.3)$$

Equations (5.2) and (5.3) imply similar scattering strengths for xx and yy polarizations, while in the D_{2h} case of an *untwinned* RE-123 crystal the corresponding strengths should be different. A 50% twinned crystal, however, should also lead to the same intensities under xx and yy polarizations. Note that for $D_{2h}(45°)$ symmetry the tensor of (5.3) remains exactly valid for the B_{1g} mode. An additional constraint must also be introduced in (5.1) if the symmetry is raised to D_{4h}: the diagonal elements α and β must be equal and therefore the xx and yy spectra must have the same intensity for A_{1g} modes. The same property should hold for the B_{1g} phonons. This will also be true, on the average, for any 50% twinned crystals, a fact that was early used to identify the first untwinned Y-123 HTSC samples [41] (see Fig. 5.3). We conclude the discussion of the A_g modes of RE-123 by mentioning that (5.3) remains approximately correct even for D_{2h} symmetry (the orthorhombicity is produced by the chains, which are far from the vibrating planes). It can then be used to identify unambiguously the phonons of quasi-B_{1g} symmetry (left part of Fig. 5.4): They appear for $\hat{\mathbf{e}}_\mathrm{L} \sim (x+y)$ and $\hat{\mathbf{e}}_\mathrm{S} \sim (x-y)$. For the case $\hat{\mathbf{e}}_\mathrm{L} \parallel \hat{\mathbf{e}}_\mathrm{S}$, with $\hat{\mathbf{e}}_\mathrm{L}$ at an angle φ respect to the x-axis, the scattering intensity should be proportional to $\cos^2 2\varphi$, whereas for $\hat{\mathbf{e}}_\mathrm{L} \perp \hat{\mathbf{e}}_\mathrm{S}$ that intensity should be proportional to $\sin^2 2\varphi$. These cloverleaf-like patterns provide an unambiguous signature for the B_{1g} modes. For B_{2g} modes (xy symmetry), the $\sin^2 2\varphi$ pattern corresponds to $\hat{\mathbf{e}}_\mathrm{L} \parallel \hat{\mathbf{e}}_\mathrm{S}$ and $\cos^2 2\varphi$ to $\hat{\mathbf{e}}_\mathrm{L} \perp \hat{\mathbf{e}}_\mathrm{S}$.

Next we discuss briefly the phonon Raman spectrum of the RE-124 HTSC, in particular Y-124 ($YBa_2Cu_4O_8$). This compound has a fixed oxygen stoichiometry and its single crystals are also *untwinned* (possibly because of the high stability of the double chains). The $xx-yy$ anisotropy, corresponding to the fact that $\alpha \neq \beta$ in (5.1), can be clearly seen in Fig. 5.5. This spectrum exhibits two additional A_g modes when compared to that of Y-123:

The O1 mode at $603\,cm^{-1}$ and the Cu1 mode at $247\,cm^{-1}$. These modes become Raman active because of the existence of two chains per PC. The other phonon modes in Fig. 5.5 peak at $103\,cm^{-1}$ (mainly Ba vibrations), $150\,cm^{-1}$ (mainly Cu2), $340\,cm^{-1}$ (O2-O3, B_{1g}-like), $439\,cm^{-1}$ (O2 + O3), and $502\,cm^{-1}$ (mainly vibrations of the *apical* oxygens O4). Note the large differences in the xx and yy intensities of some of these A_g modes, which are a signature of the orthorhombic, untwinned structure of the material.

Before closing this brief survey of the phonon Raman spectra of HTSC we consider Nd-214, the prototype of the n-type superconductors, whose PC is shown in Fig. 5.2 [49]. This material has tetragonal D_{4h} symmetry. The Cu ion is at a center of inversion and does not generate Raman active phonons. The two Nd atoms of a PC are connected by the inversion and give rise to a Davidov pair, one of whose components (at $228\,cm^{-1}$) has the Raman-active A_{1g} symmetry. The two oxygen ions placed between Nd-planes are not at a center of inversion and therefore they also generate a Davidov doublet. It is easy to see that the vertical (along z) vibrations of these oxygens, *even* with respect to an inversion center (e.g., copper), have the same pattern as the B_{1g} mode of Fig. 5.4. The peak observed at $328\,cm^{-1}$ obeys the selection rules (cloverleaf pattern vs. φ) which correspond to the B_{1g} symmetry and therefore its identification is definitive.

We discuss next disorder induced vibrational Raman spectra and illustrate them with the case of the Raman forbidden phonons of Cu1 and O1 ions in $YBa_2Cu_3O_{7-\delta}$. The corresponding vibrations become allowed when the

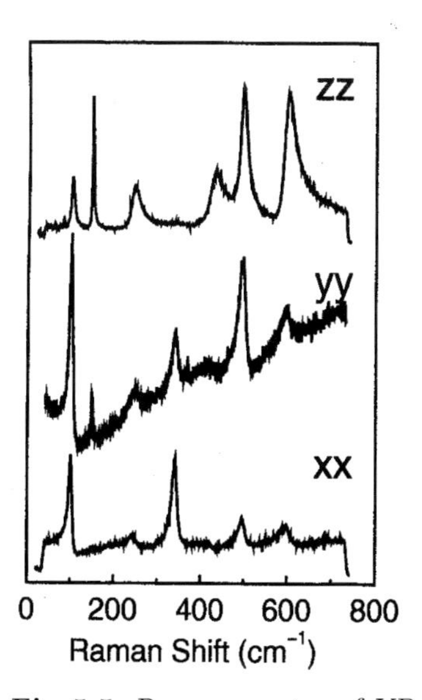

Fig. 5.5. Raman spectra of $YBa_2Cu_4O_8$ for incident and scattered polarizations parallel to the $x(xx)$, $y(yy)$, and $z(zz)$ axes, measured at 300 K with the 514 nm argon ion laser line [48]

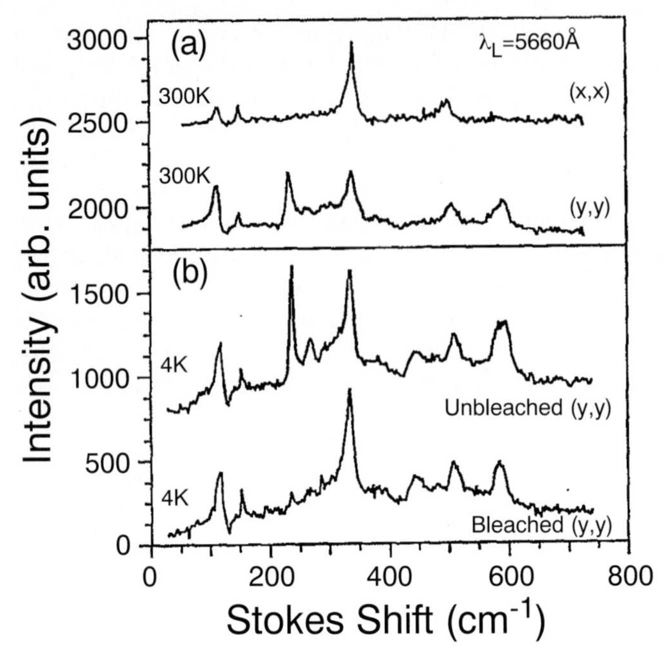

Fig. 5.6. (**a**) Raman spectra of a $YBa_2Cu_3O_7$ taken at 300 K with a laser frequency of 2.19 eV for polarizations perpendicular to the chains (xx), and parallel to the chains (yy). The sample was a nearly optimally doped ($T_c \simeq 90$ K) untwinned crystal. Graphs are vertically offset for clarity. The labels x and y correspond to the crystalline $\hat{\boldsymbol{a}}$ and $\hat{\boldsymbol{b}}$ directions, respectively. (**b**) Comparison of spectra of the dark-cooled state with the light-soaked state at 4 K (yy). The metastable state of the bottom spectrum appears after extended exposure to light with $\hbar\omega > 2.2$ eV. From [51]

chains are oxygen deficient, i.e., for $\delta \neq 0$ [50]. This fact manifests itself through the appearance of two rather sharp peaks at $\sim$ 230 and 600 cm^{-1} (see Fig. 5.6).[4]

It has been shown by Wake et al. [51] that these defect-induced modes exhibit a very rich phenomenology:

(1) First of all, they are strongly resonant (by a factor of 10) for $\hbar\omega \simeq 2.2$ eV, close to the yellow line of the Kr^+-laser.
(2) They are much stronger in the yy than in the xx configuration. See also [50].[5]
(3) The peak at 230 cm^{-1} can be *bleached* at $T \leq 100$ K (see Fig. 5.6) by *soaking* the measured spot with light of photon energy larger than 2.2 eV.

[4] Note that these frequencies are very similar to those of the A_g modes of Y-124 corresponding to O1 (605 cm^{-1}) and Cu1 (250 cm^{-1}). This supports the assignment made above.

[5] This is also the case for the corresponding peaks of Y-124 seen in Fig. 5.5.

The *bleached* peaks reappear when the sample is warmed up to temperatures higher than 180 K.

Possible explanations of these fascinating phenomena have been presented in [51] but the details are still not understood. The metastability and quenching properties may be related to the tendency to separate into different phases found in $YBa_2Cu_3O_{7-\delta}$ depending on treatment [52,53]. When $\delta = 0.5$, for instance, the O1 ions can be disordered or adopt a fully ordered stoichiometric orthorhombic (so-called OII) structure in which every second chain in the $x - y$ plane is missing.

The n-type HTSC $RE_{2-x}Ce_xCuO_4$ (RE = Gd, Sm, Eu, Nd, Pr, structure T' in Fig. 5.2) also shows, for $x \neq 0$, a sharp Raman peak at $\approx 580\,\text{cm}^{-1}$. This peak, which is strongly polarized in the zz configuration, is not present for $x = 0$. It has been conjectured that it is activated by the Ce-related disorder [54,55]. Its high frequency suggests that it is related to oxygen vibrations. This conjecture has been demonstrated through isotopic substitution $^{16}O \rightarrow {}^{18}O$: the peak shifts like the square root of the oxygen mass. Since this vibrational mode is incompatible with the stoichiometric T' structure, it has to be attributed to the perturbation produced by the cerium on the oxygen vibrations. A Ce ion close to one of the O_2 planes of Fig. 5.2 (T' structure) makes Raman active the mode in which all O_2 vibrate in the same direction along $\hat{\mathbf{c}}$, although it is not likely that this local, bond-bending-like mode could reach frequencies as high as $580\,\text{cm}^{-1}$. Another possibility is interstitial oxygen placed at midpoints between the Nd (or Ce) and Cu ions along z when Ce is present. Different types of local environments, some of them possibly related to the $580\,\text{cm}^{-1}$ mode, are also observed in the crystal field transitions of the Nd^{3+} ions in $Nd_{2-x}Ce_xCuO_4$ [56].

We have discussed, in connection with Fig. 5.4, the effect of symmetry on the phonon eigenvectors that correspond to the vibrations of the O2 and O3 atoms. In the case of tetragonal RE-123 (and also for orthorhombic Bi-1212) there is only one phonon mode of B_{1g} symmetry. Correspondingly, this mode does not mix with any others; its eigenvector, illustrated in Fig. 5.4, is determined solely by symmetry. There are, however, four A_{1g} modes (five A_g modes for orthorhombic Y-123 where both A_{1g} and B_{1g} become A_g, see the representations of the D_{2h} group in Table 5.1) for which the vibrational patterns of the different atoms involved are expected to mix to some degree. Actually, since their frequencies are well separated (115, 150, 440, $500\,\text{cm}^{-1}$) the admixture is expected to be weak.

The available knowledge concerning phonon eigenvectors, especially for complex crystals such as the high-T_c superconductors, is relatively scarce. It is nevertheless important for the calculation of physical properties such as isotope effects and electron–phonon interactions. We have assigned the $115\,\text{cm}^{-1}$ mode predominantly to Ba vibrations and the $150\,\text{cm}^{-1}$ mode to vibrations of Cu2 atoms. This assignment has been confirmed by isotopic substitution experiments [57]: If the $150\,\text{cm}^{-1}$ mode is due only to Cu it should shift like the inverse square root of the average copper mass when replacing the copper isotope ^{63}Cu by ^{65}Cu. It should, however, remain invariant when replacing barium isotopes. By performing various isotopic replacements it is possible to determine quite accurately the eigenvectors of these modes. As an example we show in Fig. 5.7 the effect of the substitution of ^{138}Ba by ^{134}Ba [57]. Here the 115 cm^{-1} mode shifts as corresponds to the Ba mass while the 150 cm^{-1} mode barely shifts, thus confirming the nearly unmixed nature of these modes. Careful fits to the measured lineshapes yield the eigenvector amplitudes listed under "experiment" in Table 5.2.

During the past decade several methods have been developed for calculating ab initio the phonon frequencies and eigenvectors from the theoretical electronic band structure. They involve the evaluation of the *total energy*

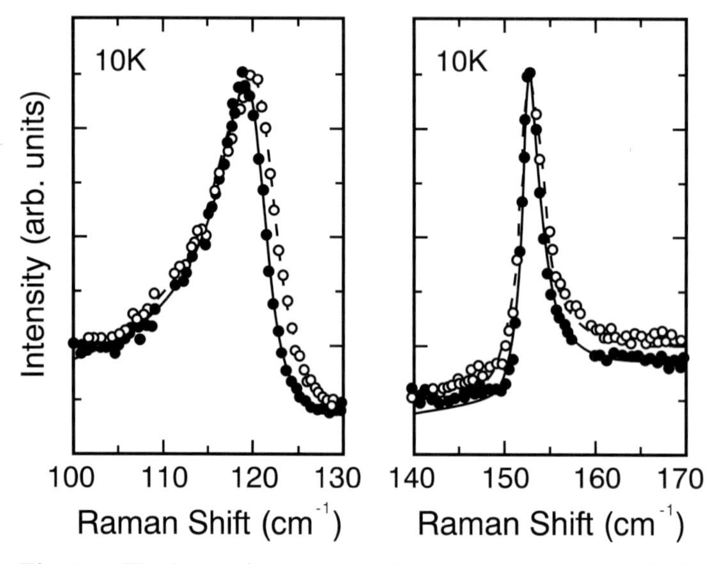

Fig. 5.7. The lowest frequency A_g Raman phonon modes of $YBa_2Cu_3O_7$ for two different Ba isotopes (• ^{138}Ba and ○ ^{134}Ba). The solid and dashed lines represent fits with Fano lineshapes. From [57]

of a crystal as a function of a static phonon displacement (frozen phonon) and thus determining the equilibrium position (corresponding to the minimum energy) and the restoring force constant (the coefficient of the quadratic term of the expansion of energy vs. displacement). Table 5.2 shows the frequencies calculated by different authors for the 115 and 150 cm^{-1} A_g phonons of Y-123, together with the eigenvectors that express the Ba-Cu admixture in these modes. We note that while the experiment indicates nearly pure atomic vibrational modes (as expressed by the square of the eigenvector components of Table 5.2) the calculations based on the local density approximation (LDA) to the exchange and correlation potentials yield almost equal

Table 5.2. Calculated eigenvectors of the two lowest A_g Raman phonons of $YBa_2Cu_3O_7$ compared with experimental results

cm^{-1}	Ba	Cu2	
116	0.93	$\mp$0.33	experiment [57]
95	0.65	0.75	LMTO-LDA [58]
105	0.81	0.59	LAPW-LDA [59]
103	0.92	0.40	LAPW-LDA [60]
115	0.93	0.36	LAPW-GGA [60]
150	0.33	$\pm$0.93	experiment [55]
130	0.76	−0.65	LMTO-LDA [58]
127	0.59	−0.80	LAPW-LDA [59]
130	0.40	−0.92	LAPW-LDA [60]
144	0.36	−0.93	LAPW-GGA [60]

admixture of Ba and Cu vibrational amplitudes. More recent calculations, in which a more sophisticated ansatz for exchange and correlation was used (the so-called *generalized gradient approximation*, GGA), agree quite well with the experimental results (see Table 5.2). The LDA-based calculations apparently slightly overestimate the off-diagonal elements of the dynamical matrix, which then leads to an admixture of the eigenvectors that is too large [57,61].

The eigenvectors of the Raman phonons in Y-123 are fairly simple. However, they become more complex (through stronger admixture of atomic vibrational patterns) for crystals with a larger number of atoms in the primitive cell. An example is provided by Y-124, a material with two parallel CuO chains per PC, staggered with respect to the constituent atoms (Cu on top of O and vice versa) [62]. As already mentioned, the inversion centers that for Y-123 were at O1 and Cu1 are shifted, for Y-124, to the midpoints of the Cu−Cu and O−O atoms in the double chains.

The even vibrations corresponding to the additional chains are Raman active, while their odd counterparts remain IR active (these modes were Raman inactive for the single chains of Y-123). Extensive isotopic substitution experiments have yet to be performed for Y-124, although its phonon frequencies have been determined by Raman spectroscopy [63]. The only information available on eigenvectors was obtained for the A_g modes by an ab initio method [64]. It is summarized in Table 5.3 together with the calculated and measured frequencies. Note that for this material the calculated Ba and Cu(2) phonon eigenvectors are nearly unmixed. However, the "additional" Cu1 and O1 (chain) modes mix quite considerably. A stronger mixing is expected simply on the basis of the larger number of atoms per PC.

Implicit in the above discussion of eigenvectors of phonons with $\boldsymbol{k} = 0$ (Γ-point of the BZ) is the fact that they must preserve the center of mass (otherwise the crystal flies away as a whole!). This happens automatically for

Table 5.3. Calculated frequencies and eigenvectors of the A_g Raman-active modes of Y-124 [64] For comparison, the measured A_g frequencies are also given in [63]. Underlined entries identify the strongest component of each eigenvector

Frequencies (cm^{-1}) Calculated	Measured	Predominant mode	Ba	Cu(2)	Cu(1)	O(3)–O(2)	O(3)+O(2)	O(4)	O(1)
89	103	Ba	−0.92	−0.29	−0.21	+0.03	−0.07	+0.06	−0.13
169	150	Cu(2)	+0.33	−0.93	−0.08	+0.04	−0.11	+0.01	−0.02
270	247	Cu(1)	−0.22	−0.15	+0.83	+0.03	−0.03	−0.31	+0.39
393	340	O(3)–O(2)	+0.01	+0.01	−0.03	+0.94	+0.27	+0.16	+0.14
426	439	O(3)+O(2)	+0.05	+0.13	−0.08	+0.33	0.77	−0.47	−0.22
489	502	O(4)	+0.01	−0.06	−0.13	+0.01	+0.56	−0.77	−0.29
576	603	O(1)	−0.02	+0.01	−0.49	−0.08	−0.06	−0.27	+0.82

the Raman modes of a pair of equivalent atoms since they move in opposite directions. For the IR modes, however, the atoms of a "Davidov" pair both move in the same direction and therefore there must be other nonequivalent atoms moving in the opposite direction so as to leave the center of mass at rest. This results in a contribution of more atoms to a given IR mode than in the case of Raman-active modes. Thus far no ab initio calculation of IR-active eigenvectors of Y-123 is available. Nonetheless, we have at our disposal a number of semiempirical lattice dynamical calculations based on, e.g., the shell model. We display what seem to be at this point the most reliable calculated eigenvectors for the IR-active modes, polarized along z (B_{1u} symmetry) of Y-123 [46,57]. Their reliability has been assessed by means of several isotopic substitutions [57].

The B_{1u} eigenvectors (IR-active along z according to Table 5.1 in D_{2h} notation) are important for the analysis of the IR spectra, specifically those measured with a z-polarized electric field. They enable one to determine transverse effective charges $e^*_{\mathrm{T}i}$, associated with each atom i, from the dimensionless oscillator strengths S_j extracted for the various B_{1u} phonons (labeled by j) from the IR spectra, using the expression:

$$S_j = \frac{4\pi}{V} \frac{\left[\Sigma e^*_{\mathrm{T}i} u_{ij}\right]^2}{\omega^2_{\mathrm{TO}_j} \Sigma_i m_i u^2_{ij}} , \tag{5.4}$$

where ω_{TO_j} is the jth TO-vibrational frequency [65] (note that the IR-active phonons split into TO components, vibrating perpendicular to $\boldsymbol{k}$ and an LO component, parallel to $\boldsymbol{k}$, at a higher frequency) and u_{ij} is the corresponding vibrational amplitude of atom i related to the eigenvector component $\boldsymbol{e}_{ij}$ through:

$$\boldsymbol{u}_{ij} = \sqrt{\frac{\hbar}{2M_i\omega_j}} \boldsymbol{e}_{ij} . \tag{5.5}$$

In (5.4) the quantity V represents the volume of the PC. It has been shown [57] that the eigenvectors of Table 5.4 lead, when combined with experimental oscillator strengths S_j using (5.4), to values of $e^*_{\mathrm{T}i}$ close to the nominal valence charges, a fact that adds some credibility to those eigenvectors.

One comment with respect to the even counterparts of the odd eigenvectors of Fig. 5.4 is in order. The A_{1g} mode has as IR-active (B_{1u}) counterpart (i.e., the Raman mode with the closest vibrational amplitudes) the 370 cm^{-1} mode of Table 5.4. Note that, as discussed above, the corresponding B_{1u} eigenvector also involves the motion of yttrium, in a direction opposite to that of O2, O3 so as to conserve the center of mass. The B_{1g} mode (D_{4h} notation), however, has as B_{1u} counterpart an eigenvector in which basically only the O1-O2 oxygens move. Since they already move in opposite directions in each plane, the center of mass is automatically preserved, exactly if the symmetry is D_{4h}, in which case the mode is neither IR (no dipole moment is associated with the displacement pattern) nor Raman active. In this case one speaks of a *silent mode*. When the D_{4h} symmetry is lowered to D_{2h} the O2 and O3 amplitudes do not have to cancel exactly and a small, residual IR activity remains. However, according to (5.4), the small resulting sum of the O2-O3 amplitudes must be squared. Therefore the optical activity of the B_{1u}-like mode (D_{4h} notation) becomes nearly two orders of magnitude smaller than that of the 370 cm^{-1} (O2+O3) B_{1u} mode (D_{2h} notation): the "silent" O2-O3 mode has thus never been seen in the corresponding IR spectra. Inelastic neutron scattering has placed it at about 190 cm^{-1} [66].

While IR and Raman activities are mutually exclusive (within the dipole approximation to the photon–electron interaction in centrosymmetric crystals), IR-active modes can be observed in Raman spectra under special circumstances [67]. This occurs for the LO components of poorly conducting (such as Y-123 for $\boldsymbol{e}_{\mathrm{L}} \parallel \hat{c}$) or semiconducting samples ($YBa_2Cu_3O_6$)

Table 5.4. Normalized eigenvector components e_{ij} for the infrared-active B_{1u} modes in $YBa_2Cu_3O_{7-\delta}$ determined from a shell-model calculation. The frequencies of the modes not observed in the spectra, most likely because of their weak oscillator strength, are given in brackets. The dominant contributions to the eigenvectors are underlined [57]

$\omega_{\mathrm{TO}}^{\mathrm{calc.}}$ (cm^{-1})	e_{Y}	e_{Ba}	$e_{\mathrm{Cu(1)}}$ "Chain"	$e_{\mathrm{Cu(2,3)}}$ "Plane"	$e_{\mathrm{O(1)}}$ "Chain"	$e_{\mathrm{O(2)}}$	$e_{\mathrm{O(3)}}$	$e_{\mathrm{O(4)}}$ "Apex"
(94.7)	0.05	$\underline{-0.263}$	−0.165	$\underline{0.615}$	−0.149	−0.090	−0.113	−0.075
150.9	0.134	$\underline{-0.450}$	$\underline{0.567}$	−0.022	0.131	0.141	0.163	$\underline{0.269}$
196.2	$\underline{0.675}$	−0.138	$\underline{-0.449}$	−0.099	−0.053	$\underline{0.222}$	$\underline{0.241}$	−0.183
(307.5)	−0.119	0.001	−0.073	0.039	0.096	$\underline{-0.446}$	$\underline{0.534}$	−0.002
317.4	−0.196	−0.057	$\underline{-0.285}$	0.068	$\underline{0.912}$	0.126	−0.023	−0.014
370.2	$\underline{0.574}$	−0.019	0.180	−0.093	$\underline{0.291}$	$\underline{-0.422}$	$\underline{-0.296}$	−0.041
531.1	0.092	0.007	$\underline{-0.485}$	−0.028	−0.091	−0.074	−0.072	$\underline{0.602}$

near an electronic interband resonance (i.e., when the laser photon energy equals that of strong interband electronic excitations). In this case the electrostatic field that accompanies the LO phonons can couple to the initial and final electronic states involved in the resonant excitations (this is the so-called Fröhlich interaction). This results in quadrupole-like Raman scattering since the *scattering efficiency* is proportional to $|\boldsymbol{k}_p|^2$. As we have already mentioned, this quantity is very small for normal visible lasers. It can, however, lead to considerable scattering efficiencies if an electronic resonance is present [65]. Such resonance is found in insulating $YBa_2Cu_3O_6$ for $\lambda_L = 676.4$ nm ($\hbar\omega_L = 1.83$ eV). In $YBa_2Cu_3O_6$, LO phonons at 206, 266, 417, and 654 cm^{-1} have been observed. They correspond to in-plane LO-like IR-active vibrations. Their strengths are proportional to the square of the corresponding transverse effective charges, as predicted by theory [65,67].

We show in Fig. 5.8 Raman spectra of nonsuperconducting $PrBa_2Cu_3O_7$. The two arrows point at two nominally Raman-forbidden, IR-active modes that become Raman allowed near resonance through the Fröhlich interaction mechanism (for $\hat{\mathbf{e}}_L \parallel \hat{\mathbf{e}}_S$, in-plane vibrations) [48]. Similar work has been recently performed for La_2CuO_4 using a resonance that appears close to the 1.16 eV line of a Nd-YAG laser [68].

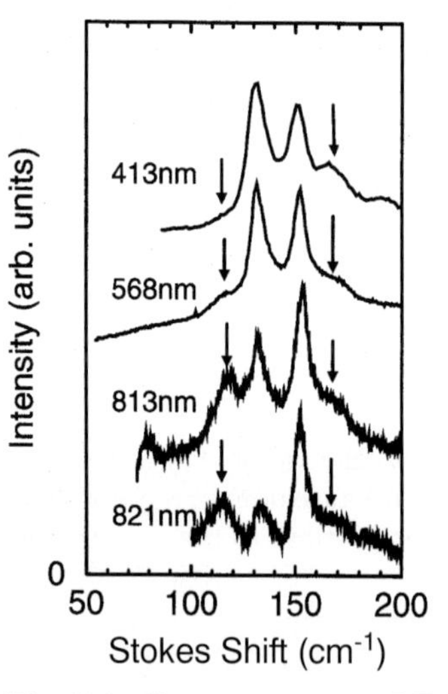

Fig. 5.8. Raman spectra of $PrBa_2Cu_3O_{7-\delta}$ for different laser wavelengths and $\hat{\mathbf{e}}_L \parallel \hat{\mathbf{e}}_S$ perpendicular to the z-axis. The arrows point to the Raman forbidden modes that become Raman active near resonance through Fröhlich interaction [48]

5.2.2 Raman Intensities, Raman Tensors

As discussed in Chap. 1, the scattering cross section can be expressed in terms of *Feynman diagrams.* Simple mathematical rules enable us to transform these diagrams into integrals of algebraic expressions corresponding to the scattering efficiency for the process under consideration [69]. Resonances (i.e., maxima) in the scattering efficiency appear for $E_\mu = \hbar\omega_L$ and for $E_\nu = \hbar\omega_S$, where E_μ and E_ν are the energies of the virtual intermediate electronic excitations. The ratio of the scattered photon flux (per unit solid angle Ω and unit path length) can be written as [65].

$$\frac{\mathrm{d}S}{\mathrm{d}\Omega} \simeq \frac{\omega_L^3 \omega_S V}{c^4} \left|\hat{\boldsymbol{e}}_L \cdot \boldsymbol{R}(\omega_L) \cdot \hat{\boldsymbol{e}}_S\right|^2 , \tag{5.6}$$

where

$$\boldsymbol{R}_j(\omega_L) = \sum_i \frac{\partial \boldsymbol{\chi}(\omega_L)}{\partial \boldsymbol{u}_i} \boldsymbol{u}_i \tag{5.7}$$

is the second rank Raman tensor for scattering by the jth Raman phonon and $\boldsymbol{\chi}$ is the susceptibility tensor. These equations (5.6) and (5.7) have been successfully used for evaluating scattering efficiencies for phonon scattering in rather simple crystals (e.g. diamond, silicon [65]). In spite of the complex PCs of high-T_c superconductors, electronic band structures were computed early and applied to calculating $\boldsymbol{\chi}(\omega)$. As early as 1990 these calculations were extended to evaluate the Raman tensors $\boldsymbol{R}_j(\omega_L)$ of the five A_g-like Raman phonons of $YBa_2Cu_3O_7$ (i.e., Y-123) [70]. These calculations yielded values of the scattering efficiency $\mathrm{d}S/\mathrm{d}\Omega$ which were in agreement with experimental data, including resonances in $\mathrm{d}S/\mathrm{d}\Omega$ vs. laser frequency ω_L.

The evaluation of (5.7) requires knowledge of the components of $\boldsymbol{u}$ on the various atoms labeled by i ($\boldsymbol{u}_i$). For the A_g modes of Y-123 these eigenvectors are fairly well known (see Table 5.2). By adjusting these eigenvectors somewhat, so as to bring the calculated $\mathrm{d}S/\mathrm{d}\Omega$ into as good an agreement as possible with the measured efficiencies vs. ω_L (the so-called resonance profiles), it is possible to improve the quality of the eigenvectors. In this manner, the authors of [70] were able to confirm that the Cu and Ba A_g modes are indeed nearly unmixed and also that the $435\,\mathrm{cm}^{-1}$ (O2+O3) modes must have considerable admixture of the $500\,\mathrm{cm}^{-1}$ mode of the apical oxygens.

Ab initio calculations for Raman tensors have also been performed for Y-124 [48]. Results for the scattering efficiency vs. ω_L calculated using the eigenvectors of Table 5.3 are shown in Fig. 5.9. They correspond reasonably well to the experimental data. Particularly rewarding is the agreement found for the *absolute* scattering efficiencies.

5.2.3 The Phases of the Raman Tensors

The Raman tensor $\boldsymbol{R}_j(\omega)$ for the jth Raman phonon of an insulator is real below the lowest absorption edge. Above this edge, and also everywhere in

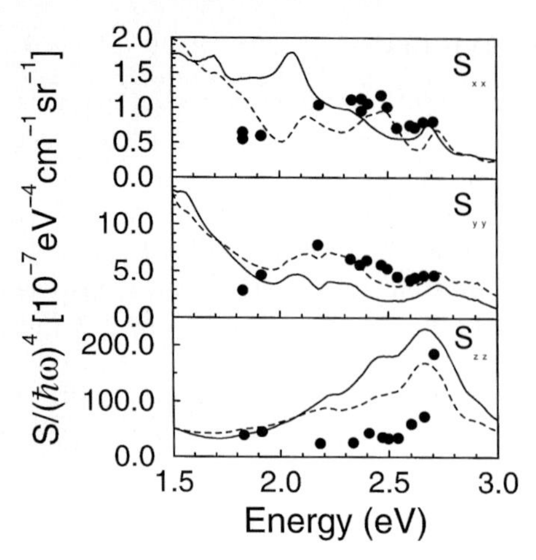

Fig. 5.9. Raman scattering efficiencies for the A_g apical O4 phonons of Y-124 versus $\omega_{\rm L}$ [normalized to $(\hbar\omega)^4$] calculated from the LMTO-LDA band structure (see text) for the (xx), (yy), and (zz) configurations. The solid lines represent the efficiency for the unmixed modes of O4 while the dashed lines give the efficiencies calculated for the corresponding mixed eigenvector of Table 5.2. The points are experimental [48]

the case of a metal, it is complex, corresponding to the fact that the susceptibility tensor $\boldsymbol{\chi}(\omega)$ is complex. For a tetragonal (or orthorhombic) crystal, $\boldsymbol{R}_j(\omega_{\rm L})$ has two (three) independent complex diagonal components at each laser frequency $\omega_{\rm L}$, each of which can be represented by a magnitude R and a phase φ. For polarizations $\hat{\mathbf{e}}_{\rm L}$ and $\hat{\mathbf{e}}_{\rm S}$ parallel to the tensor axes, only the corresponding magnitude of the Raman tensor can be determined by spontaneous Raman scattering, which is an *incoherent* process. It is possible, however, to determine the relative phases of two diagonal components of $\boldsymbol{R}(\omega_{\rm L})$ by using polarizations tilted with respect to the crystal axes. This has been recently done for Sm-123 by Strach et al. [71]. The samples were twinned thin films and thus had average D_{4h} symmetry. The phase difference between $R_{\bar{x}\bar{x}}(\omega)$ ($\bar{x}$ is the average of x and y) and $R_{zz}(\omega)$ was determined for the 114, 146, 428, and 506 cm^{-1} z-polarized phonons as follows.

A thin film of Sm-123, oriented with the z-axis in the plane, was deposited on a (110) surface of a $\mathrm{SrTiO_3}$ crystal. The Raman efficiencies for the four A_{1g} (in D_{4h} notation) phonons were measured as a function of the angle Θ between $\hat{\mathbf{e}}_{\rm L}$ and $\hat{z}$, both for $\hat{\mathbf{e}}_{\rm L} \parallel \hat{\mathbf{e}}_{\rm S}$ and $\hat{\mathbf{e}}_{\rm L} \perp \hat{\mathbf{e}}_{\rm S}$. Results for $\hat{\mathbf{e}}_{\rm L} \parallel \hat{\mathbf{e}}_{\rm S}$ are shown by the dots in Fig. 5.10. They were fitted with the equation (obtained from (5.2)):

$$I_{\|}(\omega,\Theta) = \alpha^2 \sin^4\Theta + \gamma^2\cos^4\Theta + 2\alpha\gamma\sin^2\Theta\cos^2\Theta\cos(\varphi_\alpha - \varphi_\gamma)\,, \tag{5.8}$$

where α and γ represent the magnitude of the $\overline{xx}$ and zz components of the Raman tensor and φ_α, φ_γ the corresponding phase angles. The solid lines in Fig. 5.10 display the best fit with (5.8) which yields the values of $\varphi_{\alpha\gamma} = \varphi_\alpha - \varphi_\gamma$ given in the figure. The dashed (dotted) curves represent the fits obtained when setting $\varphi_{\alpha\gamma} = 0$ ($\varphi_{\alpha\gamma} = \pi$).

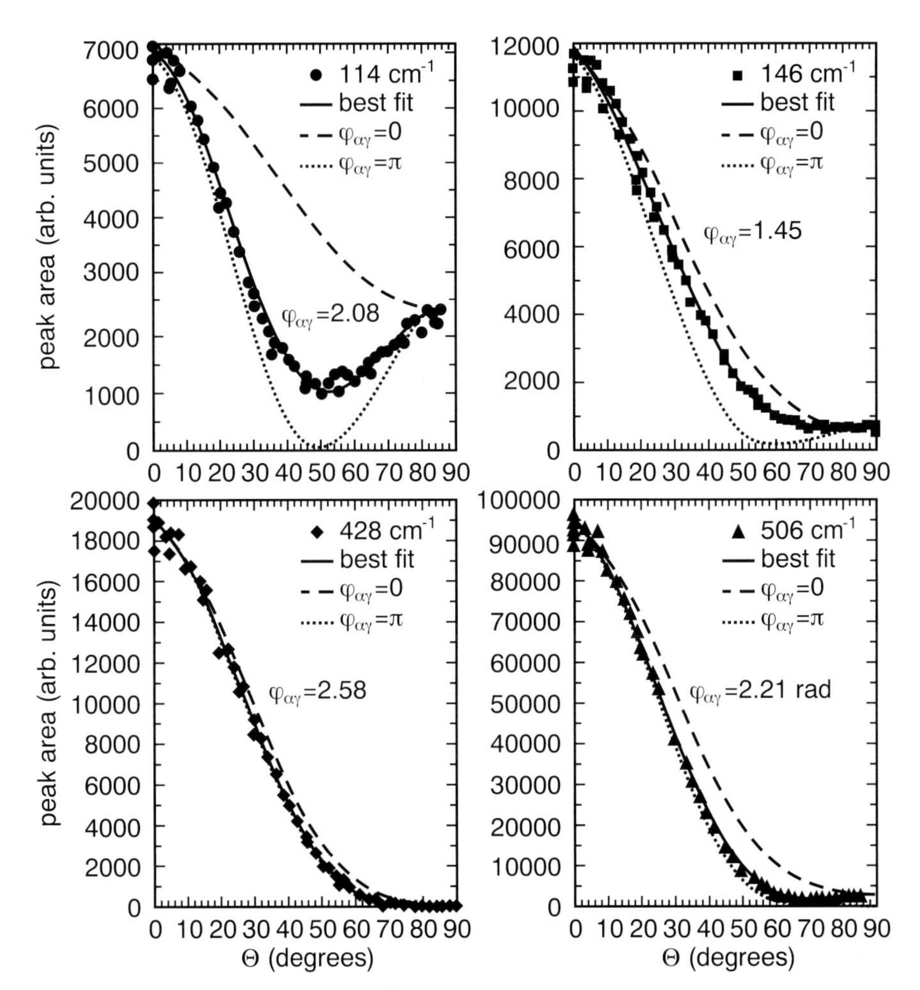

Fig. 5.10. Raman intensities $I_{\|}(\Theta)$ of the four A_{1g}-like phonons of $SmBa_2Cu_3O_{7-\delta}$ at room temperature measured on a (110) surface in parallel polarization as a function of angle Θ. *Solid symbols* correspond to experimental values determined from the spectra by numerical fits. *Solid lines* represent the results of a least squares fit of (5.8) to the data. The best fit values for $\varphi_{\alpha\gamma} = \varphi_\alpha - \varphi_\gamma$ are given in the figures. The *dashed* and *dotted lines* show intensity functions calculated by using the same values $|\alpha|$ and $|\gamma|$ as determined for the best fit, but phase shifts $\varphi_{\alpha\gamma} = 0$ and $\varphi_{\alpha\gamma} = \pi$, respectively [71]

The dependence of $I_{\|}$ on Θ is expected to be different for each laser frequency. We show in Fig. 5.11 the anisotropy ratio $|a/c|$ and the phase difference φ_{ac} for the $146\,\mathrm{cm}^{-1}$. Cu2 modes vs. ω_{L}. The solid curves represent the results of calculations for Y-124 [48]: when performing this work calculated values for φ_{ac} were not available for Sm-123.

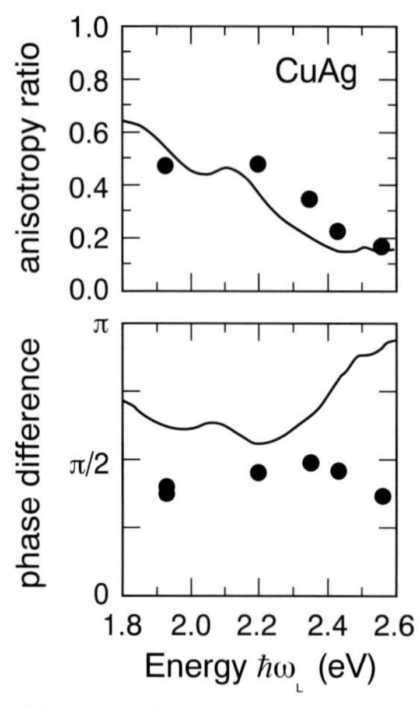

Fig. 5.11. Anisotropy ratio and phase difference between the Raman in-plane tensor components of Sm-123 parallel and perpendicular to the planes. The *points* are experimental. The *lines* were calculated from the Raman tensors of Y-124 [48,71]

5.3 Scattering by Intraband Electronic Excitations

5.3.1 Normal Metals

Two types of Raman scattering by electronic excitations can be observed in crystals

(a) *Interband excitations* from an occupied (at least partly) to an unoccupied band. These transitions are possible in insulators (or semiconductors) as well as in metals. If parity is a good quantum number (i.e., in the presence of inversion symmetry) the scattering Hamiltonian couples, in the dipole approximation, only electronic states of the same parity.
(b) *Intraband excitations* in the case of metals or doped semiconductors. The resulting excitations are automatically *even* since the initial and final

Bloch functions are nearly the same (at least for low energy excitations). Nevertheless, low frequency excitations in *clean* metals cannot be produced by optical techniques because of the small transfer wavevectors $\boldsymbol{k}$ associated with the optical excitations. The problem is schematically illustrated in Fig. 5.12a: For any given initial states below E_{F} there are no final states with energy $\hbar\omega = \hbar\omega_{\mathrm{L}} - \hbar\omega_{\mathrm{S}}$ and $\boldsymbol{k} = \boldsymbol{k}_{\mathrm{L}} - \boldsymbol{k}_{\mathrm{S}}$. Nor are such excitations found in IR spectroscopy, where $\boldsymbol{k}$ is even smaller. This problem is related to the so-called Landau damping: in order to obtain excitations in a *clean* metal at a frequency ω one must have a minimum transfer of wavevector:

$$\frac{k}{k_{\mathrm{F}}} = \frac{\omega}{2\omega_{\mathrm{F}}}, \tag{5.9}$$

where $\hbar\omega_{\mathrm{F}}$ represents the Fermi energy ($\approx 1\,\mathrm{eV}$) and k_{F} is the magnitude of the Fermi wavevector, which in a metal is of the order of π/a. For $\hbar\omega \simeq 0.05\,\mathrm{eV}$, equation (5.9) requires $k \simeq 0.03 k_{\mathrm{F}}$, a value too high to be reached with standard optical photons (see Fig. 5.12a).

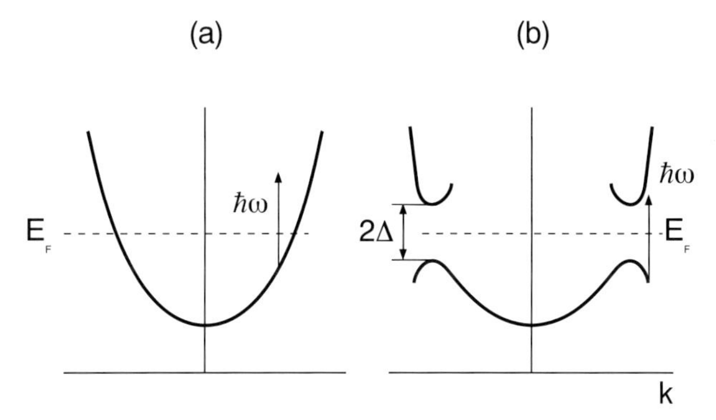

Fig. 5.12. (**a**) Schematic diagram of the conduction band structure and the Fermi surface of a metal in the normal state. E_{F} represents the Fermi energy. The arrow labeled $\hbar\omega$ symbolizes intraband excitations which are optically forbidden in the electric dipole approximation (no real final state). (**b**) Corresponding diagram for the superconducting state. Note that the pair-breaking transitions become allowed

We have discussed, so far, the *clean limit*. In the so-called *dirty limit*, impurities, phonons, surfaces and other crystal "defects" or higher order many-body interactions produce a short electron mean free path l which results in a smearing of the scattering $\boldsymbol{k}$-vector by an amount $1/l$. For l sufficiently short, (5.9) can then be satisfied with optical spectroscopies: both Raman scattering and IR absorption can induce intraband (so-called *pair breaking*) electronic excitations. The exact clean limit can never be reached since, in the best of the cases, we have scattering by acoustic phonons and also by many-body electron–electron interactions. Correspondingly, metals always absorb somewhat in the IR: except as very thin films, they never become transparent. Actually, below the *screened* plasma frequency Ω_p, the reflectivity equals *one* in the clean limit, so that even if the metal is not lossy, it completely reflects the light and, thus, becomes opaque. Consequently, infrared spectra are determined by both, the real and the imaginary parts of the susceptibility $\chi_1(\omega)$ and $\chi_2(\omega)$, respectively. A very precise novel technique, that of FT-ellipsometry [72], has been recently successful for the determination of $\chi_1(\omega) + \mathrm{i}\chi_2(\omega)$ in high-T_c superconductors [73], especially using synchrotron radiation as a source [29].

Returning to Raman scattering, the scattering Hamiltonian is obtained from that of a free electron:

$$H = \frac{1}{2m}\left(\boldsymbol{p} + \frac{e}{c}\boldsymbol{A}\right)^2 \tag{5.10}$$

by replacing $\boldsymbol{A}$ by the sum of the incident and scattered vector potentials $\boldsymbol{A} = \boldsymbol{A}_\mathrm{L} + \boldsymbol{A}_\mathrm{S}$. This leads to the Hamiltonian for *Thompson scattering*:

$$H_\mathrm{R} = \frac{e^2}{c^2}\frac{1}{m}\boldsymbol{A}_\mathrm{L} \cdot \boldsymbol{A}_\mathrm{S} \ . \tag{5.11}$$

This Hamiltonian is equivalent to a static (longitudinal) electric potential. It produces, as a response, a charge density fluctuation which, at low frequencies, becomes screened by the dielectric response of the metal. The resulting scattering efficiency is

$$\frac{\partial^2 S}{\partial\Omega\partial\omega} \propto -\mathrm{Im}\left[\varepsilon_0 + 4\pi\chi(\omega)\right]^{-1} , \tag{5.12}$$

where $\omega = \omega_\mathrm{L} - \omega_\mathrm{S}$, $\chi(\omega)$ is the electric susceptibility of the free electron gas and ϵ_0 the remaining background dielectric constant (due mainly to electronic interband excitations). Using for $\chi(\omega)$ the Drude expression:

$$\chi = -\varepsilon_0 \frac{\Omega_\mathrm{p}^2}{\omega(\omega + \mathrm{i}\Gamma)} , \tag{5.13}$$

we find from (5.12) a scattering spectrum that vanishes for $\omega \to 0$ but peaks at the *screened plasma frequency* Ω_p:

$$\Omega_\mathrm{p} = \left(\frac{4\pi N e^2}{\varepsilon_0 m}\right)^{1/2} , \tag{5.14}$$

where N is the concentration of free carriers (per unit volume). In Y-123, $\hbar\Omega_\mathrm{p} \simeq 1\,\mathrm{eV}$ for in-plane electric fields. This frequency, which is quite large (for Raman spectroscopy) and overdamped ($\Gamma \simeq \omega$), has not been seen in the Raman spectra. It can be indirectly estimated from the frequency at which $\varepsilon_1(\omega) = \varepsilon_0 + 4\pi\chi_1(\omega)$ vanishes.

Equation (5.11) represents the behavior of free electrons. In ordinary metals, however, free electrons are subject, in addition, to a periodic crystal potential $V(\boldsymbol{r})$ which leads to folding of the electron energy bands into the first Brillouin zone and the appearance of interband transitions between the occupied and the unoccupied folded bands. These transitions are induced by the $\boldsymbol{p}\cdot(\boldsymbol{A}_\mathrm{L}+\boldsymbol{A}_\mathrm{S})$ terms of (5.10). Neglecting resonance effects, i.e., assuming that ω_L and ω_S are quite far from resonant interband electronic transitions, it is possible to use $\boldsymbol{k}\cdot\boldsymbol{p}$ theory [74–76] and eliminate the crystal potential by introducing an inverse *tensor* effective mass $\boldsymbol{\mu}^{-1}$. Equation (5.11) can thus be generalized to read:

$$H_\mathrm{R}=\frac{e^2}{c^2}\left(\boldsymbol{A}_\mathrm{L}\cdot\boldsymbol{\mu}^{-1}\cdot\boldsymbol{A}_\mathrm{S}\right) . \tag{5.15}$$

The accuracy of the non-resonance assumption that leads to (5.15) is hard to verify. Calculations of scattering efficiencies without this assumption suggest that (5.15) is semiquantitatively correct for high-T_c superconductors when using ordinary laser frequencies [75]. Strong deviations from the predictions of (5.15) are found at the direct gaps of semiconductors [74,77,78].

For a metal in which all electrons are described by the same $\boldsymbol{\mu}^{-1}$, with the same orientation of the axes of the mass tensor, the perturbation Hamiltonian (5.15) also leads to screened low energy excitations and a peak at the plasma frequency, which depends on the directions of $\boldsymbol{A}_\mathrm{L}$ and $\boldsymbol{A}_\mathrm{S}$ if $\boldsymbol{\mu}^{-1}$ is anisotropic (i.e., in noncubic crystals). A more interesting case is that in which there are two or more groups of carriers with different masses (this includes HTSC with two or more sheets of the Fermi surface, such as for Y-123). It can even happen in cubic crystals such as silicon [74,77]. It also applies to a single Fermi surface sheet provided $\boldsymbol{\mu}^{-1}$ varies significantly around that sheet. In these cases it is possible to scatter via partly unscreened low frequency excitations: two (or more) terms induced by (5.15) with different masses can combine to yield carrier fluctuations with zero charge but nonzero expectation values of the Hamiltonian (5.15) [79]. Such unscreened multicarrier Raman processes have been profusely studied for semiconductors [74,77,78]. They also appear in the normal state of high-T_c superconductors [79]. In the latter case the frequency dependence of the scattering efficiency (actually the lack of it!) led Varma et al. [80] to postulate the existence of a marginal Fermi liquid.

The following simple phenomenological expression has been often used to account for the Raman efficiency of electronic scattering in a metallic system with a distribution of carrier masses [77,79]:

$$\frac{\partial^2 S}{\partial\omega\partial\Omega}=\left[1+n_\mathrm{B}(\omega,T)\right]\frac{\omega\Gamma B}{\omega^2+\Gamma^2} , \tag{5.16}$$

where Γ represents the scattering rate and B (dimensions: $L^{-1}T$) is given by:

$$B = \frac{\hbar e^4}{\pi c^4} N(E_{\mathrm{F}}) \left\langle \left| \hat{e}_{\mathrm{L}} \cdot \left(\frac{1}{\boldsymbol{\mu}} - \left\langle \frac{1}{\boldsymbol{\mu}} \right\rangle_{\mathrm{F}} \right) \cdot \hat{e}_{\mathrm{S}} \right|^2 \right\rangle_{\mathrm{F}} , \tag{5.17}$$

where $\langle \ldots \rangle_{\mathrm{F}}$ represents an average over the Fermi surface.

The components of the $\boldsymbol{\mu}^{-1}$ tensor are given by:

$$\mu_{ij}^{-1} = \hbar^{-2} \left(\frac{\partial^2 E}{\partial k_i \partial k_j} \right)_{E_{\mathrm{F}}} . \tag{5.18}$$

The derivatives in (5.18) have been evaluated for Y-123 and Y-124 from the electron energy bands calculated ab initio with the LMTO-LDA method [81,82] Using reasonable values of the loss parameter $\Gamma(\omega, T)$ obtained within the marginal Fermi liquid model (i.e., basically proportional to ω or to T, whatever the larger, or, in analytical form $\Gamma \propto \sqrt{[\alpha\omega]^2 + T^2}$, with α an adjustable parameter of the order of unity), it was possible to calculate absolute values of the electronic scattering efficiencies of high-T_{c} materials in the normal state that are in semiquantitative agreement with experimental data. Such data are obtained in the form of different symmetry components of the Raman spectra generated by using different configurations for the incident and scattered polarizations. The different irreducible symmetry components correspond to different symmetrized components of $\boldsymbol{\mu}^{-1}$: a *symmetric* second rank tensor possesses, within the D_{4h} point group, two components of A_{1g} symmetry (μ_{zz}^{-1} and $\mu_{xx}^{-1} + \mu_{yy}^{-1}$), one of B_{1g} symmetry ($\mu_{xx}^{-1} - \mu_{yy}^{-1}$), one of B_{2g} symmetry (μ_{xy}^{-1}), and one of E_g symmetry (doubly degenerate, $\mu_{xz}^{-1} = \mu_{yz}^{-1}$). Since the electronic band structure of high-T_{c} materials is nearly two dimensional (it may be exactly two dimensional if carrier *hopping* between CuO_2 planes along z is incoherent, as often assumed), all inverse masses with one or two indices z can be safely set to zero (i.e., Raman scattering for the corresponding polarizations is very weak [83]). We are then left with the mass components $\mu_{xx}^{-1} + \mu_{yy}^{-1}$, $\mu_{xx}^{-1} - \mu_{yy}^{-1}$ and μ_{xy}^{-1}, which must be contracted with $\hat{\mathbf{e}}_{\mathrm{L}}$ and $\hat{\mathbf{e}}_{\mathrm{S}}$, squared, and averaged over the Fermi surface (or surfaces) in order to obtain the scattering efficiency. The so-called B_{1g} component of the spectra ($x^2 - y^2$ symmetry) is obtained for $\hat{\mathbf{e}}_{\mathrm{L}} \perp \hat{\mathbf{e}}_{\mathrm{S}} \parallel (x+y)$ while B_{2g} (xy symmetry) is obtained for $\hat{\mathbf{e}}_{\mathrm{L}} \perp \hat{\mathbf{e}}_{\mathrm{S}} \parallel (x)$. In order to determine the A_{1g} component, we must calculate linear combinations of spectra for $\hat{\mathbf{e}}_{\mathrm{L}} \parallel \hat{\mathbf{e}}_{\mathrm{S}}$ and the B_{1g} or B_{2g} components measured for $\hat{\mathbf{e}}_{\mathrm{L}} \perp \hat{\mathbf{e}}_{\mathrm{S}}$ (this procedure, unfortunately, introduces errors). For instance we find for $\hat{\mathbf{e}}_{\mathrm{L}} \parallel \hat{\mathbf{e}}_{\mathrm{S}} \parallel x$ the symmetry components:

$$x^2 = \frac{1}{2}\left(x^2 + y^2\right) + \frac{1}{2}\left(x^2 - y^2\right) = A_{1g} + B_{1g} , \tag{5.19}$$

and for $\hat{\mathbf{e}}_{\mathrm{L}} \parallel \hat{\mathbf{e}}_{\mathrm{S}} \parallel (x + y)$:

$$\frac{1}{2}(x + y)^2 = \frac{1}{2}\left(x^2 + y^2\right) + (xy) = A_{1g} + B_{2g} . \tag{5.20}$$

In spite of the error bars inherent to the determinations of absolute cross sections, the agreement between measured and calculated quantities is rather remarkable [81,82].

As already discussed, Y-123 and Y-124 do not belong to the D_{4h} tetragonal point group but to the orthorhombic D_{2h} (although for the CuO_2 planes D_{4h} should be a reasonable approximation). In D_{2h} notation the (x, x) and (y, y) polarization configurations both have A_g symmetry but need not lead to the same scattering efficiency: scattering in the (y, y) configuration is usually twice as strong as (x, x) for Y-123 and Y-124. For these materials $(x+y)(x-y) = (x^2-y^2)$ becomes fully symmetric, so that three spectra of the same symmetry (A_g) are found [corresponding to (y, y), (x, x), and $(x+y)(x-y)$]. Only the (xy) polarization configuration leads to a non-fully-symmetric spectrum (B_{1g} in D_{2h} notation for Y-123 and Y-124). This is the case treated in [81,82]. We recall the fact, already discussed in Sect. 5.1, that there are two possible ways of lowering the D_{4h} symmetry to D_{2h}. One of them applies to Y-123, where the symmetry is broken by the presence of chains along y. The other obtains in the bismuth superconductors and corresponds to a distortion along $x+y$ (we keep the x and y axes along the Cu-O2, Cu-O3 bonds). In this case the (xy) polarization configuration becomes fully symmetric (A_g) while $(x+y)(x-y) = (x^2-y^2)$ reverses sign upon reflection against the $(x+y)$ plane. Its symmetry remains B_{1g} in the D_{2h} (45°) group (see Table 5.1). These symmetry considerations will be important in the next subsection.

At this point, a few remarks about the *screening* A_g spectra are needed. The screened scattering efficiencies are proportional to the mass variance [77,79]:

$$\left\langle |\mathbf{e}_{\mathrm{L}} \cdot \boldsymbol{\mu} \cdot \mathbf{e}_{\mathrm{S}}|^2 \right\rangle_{\mathrm{F}} - |\langle \hat{\mathbf{e}}_{\mathrm{L}} \cdot \boldsymbol{\mu} \cdot \hat{\mathbf{e}}_{\mathrm{S}} \rangle_{\mathrm{F}}|^2 \,, \tag{5.21}$$

where F represents an average over the Fermi surface. The second term in (5.21) vanishes for nonsymmetric polarization configurations (e.g., B_{1g} and B_{2g}) since a nonsymmetric function yields a zero average upon application of all symmetry operations of a point group. This term represents the screening of the charge density fluctuations. Hence, *B_{1g} and B_{2g} spectra are not screened.* The A_g spectra, however, are fully screened if all carriers have the same mass *tensor*, as can be easily shown by taking the constant matrix elements out of the $\langle \ldots \rangle_{\mathrm{F}}$ average: mass tensors varying around the Fermi surface are required if the symmetric spectra are not to vanish at low frequencies. Otherwise we are back to the fully screened case in which only a peak at Ω_{p} is observed.

Note that at high temperatures the statistical factor $[1+n_{\mathrm{B}}(\omega, T)]$ in (5.2) is equal to $k_{\mathrm{B}}T/\hbar\omega$, and the marginal Fermi liquid model gives $\Gamma \propto T$. Replacing these values into (5.16) we find a scattering efficiency independent of T and ω, in agreement with experimental observations. This was, in fact, the motivation of Varma et al. [80] for introducing the marginal Fermi liquid

(MFL) model. Within this model $\Gamma \propto \omega$ for $k_B T \ll \hbar\omega$. This and (5.16) also yields a frequency independent scattering efficiency.

In the past few years, some effort has been made to put the theory of the scattering parameter Γ on a more microscopic basis [84–86]. Usually, the general idea is to relate $\Gamma(\omega, T)$ to the imaginary part of an electron self-energy induced by emission and reabsorption of a virtual boson. Since the nature of this boson is not well known, however, one must accept some degree of phenomenology in this type of approach. The most logical choice for this boson is an antiferromagnetic fluctuation of the type considered as likely candidate for effecting the superconducting pairing.[6]

In most of the theoretical publications the scattering efficiency of (5.16) is recast into the form:

$$\frac{\partial^2 S}{\partial\omega\partial\Omega} = \frac{r_0^2}{\pi}\left[1 + n_B(\omega, T)\right] \mathrm{Im}\chi_R(\omega, T) , \tag{5.22}$$

where $\chi_R(\omega, T)$ is the appropriate component of the so-called Raman susceptibility tensor, which includes the mass vertex of (5.17), $r_0 = e^2/mc^2$ is the Thomson radius and the wavevector transfer has been set equal to zero as corresponds to the dipole approximation (long laser wavelength). Within the phenomenological model of (5.16) we have:

$$\chi_R(\omega, t) = \chi_R' + \chi_R'' \propto \frac{\omega B}{\omega - i\Gamma} . \tag{5.23}$$

The aim of the more recent calculations [84–86] has been to obtain analytical or numerical values for $\Gamma(\omega, T)$ involving a physically transparent mechanism. As already mentioned, the usual ansatz corresponds to antiferromagnetic fluctuations having wavevectors within a narrow range centered around $\frac{\pi}{a}(\pm 1, \pm 1, 0)$. Such wavevectors couple strongly points of the Fermi surface close to $\frac{\pi}{a}(\pm 1, 0, 0)$ with those around $\frac{\pi}{a}(0, \pm 1, 0)$. Electrons at these points (the so-called *hot spots*) are, therefore, heavily damped. According to (5.16) the scattering efficiency will be small for small ω ($\sim \frac{\omega}{\Gamma} \approx \frac{\omega}{T}$, $T < T_c$) and large for large ω ($\sim \frac{\Gamma}{\omega} \approx 1$), with a crossover at $\omega \approx \Gamma$. This crossover can be seen in Fig. 5.13 for B_{1g} and B_{2g} components of $\chi''(\omega, T)$ calculated and measured for an overdoped Bi-2212 sample.[7] Note that the temperature dependence of the B_{2g} spectra in Fig. 5.13 is considerably weaker than that of the B_{1g} spectra. This is related to the fact that the *hot spots* contribute strongly to B_{1g} spectra (the B_{1g} mass vertex is strongest at $\approx \frac{\pi}{a}(\pm 1, 0, 0)$ and $\frac{\pi}{a}(\pm 1, 0, 0)$, i.e., at the *hot spots* while the B_{2g} mass vertex (xy symmetry) is strongest along the lines $\frac{\pi}{a}(\pm\xi, \pm\xi, 0)$ where only *colder spots* appear).

[6] For an analytical expression corresponding to this interaction see (13) in [85] or (16) of [84].

[7] Note that most recent semimicroscopic calculations treat only B_{1g} and B_{2g} spectra so as to avoid the complications associated with screening and with the experimental uncertainties [84–86].

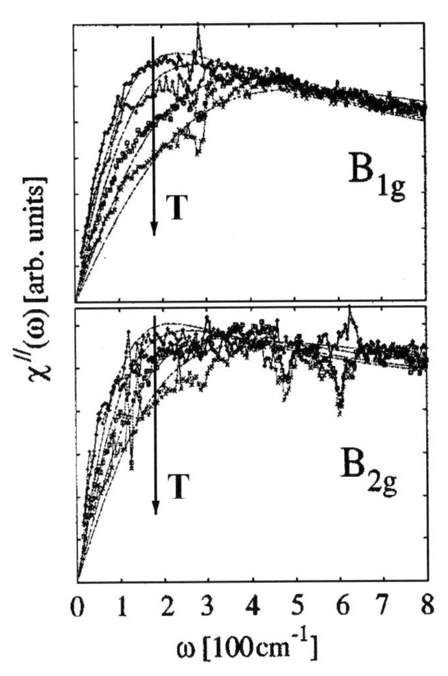

Fig. 5.13. Fit of the theory of [85] to the B_{1g} and B_{2g} spectra reported in [87] for *overdoped* $Bi_2Sr_2CaCu_2O_{8+\delta}$ ($T_c = 55$ K) at 60, 90, 150, and 200 K from top to bottom. From [85]

It is of interest to consider the slope of $\chi''(\omega, T)$ for $\omega \to 0$, which according to (5.23) should be proportional to Γ^{-1}. The inverse slopes Γ corresponding to the calculations of Fig. 5.13 are plotted in Fig. 5.14 versus temperature for the B_{1g} and B_{2g} (D_{4h} notation) scattering channels of overdoped Bi2212. For the B_{1g} channel, a larger inverse slope is found at all temperatures, as expected from the hot spot argument given above. Note also that in the B_{1g} case, Γ is approximately proportional to T, as expected for a marginal Fermi liquid. When multiplying this T-dependence of Γ^{-1} by $\frac{T}{\omega}$, the high temperature, low frequency limit of $[1 + n(\omega, T)]$, a temperature independent scattering efficiency is found, in agreement with the experimental observations for optimally doped samples. Figure 13 of [85] displays results similar to those in our Fig. 5.14 but for an underdoped sample. For the B_{2g} channel the *inverse slope* turns out to be somewhat larger than for the B_{1g} channel, the latter being approximately given by:

$$\Gamma(T) \simeq \Gamma_0 \left(1 + 10^{-3} T\right) . \tag{5.24}$$

When multiplying the result of (5.24) by the high temperature Bose–Einstein factor we predict a considerable increase of $\partial^2 S / \partial\omega \partial\Omega$ with increasing temperature which has been observed experimentally for Y-124, an

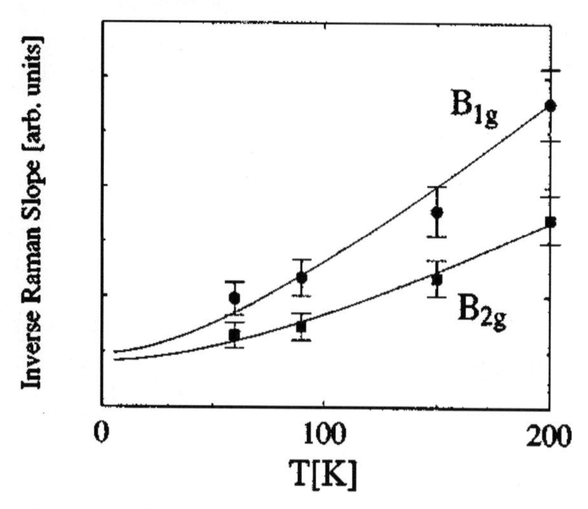

Fig. 5.14. Fit of the calculated inverse slopes of the Raman response for the two symmetry channels B_{1g} and B_{2g}. The inverse slope corresponds to the Γ of (5.16). From [85]

intrinsically underdoped HTSC [27]. The authors of [27] performed measurements for a Y-124 crystal under pressure: it is known that pressure increases doping in Y-124 which becomes optimally doped under a pressure of about 10 GPa [88]. The Raman spectra measured as a function of pressure (both B_{1g} and B_{2g} components) yield spectra that depend on temperature following the Bose–Einstein factor at atmospheric pressure while they become temperature independent at pressures of 10 GPa ($\approx$100 kbar) at which Y-124 becomes optimally doped [88]. This type of investigation should be pursued in more detail, especially side-by-side with the theoretical work now in progress.

5.3.2 Scattering in the Superconducting State

Probably the first observation of a superconducting *pair breaking* gap by Raman spectroscopy was made by Lyons et al. [2]. We show in Fig. 5.15 a spectrum measured in 1987 with unpolarized light for an untwinned crystal of Y-123 above and below T_c [41]. From this spectrum, which was taken on an (x, y) plane but with unpolarized light, the B_{1g} phonon that appears at 340 cm^{-1} has been subtracted. The polarized spectra of the superconducting phases, different for different polarizations, suggest the presence of an anisotropic gap with nodes around the Fermi surfaces (a spectral weight is observed down to the lowest Raman frequency) [89]. The maximum gap revealed by Fig. 5.15 is $2\Delta_0 \simeq 350\,\mathrm{cm}^{-1}$; for $T_c \simeq 90\,\mathrm{K}$ it leads to $(2\Delta_0/k_\mathrm{B}T_c) \simeq 5.6$, much larger than the BCS value ($2\Delta_0/k_\mathrm{B}T_c \simeq 3.6$). These high values are typical for HTSC.

Within the framework of a BCS-like theory the pair breaking energy can be written as:

$$\Delta E_{\boldsymbol{k}} = 2\sqrt{\Delta_{\boldsymbol{k}}^2 + \epsilon_{\boldsymbol{k}}^2}\,, \tag{5.25}$$

where $\epsilon_{\boldsymbol{k}}$ is the normal state energy of electrons and holes measured from the Fermi energy. Equation (5.25) indicates that in the superconducting state finite excitation energies $\Delta E_{\boldsymbol{k}} = 2\Delta_{\boldsymbol{k}}$ become possible even for $\boldsymbol{k} \simeq 0$ (see Fig. 5.12b): These excitations are not possible in the normal state, at least in the clean limit. Calculations for $T < T_\mathrm{c}$ and $\Gamma \to 0$ should, therefore, yield a peak at the maximum value of $2\Delta_{\boldsymbol{k}}$ which we call Δ_0. Above this energy, however, the scattering efficiency calculated for $\Gamma \to 0$ will tend to that calculated for the normal state [3,90,91], i.e., to zero. Calculations for $T < T_\mathrm{c}$ which include the effects of Γ and thus lead to a finite scattering efficiency even for $\hbar\omega \gg 2\Delta_0$, have been published recently [84–86]. We discuss first the calculations for $\Gamma \to 0$ and compare them with experimental results.

When coupling spectroscopic Hamiltonians to pair breaking excitations one must include the so-called coherence factors which depend on the parity of that Hamiltonian [92]. It is easy to show that these coherence factors lead, in the case of IR spectra, to vanishing IR coupling in the clean limit ($\Gamma \to 0$). Hence only for finite carrier mean-free-path can one observe the gap 2Δ in IR spectra. For Raman excitations, however, the coherence factors interfere constructively and a gap, similar to that in Fig. 5.15, should appear in the measured spectra.

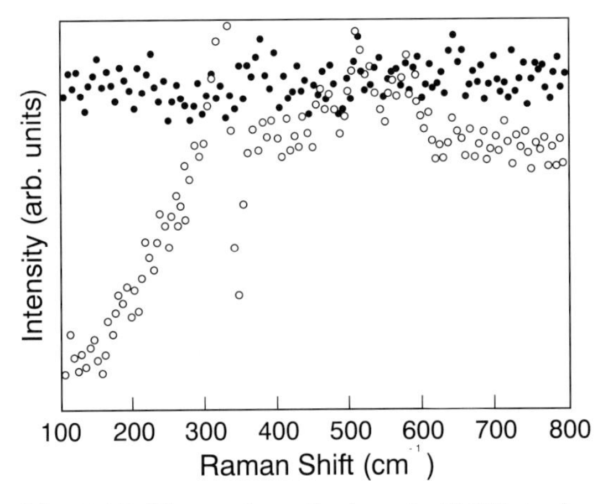

Fig. 5.15. Electronic scattering of a Y-123 single crystal for two temperatures. The *points* were obtained by subtracting the experimental data measured at $T = 100\,\mathrm{K}$ from those measured at $4\,\mathrm{K}$ (*open circles*) and at $89\,\mathrm{K}$ (*solid circles*). From [41]

The lineshape for the pair-breaking spectra was early calculated by Klein and Dierker [93]. The corresponding scattering efficiency is given by

$$\frac{\partial^2 S}{\partial\omega\partial\Omega} = \frac{\hbar e^4}{c^4}\mathrm{Im}\left\{\langle\mu^{-1}F\rangle_\mathrm{F} - \frac{\langle\mu^{-1}F\rangle_\mathrm{F}^2}{\langle F\rangle_\mathrm{F}}\right\}, \tag{5.26}$$

where the inverse mass μ^{-1} represents $\hat{\mathbf{e}}_\mathrm{L}\cdot\boldsymbol{\mu}^{-1}\cdot\hat{\mathbf{e}}_\mathrm{S}$, and F is the so-called Tsuneto function, an analytic function whose imaginary part is

$$\mathrm{Im}\,F(\omega,\boldsymbol{k}) = \pi N_\mathrm{F}\frac{4\,|\Delta_{\boldsymbol{k}}|^2}{\omega\sqrt{\omega^2 - 4\,|\Delta_{\boldsymbol{k}}|^2}}, \tag{5.27}$$

N_F being the density of electronic states at the Fermi surface. Note the close correspondence between the term in the curly brackets of (5.26) and (5.21): The second term inside the curly brackets of (5.26) represents the screening. Since F is fully symmetric with respect to the point group, the screening term vanishes if μ^{-1} does not contain a fully symmetric component (of A_g or A_{1g} symmetry). However, if μ is constant around the Fermi surface (5.26) also vanishes regardless of the $\boldsymbol{k}$-dependence of F.

Equation (5.26) has been evaluated by several authors, with different theoretical models for $\boldsymbol{\mu}^{-1}$ obtained either from semiempirical expressions for the electronic bands (tight binding method) [3] or from ab initio LMTO-ASA band structures [82,90,91]. In some of these calculations the averages over the Fermi surfaces implicit in (5.26) were strictly performed (similar to the procedure required to evaluate (5.21)) while others take into consideration the fact that $2\Delta_0$ can be a sizeable fraction of the Fermi energy, in which case a volume integration over the whole Brillouin zone may be more appropriate for handling the scattering efficiency than the Fermi surface integrations implicit in (5.26) [91]. Several assumptions are made for the variation of $\Delta_{\boldsymbol{k}}$ with $\boldsymbol{k}$, the most popular ones being nowadays the assumption of a gap of B_{1g} symmetry (in D_{4h} notation), i.e., either $k_x^2 - k_y^2$, $\cos k_x^2 - \cos k_y^2$, or $\cos 2\varphi$, where φ is the angle of $\boldsymbol{k}$ with the x-axis. In [91] calculations were carried out for several scattering configurations using LMTO band structures and performing both Fermi surface and Brillouin zone integrations. The results were shown not to be qualitatively different, the differences between the absolute scattering efficiencies calculated in the two ways just described only being important when there is a *van Hove singularity* near the Fermi surface.

We show in Fig. 5.16 the efficiencies for various scattering configurations calculated for optimally doped Y-123 using a three-dimensional Brillouin zone integration [91]. In all cases the unscreened and the screening component, plus their difference (i.e., the total efficiency to be compared with the measured one) are given. The (xy) configuration is nonsymmetric and, as expected, the screening term vanishes. The $(x'y')$ configuration $[x' = (x+y),\ y' = (x-y)]$ would be nonsymmetric for the exact D_{4h} symmetry (B_{1g}). The symmetry

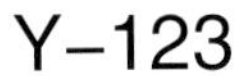

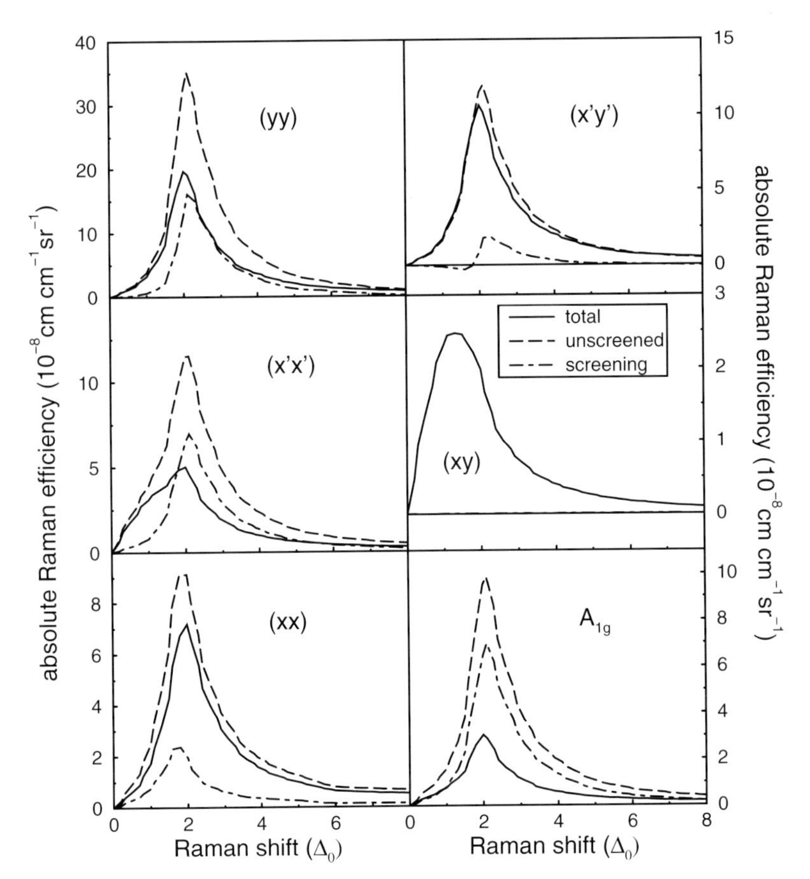

Fig. 5.16. Theoretical electronic scattering efficiencies obtained by BZ integration for Y-123 in the superconducting state. Each of the six panels contains the calculated total absolute Raman efficiency for electronic Raman scattering and its two constituents, the unscreened and the screening part evaluated according to (5.26). From [91]

of the calculated structure, however, is D_{2h} for which B_{1g} becomes A_g (fully symmetric): Hence a small amount of screening is present. The screening is, however, large for the parallel scattering configurations, which contain a large A_g component.

A comparison of the calculations of Fig. 5.16 with the experimental results of Fig. 5.17 leads to interesting observations. First of all note the expected fact that in the calculation of Fig. 5.16 the scattering efficiency tends to zero for $\hbar\omega \gg 2\Delta_0$, while this is not the case for the experimental spectra of Fig. 5.17. The most remarkable observation, however, is the close agreement between calculated and measured scattering efficiencies (absolute efficiencies

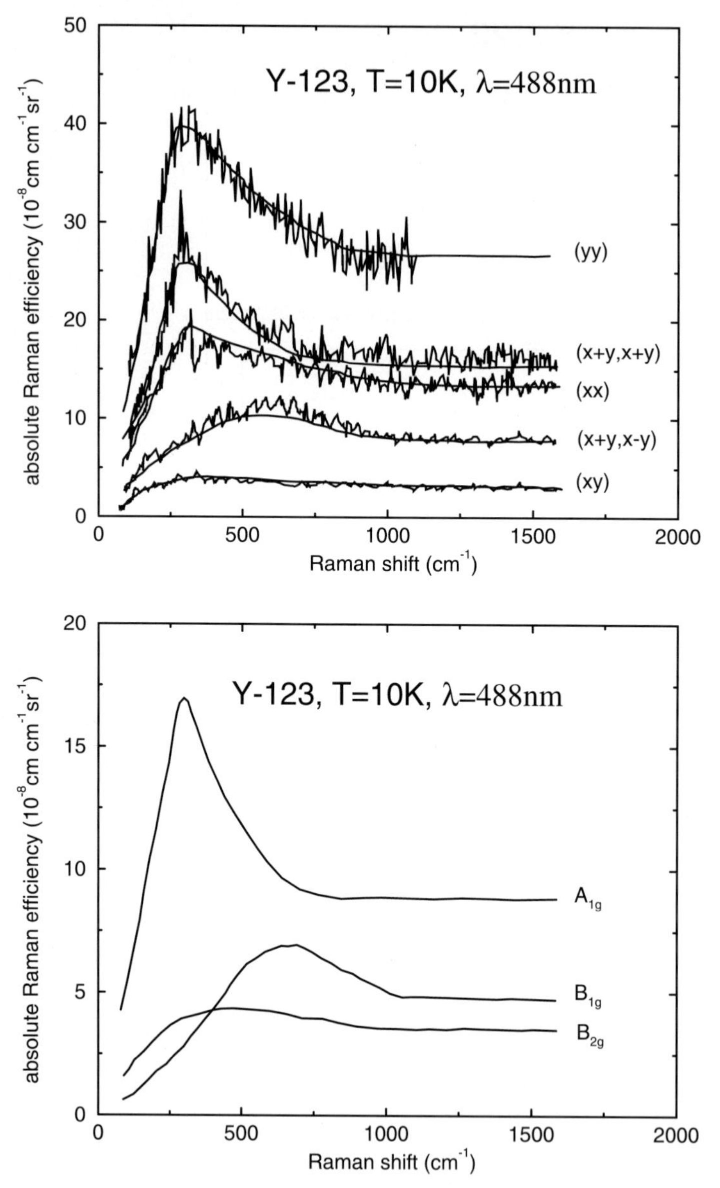

Fig. 5.17. Experimental Raman-scattering efficiencies for Y-123 from [91]. The vertical scales are absolute Raman efficiencies, measured at $T = 10\,\mathrm{K}$ and an exciting laser wavelength of $\lambda_\mathrm{L} = 488$ nm. The A_{1g} component extracted according to $I_{A_{1g}} = (I_{xx} + I_{yy})/2 - I_{x'y'}$ is plotted in the lower panel together with the quasitetragonal B_{1g} and B_{2g} components [90]

are only rarely found [94] in the Raman literature for solids). Of particular interest is the (relatively) small value of the maximum efficiency for the (xy) configuration (B_{2g} for the D_{4h}, B_{1g} for the D_{2h} point group of RE-123) and the fact that the maximum occurs well below $2\Delta_0$, the maximum gap used in the calculation which was performed with the ansatz $2\Delta_k \simeq 2\Delta_0 \cos 2\varphi$. The reason is that the $2\Delta_0$ gap is found at $\varphi = 0, \pi$ and $\pm\pi/2$. At these angles any function of xy symmetry vanishes; hence the small value of the (xy) efficiency (in spite of the lack of screening!) and the concomitant fact that the spectral maximum occurs well below $2\Delta_0$. The maximum efficiency occurs for the (y, y) configuration, both as calculated and measured. In the calculations this fact results from having implicitly included carriers in the chains along y since they cannot be easily separated from those in the planes in the 3D band calculation used to determine $\boldsymbol{\mu}^{-1}$. These chains, when decoupled from the planes, have a 1D band structure and thus only contribute to μ_{yy}^{-1}.

The most striking discrepancy between theory (Fig. 5.16) and measurements (Fig. 5.17) refers to the A_{1g} component, the experimental one being extracted as a linear combination of two spectra (given in Sect. 5.3.1). The calculated A_{1g} is, due to screening, considerably weaker than the B_{1g} (D_{4h}) component. In the measured spectra, however, the A_{1g} component appears to be the largest (see Fig. 5.17). The reason for this discrepancy is not known, although we know that the magnitude of A_{1g} depends rather critically on the details of the mass fluctuations around the Fermi surface. The presence of two or more CuO_2 planes is likely to enhance mass fluctuations and thus, by lowering the screening, to enhance the A_{1g} spectra (see, e.g., Fig. 17 of [90]). Strong fluctuations of $\boldsymbol{\mu}^{-1}$ can also be introduced ad hoc for a single Fermi surface so as to enhance the A_{1g} scattering by decreasing the screening [95,96]. This effect, however, is not at all *robust* [97,98] while the predominance of the A_{1g} component is rather general, at least in cases in which two or more CuO_2 planes are present.

Another important difference between the calculated and measured A_{1g}- and B_{1g}-like Raman pair-breaking spectra concerns the position of their maxima in the energy scale, which in the unscreened case should be $2\Delta_0$ (or slightly below if the Fermi surface is not continuous vs. φ, i.e., if pieces are missing around $k_x = 0$ or $k_y = 0$). In the calculated A_{1g} spectra it is possible to lower the peak through screening, but only by a small amount (see Fig. 10 of [90]). The experimental spectra of optimally doped Y-123 (see Fig. 5.17) show peaks in the A_{1g} component at about 0.6 times the frequency of the B_{1g} peaks (we have been implicitly considering *optimally doped* samples; for under or over-doped samples see [99]). Note that the measured ratio $2\Delta_0/k_\mathrm{B}T_\mathrm{c}$ increases when the doping decreases.

In [96] an attempt was made to bring the calculated spectra into better agreement with experimental ones by introducing vertex corrections to the scattering efficiency (5.26). Unfortunately, an algebraic error was made that, once corrected, led to a fully screened (i.e., with vanishing efficiency)

A_{1g} scattering component [96,97], a fortuitous result which is in complete disagreement with the experimental spectra. While there is universal agreement among workers in the field concerning the experimental results (at least for materials with two or more CuO_2 planes), the theoretical situation has been muddled by computational errors and delicate assumptions [95–97]. It is therefore difficult to decide which, if any, of the two peaks, that in the A_{1g} or that in the B_{1g} spectrum, should be assigned to the pair breaking gap $2\Delta_0$. The assignment to the A_{1g} peaks in most optimally doped cases leads to $2\Delta_0/k_B T_c \simeq 5$ while the B_{1g} assignment leads to $2\Delta_0/k_B T_c \simeq 8$. This difference, however, becomes smaller in the overdoped case [99]. The presence of two CuO_2 planes allows, in principle, the introduction of two different gaps, one related to the bonding and the other to the antibonding sheet of the Fermi surface [90]. One should, however, mention that the B_{1g} peak has been reported to persist *above* T_c, especially in *underdoped* materials, a property typical of the so-called *pseudogap*, which may be also related to the presence of *hot spots* for $\varphi = 0, \frac{\pi}{2}, \pi$, and $\frac{3\pi}{2}$ [100]. Note that in [86] it is pointed out that the discrepancy between the frequencies of the peaks in the B_{1g} and the A_{1g} spectra may be even more serious than generally believed, in particular in the underdoped region. These authors convincingly argue that the B_{1g} peak does not correspond to $2\Delta_0$ but to a quasiparticle bound state which, in the strongly underdoped region, peaks at a frequency $\approx 1.5\Delta_0$. Unfortunately, the recent theoretical papers [84–86] do not treat the A_{1g} spectra (possibly because of the experimental and theoretical difficulties mentioned in Sect. 5.3.1). If the B_{1g} spectra were to peak at $1.5\Delta_0$, a rather disturbing situation that would demand an explanation would arise. Moreover, $2\Delta_0/T$ would be as high as 12.

In Fig. 5.18 we reproduce results from [86] showing experimental B_{1g} Raman spectra of Bi-2212 in four doping regimes [101], compared with theoretical fits to the calculation mentioned above. These calculations include vertex corrections (i.e., interaction between electron-like and hole-like quasiparticles) which, in the underdoped case, lead to a quasibound state (a band resonance since this state overlaps with a weak scattering continuum always present down to $\omega \to 0$ for a $(k_x^2 - k_y^2)$-like gap). Figure 5.18 clearly shows that the *binding energy* of the quasibound state vanishes in the overdoped case (i.e., no quasibound state exists), is small for optimal doping and rather large ($\gtrsim 0.5\Delta_0$) in the underdoped samples. We should mention that the value of the gap Δ_0 used to normalize the frequency scale of Fig. 5.18 has been extracted from separate photoemission [102] and tunneling experiments [103] on different samples.

In spite of the controversy concerning the peak positions and the strengths of A_{1g} spectra, there seems to be a rather general agreement about the information contained in the asymptotic behavior of the scattering efficiency for

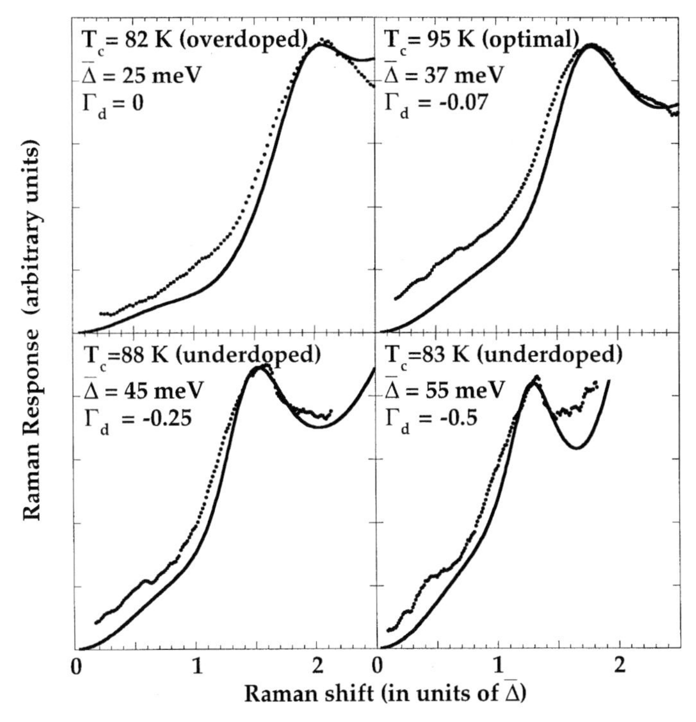

Fig. 5.18. Fits of the theoretical Raman intensity (B_{1g} component) with final state interaction to the experimental data for slightly overdoped, optimally doped and underdoped Bi-2212 materials. In the absence of the final state interaction, the peak frequency in the Raman intensity, even at strong coupling, is almost exactly twice the peak frequency in the density of states. The parameter Γ_d measures the strength of the vertex corrections which according to [86] give rise to a pseudo-resonance in the scattering efficiency. With underdoping, the peak in the Raman response progressively deviates down from $2\bar{\Delta}$

$\omega \to 0$. Such behavior can be represented by a power law [3]:

$$\left[\frac{\partial^2 S}{\partial\omega\partial\Omega}\right] \propto \omega^s \, . \tag{5.28}$$

The exponent s contains information about the symmetry properties of the gap function $\Delta_{\boldsymbol{k}}$ [3,97,98]. For a gap with $k_x^2 - k_y^2$, i.e., B_{1g} symmetry (D_{4h} notation) there are nodes of $\Delta_{\boldsymbol{k}}$ along the $k_x + k_y$ and $k_x - k_y$ diagonals of the Fermi surface. If the Raman vertex, i.e., the mass $\mu^{-1} = \hat{\mathbf{e}}_\mathrm{L} \cdot \boldsymbol{\mu}^{-1} \cdot \hat{\mathbf{e}}_\mathrm{S}$, does not vanish along these lines (e.g. for A_{1g} and B_{2g} spectra), a straightforward, density-of-states-type integration leads in this case to $s = 1$. If μ^{-1} also has nodes at the $x + y$ and $x - y$ diagonals (i.e., for a B_{1g}-scattering configuration) the Raman efficiency vanishes faster than linearly for $\omega \to 0$: the corresponding integration leads to $s = 3$ [3]. There is considerable evidence supporting this type of behavior, which can be translated into support for a B_{1g}-like [i.e., $\Delta_{\boldsymbol{k}} \sim \left(k_x^2 - k_y^2\right)$] gap. A rather striking example is shown in Fig. 5.19 for Hg-1223 [98], the superconductor with the highest T_c known ($T_c \simeq 130\,\mathrm{K}$) which has tetragonal symmetry (D_{4h} point group). The $\sim \omega^3$ behavior exhibited by the scattering efficiency for $\omega \to 0$ supports the presence of a gap with B_{1g} symmetry. We should recall, however, that Raman spectra are only sensitive to $|\Delta_{\boldsymbol{k}}|$, not to the phase of $\Delta_{\boldsymbol{k}}$ (the determination of the phase requires quantum interference experiments, see [104]). Hence, the $s = 3$ behavior of Fig. 5.19 would also be compatible with the A_{1g}-like $\left|k_x^2 - k_y^2\right|$ gap dependence. The spectra taken for the A_{1g} and B_{2g} scattering configurations reveal a linear behavior ($s = 1$) for $\omega \to 0$, compatible also with either a $(k_x^2 - k_y^2)$- or a $\left|k_x^2 - k_y^2\right|$-type of gap [3,98].

Of particular interest is the measured behavior of $\partial^2 S/\partial\omega\partial\Omega$ for $\omega \to 0$ (i.e., the fitted values of s) for orthorhombic high-T_c materials in the superconducting state. As mentioned in Sect. 5.1, there are two possible types of distortion that lead from D_{4h} to D_{2h} symmetry: a perturbation along the CuO bonds of the planes (e.g. that produced by the presence of chains) and a perturbation along the $x+y$ diagonal. The former (the case of Y-123) trans-

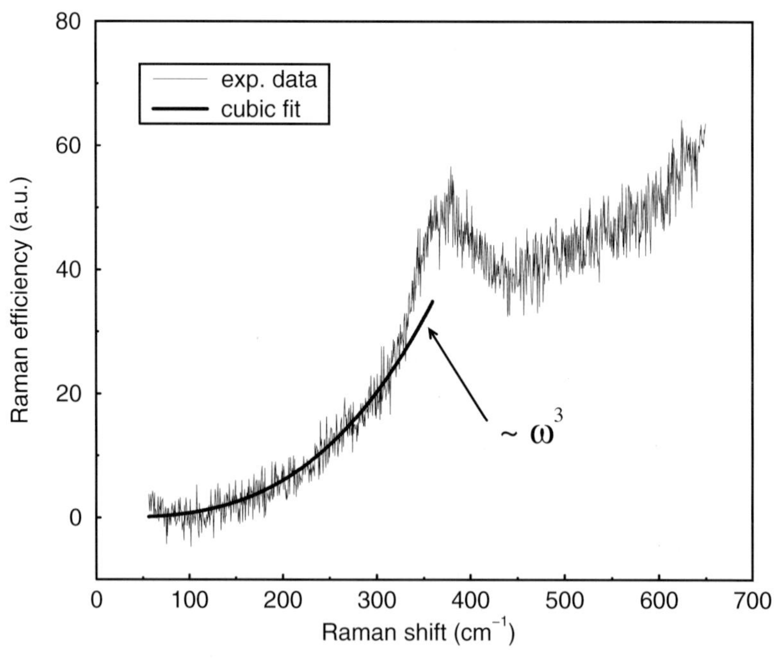

Fig. 5.19. Measured B_{1g} spectrum of $HgBa_2Ca_2Cu_3O_{8-\delta}$ ($T = 130\,\mathrm{K}$) in the superconducting state showing the asymptotic dependence $\propto \omega^3$ predicted and found for small ω (*solid line*) [98]

forms the assumed B_{1g} symmetry of $\Delta_{\boldsymbol{k}}$ into A_g, while the latter (Bi-2212) leaves a $(k_x^2 - k_y^2)$-like gap B_{1g}-like. Hence, in the case of Bi-2212 one expects exact ω^3 behavior of $\partial^2 S/\partial\omega\partial\Omega$ for $\omega \to 0$ in the clean limit. This is confirmed by the data [105] and the cubic fit of Fig. 5.20a. In the case of Y-123 and Y-124, a small linear component should appear in the spectra for the $(x+y, x-y)$ scattering configuration as a result of the orthorhombic distortion along $\hat{\boldsymbol{y}}$ induced by the chains. This is confirmed by the fit to the data of Hackl et al. [87,106] shown in Fig. 5.20b [4].

The D_{2h} symmetry of Y-123, while adding an A_g-like term (e.g. a constant) to $k_x^2 - k_y^2$, shifts the nodes of *both* $\Delta_{\boldsymbol{k}}$ and $\mu_{xx}^{-1} - \mu_{yy}^{-1}$ away from the $x \pm y$ diagonals. It is therefore not possible to extract, from the linear component of the fit in Fig. 5.20b, quantitative information about the A_g-B_{1g} admixture in $\Delta_{\boldsymbol{k}}$. This can, however, be done if independent information is available for the corresponding admixture in $(\mu_{xx}^{-1} - \mu_{yy}^{-1})$ obtained from LMTO-LDA calculations. In [4] this information was combined with experimental data so as to extract information about B_{1g}-A_g admixture in $\Delta_{\boldsymbol{k}}$: a shift of the gap nodes from $\varphi = 45°$ to $36°$ was found.

Other than the reliability and validity of the LMTO-LDA one-electron band structures, the estimate given above suffers from another drawback. It has been shown in [107] that the presence of impurities or other defects can also introduce linear components (i.e., with $s = 1$) in the B_{1g} scattering for $\omega \to 0$. One may, however, surmise that impurities do not play an important role in the cases just mentioned, from the fact that Y-123 ($YBa_2Cu_3O_7$), in which a linear term clearly appears (Fig. 5.20b), is basically a stoichiometric material. The HTSC Bi-2212 and Hg-1223, in which they do not appear (Figs. 5.19 and 5.20a), have intrinsic oxygen defects in their charge reservoirs (the Bi-O and the Hg-O planes). We should also point out that nonperturbative calculations, of the type displayed in Fig. 5.18, also have a tendency to modify the asymptotic ω^3 law characteristic of the spectra of B_{1g} symmetry for B_{1g} symmetric gaps.

5.4 Electron–Phonon Interaction

In the previous sections we discussed Raman active phonons and intraband electronic excitations, the latter taking place in the normal state (induced by the imaginary part of a self-energy) and in the superconducting state (pair breaking excitations appearing even in the clean limit). For a certain scattering configuration, belonging to a specific irreducible representation of the point group, one can observe both, phonons and electronic excitations. If both types of excitations have the same symmetry, it should be possible to annihilate one of them while creating the other. These processes are represented by matrix elements of the electron–phonon interaction Hamiltonian d_{ep}. The process of annihilation of a phonon followed by creation of an electronic excitation is isomorphic to the process of Raman scattering by elec-

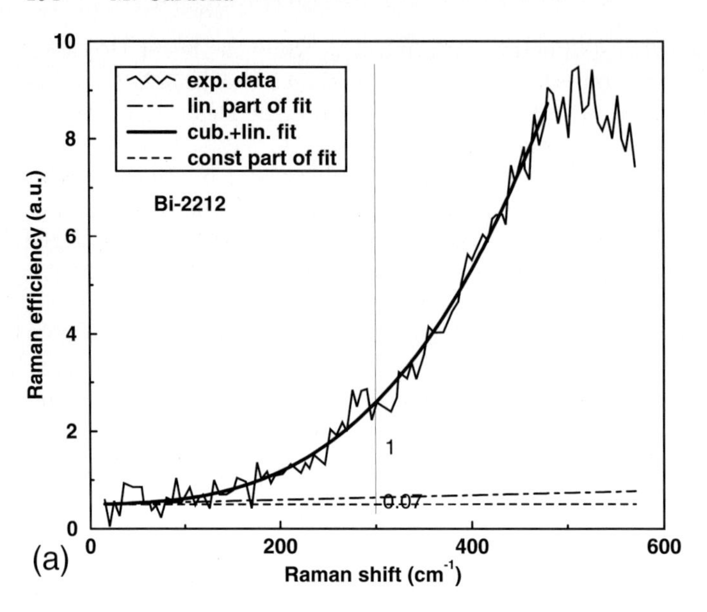

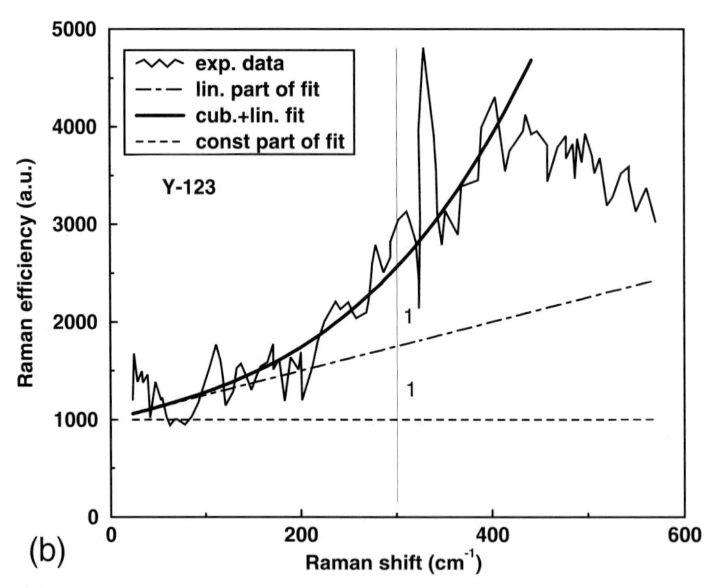

Fig. 5.20. (**a**) Noisy curve: experimental Raman efficiency reported in [105] for Bi-2212. The *solid line* represents a fit with a linear ($\propto \omega^s$, $s = 1$) plus a cubic ($s = 3$) function plus a constant background (*short dashed line*) [98]. (**b**) Noisy curve: experimental Raman efficiency of Y-123 reported in [87,106]. The *solid line* represents a fit with a linear ($s = 1$) plus a cubic ($s = 3$) ω^s function plus a constant background [4]

tronic excitations: $\boldsymbol{A}_S^+ \cdot \boldsymbol{A}_L$ corresponds to the phonon amplitude while the scattering vertex μ^{-1} corresponds to d_{ep} [77,108,109]. The coupling of symmetric (A_g- or A_{1g}-like) phonons to intraband electronic excitations is thus screened by the dielectric response of the carriers while that of nonsymmetric phonons is not screened. The screening of the coupling to a symmetric phonon is complete if d_{ep} is constant around the Fermi surface (FS). As in the case of electronic Raman scattering by mass fluctuations, fluctuations of d_{ep} around the FS allow some measure of unscreened coupling, even for fully symmetric phonons (see (5.17) and (5.26) where μ^{-1} must be replaced by d_{ep}). The effect of the electron–phonon interaction on the phonon spectra is characterized by three phenomena:

(1) A renormalization of the phonon frequency corresponding to the virtual (i.e., energy non-conserving) creation and annihilation of electronic excitations.
(2) If the phonon frequencies overlap with the continuum of electronic excitations (or with pair breaking excitations across the gap in the superconducting state) the processes in which a phonon is destroyed through creation of an electronic excitation can be real (i.e., energy conserving). They then decrease the phonon lifetime τ_p while increasing its linewidth Γ_p according to:

$$\Gamma_p \cdot \tau_p \simeq 1\,, \tag{5.29}$$

where Γ_p is the full-width-at half maximum (FWHM) in rad $\cdot$ s^{-1}. When crossing T_c, Γ_p can either increase or decrease. An increase is found when the phonon frequency is close to the peak of the pair breaking electronic excitations (since the decay into electronic excitations is forbidden in the *clean* limit for $\boldsymbol{k}_p \simeq 0$) while a decrease can take place (outside the *clean* limit) when the frequency is well within the gap that opens up in the electronic excitations below T_c. The exact crossover point between broadening and sharpening is hard to predict since the broadening in the normal state is induced by either defects or other many-body interactions while that below a $(k_x^2 - k_y^2)$-like gap in the superconducting state has an intrinsic component [110]. In the language of many-body perturbation theory one says that $\Gamma_p = -2\Sigma_2$, where Σ_2 is the imaginary part of the *self-energy* of interaction. The width $\Gamma_p = \tau_p^{-1}$ in the superconducting state is equivalent (to multiplicative constants and after replacing μ^{-1} by d_{ep}) to the Raman scattering efficiency of (5.26). The corresponding real part of the self energy, Σ_1, represents the phonon self-energy shift. Its counterpart in the Raman scattering process represents a minute renormalization of the speed of light in the medium and is usually negligible.
(3) Renormalization of the phonon scattering efficiency due to electron–phonon interaction. Virtual transitions produce not only a renormalization of the phonon frequency (as discussed in (1)) but also an admixture

to the phonon of the interacting electronic excitations. To first order in d_{ep} this admixture can be expressed as:

$$\overline{|p\rangle} = |p\rangle + \sum_{e} \frac{d_{\mathrm{ep}}}{\overline{E}_p - E_e} |e\rangle , \tag{5.30}$$

where $\overline{|p\rangle}$ and $|p\rangle$ represent the eigenvectors of the renormalized and unrenormalized phonons of energy E_p and $|e\rangle$ those of the coupling electronic excitations of energy E_e. The scattering efficiency is obtained by calculating the matrix elements of the scattering operator between $\overline{|p\rangle}$ and the ground state. From (5.30) we surmise that an interference occurs between the scattering amplitudes of the unrenormalized excitations $|p\rangle$ and $|e\rangle$. Depending on the product of the signs of R_e and R_p (the unrenormalized Raman tensors for electronic and phonon scattering) and d_{ep}, the interference is either constructive or destructive for $E_e > \overline{E}_p$ (and vice versa for $E_e < \overline{E}_p$). This gives rise to asymmetric, so-called Fano–Breit–Wigner-type line shapes (see curves in Fig. 5.21 for $T > 200\,\mathrm{K}$) [69]. If the unrenormalized electronic continuum is flat, one can represent such line shapes by the expression [12,13]

$$\frac{\partial^2 S}{\partial\omega\partial\Omega} = \frac{|q+\mathcal{E}|^2}{1+\mathcal{E}^2} , \tag{5.31}$$

where $\mathcal{E} = 2(\hbar\omega - \overline{E}_p)/\Gamma_{\mathrm{p}}$ and q is the dimensionless Fano asymmetry parameter.

The effect of the change in Σ_1 on the phonon frequency when crossing T_{c} is represented in Fig. 5.21 for the B_{1g} phonon of an Y-123 sample. A significant softening of about $6\,\mathrm{cm}^{-1}$ is found below T_{c}. The relationship between the self-energy change at T_{c} shown in Fig. 5.21 and the superconducting transition was verified by Ruf and Cardona in [111]. These authors showed that a similar change in Σ_1 occurred when the transition was induced by a magnetic field.

Zeyher and Zwicknagl were the first to investigate theoretically the self-energies Σ_1 and Σ_2 in detail [112]. They assumed a strong coupling superconductor but with a fully symmetric gap.[8] This calculation was generalized in [113] to include a B_{1g}-like [i.e., $(k_x^2 - k_y^2)$-like] $\Delta_{\boldsymbol{k}}$. Using these theoretical approaches it is possible to obtain an average value of d_{ep}, and the corresponding dimensionless parameter λ as defined by McMillan [114]. In order to get a feeling for the strength of the electron–phonon interaction, which might be responsible for BCS-like superconductivity, it is convenient to estimate the parameter λ that one would obtain if all phonons would couple to the electronic excitations as much as the B_{1g} phonon whose effect is illustrated in Fig. 5.21. This value was estimated to be $\lambda = 1$ [115], a value

[8] At that time the B_{1g}-like gap had not yet come to the fore.

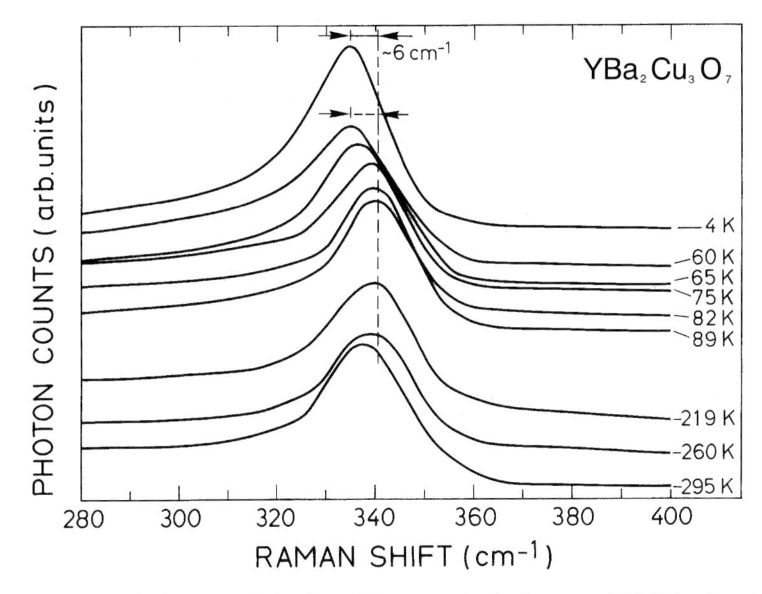

Fig. 5.21. Softening of the B_{1g} (D_{4h} notation) phonon of Y-123 when lowering the temperature below T_c. Because of heating by the laser power in these early data, the nominal temperatures given in the figure are only approximate. The measured spectra correspond to those reported in [41]

that corresponds to strong coupling and would, by itself, lead to a fairly high-T_c (but, unfortunately, also large isotope effects on T_c, which are not observed [116,117]!). It is, however, not very likely that all phonons couple as much as that B_{1g} mode [45]. Neutron scattering experiments indicate that d_{ep} of the B_{1g} phonon branch decreases monotonically with increasing $\boldsymbol{k}$ [118].

A specially strong case of electron–phonon coupling has been observed for Hg-1234 [5]. In Fig. 5.22 we show Raman spectra for this material below and above $T_c = 123\,\mathrm{K}$. Three phonons are observed at about 220, 375 and $400\,\mathrm{cm}^{-1}$. They can be assigned to z-polarized A_{1g}-type vibrations of the four CuO_2 planes admixed with vibrations of two of the three Ca spacer planes (a total of three A_{1g} modes). The 220 and $375\,\mathrm{cm}^{-1}$ modes soften considerably below T_c, especially the latter (from 395 to $348\,\mathrm{cm}^{-1}$). This softening is the largest observed thus far for any high-T_c superconductor; its magnitude is similar to that estimated from changes in the vibrational amplitudes in some channeling and neutron absorption experiments [119], which remain to be understood. It would correspond to $\lambda \simeq 6$ if all phonons would couple as strongly (a rather unlikely situation). Such an average coupling constant $\lambda \simeq 6$ would suffice to explain the high-T_c simply on the basis of electron–phonon interaction but be incompatible with the measured isotope effect [116,117].

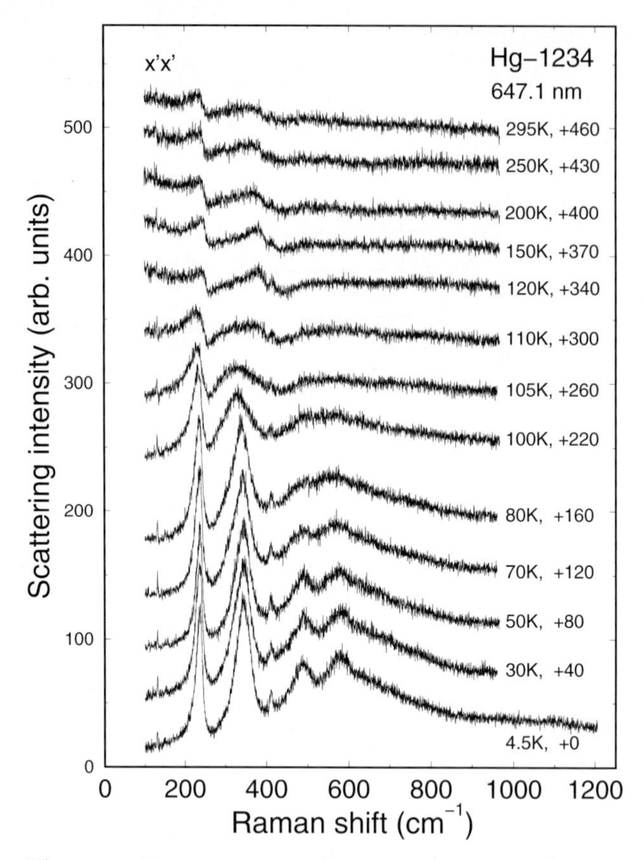

Fig. 5.22. Raman spectra of Hg-1234 ($T_c = 123\,K$) measured at various temperatures between RT and 4.5 K in $x'x'$ polarization. The numbers in the right column give the offset of the spectra with respect to that at the bottom [5]

Of particular interest is the strong increase in the intensities of the 220 and 375 cm^{-1} phonons in the superconducting phase. It has been attributed [5] to the admixture to the phonon of pair breaking electronic excitations. The latter have, according to Fig. 5.22, a very high Raman efficiency. The value of d_{ep} extracted from the softening below T_c suffices to explain the strong intensity increase. Early observations of changes of scattering efficiencies at T_c were reported by Friedl et al. [120] for Y-123.

Phonon anomalies *above* T_c have been observed for underdoped materials (e.g. Y-124), in particular in the IR-spectra [121]. They seem to be related to the presence of pseudogaps which are now generally believed to remain open above T_c. One often observes broad phonon anomalies with decreasing temperature, starting at the temperature T^* at which the pseudogap opens. These anomalies become sharper when crossing T_c. Typical effects of this kind are illustrated in Fig. 5.23 for an $YBa_2Cu_3O_{6.5}$ crystal with $T_c = 53\,K$ [122].

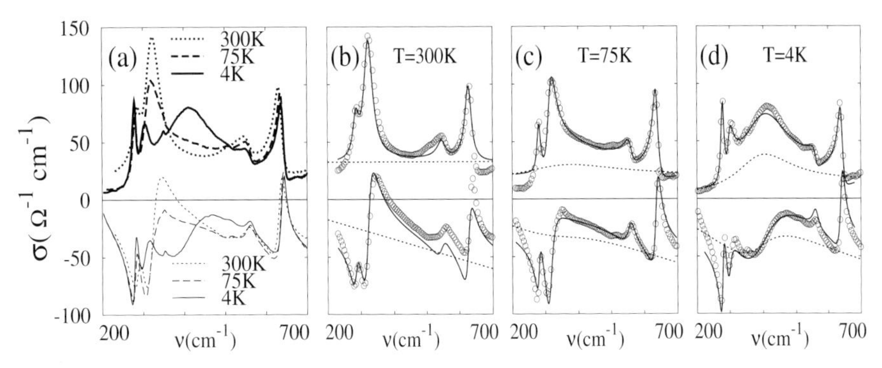

Fig. 5.23. (**a**) Measured spectra of the c-axis conductivity, $\sigma = \sigma_1 + i\sigma_2$, of $YBa_2Cu_3O_{6.5}$ with $T_c = 53\,K$ together with the fits obtained by using the model of [122] for (**b**) $T = 300\,K$, (**c**) $T = 75\,K$, and (**d**) $T = 4\,K$. The *dotted lines* represent the bare electronic contributions

The figure displays the spectra of the real and imaginary parts of the out-of-plane conductivity

$$\sigma_{zz} = \frac{\omega i}{4\pi}(1 - \epsilon_{zz}) \tag{5.32}$$

of this crystal measured in the far infrared by means of ellipsometry using synchrotron radiation as a light source [29,72,73]. The following dramatic changes of the measured spectra are observed when lowering the temperature, starting even above T_c but becoming particularly strong below T_c:

(1) A broad band appears in the 400–500 cm^{-1} region. This band has been observed by a number of other authors and for other undoped materials with at least two CuO_2 planes per PC [123,124]. It has been attributed to the *transverse plasmons* that become possible in a composite medium consisting of two different alternating conducting layers [125].
(2) Beside the broad transverse plasmon band, sharp phonon peaks appear at around 300, 550 and 650 cm^{-1}. These peaks show striking changes in position, width and strength concomitant with the changes in the plasmon band with temperature.
The authors of [122] were able to interpret the phenomena displayed in Fig. 5.23 on the basis of a simple electrostatic local field model involving the conducting CuO_2 layers and the IR-active phonons [122].

5.5 Crystal Field Transitions Between f-Electron Levels

Many of the high-T_c superconductors contain rare earth ions in trivalent states. These ions possess $4f$-electrons in the core (exception: La^{3+}, Ce^{4+} and, of course, Y^{3+}) that are strongly correlated and must be treated as a many-electron eigenstate characterized, within the $L \cdot S$ coupling scheme, by the total orbital (S, P, D, F, G, H, I) and spin (S) angular momenta and the resulting total angular momentum J. The quantum numbers for the ground state are easily determined using the Pauli principle and Hund's rules [126,127]. The ground state of the Nd^{3+} ion, containing three f-electrons, is thus found to have $S = \frac{3}{2}$, $L = 6$, and $J = \frac{9}{2}$ ($^4I_{9/2}$ in the standard notation). The other two rare earth ions that have been studied by Raman spectroscopy in high-T_c superconductors are Pr^{3+} ($^3H_4$) and Sm^{3+} ($^6H_{5/2}$). These ions fall into two categories: those with an odd number of f-electrons N_f, for which all states are at least doubly degenerate in the absence of a magnetic field (Kramers degeneracy, e.g. Nd^{3+} and Sm^{3+}) and those with an even number of f-electrons (e.g., Pr^{3+}), for which the states can be nondegenerate. In the presence of a crystal field (generated by the ionic charges of all other atoms), the ionic states just mentioned split. The resulting states belong to one of the reps of the point group of the site in which the ion is placed. This gives a single group representation if the number of f-electrons is even and a double group representation if it is odd (so as to preserve the Kramers degeneracy). Since the f-electron states possess magnetic moments, transitions between the ground and the excited crystal field states can be investigated by inelastic neutron scattering [128]. Many of these transitions are also Raman allowed since excitations from an f to another f-electron level have even parity (note that if the rare earth ions are at a center of inversion, as in the case of Nd-123 but not for Nd_2CuO_4, the crystal-field (CF) transitions are IR forbidden). Nevertheless, in high-T_c superconductors CF transitions have only been observed by Raman spectroscopy for Pr^{3+}, Nd^{3+} and Sm^{3+}. The reason probably is that these transitions usually take place via excited $4f$ multiplets as intermediate states and the $f \rightarrow f$ virtual transitions are even, and therefore dipole forbidden by parity, i.e., intermediate states of either d- or g-symmetry, at least $\simeq 6$ eV above the ground state, would be required for allowed transitions. If the rare earth ions are at a center of inversion, no intra-atomic $f + d$, or $f + g$ admixture is possible. Sufficient hybridization with neighboring ions (whose electronic wavefunctions can be d- or g-like with respect to the site of the rare earth) may be necessary to allow, in this case, observation of CF transitions by Raman spectroscopy. This does not seem to happen for most rare earth ions when placed at inversion centers. The CF transitions have, nevertheless, been observed for Nd-123 [129] and Sm-123 [130]; in most of these cases they become observable thanks to the admixture of Raman active phonon eigenstates induced by electron–phonon interaction [129–131].

Inelastic neutron scattering is a more general method of observing CF transitions: their coupling is related to a magnetic dipole moment. Raman scattering has a more complicated coupling mechanism involving electric dipole transitions to electronic intermediate states, which are forbidden or nearly forbidden. Whenever the CF transitions appear in the Raman spectra, however, the Raman technique is simpler than neutron scattering and provides higher resolution and accuracy in the frequency determination. Moreover, it enables investigation of the coupling of CF transitions to phonons [129] and the determination of the corresponding electron–phonon coupling constant [131]. Raman scattering is particularly useful for performing measurements in a magnetic field [132].

We have conjectured that the difficulty of observing Raman scattering by crystal field excitations in RE-123 materials is due to the fact that the rare earth ion (RE) is located at an inversion center. In the n-type superconductors of the RE_2CuO_4 family [133–135] there are two equivalent rare earth sites connected by an inversion center at the midpoint between them. The crystal field excitation can thus be either an odd or an even combination of excitations in each of the two equivalent ions. The lack of inversion symmetry allows, in principle, admixture of intraionic f and (d, g) levels, a fact that may enhance the virtual dipole transitions to the intermediate states. A quantitative evaluation of the corresponding scattering efficiencies for RE_2CuO_4 systems has not yet been performed but the experimental spectra show CF transitions that are allowed in their own right, without admixture of phonons. These transitions take place inside the lowest CF multiplet [134] and also from this multiplet to higher ones [135]. For a study of intermultiplet CF transitions by neutron spectroscopy in $SmBa_2Ba_2Cu_3O_7$, using a spallation source, see [136].

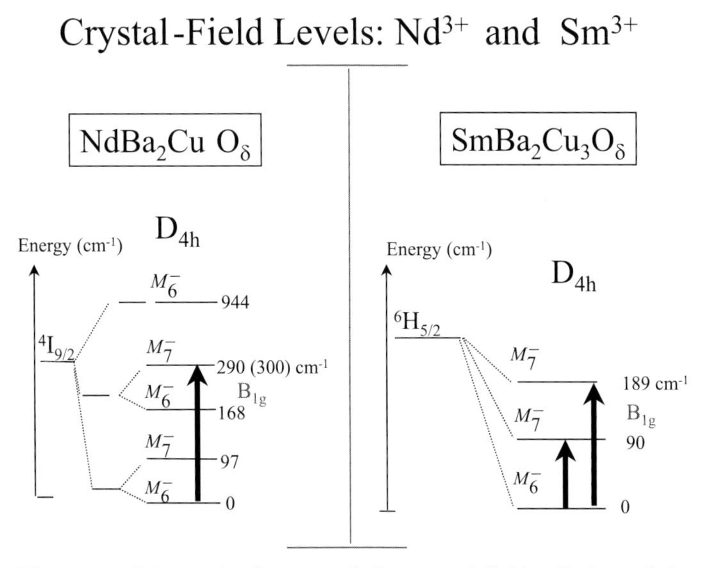

Fig. 5.24. Schematic diagram of the crystal field splitting of the ground states of the Nd^{3+} and Sm^{3+} ions in RE-123. In both cases Kramers doublets are present (courtesy of A.A. Martin)

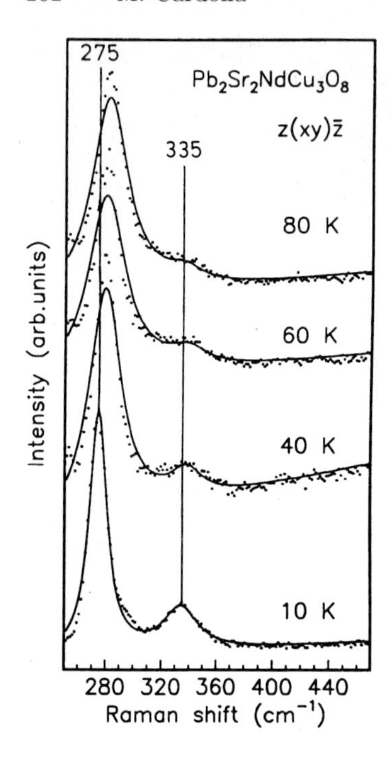

Fig. 5.25. Cross-polarized Raman spectra of the cf plus B_{1g}-phonon doublet of $Pb_2Sr_2NdCu_3O_8$ at 10, 40, 60, and 80 K together with least square fits to a theory which includes the coupling of the two excitations [137]. Note the decrease in the strength of the component $\simeq 335\,cm^{-1}$, due mainly to the CF transitions, with increasing T

As an illustration of the concepts just presented we show in Fig. 5.24 a diagram of the CF splittings of the ground state CF multiplets of Nd-123 ($^4I_{9/2}$) and Sm-123 ($^6H_{5/2}$). For Nd-123 the tenfold degeneracy of the $^4I_{9/2}$ multiplet is split by the *cubic* component of the CF into two quadruplets and a doublet. The addition of a tetragonal component, as required by D_{4h} symmetry, splits each of the quadruplets into two Kramers doublets, with M_6^- and M_7^- symmetry, respectively. The $M_6^- \rightarrow M_7^-$ quadruplet excitations indicated by the arrows in Fig. 5.24 have a component of B_{1g} symmetry, at a frequency of about $300\,cm^{-1}$, which is very close to that of the B_{1g} phonon of Fig. 5.4: the two excitations mix and, at low temperatures, become nearly fully hybridized, although the lower frequency component has predominant CF character in $NdBa_2Cu_3O_{7-\delta}$. Both excitations become observable in Raman scattering because of their coupling to the strongly Raman active phonon component.

Raman spectra illustrating the B_{1g} coupled phonon plus CF transitions of the Nd^{3+} ion in $Pb_2Sr_2NdCu_3O_8$ are shown in Fig. 5.25 for several temperatures [137].[9] Instead of the single B_{1g} phonon peak characteristic of materials with two CuO_2-planes around $300\,cm^{-1}$ we observe, at 10 K, a doublet the upper component decreasing with increasing temperature to disappear nearly completely at 300 K (not shown in Fig. 5.25). Isotopic substitution of ^{16}O by

[9] Note that for this material the bare B_{1g} phonon is at lower frequency than its CF counterpart.

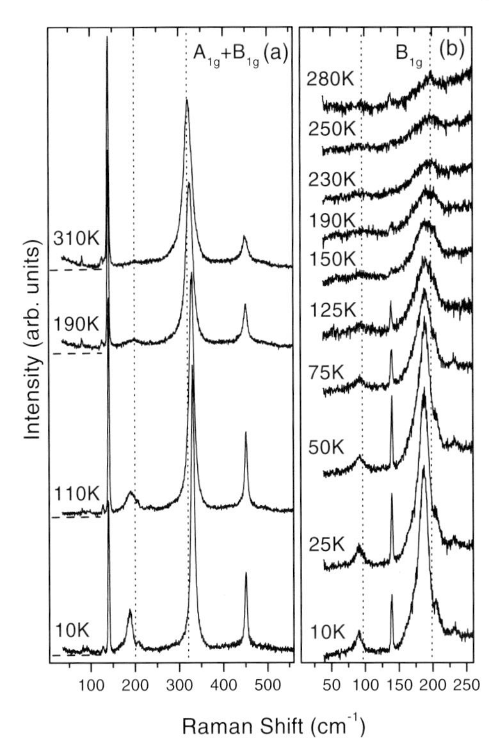

Fig. 5.26. Polarized Raman spectra of (**a**) $A_{1g} + B_{1g}$ and (**b**) B_{1g} for a non-superconducting $SmBa_2Cu_3O_6$ crystal at different tempertures. Crystal field excitations are seen at 90 and $190\,cm^{-1}$. They disappear with increasing temperature because of the equalization of the population of the initial and the final states [130]

^{18}O indicates that both peaks correspond to mixed phonon and electronic excitations [137]. The CF excitation at hand is equivalent to that labeled as B_{1g} in the left side diagram of Fig. 5.24. Increasing the temperature results in the growth of the population of the *final state* M_7^- of Fig. 5.24, at the expense of the ground state population. At 300 K, both states become nearly equally populated and the corresponding spectral component disappears, resulting in an increase of the frequency of the phonon-like peak (it was pushed down at low temperatures by the interaction with the CF transitions). A fit to the coupling Hamiltonian yields the energies of the uncoupled excitations and the coupling constants [137].

The other example we present involves the ground state CF multiplet of Sm-123 (see Fig. 5.26). In Fig. 5.26 we show the corresponding Raman spectrum measured both, in $A_{1g} + B_{1g}$ (a) and in pure B_{1g} configuration (b). The weak peak at $90\,cm^{-1}$ and the stronger one at $190\,cm^{-1}$ correspond to the two CF excitations shown by arrows in Fig. 5.24. These excitations are also expected to mix with the B_{1g} phonon, but considerably less than for

Nd-123 because of the larger frequency separation (and the correspondingly larger energy denominators). Actually, the $90\,\text{cm}^{-1}$ peak seems to be mainly due to the Raman activity of the CF excitation, while only 20% of the activity is due to the CF component for the $190\,\text{cm}^{-1}$ peak [130].

5.6 Light Scattering by Magnons in HTSC and Their Antiferromagnetic Parent Compounds

5.6.1 Antiferromagnetic Structures in the Underdoped Parent Compounds

Most of the light scattering phenomena in HTSC discussed so far have been interpreted on the basis of one-electron band structure theories [70,138]. On several occasions we have mentioned, however, the presence of *antiferromagnetic* (AF) *fluctuations* which signal strong electron correlation lying beyond the one-electron ansatz. Ab initio one electron band structure calculations are usually based on local density functionals which describe, in an average manner, exchange and correlation effects. As mentioned in connection with the theory of phonon frequencies and eigenvectors (Sect. 5.2.1), efficiencies for Raman scattering by phonons (Sect. 5.2.2) and by electronic excitations (Sect. 5.3), ab initio one electron calculations have reached a high degree of sophistication and success in treating those questions. The phenomena related to AF in HTSC and their parent compounds, however, lie beyond the scope of one-electron theory. They are usually treated theoretically by assuming a simplified one-electron band structure (e.g. tight binding) and adding to it, in as sophisticated a manner as possible and necessary, the appropriate ingredients of electronic correlations. This field has become a splendid playground for theories, both of the numerical and the analytical variety. Also, experimentalists have delved at length into the consequences of the strong correlations that affect the $3d$ electrons of the Cu^{2+} ions in the CuO_2 planes: Néel-type antiferromagnetic order (as revealed by elastic neutron scattering, $T_N = 420\,\text{K}$ for $YBa_2Cu_3O_6$ [139]) and spin wave excitations (i.e. magnons), evidenced by inelastic neutron scattering [140] and by Raman spectroscopy [9]. In spite of the rather unstructured, to some experimentalists boring, features that appear in the Raman spectra of magnons (a single broad band, $\approx 800\,\text{cm}^{-1}$ wide, centered at $\approx 2600\,\text{cm}^{-1}$, see Fig. 5.27) the amount of experimental work published in the field is overwhelming. The reason is to be sought in the robustness of this broad structure, found to be almost the same in all insulating partners of high-T_c superconductors, and persisting even in their underdoped variety well above T_N [10], interesting polarization selection rules (see Fig. 5.28) [141] and its unusually large width. The persistence of these scattering spectra above T_N, and in doped metallic materials, suggests the wording of "scattering by AF fluctuations" [142] which, as already mentioned, play nowadays a prominent role among the "intermediate

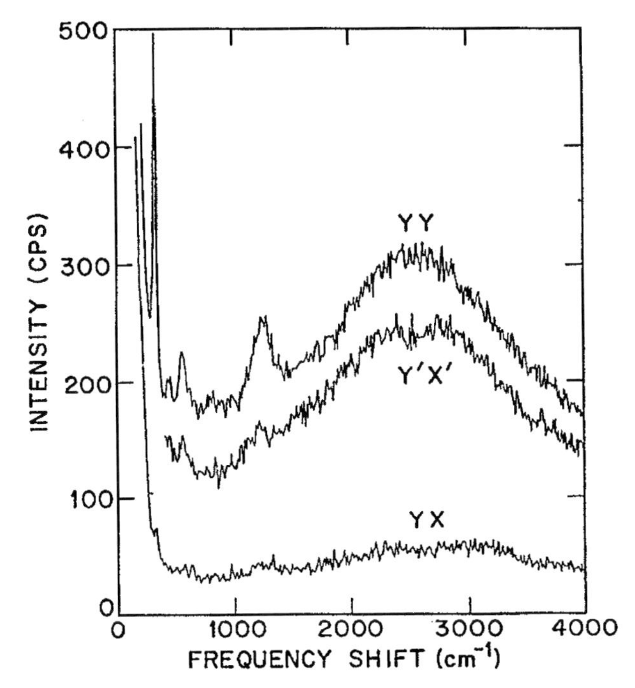

Fig. 5.27. Spectra obtained for $YBa_2Cu_3O_6$ (nonsuperconducting) in three polarization geometries relative to the basal plane. The *yy* trace is shifted 50 cps upwards, otherwise the scales are the same. The laser wavelenght used was 4880 Å (2.54 eV) [9]

bosons" considered as possible candidates for effecting the superconducting pairing. Hence, there is also great theoretical interest in the scattering by magnons and AF fluctuations in HTSC and parent compounds [143–152]. In previous sections of this chapter we have not been able to give an exhaustive list of relevant literature references. This is also going to be the case in the present one. We hope, however, to have chosen a representative cross section of the published work. In the cited publications, the interested reader will find most of the literature references to work performed to date on light scattering by magnons in HTSC.

Let us consider the typical insulating compounds, La_2CuO_4 and Nd_2CuO_4 which have (see Fig. 5.2) one CuO_2 plane per tetragonal PC. We have mentioned in Sect. 5.1 that these stoichiometric materials have a valence charge in the cations (+8) which equals, with the opposite sign, that in the anions. The charge compensation explains, at least in a naive qualitative way, why these materials are insulators. This conclusion, however, contradicts the predictions one may draw from the corresponding one-electron band structure: the Cu^{2+} ions have nine $3d$ electrons, i.e., one hole in an otherwise filled set of one-electron $3d$ bands. Each of these bands can usually accommodate two electrons per PC. Therefore the half-filled $3d$ band would be expected to lead

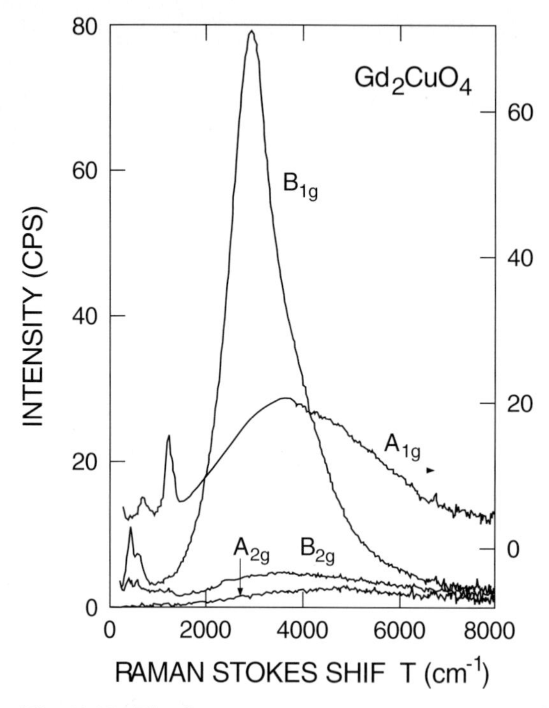

Fig. 5.28. The four pure symmetry components of the Raman scattering intensity vs energy shift of Gd_2CuO_4 taken with an incident laser wavelength of 4880 Å (2.54 eV). The A_{1g} spectrum has been vertically offset for clarity [141]

to electronic conduction. The paradox is solved when one considers the large Coulomb repulsion energy needed to fill the half-empty $3d$ band. Such energy is much larger for double (~9 eV, see [152]) than for single occupancy. This correlation energy for double occupancy is usually called the "Hubbard U". In the standard one-electron band structures it is treated in an average manner, as part of the LDA functional, without distinguishing whether the relevant states are unoccupied, singly occupied, or doubly occupied. A half-filled band is then able to conduct electricity. This will be a reasonable approximation if the one-electron band width is larger than U, a condition that does not hold for the $3d$ electrons of Cu in the cuprates under consideration.

Materials with a half-filled $3d$ band, such as La-123, Nd-123 or, approximately, $YBa_2Cu_3O_{6.5}$, are thus expected to have the lowest Coulomb energy (ground state) when each of the Cu atoms in the CuO_2 planes contains one single missing $3d$ electron (i.e., one hole). A charge fluctuation leading to a $3d^{10}$ plus a $3d^8$ configuration (needed for electrical conduction) requires an energy of more than 9 eV and is therefore strongly suppressed, even at high temperatures (the melting temperature T_M lies below 0.1 eV!). We shall see in Sect. 5.6.3 that the half-occupied $3d$ orbitals have a tendency to order their spins in a 2D antiferromagnetic fashion which is illustrated in Fig. 5.29 and follows from the 2D Heisenberg Hamiltonian:

$$H_A = -J \sum_{(i,j)} \boldsymbol{S}_i \cdot \boldsymbol{S}_j \ . \tag{5.33}$$

Typical values of the nearest neighbor antiferromagnetic exchange constant J for HTSC parent compounds, are 0.15 eV (1200 cm^{-1}). In (5.33) $\boldsymbol{S}_i$ and $\boldsymbol{S}_j$ are spin operators acting on the spins of either sublattice i (spin up ↑) or sublattice j (spin down ↓ for the AF state) and the sum over (i, j) represents a sum over four nearest neighbor Cu-Cu bonds j for each i (see Fig. 5.29). Because of the large values of J, the antiferromagnetic ground state may be expected to persist up to relatively high temperatures. T_{N}, however, is not very high ($\approx$ 400 K). The 2D nature of H_{A} is responsible for the low T_{N} [153] and for the presence of strong antiferromagnetic fluctuations at temperatures higher than T_{N}, which would be suppressed in the 3D case [154].

For future reference we show in Fig. 5.29 the AF primitive cell (PC) and the corresponding Brillouin zone of a CuO_2 plane with the spin orientations appropriate to the HTSC-like cuprates. Note that materials with two or

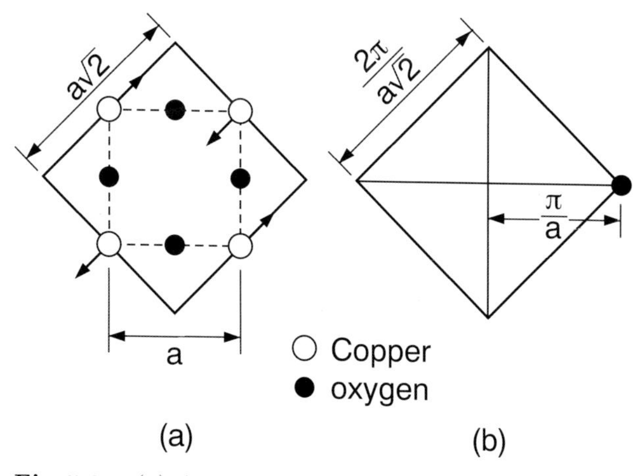

Fig. 5.29. (**a**) AF Primitive cell of a 2D CuO_2 plane indicating the directions of the AF-ordered Cu-spins (*solid lines*). The *dashed lines* represent the PC of the spin-disordered material $a \simeq 3.8$ Å, see Fig. 5.1. (**b**) Brillouin zone of the 2D AF CuO_2 planes

more CuO_2 planes per PC usually exhibit antiferromagnetic exchange coupling between the nn $3d^9$ copper ions of adjacent planes. The corresponding *interplane* J is considerably smaller than its *intraplane* counterpart. The observation of AF order along z within the CuO_2 bilayers, illustrated in Fig. 5.1, has been reported in [155].

5.6.2 Introduction to Light Scattering by Magnons in Antiferromagnets

The Hamiltonian (5.33) yields, in its ground state, the AF order sketched in the PC of Fig. 5.29a. When the AF structure in Fig. 5.29a is periodically repeated in two dimensions, this 2D AF can carry Bloch-like excitations, characterized by a wavevector $\boldsymbol{k}$, corresponding to spin fluctuations in either one or the other of the spin-up and spin-down sublattices i, j. The sketchy theoretical background given in this section will follow closely the pioneering work of Fleury and Loudon [8].

In order to derive from (5.33) Bloch-like excitations, one introduces magnon creation and annihilation operators acting on the spins of sublattices $i(\uparrow)$ and $j(\downarrow)$ $\alpha^+_{\uparrow\boldsymbol{k}}$, $\alpha_{\uparrow\boldsymbol{k}}$, $\alpha^+_{\downarrow\boldsymbol{k}}$, $\alpha_{\downarrow\boldsymbol{k}}$. The connection between the operators $\boldsymbol{S}_i$, $\boldsymbol{S}_j$ and the magnon operators is found through the so-called Holstein–Primakov transformation [8,156].[10] The transformed Hamiltonian is:

$$H = \sum_k \left[E_{\uparrow\boldsymbol{k}} \alpha^+_{\uparrow\boldsymbol{k}} \alpha_{\uparrow\boldsymbol{k}} + E_{\downarrow\boldsymbol{k}} \alpha^+_{\downarrow\boldsymbol{k}} \alpha_{\downarrow\boldsymbol{k}} \right] \tag{5.34}$$

where the sum over $\boldsymbol{k}$ extends over the AF Brillouin zone (Fig. 5.29b). This Hamiltonian is diagonal in the magnon base, with doubly degenerate eigenvalues corresponding to the fact that both sublattices are equivalent:

$$\begin{aligned} E_{\boldsymbol{k}} &= \hbar\omega_{\boldsymbol{k}} = 4JS\left[1 - \gamma^2_{\boldsymbol{k}}\right]^{1/2} \\ \gamma_{\boldsymbol{k}} &= \frac{1}{2}\left[\cos k_x a + \cos k_y a\right] . \end{aligned} \tag{5.35}$$

For the case under consideration, $S = \frac{1}{2}$. The maximum value of $E_{\boldsymbol{k}}$, found at the edge of the AF Brillouin zone (Fig. 5.29b) $\boldsymbol{k}_x = \left(\frac{\pi}{a}, 0\right)$ or $\boldsymbol{k}_y = \left(0, \frac{\pi}{a}\right)$, amounts to

$$E_{\mathrm{M}} = \hbar\omega_{\mathrm{M}} = 2J . \tag{5.36a}$$

This value can be heuristically obtained by simply flipping one up-spin with respect to its four down-neighbors:

$$\Delta E = \hbar\omega_{\mathrm{M}} = 4 \times \frac{1}{2} \times J = 2J . \tag{5.36b}$$

The degeneracy associated with the two sublattices can be split by a magnetic field. One may conjecture that this property can be used to identify magnon structures in Raman spectra. However, the splitting that can be achieved with the highest static fields available ($\lesssim 5\,\mathrm{cm}^{-1}$) is too small to be detected in view of the large widths of the spectral structures observed ($\approx 2000\,\mathrm{cm}^{-1}$).

[10] Note that in [145] the so-called Dyson–Maleev transformation, more appropriate to the treatment of nonlinear phenomena, is used.

Scattering by one magnon is not easy to observe either. The $\boldsymbol{k}$-conservation selection rule implies scattering by one magnon with $\boldsymbol{k} \simeq 0$. In cuprates with only one CuO_2 plane per unit cell, (5.35) leads to magnon frequencies $\omega_{\boldsymbol{k}}$ tending to zero for $k \to 0$. If two CuO_2 planes are present (like in Fig. 5.1), their symmetric and antisymmetric combinations should generate acoustic- and optic-like magnons,[11] the latter with $\omega_{\boldsymbol{k}} \neq 0$. Two different mechanisms have been proposed for the scattering by magnons with $k \simeq 0$. The first one is purely magnetic scattering, induced by the coupling of the *magnetic field* (or the vector potential) of the electromagnetic radiation. Because of the weakness of such coupling, the estimated scattering efficiency is $10^{-13}\,\mathrm{cm^{-1}sr^{-1}}$, negligibly small compared with the observed values for two-magnon scattering $10^{-4}\,\mathrm{sr^{-1}cm^{-1}}$, see [158].

The other proposed one-magnon scattering process is of electric dipole nature, using as intermediate states electronic states split by spin–orbit interaction. These states contain mixtures of spin-up and spin-down components. It is thus possible to raise a spin-up electron to one of these excited states via the $\boldsymbol{p} \cdot \boldsymbol{A}_\mathrm{L}$ electric dipole Hamiltonian and to return it to a spin-down state, using the spin-down component of the excited state, via the $\boldsymbol{p} \cdot \boldsymbol{A}_\mathrm{S}$ Hamiltonian. Much larger efficiencies (between 10^{-5} and $10^{-10}\,\mathrm{cm^{-1}sr^{-1}}$, the broad range is due to the possibility of resonance with the intermediate state) have been estimated for these processes than for the magnetic dipole counterparts. However, they may still be too small to be observed in high-T_c parent compounds.[12]

The spectral bands shown in Fig. 5.27 and 5.28, centered around $2600\,\mathrm{cm^{-1}}$, have been identified as representing scattering by two magnons. The density of states corresponding to (5.35) has maxima for $\boldsymbol{k}_x = (\pm\frac{\pi}{a}, 0)$ and $\boldsymbol{k}_y = (0, \pm\frac{\pi}{a})$, i.e. at the edge of the AF Brillouin zone, where ω_k reaches its maximum value [160]. Using (5.36a) and (5.36b) one may naively surmise that the corresponding peak frequency is $4J$. However, this guess is incorrect: The reason is given next.

We shall see in Sect. 5.6.4 that the process resulting in two magnon scattering involves two electric dipole transitions ($\boldsymbol{p} \cdot \boldsymbol{A}_\mathrm{L}$, $\boldsymbol{p} \cdot \boldsymbol{A}_\mathrm{S}$) between nearest neighbor copper atoms and a double spin flip, the latter induced by an exchange operator similar to that of (5.33). The resulting excited state is shown schematically in Fig. 5.30a. The scattering by two magnons just described corresponds to two spin flips in nearest neighbor Cu ions: Each of the two spins flips with respect to only *three* Cu neighbors thus leading to a total excitation energy $\frac{1}{2}J \times 6 = 3\,J$, not $4\,J$. Quantum fluctuation corrections lower the value of the excitation energy at the spectral maximum down to $2.76\,J$ for $S = \frac{1}{2}$ [145]. From the peak energy of $2600\,\mathrm{cm^{-1}}$ (Fig. 5.27) we thus

[11] Optic magnons have been observed near $\boldsymbol{k} = 0$ at $524\,\mathrm{cm^{-1}}$ (67 meV) in $YBa_2Cu_3O_{6.3}$ by inelastic neutron scattering [157].

[12] A similar form of spin flip scattering is observed, under resonance condititions, in doped semiconductors. See Fig. 5.27 in [159].

obtain the AF exchange constant $J = 940\,\text{cm}^{-1} = 0.12\,\text{eV}$ for $YBa_2Cu_3O_6$ and $J = 1100\,\text{cm}^{-1} = 0.135\,\text{eV}$ for La_2CuO_4 [9].

In order to confirm the assignment of the Raman peak observed at $2600\,\text{cm}^{-1}$ for $YBa_2Cu_3O_6$, we compare the exchange constant $J = 0.12\,\text{eV}$ obtained from it with the slope c_M of the magnon dispersion relation measured for $YBa_2Cu_3O_{6.15}$ by inelastic neutron scattering. From (5.35) we find:

$$c_M = \sqrt{2}Ja = \sqrt{2} \times 0.12 \times 3.9\,\text{eVÅ} = 0.7\,\text{eVÅ}\,. \tag{5.37}$$

This value is in acceptable agreement with the measured one (1 eVÅ) [140], especially in view of the simplicity of the theoretical ansatz used. For a state-of-the-art discussion of the magnon spectrum of high-T_c superconductors see [161].

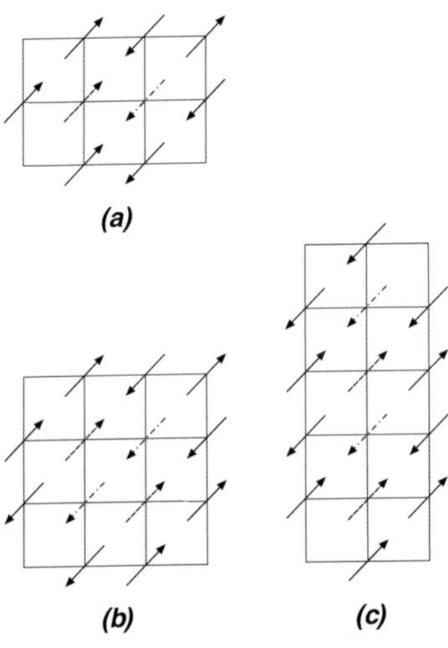

Fig. 5.30. (**a**) Schematic diagram of the excited state that results after scattering by *two magnons* in a CuO_2 plane. The *dashed arrows* represent two spins flipped in the scattering process, each against three of its neighbors. Therefore, the excitation energy is $E = 3\,J$. A quantum correction lowers this value to $2.76\,J$ [145]. (**b**) and (**c**) Schematic diagram of the final state after scattering by *four magnons*, (**b**) a plaquette-type excitation corresponding to $E = 4\,J$, while the linear excitation in (**c**) corresponds to $E = 5\,J$

5.6.3 Electronic Structure of the CuO_2 Antiferromagnetic Insulator and the Mechanism of Scattering by Two Magnons

The large efficiency measured for the scattering by two magnons ($\approx 10^{-4}\,\text{cm}^{-1}\text{sr}^{-1}$ [158]) demands an electric dipole coupling mechanism for the incident and scattered photons. Such a mechanism is not explicitly apparent in the phenomenological Hamiltonian of (5.33). This Hamiltonian appears naturally in the standard treatments of one-electron interaction plus Hubbard repulsion, such as the so-called three-band Hubbard model [151,152]. The three bands are the $(x^2 - y^2)$-like copper $3d$ bands, the p_x band of the O2 oxygen and the p_y band of O3 (see Fig. 5.1). A nearest neighbor hopping integral equal to $-t$ for the $pd\sigma$ coupling and a smaller, next nearest neighbor O2-O3 hopping integral t' define the one-electron problem, which is then completed by the Hubbard double occupancy energy U_{dd} plus, possibly, two additional smaller Hubbard energies U_{pd} and U_{pp}. The model leads to the density of states sketched in Fig. 5.31 [148].

We have also introduced in Fig. 5.31 the so-called Zhang–Rice singlet band. After some calculations, Zhang and Rice [162] concluded that the physics of Fig. 5.31 is basically contained in the so-called t-J model, in which the oxygen electrons do not appear explicitly. Instead, one has a singlet ZR spin and a hole (absence of spin) at each of the copper sites. The spins are coupled antiferromagnetically following (5.33). Using perturbation theory, the *superexchange* constant J can be written as a function of the parameters of the 3

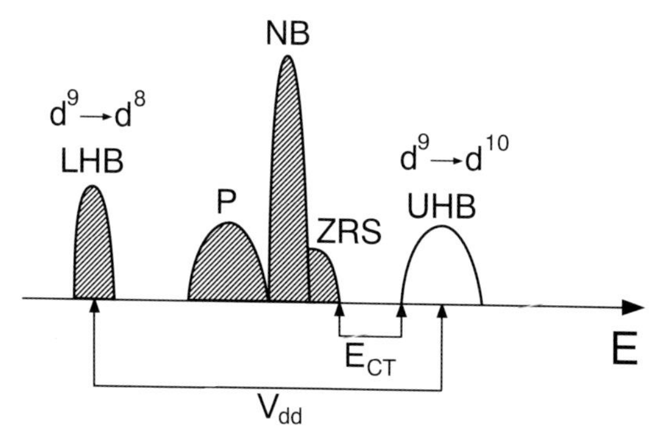

Fig. 5.31. Schematic density of states diagram of $YBa_2Cu_3O_{6.5}$ as a change transfer insulator (E_{CT} = charge transfer gap) in the 3-band Hubbard model. The shaded states are electron occupied (hole empty) whereas the upper Hubbard band (UHB) is electron empty (hole occupied). ZRS represents the Zhang–Rice singlet band [148]

band Hubbard model:

$$J \simeq \frac{4t^4}{\Delta^2}\left(\frac{1}{U_{dd}} + \frac{1}{\Delta}\right) , \tag{5.38}$$

where $\Delta \simeq 3.5\,\text{eV}$ represents the difference between the atomic p and d energies, $U_{dd} \simeq 8.8\,\text{eV}$ and $t = 1.3\,\text{eV}$. Using these energies we find $J = 0.09\,\text{eV}$, in reasonable agreement with the values derived from the peak in the *two-magnon spectra*.

It is easy to construct the Hamiltonian for Raman scattering by two magnons, in lowest (third) order of perturbation theory, following a path similar to that which led from the three-band Hubbard model to (5.33) and (5.38). The relevant steps are illustrated in Fig. 5.32 for the case of incident and scattered fields $\boldsymbol{E}_{\mathrm{L}}$, $\boldsymbol{E}_{\mathrm{S}}$ directed along the $\hat{x}$-axis (Fig. 5.1). The Hamiltonian expressing the incident field $(\boldsymbol{p}_{\mathrm{L}} \cdot \boldsymbol{E}_{\mathrm{L}})$ can excite an electron of the "antibonding" oxygen O2 to the UHB, of the copper on the left in step (a), the corresponding dipole matrix element being proportional to t. In step (b), the Hamiltonian (5.33) flips the up-spin of O2 and the down-spin of the copper to the right. The last step (c) involves the deexcitation of the UHB of the copper on the left through a transition to the empty up-spin O2. The deexcitation energy is taken up by the emitted photon. We take both participating photons to be polarized with $\boldsymbol{E}$ along $\hat{\boldsymbol{x}}$. The combined scattering Hamiltonian is proportional to [8,143]

$$H_{\mathrm{R}} \propto \frac{t^2}{U - \omega_{\mathrm{L}}} \sum_{(i,j_x)} \boldsymbol{S}_i \cdot \boldsymbol{S}_{j_x} \left(\boldsymbol{E}_{\mathrm{L}} \cdot \boldsymbol{\mu}_x\right) \left(\boldsymbol{E}_{\mathrm{S}} \cdot \boldsymbol{\mu}_x\right) , \tag{5.39}$$

where the effective Hubbard energy U is, approximately, the average excitation energy from the ZRS (see Fig. 5.31) to the UHB ($\approx$ 3.5 eV [148]). The sum over (i, j_x) in (5.39) is extended for each i of the up-spin sublattice to the two j nearest neighbors along $\hat{\boldsymbol{x}}$. The scattering described by (5.39) corresponds to $\boldsymbol{E}_{\mathrm{L}} \parallel \boldsymbol{E}_{\mathrm{S}} \parallel \hat{\boldsymbol{x}}$.

Let us assume $|\boldsymbol{\mu}_x| = |\boldsymbol{\mu}_y| = \mu$ and add to (5.39) the corresponding Hamiltonian for $\boldsymbol{E}_{\mathrm{L}} \parallel \boldsymbol{E}_{\mathrm{S}} \parallel \hat{y}$. In this manner we obtain the scattering Hamiltonian for a polarization configuration of A_{1g} symmetry:

$$H_{\mathrm{R}}^{(A_{1g})} \propto \frac{t^2 \mu^2}{U - \omega_{\mathrm{L}}} \left[\sum_{(i,j)} \boldsymbol{S}_i \cdot \boldsymbol{S}_j \right] \frac{1}{2} \left(E_{\mathrm{L}x} E_{\mathrm{S}x} + E_{\mathrm{L}y} E_{\mathrm{S}y}\right) , \tag{5.40}$$

where j is summed over the four nearest neighbors of i. Note that the Hamiltonian (5.40) commutes with the unperturbed one given in (5.33). It cannot, therefore, induce any excitations when applied to the ground state of (5.33). Thus, to this order in $t^2/(U - \omega_{\mathrm{L}})$, scattering of A_{1g} symmetry should not take place. This is not the case for the Hamiltonian (5.39), which can be decomposed into the sum of (5.40) and:

$$H_{\mathrm{R}}^{(B_{1g})} \propto \frac{t^2}{U - \omega_{\mathrm{L}}} \sum_{i,j_{x,y}} \frac{1}{2} \boldsymbol{S}_i \cdot \left[\boldsymbol{S}_{j_x} E_{\mathrm{L}x} E_{\mathrm{S}x} - \boldsymbol{S}_{j_y} E_{\mathrm{L}y} E_{\mathrm{S}y}\right] , \tag{5.41}$$

where the sum over j is split into nearest neighbors along $\hat{\boldsymbol{x}}$ and along $\hat{\boldsymbol{y}}$. The Hamiltonian (5.41) corresponds to scattering of B_{1g} symmetry by two magnons, which, contrary to that of A_{1g} symmetry, should not vanish. This is the reason why the B_{1g} scattering is usually dominant, at least away from

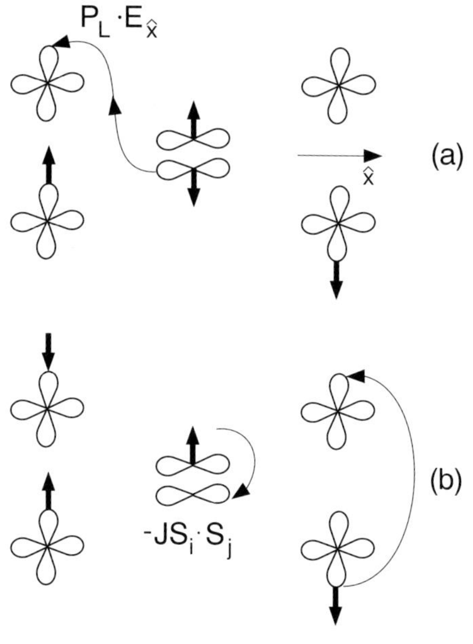

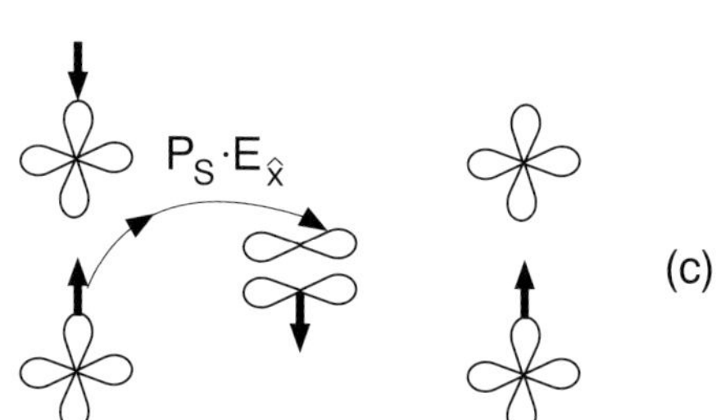

Fig. 5.32. Schematic diagram of the three steps leading to light scattering by two magnons in CuO_2 AF planes. (**a**) Electric dipole virtual excitation to a UHB. (**b**) Exchange induced spin flip in the oxygen and the copper. (**c**) Deexcitation from an UHB by dipole emission of the scattered photon. The Stokes shift should be $\simeq 3\,J$ according to Fig. 5.30

resonance ($\omega_{\mathrm{L}} \ll U \simeq 3.5\,\mathrm{eV}$) (see Figs. 5.27 and 5.28). Scattering of A_{1g} symmetry becomes possible to higher order of perturbation theory. The next higher order terms in the $t/(U-\omega_{\mathrm{L}})$ are [143]:

$$H_{\mathrm{R}'}^{(A_{1g})} \propto \frac{t^4}{(U-\omega_{\mathrm{L}})^3} \sum_{i,j} \left(S_i \cdot S_{j_y}\right) \left(S_{i_{x+y}} \cdot S_{j_x}\right) , \tag{5.42}$$

where, for simplicity, we have only included terms which correspond to nearest neighbor exchange interaction. The Hamiltonian of (5.42) flips four spins in a plaquette such as that of Fig. 5.30 and should result in a Raman shift equal to $4J$. In [143] the reader can also find terms of B_{2g} symmetry (i.e. $E_x E_y$) and A_{2g} symmetry (that of an antisymmetric Raman tensor) which become nonvanishing to order $t^4/(U-\omega_{\mathrm{L}})^3$. The B_{2g} terms also require second neighbor exchange interaction:

$$H_{\mathrm{R}'}^{(B_{2g})} \propto \frac{t^4}{(U-\omega_{\mathrm{L}})^3} \sum_{i,j} \left(S_i S_{i_{x+y}} - S_j S_{j_{x-y}}\right) \tag{5.43}$$

Therefore scattering in the B_{2g} (E_xE_y) symmetry configuration should be very weak, weaker than A_{1g} scattering, in agreement with the results of Fig. 5.28. The corresponding A_{2g} scattering operator should produce scattering involving three spin flips which requires that either the ground state or the excited state be chiral [37].

5.6.4 Lineshape of Two-Magnon Raman Scattering in the Insulating HTSC Phases

The lineshape of light scattering by two magnons in AF materials was discussed well before the discovery of HTSC. We mention the pioneering work of Elliott and Thorpe [163] who developed a Green's function method for calculating the two-magnon Raman lineshape in 3D antiferromagnets, and Parkinson [160], who used the method to calculate the spectrum of the 214-like material K_2NiF_4, a 2D antiferromagnet with two holes in the d-shells of Ni (d^8), i.e. according to Hund's rule, with $S = 1$. The basic idea behind this work is the assumption of isotropy of the magnon dispersion relation (5.35) which, at the edge of the magnetic BZ leads to a typical square root singularity in the density of magnon states $\sim (\omega_\mathrm{M} - \omega)^{-1/2}$. This functional dependence also applies to the density of two magnons with the constraint $\boldsymbol{k}_1 + \boldsymbol{k}_2 \simeq 0$, imposed by Raman spectroscopy. The $(\omega_\mathrm{M} - \omega)^{-1/2}$ singularity is lifted by magnon–magnon interaction and a spectral shape similar to the one observed for K_2NiF_4 and also Rb_2MnF_4 ($S = \frac{5}{2}$) is found. Analogous calculations have been performed by Weber and Ford for the cuprates [164]: They reveal an essential difference between the two-magnon band calculated on the basis of the Gaussian-broadened dispersion relation (5.35) and the experimental spectra: while the former are asymmetrically broadened (with a sharp edge at the high frequency side corresponding to $(\omega_\mathrm{M} - \omega)^{-1/2}$) the latter have a long tail towards high frequencies (see Fig. 5.33). Figure 5.35 also indicates that the two-magnon spectra broaden considerably with increasing temperature. This broadening is not likely to result from magnon–magnon interactions since the magnon frequencies involved are considerably larger than those that correspond to the temperatures of the measurement. It must, therefore, be related to the thermal excitation of phonons which broaden the magnons through phonon–magnon interaction. Such interaction arises naturally from the modulation of the p-d first neighbor hopping parameter t when modulating the copper–oxygen distance r: According to Harrison [166] t should be proportional to $r^{-3.5}$. The authors of [165] fitted the spectra of Fig. 5.33 with the sum of two temperature dependent broadening mechanisms: magnon–magnon and magnon–phonon interaction. The Lorentzian broadening parameter (HWHM) at 600 K calculated from these mechanisms, is $\sim 600\,\mathrm{cm}^{-1}$, about $300\,\mathrm{cm}^{-1}$ less than that needed to fit the spectrum measured at 600 K.

This discrepancy and the asymmetric lineshape, with a long tail toward higher frequencies, have triggered a large amount of theoretical work, most

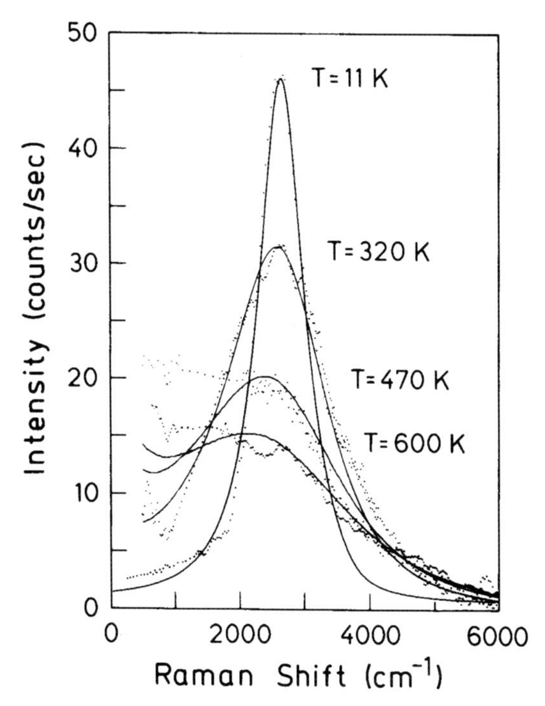

Fig. 5.33. Measured (*dots*) and calculated (*solid line*) B_{1g}-like two-magnon spectra of $EuBa_2Cu_3O_6$ for various temperatures. The experimental high-energy tails of the spectra taken at 11 K and 320 K spectra merge with each other [165]

of which is reviewed in [146]. Three causes have been suggested for these anomalous spectral shapes:

(1) Quantum fluctuations due to the small value of S ($S = \frac{1}{2}$) in the cuprates (as opposed to K_2NiF_4 and Rb_2MnF_4) [154]. This suggestion has been rejected in [167] on the basis that similar anomalies appear for an antiferromagnet with $S = 1$: $NiPS_3$.
(2) Electron phonon interaction. Besides the heuristic, perturbative calculations in [165], a number of cluster calculations with the phonons represented by static distortions of the atomic sites in the cluster have been presented [146,168,169]. The authors make use of the $r^{-3.5}$ dependence of the hopping parameter t in order to simulate the effects of the phonons. In calculations using finite clusters, it is, however, difficult to separate the coherent components of the spectral functions from the incoherent background due to the frozen phonon displacements. Nevertheless, the calculated curves describe well the measured B_{1g} spectra except, possibly, for the high frequency tails [168].
(3) The high frequency tails can be explained by including contributions of scattering by more than two magnons, such as the processes depicted in

Fig. 5.3c,d which peak at $4\,J$ and $5\,J$. These processes are essential for understanding the scattering of A_{1g} symmetry which, according to (5.42) should be weak and should peak at $\omega \approx 4\,J$ (see Fig. 1 of [168]).

We close this section by mentioning that spectra which are likely to be due to the creation of two magnons plus one phonon have been observed in the IR absorption of high-T_c superconductors [170,171].

5.6.5 Resonant Raman Scattering by Magnons

The expressions (5.40)–(5.43) indicate the presence of resonance effects in the scattering by magnons. The dominant scattering, i.e. that of B_{1g} symmetry, is represented by the Hamiltonian (5.41) which contains a single resonant energy denominator (note, however, that (5.41) must be squared in order to obtain the resonant scattering efficiency). This dominant B_{1g} scattering corresponds to scattering by two magnons. Scattering of A_{1g} symmetry involves, as discussed in Sect. 5.6.2, four or more magnons and is more strongly resonant than the dominant two-magnon scattering of B_{1g} symmetry (note that (5.42) has *three* energy denominators); it should display peaks in the scattered spectra at $E_s \approx 4\,J$ and $5\,J$ (see Fig. 5.31). Such peaks have been reported by several workers. We mention here the measurements of Rübhausen et al. [10] performed on the nonsuperconducting material $PrBa_2Cu_{2.7}Al_{0.3}O_7$: a broad band centered at around $E_s = 4\,J$, displaying a long tail ($E_s \simeq 5\,J$, $6\,J$) towards higher energies in the A_{1g} scattering geometry when the laser approaches the resonant energy of $\simeq 2.8\,eV$. These authors also observed a resonance at nearly the same laser frequency in B_{1g} configuration for which the scattering spectrum is peaked at $E_s \simeq 3\,J$. Close to resonance, the long energy tail also contains some evidence for excitations at $E_s = 4\,J, 5\,J, 6\,J, \ldots$ which have been predicted in [143].

A detailed investigation of the resonance in the B_{1g} two-magnon scattering has been performed by Brenig et al. [94] using six different laser lines. Figure 5.34 displays the integrated absolute Raman efficiencies of the spectra measured for four different cuprate samples vs. laser photon energy. The solid curve shows the results of a calculation (in absolute efficiency units), similar to that which led to (5.41), requiring a few adjustable band structure and correlation parameters which take physically reasonable values. The fact that the calculation can reproduce the large measured scattering efficiencies is quite remarkable. The lower edge of the resonance curve and the few available experimental points, peak at $\hbar\omega_{\mathrm{L}} \simeq 2.7\,\mathrm{eV}$. This photon energy corresponds to that of a structure observed in the dielectric function of $PrBa_2Cu_2Cu_3O_7$ for in-plane polarization [172].

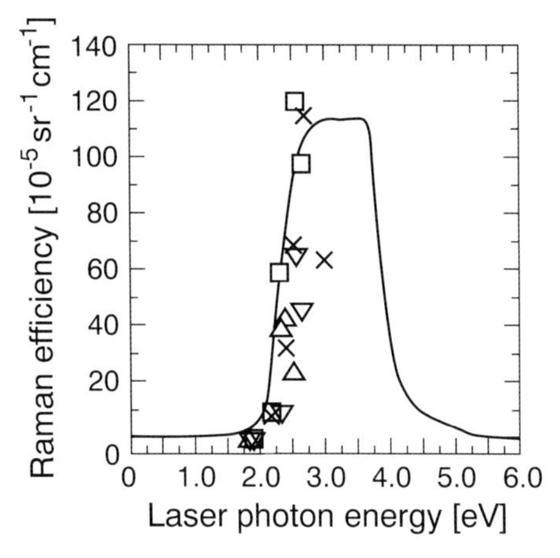

Fig. 5.34. Observed B_{1g} resonant Raman efficiency for various cuprates (*markers*) compared with the calculated efficiencies (*solid line*): (□) $YBa_2Cu_3O_{6.0}$; (×) $YBa_2Cu_3O_{6.3}$, scaled to $O_{6.0}$; (△) Nd_2CuO_4; (▽) La_2CuO_4 [94]

5.6.6 Scattering by Magnetic Fluctuations in Doped (Superconducting) Cuprates

Figures 5.27, 5.28, 5.33, and 5.34 demonstrate the robustness of scattering by two magnons in the insulating partners of HTSC cuprates. In spite of a natural reluctance to admit the simultaneous appearance of antiferromagnetism and superconductivity, it was early realized that two-magnon-like features persisted in the doped phonons of those cuprates [142] even in their superconducting state. Their strength has a tendency to decrease with increasing doping (see Fig. 5.35), becoming rather weak in strongly overdoped samples (for spectra of overdoped Bi-2212 see [173]). Actually, for some highly overdoped samples, two-magnon structures appear clearly only below T_c (see Fig. 1a of [173]). For underdoped samples, the two-magnon features are enhanced when the laser frequency approaches resonance (around 3 eV, see Fig. 1 of [101]). This resonance effect is much weaker for overdoped samples. Although two-magnon features can be seen for samples with any degree of doping, the following facts are generally observed when the doping is increased.

(1) The intensity of the "two-magnon" features decreases (see Fig. 5.35 and [174]).
(2) The corresponding band broadens [174].
(3) Its peak shifts slightly to lower frequencies [174].

It is generally believed that the AF long-range order disappears upon doping. The fact that AF-like scattering can be observed in doped samples, in which the AF order has been destroyed, suggests that the scattering is due to spin-flip transitions in short-range antiferromagnetic fluctuations. The

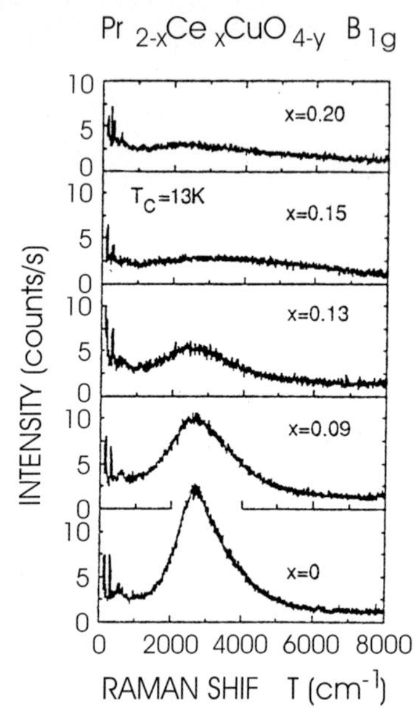

Fig. 5.35. Concentration dependence of the B_{1g} Raman spectra in the n-type superconductor $Pr_{2-x}Ce_xCuO_{4-y}$ at 300 K. Crystals with $x = 0$, 0.09, and $x = 0.13$ belong to the AF phase at 0 K. The $x = 0.15$ crystal exhibits superconductivity at 13 K. The $x = 0.2$ crystal belongs to the metallic phase [142]

near disappearance of "two-magnon" scattering in heavily overdoped cuprates reveals that for these materials the coherence length of the antiferromagnetic fluctuations is no more than about two lattice constants. One may conjecture that the decrease in the coherence length of the AF fluctuations is responsible for the decrease of T_c in the overdoped region, under the assumption that the superconducting pairing is effected by the AF fluctuations.

A few other facts suggest a close correlation between the AF fluctuations (as observed by Raman scattering) and the superconductivity. It has been observed for underdoped samples (e.g. $Y_{0.8}Pr_{0.2}Ba_2O_7$) that the normal-to-superconducting transition is accompanied by a significant transfer of scattering strength from the A_{1g} to the B_{1g} spectrum in the region from 1000 to 3000 cm^{-1} [175].

The ubiquitous multimagnon Raman bands discussed in this section exhibit a considerable number of interesting features, some of which, e.g. resonance effects, are just beginning to be investigated. These investigations may provide clues for the definitive elucidation of the mechanism responsible for the superconductivity in these intriguing materials.

Acknowledgments. The work described in this chapter would not have been possible without the contributions of innumerable collaborators whose names appear in the roster of references. Special recognition is deserved by the experts in crystal growth who patiently prepared the samples used for Raman- and IR-spectroscopy. Although the list is too long to allow explicit mention here, I would like to express my gratitude, in particular, to my long-time associates Tobias Ruf, Christian Thomsen, and Christian Bernhard. O.K. Andersen and N.E. Christensen also deserve my thanks for having introduced me to the wonders of the LMTO-LDA band structure calculations. Financial support from the Fonds der Chemischen Industrie is also acknowledged. Last but not least, I would like to thank Sabine Birtel for typing and masterfully setting up the final version of the manuscript.

References

1. M. Schwoerer-Bohning, A.T. Macrander, D.A. Arms: Phys. Rev. Lett. **80**, 5572 (1998)
2. K.B. Lyons, S.H. Liou, M. Hong, H.S. Chen, I. Kwo, T.J. Negran: Phys. Rev. B **36**, 5592 (1987)
3. T. Devereaux, D. Einzel: Phys. Rev. B **51**, 16336 (1995); ibid **54**, 15547 (1996)
4. T. Strohm, M. Cardona: Solid State Commun. **104**, 233 (1997)
5. V.G. Hadjiev, X.J. Zhou, T. Strohm, M. Cardona, Q.M. Lin, C.W. Chu: Phys. Rev. B **58**, 1043 (1998)
6. X.J. Zhou, M. Cardona, C.W. Chu, Q.M. Lin, S.M. Loureiro, M. Marezio: Phys. Rev. B **54**: 6137 (1996)
7. M. Guillaume, W. Henggeler, A. Furrer, R.S. Eccleston, V. Tromov: Phys. Rev. Lett. **74**, 3423 (1995)
8. P.A. Fleury, R. Loudon: Phys. Rev. **166**, 514 (1968)
9. K.B. Lyons, P.A. Fleury, L.F. Schneemeyer, J.V. Waszczak: Phys. Rev. Lett. **60**, 732 (1988)
10. M. Rübhausen, C.T. Rieck, N. Diekmann, K.-O. Subkee, A. Bock, U. Merkt: Phys. Rev. B **56**, 14797 (1997)
11. J.P. Carbotte: Rev. Mod. Phys. **62**, 1027 (1990); D.J. Scalapino et al: Phys. Rev. B **34**, 8190 (1986); P. Monthoux and D. Pines: Phys. Rev. B **47**, 6068 (1993)
12. C. Thomsen, M. Cardona: in *Physical Properties of High Temperature Superconductors I*, ed. by D.M. Ginsberg (World Scientific, Singapore 1989) p. 409
13. C. Thomsen: in *Light Scattering in Solids VI*, ed. by M. Cardona, G. Güntherodt (Springer, Heidelberg 1991) p. 285
14. R. Feile: Physica C **159**, 1 (1989)
15. M. Cardona: Physica C **317–318**, 30 (1999)
16. E. Faulques: *Materials Synthesis and Characterization* (Plenum, New York 1996)
17. J.R. Ferraro, V.A. Maroni: Appl. Spectrosc. **44**, 351 (1990)
18. *Physical Properties of High Temperature Superconductors I – V*, ed. by D.M. Ginsberg (World Scientific, Singapore 1989–1998)
19. M. Cyrot, D. Pavuna: *Introduction to Superconductivity and High-T_c Materials* (World Scientific, Singapore 1992)
20. L. Gao et al.: Phys. Rev. B **50**, 4260 (1994)

21. M.K. Wu, J.R. Ashburn, C.J. Torng, P.H. Hor, R.L. Meng, L. Gao, Z.J. Huang, Y.Q. Wang, C.W. Chu: Phys. Rev. Lett. **58**, 908 (1987)
22. A. Leggett: Phys. Rev. Lett. **83**, 392 (1999)
23. For an exhaustive pictorial description of the structure of Nd-214 and other HTSC see: H. Shaked et al.: *Crystal Structures of the High-T_c Superconducting Copper Oxides* (Elsevier Science, Amsterdam 1994)
24. Y. Tokura, H. Takagi, S. Uchida: Nature **337**, 345 (1989)
25. K. Oka, Z. Zou, J. Ye: Physica C **300**, 200 (1998)
26. O. Fischer, C. Renner, I. Maggio-Aprile: in *The Gap Symmetry and Fluctuations in High-T_c Superconductors*, ed. by J. Bok, G. Deutscher, D. Pavune, S.A. Wolf: NATO ASI, Series B: Physics **371**, 487 (1998)
27. T. Zhou, K. Syassen, M. Cardona, J. Karpinski, E. Kaldis: Solid State Commun. **99**, 669 (1996)
28. E.T. Heyen, R. Liu, C. Thomsen, R. Kremer, M. Cardona, J. Karpinski, E. Kaldis, S. Rusiecki: Phys. Rev. B **41**, 11058 (1991)
29. C. Bernhard, R. Henn, A. Wittlin, M. Kläser, Th. Wolf, G. Müller-Vogt, C.T. Lin, M. Cardona: Phys. Rev. Lett. **80**, 1762 (1998)
30. J.G. Bednorz, K.A. Müller: Z. Phys. B **64**, 189 (1986)
31. B. Morosin, D.S. Ginley, P.F. Hlava, M.J. Car, R.J. Baughman, J.E. Shirbe, E.L. Venturini, J.F. Kwak: Physica C **152**, 413 (1988)
32. D.E. Cox, C.C. Torardi, M.A. Subramanian, J. Gopalakrishnan, A.W. Sleight: Phys. Rev. B **38**, 6624 (1988)
33. H. Maeda, Y. Tanaka, M. Fuksutumi, T. Asano: Jpn. J. Appl. Phys. **27**, L209 (1988)
34. H.G. von Schnering et al.: Angew. Chem. Int. Ed. Engl. **27**, 574 (1988)
35. J.D. Jorgensen, B.W. Veal: Phys. Rev. B **41**, 1863 (1990)
36. Z.-X. Shen, D.S. Dessau: Physics Reports **253**, 1 (1995)
37. B. Gegenheimer: private communication
38. T. Haage, J. Zegenhagen, J.Q. Li, H.-U. Habermeier, M. Cardona: Phys. Rev. B **56**, 8404 (1997)
39. H.-U. Habermeier, A.A.C.S. Lourenço, B. Friedl, J. Kircher, J. Kohler: Solid State Commun. **77**, 683 (1991)
40. C. Thomsen, R. Wegerer, H.-U. Habermeier, M. Cardona: Solid State Commun. **83**, 199 (1992)
41. C. Thomsen, M. Cardona, B. Gegenheimer, R. Liu, A. Simon: Phys. Rev. B **37**, 9860 (1988)
42. B. Dam, J.M. Huijbregtse, F.C. Klaassen, R.C.F. van der Geest, G. Doornbos, J.H. Rector, A.M. Testa, S. Freisem, J.C. Martinez, B. Stäuble-Pümpin, R. Griessen: Nature **399**, 439 (1999)
43. J.L. McMannus-Driscoll: Ann. Rev. Mater. Sci. **28**, 421 (1998)
44. R. Kossowsky, S. Bose, V. Pan, Z. Durusoy: NATO Conf. Proc. Series E, Vol. 356 (Kluwer, Dordrecht 1999); *Proc. of Int. Workshop on Critical Current in Superconductors for Practical Applications*, ed. by L. Zhou, W.H. Weber, E.W. Collins (World Scientific, Singapore 1977)
45. B. Friedl, C. Thomsen, E. Schönherr, M. Cardona: Solid State Commun. **76**, 1107 (1990)
46. J. Humlíček, J. Kircher, H.-U. Habermeier, M. Cardona, R. Röseler: Physica C **190**, 383 (1992)

47. Ya.G. Ponomarev, E.B. Tsokur, M.V. Tsudakova, S.N. Tschesnokov, M.E. Shabalin, M.A. Lorenz, M.A. Hein, G. Müller, H. Piel, B.A. Arminov: Solid State Commun. **111**, 513 (1999)
48. B. Lederle: Doctoral Dissertation, University of Stuttgart, 1997.
49. E.T. Heyen, G. Kliche, W. Kress, W. König, M. Cardona, E. Rampf, J. Prade, U. Schröder, A.D. Kulkarni, F.W. de Wette, S. Piñol, D.McK. Paul, E. Morán, M.A. Alario-Franco: Solid State Commun. **74**, 1299 (1990)
50. C. Thomsen, M. Cardona, B. Gegenheimer, R. Liu, A. Simon: Phys. Rev. B **37**, 9860 (1988)
51. D.R. Wake, F. Slakey, M.V. Klein, J.P. Rice, D.M. Ginsberg: Phys. Rev. Lett. **67**, 3728 (1991)
52. M.N. Iliev, H.-U. Habermeier, M. Cardona, V.G. Hadjiev, R. Gajić: Physica C **279**, 63 (1997)
53. M.N. Iliev, P.X. Zhang, H.-U. Habermeier, M. Cardona: J. of Alloys and Compounds **251**, 99 (1997)
54. E.T. Heyen, R. Liu, M. Cardona, S. Piñol, D. McK. Paul, E. Morán, M.A. Alario-Franco: Phys. Rev. B **43**, 2857 (1991)
55. V.M. Orera, M.L. Sanjuán, R. Alcalá, J. Fontcuberta, S. Piñol: Physica C **168**, 161 (1990)
56. S. Jandl, P. Dufour, T. Strach, T. Ruf, M. Cardona, V. Nekvasil, C. Chen, B.M. Wanklyn, S. Piñol: Phys. Rev. B **53**, 8632 (1996)
57. R. Henn, T. Strach, E. Schönherr, M. Cardona: Phys. Rev. B **55**, 3285 (1997)
58. C.O. Rodriguez A.I. Liechtenstein, I.I. Mazin, O. Jepsen, O.K. Andersen, M. Methfessel: Phys. Rev. B **42**, 2692 (1990)
59. R.E. Cohen, W.E. Pickett, H. Krakauer: Phys. Rev. Lett. **64**, 2575 (1990)
60. R. Kouba, C. Ambrosch-Draxl: Physica C **282–287**, 1635 (1997)
61. M. Cardona, X.J. Zhou, T. Strach: *Proceedings of the 10th Anniversary HTS Workshop on Physics, Materials and Applications*, ed. by B. Batlogg, C.W. Chu, W.K. Chu, D.U. Gubser, K.A. Müller (World Scientific, Singapore 1996), p. 72
62. E. Kaldis, P. Fischer, A.W. Hewat, J. Karpinski, S. Rusieki: Physica C **159**, 668 (1989)
63. E.T. Heyen, R. Liu, C. Thomsen, R. Kremer, M. Cardona, J. Karpinski, E. Kaldis, S. Rusieki: Phys. Rev. B **41**, 11058 (1990)
64. F. Paulsen, A.I. Lichtenstein, O.K. Andersen: private communication
65. M. Cardona: in *Light Scattering in Solids II*, ed. by M. Cardona, G. Güntherodt (Springer, Heidelberg 1982), p. 19
66. L. Pintschovius, W. Reichardt: in *Physical Properties of High-T_c Superconductors IV*, D.M. Ginsberg ed. (World Scientific, Singapore 1994), p. 295
67. E.T. Heyen, J. Kircher, M. Cardona: Phys. Rev. B **45**, 3037 (1992)
68. Y. Lin, J.E. Eldridge, J. Sichelschmidt, S.-W. Cheong, T. Wahlbrink: Solid State Commun. **112**, 315 (1999)
69. V.I. Belitsky, A. Cantarero, M. Cardona, C. Trallero-Giner, S.T. Pavlov: J. Phys.: Condens. Matter **9**, 5965 (1997); V.I. Belitsky, M. Cardona, I.G. Lang, S.T. Pavlov: Phys. Rev. B **46**, 15767 (1992)
70. E.T. Heyen, S.N. Rashkeev, I.I. Mazin, O.K. Andersen, R. Liu, M. Cardona, O. Jepsen: Phys. Rev. Lett. **65**, 3048 (1990)
71. T. Strach, J. Brunen, B. Lederle, J. Zegenhagen, M. Cardona: Phys. Rev. B **57**, 1292 (1998)

72. A. Röseler: *Infrared Spectroscopic Ellipsometry*, (Akademie Verlag, Berlin 1990)
73. K. Kamarás, K.-L. Barth, F. Keilmann, R. Henn, M. Reedyk, C. Thomsen, M. Cardona, J. Kircher, P.L. Richards, J.-L. Stehlé: J. Appl. Phys. **78**, 1235 (1995)
74. G. Abstreiter, M. Cardona, A. Pinczuk: *Light scattering by electronic excitations in semiconductors*, in *Light Scattering in Solids IV*, (Springer, Heidelberg 1984), p. 5
75. S.N. Rashkeev, G. Wendin: Phys. Rev. B **47**, 11603 (193); Z. Phys. B **33** (1993)
76. P.Y. Yu, M. Cardona: *Fundamentals of Semiconductors*, (Springer, Heidelberg 1995, 2nd Edn. 1999)
77. M. Cardona, I.P. Ipatova: in *Elementary Excitations in Solids*, ed. by J.L. Birman, C. Sébenne, R.F. Wallis (Elsevier, Amsterdam 1992) p. 237
78. S. Das Sarma, D.-W. Wang: Phys. Rev. Lett. **83**, 816 (1999)
79. A. Zawadowski, M. Cardona: Phys. Rev. B **42**, 10732 (1990)
80. C.M. Varma, P.B. Littlewood, S. Schmidt-Rink, E. Abrahams, A.E. Ruckenstein: Phys. Rev. Lett. **63**, 1996 (1989)
81. M.C. Krantz, I.I. Mazin, D.H. Leach, W.Y. Lee, M. Cardona: Phys. Rev. B **51**, 5949 (1995)
82. M. Cardona, T. Strohm, J. Kircher: in *Spectroscopic Studies of Superconductors*, ed. by I. Bozovic, D. van der Marel, Proc. SPIE **2696**, 182 (1996)
83. For a discussion of Raman scattering in *zz* polarization in HTSC see W.C. Wu, J.P. Carbotte: Phys. Rev. B **56**, 6327 (1997)
84. C. Jiang, J. Carbotte: Phys. Rev. B **53**, 11868 (1996)
85. T.P. Devereaux, A.P. Kampf: Phys. Rev. B **59**, 6411 (1999)
86. A.V. Chubukov, D.K. Morr, G. Blumberg: Solid State Commun. **112**, 183 (1999)
87. R. Hackl, G. Krug, R. Nemetschek, M. Opel, B. Stadlober: Proc. SPIE 2696, 194 (1996)
88. E.N. van Eenige, et al.: Physica C **168**, 482 (1990)
89. S.L. Cooper, M.V. Klein, B.G. Pazol, J.P. Rice, D.M. Ginsberg: Phys. Rev. B **37**, 5920 (1987)
90. M. Krantz, M. Cardona: J. of Low Temp. Physics **99**, 205 (1995)
91. T. Strohm, M. Cardona: Phys. Rev. B **55**, 12725 (1997)
92. J.R. Schrieffer: *Theory of Superconductivity*, (Benjamin, New York 1964)
93. M.V. Klein, S.B. Dierker: Phys. Rev. B **29**, 4976 (1984)
94. W. Brenig, P. Knoll, M. Mayer: Physica B **237–238**, 95 (1997)
95. T. Devereaux, D. Einzel: Phys. Rev. B **54**, 15547 (1996)
96. D. Manske, C.T. Rieck, R. Das Sharma, A. Bock, D. Fay: Phys. Rev. B **56**, R2940 (1997), erratum: ibid **58**, 8841 (1998)
97. T. Strohm, D. Munzar, M. Cardona: Phys. Rev. B **58**, 8839 (1998)
98. M. Cardona, T. Strohm, X.J. Zhou: Physica C **282–287**, 222 (1997)
99. C.K. Kendziora, A. Rosenberg: Phys. Rev. B **52**, 9867 (1995)
100. A. Chubukov: Europhys. Let. **44**, 655 (1998)
101. G. Blumberg, M. Kang, M.V. Klein, K. Kadowaki, C. Kandziora: Science **278**, 1427 (1997); J. Phys. Chem. Solids **59**, 1932 (1998)
102. M.R. Norman et al.: Nature **392**, 157 (1998); Z.X. Shen et al.: Science **28**, 259 (1998)
103. Ch. Renner, B. Ravaz, J.-Y. Genoud, K. Kadowaki, O. Fischer: Phys. Rev. Lett. **80**, 149 (1998)

104. D.J. van Haerlingen: Rev. Mod. Phys. **67**, 515 (1995)
105. T. Staufer, R. Nemetschek, R. Hackl, P. Müller, H. Veith: Phys. Rev. Lett. **68**, 1069 (1992)
106. R. Hackl, W. Gläser, P. Müller, D. Einzel, K. Andres: Phys. Rev. B **38**, 7133 (1988)
107. T.P. Devereaux: Phys. Rev. Lett. **74**, 4313 (1995); L.S. Borkowski, P.J. Hirschfeld: Phys. Rev. B **49**, 15404 (1994)
108. T.P. Devereaux: Phys. Rev. B **50**, 10287 (1994)
109. M. Opel, R. Hackl, T.P. Devereaux, A. Virosztek, A. Zawadowski, A. Erb, E. Walker, H. Berger, L. Forró: Phys. Rev. B **60**, 9836 (1999)
110. C. Thomsen, B. Friedl, M. Cieplak, M. Cardona: Solid State Commun. **78**, 727 (1991)
111. T. Ruf, M. Cardona: Phys. Rev. Lett. **63**, 2288 (1989)
112. R. Zeyher, G. Zwicknagl: Solid State Commun. **66**, 617 (1988); Z. Phys. B **78**, 175 (1990)
113. E.J. Nicol, C. Jiang, J.P. Carbotte: Phys. Rev. B **47**, 8131 (1993)
114. P.B. Allen, B. Mitrovic: in *Solid State Physics*, Vol. 37, ed. by H. Ehrenreich, D. Turnball (Academic, New York 1982) p. 1
115. C. Thomsen, M. Cardona, B. Friedl, I.I. Mazin, C.O. Rodriguez, O.K. Andersen: Solid State Commun. **75**, 219 (1990)
116. J. Franck: in *Physical Properties of High T_c Superconductors* Vol. IV (World Scientific, Singapore 1994) p. 189
117. P.B. Allen: Nature **335**, 396 (1988)
118. N. Pyka, W. Reichardt, L. Pintschovius, G. Engel, J. Rossat-Mignon, J.Y. Henry: Phys. Rev. Lett. **70**, 14570 (1993)
119. R.P. Sharma, T. Venkatesan, Z.H. Zhang, J.R. Liu, R. Chu, W.K. Chu: Phys. Rev. Lett. **77**, 4624 (1996); H.A. Mook, M. Mostoller, J.A. Harvey, N.W. Hill, B.C. Chakoumakos, B.C. Sales: Phys. Rev. Lett. **65**, 2712 (1990)
120. B. Friedl, C. Thomsen, H.-U. Habermeier, M. Cardona: Solid State Commun. **78**, 291 (1991)
121. A.P. Litvinchuk, C. Thomsen, M. Cardona: in *Physical Properties of High-Temperature Superconductors IV*, ed. by D.M. Ginsberg (World Scientific, Singapore 1994) p. 375
122. D. Munzar, C. Bernhard, A. Golnik, J. Humlíček, M. Cardona: Solid State Commun. **112**, 365 (1999)
123. C.C. Homes et al.: Physica C **254**, 265 (1995)
124. S. Tajima et al.: Phys. Rev. B **59**, 6631 (1999)
125. D. Van der Marel, A. Tsvetkov: Czech. Journ. of Phys. **46**, 3165 (1996)
126. G. Schaack: in *Light Scattering in Solids VII*, ed. by M. Cardona, G. Güntherodt (Springer, Heidelberg 1999)
127. M.T. Hutchings: in *Solid State Physics*, ed. by F. Seitz, D. Turnbull (Academic, New York 1964) Vol. 16, p. 227
128. J. Mesot, A. Furrer: *The crystal field as a local probe in rare earth based high-T_c superconductors*, in *Neutron Scattering in High-T_c Superconductors*, ed. by A. Furrer (Kluwer, Dordrecht 1998) p. 335; J. Mesot, A. Furrer: J. Supercond. **10**, 623 (1997)
129. E.T. Heyen, R. Wegerer, E. Schönherr, M. Cardona: Phys. Rev. B **44**, 10195 (1991)

130. A.A. Martin, V.G. Hadjiev, T. Ruf, M. Cardona, T. Wolf: Phys. Rev. B **58**, 14211 (1998); note, however, that the CF transitions recently observed by Raman scattering in Sm-123 seem to have a component that is mainly of *electronic* origin
131. T. Ruf, R. Wegerer, E.T. Heyen, M. Cardona, A. Furrer: Solid State Commun. **85**, 297 (1993)
132. T. Ruf, E.T. Heyen, M. Cardona, J. Mesot, A. Furrer: Phys. Rev. B **46**, 11792 (1992)
133. J.A. Sanjurjo, G.B. Martins, P.G. Pagliuso, E. Granado, I. Torriani, C. Rettori, S. Oseroff, Z. Fisk: Phys. Rev. B **51**, 1185 (1995)
134. P. Dufour, S. Jandl, C. Thomsen, M. Cardona, B.M. Wanklyn, C. Changkang: Phys. Rev. B **51**, 1053 (1995)
135. S. Jandl, P. Dufour, T. Strach, T. Ruf, M. Cardona, V. Nekvasil, C. Chen, B.M. Wanklyn: Phys. Rev. B **52**, 15558 (1995)
136. M. Guillaume, W. Henggeler, A. Furrer, R.S. Eccleston, V. Trounov: Phys. Rev. Lett. **74**, 3423 (1995)
137. R. Wegerer, C. Thomsen, T. Ruf, E. Schönherr, M. Cardona, M. Reedyk, J.S. Xue, J.E. Greedan, A. Furrer: Phys. Rev. B **48**, 6413 (1993)
138. W.E. Pickett: Rev. Mod. Phys. **61**, 433 (1989)
139. J.M. Tranquada et al.: Phys. Rev. Lett. **60**, 156 (1988)
140. J. Rossat-Mignaud, L.P. Regnault, C. Vettier, P. Burlet, J.Y. Henry, G. Lapertot: Physica B **169**, 58 (1991); for an early review of neutron scattering studies see R.J. Birgenau, G. Shirane: in Ref. [12], p. 152
141. P.E. Sulewski, P.A. Fleury, K.B. Lyons, S.-W. Cheong: Phys. Rev. Lett. **67**, 3864 (1991)
142. I. Tomeno, M. Yoshida, K. Ikeda, K. Tai, K. Takamuku, N. Koshizaka, S. Tanaka: Phys. Rev. B **43**, 3009 (1991)
143. B. Shastry, B.I. Shraiman: Phys. Rev. Lett. **65**, 1068 (1990), ibid: Int. J. of Modern Physics B **5**, 365 (1991)
144. A.W. Sandwik, S. Capponi, D. Poilblanc, E. Dagotto: Phys. Rev. B **57**, 8478 (1998)
145. C.M. Canali, S.M. Girvin: Phys. Rev. B **45**, 7127 (1992)
146. J. Eroles, C.D. Batista, S.B. Bacci, E.R. Gagliano: Phys. Rev. B **59**, 1468 (1999)
147. R.R. Singh, D.A. Huse: Phys. Rev. B **40**, 7247 (1989)
148. W. Brenig: Phys. Reps. **251**, 153 (1995)
149. T.C. Hsu: Phys. Rev. B **41**, 11379 (1990)
150. A.V. Chubukov, D.M. Frenkel: Phys. Rev. Lett. **74**, 3057 (1995)
151. E. Dagotto: Rev. Mod. Phys. **60**, 771 (1994)
152. W. Brenig: Phys. Reps. **251**, 153 (1995), p. 163
153. N.D. Mermin, H. Wagner: Phys. Rev. Lett. **17**, 1136 (1966)
154. R.R.P. Singh, P.A. Fleury, K.B. Lyons, P.E. Sulewski: Phys. Rev. Lett. **62**, 2736 (1989)
155. J.M. Tranquada: Phys. Rev. Lett. **60**, 1330 (1988)
156. C. Kittel: Quantum Theory of Solids (Wiley, New York 1963)
157. D. Reznik, B. Bourges, H.F. Fong, L.P. Regnauld, J. Bossy, C. Vettier, D.L. Milius, I.A. Aksay, B. Keimer: Phys. Rev. B **53**, 1474 (1996)
158. P. Knoll, M. Mayer, W. Brenig, Ch. Waidacher: J. Low Temp. Physics **105**, 283 (1996)

159. *Light Scattering in Solids IV*, ed. by M. Cardona, G. Güntherodt (Springer, Heidelberg 1984) p. 44
160. J.B. Parkinson: J. Phys. C **2**, 2012 (1969)
161. G. Aeppli, S.M. Hayden, P. Dai, H.A. Mook, R.D. Hunt, T.G. Perring, F. Dogan: Phys. Stat. Solidi. B **215**, 519 (1999) and recent unpublished work by G. Aeppli which reveals the second and third neighbor exchange constants of La_2CuO_4
162. F.C. Zhang, T.M. Rice: Phys. Rev. B **37**, 3759 (1998)
163. R.J. Elliott and M.F. Thorpe: J. Phys. C **2**, 1630 (1969)
164. W.H. Weber, G.W. Ford: Phys. Rev. B **40**, 6890 (1989)
165. P. Knoll, C. Thomsen, M. Cardona, P. Murugaraj: Phys. Rev. B **42**, 4842 (1990)
166. W.A. Harrison: *Electronic and Stucture and Properties of Solids* (Freema, San Francisco 1980)
167. S. Rosenblum, A.H. Francis, R. Merlin: Phys. Rev. B **49**, 4352 (1994)
168. F. Nori, R. Merlin, S. Haas, A.W. Sandvik, E. Dagoto: Phys. Rev. Lett. **75**, 553 (1995)
169. D.U. Saenger: Phys. Rev. B **52**, 1025 (1995)
170. J.D. Perkins, J.M. Graybeal, M.A. Kastner, R.J. Birgenau, J.P. Falck, M. Greven: Phys. Rev. Lett. **71**, 1621 (1993)
171. J. Lorenzana, R. Eder, M. Meinders, G.A. Sawatzky: J. Supercond. **8**, 567 (1995)
172. J. Kircher, M. Cardona, S. Gopalan, H.-U. Habermeier, D. Fuchs: Phys. Rev. B **44**, 2410 (1991)
173. M. Rübhausen, O.A. Hammerstein, A. Bock, U. Merkt, C.T. Riek, P. Guptasarma, D.G. Hinks, M.V. Klein: Phys. Rev. Lett. **8**, 5349 (1999)
174. M. Pressl, M. Mayer, P. Knoll, S. Lo, U. Hohenester, E. Holzinger-Schweiger: J. Raman Spectrosc. **27**, 343 (1996)
175. M. Rübhausen, N. Diekmann, A. Bock, U. Merkt, W. Widder, H.F. Braun: Phys. Rev. B **53**, 8619 (1996)

V Thoughts About Raman Scattering from Superconductors

Miles V. Klein

The prospect of inelastic scattering of light from the electronic states of a system was proposed in 1923 by Smekal [1] five years before the discovery of what is now called vibrational Raman scattering by Raman and Krishnan (from liquids) [2] and by Landsberg and Mandelshtam (from crystals) [3].

Superconductivity was discovered by Kamerlingh Onnes in 1911 [4]. In spite of extensive experimental work during the first half of the 20th century, the theoretical explanation was long in coming. That was provided by Bardeen, Cooper, and Schrieffer (BCS) in 1957 [5]. Thanks to them, we know that superconductivity is the result of pairing correlations that alter the ground state of a metal. These correlations also change dramatically the Raman-active electronic excitations, as first pointed out by Abrikosov and Fal'kovskii in 1961 [6]. A more realistic treatment was published in 1974, taking into account the anisotropy of the band structure in a real solid [7]. For spectroscopy the most important effect of the superconducting correlations is the *energy gap*, a shift in the value of the single-particle energy levels away from the Fermi energy. The maximum value of the shift is the gap Δ, which occurs right at the Fermi surface. This can be seen experimentally using single-electron spectroscopies, such as tunneling or, in the case of some cuprate high temperature superconductors (HTS), angle-resolved photoemission spectroscopy. These techniques show states at the Fermi energy shifting by Δ. The Raman scattering process creates a particle-hole pair in the normal state of the metal. Such an excitation is sometimes called an electron-hole excitation because a particle is moved from an occupied state to an unoccupied state, creating, respectively, a hole and an electron. In the superconducting state each single particle state, or quasi-particle, is a mixture of an electron and a hole; thus the particle-hole pair becomes simply a pair of quasi-particles. As a result of energy shifts, the minimum energy required to create a particle-hole excitation at temperature $T=0$ is 2Δ. This should lead to a sharp onset of Raman scattering, producing a peak at a Raman energy shift of 2Δ.

Raman scattering from metals is weak because the small optical penetration depth severely limits the scattering volume. Before the advent of HTS's, the maximum expected value of 2Δ corresponded to a Raman shift of about $50\,\mathrm{cm}^{-1}$. To detect a peak this close to the laser line requires a clean sample surface, high sensitivity, and excellent stray light rejection by the spectrometer. These conditions were not met until 1980 when Sooryakumar and Klein found Raman peaks at an energy near 2Δ in the layered transition metal dichalcogenide 2H–$TaSe_2$ [8]. This material has a micaceous structure and easily cleaves, leaving a nearly perfect surface. It undergoes a phase transition to a charge-density-wave (CDW) state at 33 K and becomes superconducting at 7 K. In the CDW state there are new Raman-active phonon modes corresponding to dynamic modulation of the amplitude of the CDW. These are near $40\,\mathrm{cm}^{-1}$ and are quite broad, whereas the gap peaks are near $18\,\mathrm{cm}^{-1}$. In fact the peaks of the CDW amplitude modes overlap the gap peaks and are believed to be the source of their Raman activity. It was suggested by Littlewood and Varma in 1981 [9] that coupling between the CDW amplitude modes and the superconducting electrons is via the amplitude of the superconducting order parameter (gap) rather than via the usual particle-hole pairs. In this picture, the Raman peaks near 2Δ are signatures of a collective mode, the amplitude mode of the superconducting gap, rather than the incoherent superposition of peaks representing quasi-particle pairs. In particle physics this collective mode is known as the Higgs mode, or the Higgs boson [10].

The Abrikosov prediction of pure electronic Raman scattering from superconducting gap excitations was realized in the A15 family of superconductors and in Nb by the Munich and Urbana groups, starting in 1983 [11–15]. To a first approximation the results could be explained by the BCS-based theory of Raman scattering [7,13], but some inconsistencies remain.

After the discovery of HTS, attention turned to them, and Raman investigators along with everyone else started wrestling with the new problems posed by these materials. Raman scattering is unique among all the spectroscopic probes of the cuprate HTS's in that it measures the long-wavelength response of the three excitations that are most directly involved in the superconducting transition. In my ordering of their relevance to high temperature superconductivity, the electronic (gap) excitations are first, the magnons second, and the optical phonons third.

The intensity, shape, and position of the electronic 2Δ gap peak depend on the polarization geometry of the experiment, which measures the symmetry of the created excitations (just as in the more familiar case of phonon Raman scattering). The origin of this symmetry-dependence lies with the anisotropy of the electronic band structure of the material and of the superconducting gap, when expressed in terms of the pseudo-momentum of the electron.

High temperature superconductors are highly unusual metals, whose properties are quite unlike those of familiar metals such as Al or Cu. The conduction electrons in the HTS interact strongly and cannot be considered to be "nearly free" as is true of Al or Cu. In many respects the HTS are "bad" metals. Their key superconducting properties are also highly unusual. Experiments that are sensitive to the phase of the superconducting gap as a function of pseudo-momentum have shown that the gap is unconventional in that it changes sign and has nodes. In the context of the Raman spectra, this strongly affects the symmetry-dependence of the gap peak.

In the metallic state, the bad-metal properties result from the fact that these materials are doped, nearly two-dimensional, spin one-half, antiferromagnetic insulators. The insulating materials are the "parent compounds" of the HTS, and their "children", the HTS, inherit some characteristics from the parents. The carriers doped into the insulator interact with the spins, breaking up the long-range antiferromagnetic order, and the carriers are in turn strongly affected by the remnant antiferromagnetism.

Magnetic Raman scattering comes from the residual antiferromagnetic fluctuations that remain in the metallic and superconducting states. This form of Raman scattering is second in importance only to the gap excitations because its presence proves that magnetic fluctuations coexist with superconductivity over much of the doping range. Magnetism was generally believed to be inimical to superconductivity, but that is not necessarily true of antiferromagnetic fluctuations. In some credible scenarios such fluctuations are believed to cause HTS in the cuprates. Within these scenarios, it is important to determine the connection between those fluctuations responsible for HTS and those that are most prominent in the magnetic Raman spectra.

Several of the Raman-active phonons in the cuprates change their shape, position, and intensity as the sample is cooled through the superconducting transition. This is evidence of strong electron-phonon coupling. The present consensus is that the coupling is not strong enough to cause HTS behavior. This coupling allows us to use the phonons as internal probes of the electronic excitations and their change with superconductivity. The application of this technique has been quite sophisticated, particularly with respect to that most studied of the HTS, $YBa_2Cu_3O_{6+x}$.

References

1. A. Smekal: Naturwiss. **11**, 873 (1923)
2. C.V. Raman, R.S. Krishnan: Nature **121**, 101 (1928)
3. G. Landsberg, L. Mandelshtam: Naturwiss. **16**, 557 (1928)
4. H.K. Onnes: Comm. Phys. Lab. Univ. Leiden, 119, 120, 122 (1911)
5. J. Bardeen, L.N. Cooper, J.R. Schrieffer: Phys. Rev. **108**, 1175 (1957)
6. A.A. Abrikosov, L.A. Fal'kovskii: Sov. Phys. JETP **13**, 179 (1961)
7. A.A. Abrikosov, V.M. Genkin: Sov. Phys. JETP **38**, 417 (1974)
8. R. Sooryakumar, M.V. Klein: Phys. Rev. Lett. **45**, 660 (1980)
9. P.B. Littlewood, C.M. Varma: Phys. Rev. Lett. **47**, 811 (1981); Phys. Rev. B **26**, 4883 (1992)
10. See the comments of P. Higgs and Y. Nambu: In *The Rise of the Standard Model*, ed. by L. Hoddeson, L. Brown, M. Riordan, M. Dresden (Cambridge University Press, Cambridge 1977), Chap. 28
11. R. Hackl, R. Kaiser, S. Schicktanz: J. Phys. C: Solid State Phys. **16**, 1729 (1983)
12. S.B. Dierker, M.V. Klein, G.W. Webb, Z. Fisk: Phys. Rev. Lett. **50**, 853 (1983)

13. M.V. Klein, S.B. Dierker: Phys. Rev. B**29**, 4976 (1984)
14. R. Hackl, R. Kaiser: W. Gläser: Physica C **162-164**, 431 (1989)
15. R. Hackl, R. Kaiser: Phys. C: Solid State Phys. **21**, L453 (1988)

VI Two-Magnon Inelastic Light Scattering

David J. Lockwood

Spin waves, or magnons, are low-lying excitations that occur in magnetically ordered materials. The concept of a spin wave was first introduced by Bloch [1] in 1930. He postulated that electron spins can deviate slightly from their ordered ground-state alignment and the resulting excitation propagates with a wavelike behaviour through the solid. Since the electron spins are described by quantum-mechanical operators, spin waves are also quantized, with the basic quantum excitation being termed the magnon.

The first observation of inelastic light scattering from magnons was made in 1966 by Fleury, Porto, and co-workers [2]. This was in the midst of a golden period in Raman scattering research when many important new discoveries were made [3] consequent upon the use of newly developed lasers as light sources. The use of laser excitation at power densities unachievable with earlier incoherent light sources, coupled with the development of the double-grating spectrometer by Mitteldorf and colleagues at Spex Industries and the use of photoelectric detection, had revolutionized and revitalized the technique of Raman spectroscopy.

Fleury et al. [2] observed a temperature-dependent light scattering from one- and two-magnon excitations in the rutile-structure antiferromagnet FeF_2. The one-magnon scattering occurred from magnons with wavevectors near the magnetic Brillouin zone centre while the two-magnon peak had different polarization characteristics and occurred at an energy nearly twice that of Brillouin zone boundary magnons. This in itself was not unusual, as first- and second-order scattering from phonons was well known in solids. What was striking, however, was the extraordinarily high intensity of the two-magnon peak – comparable to or greater than the one-magnon cross section – and the unusual line shape found in FeF_2 and other simple antiferromagnets [4]. Clearly, the two-magnon scattering was not the usual second-order effect seen, for example, in phonon Raman scattering.

Bass and Kaganov [5] first proposed in 1959 that light scattering from magnons would logically take place through the magnetic dipole interaction between the spin fluctuations and magnetic vector of the light. However, Elliott and Loudon [6] showed in 1963 that this interaction is very weak and that a much more efficient mechanism is provided through the electric dipole interaction. Nevertheless, this latter mechanism taken in second-order yields a two-magnon scattering cross section several orders of magnitude smaller than for one-magnon scattering and would be hard to observe (it has not been seen in ferromagnets) [4]. This discrepancy with experimental observations on antiferromagnets and ferrimagnets was resolved in 1968 following a suggestion from theoretical work of Moriya [7] and Loudon [8] of an alternative and more efficient mechanism now known as the exchange scattering mechanism [9].

To illustrate this mechanism we consider a simple two-sublattice antiferromagnet, where there are two magnon branches with frequencies $\omega^{\pm}(\boldsymbol{k})$. Physically these correspond to the excitations in which the total S^z spin operator changes by unity, and we denote the state functions by $|\boldsymbol{k},+\rangle$ and $|\boldsymbol{k},-\rangle$ according to whether S^z is increased (+) or decreased (−). These may be used to construct the linearly independent two-magnon states (with excitation wavevectors $\boldsymbol{k}$ and $-\boldsymbol{k}$) as follows:

$$\begin{array}{ll} \text{State} & \Delta S^z \\ |a\rangle = |\boldsymbol{k},+\rangle|-\boldsymbol{k},+\rangle & +2 \\ |b\rangle = 2^{-1/2}\left[|\boldsymbol{k},+\rangle|-\boldsymbol{k},-\rangle + |\boldsymbol{k},-\rangle|-\boldsymbol{k},+\rangle\right] & 0 \\ |c\rangle = 2^{-1/2}\left[|\boldsymbol{k},+\rangle|-\boldsymbol{k},-\rangle - |\boldsymbol{k},-\rangle|-\boldsymbol{k},+\rangle\right] & 0 \\ |d\rangle = |\boldsymbol{k},-\rangle|-\boldsymbol{k},-\rangle & -2 \end{array} \qquad \text{(VI.1)}$$

For a simple ferromagnet there is just a single (acoustic) magnon branch $|\boldsymbol{k},-\rangle$, and the only possible two-magnon state is $|d\rangle$, which corresponds to $\Delta S^z = -2$. For an antiferromagnet there are the extra possibilities of having two-magnon excitations with $\Delta S^z = 0$ and $\Delta S^z = +2$. The $\Delta S^z = \pm 2$ scattering matrix elements come from fourth-order perturbation theory [7]: second-order in the electric-dipole interactions (as usual in light scattering) and also second-order in the spin-orbit interaction. This is just the Elliott-Loudon one-magnon process [6] extended to higher order. The $\Delta S^z = 0$ scattering matrix elements come from third-order perturbation theory: second-order in the electric-dipole interaction and first-order in the exchange. The essential feature is that spin deviations are created at a pair of exchange-coupled magnetic sites *on opposite sublattices* (hence giving $\Delta S^z = 0$) through virtual electronic transitions to higher states. This results in a much more efficient mechanism for two-magnon scattering than provided by the $\Delta S^z = +2$ case. Also, since it is physically quite different from the one-magnon scattering mechanism, there is no *a priori* reason to expect two-magnon scattering to be weaker than one-magnon scattering. In (VI.1) it is the even-parity state $|b\rangle$ that corresponds to two-magnon scattering, while the odd-parity state $|c\rangle$ is important for two-magnon absorption.

Thus, in two-magnon scattering in antiferromagnets, two spin deviations with wavevectors $\boldsymbol{k}$ and $-\boldsymbol{k}$ are created in close proximity on opposite magnetic sublattices and are coupled by the exchange interaction. Because the two-magnon density of states peaks at the Brillouin zone boundary, the two-magnon Raman scattering was expected to be dominated by zone edge magnons. However, the experimental peak energies were observed to lie below the corresponding zone edge values. This is partly because of weighting factors that must be included in calculating the Raman intensity, but also because of magnon-magnon interactions [10]. The latter effect is important even at low temperatures, since the two exchange-coupled magnons are created on neighbouring lattice sites and hence they interact strongly. It should be noted that for applied magnetic fields B below the spin-flop transition point, the two-magnon scattering is insensitive to B [9]. This is because the effect of B is to increase $\omega^+(\boldsymbol{k})$ and to decrease $\omega^-(\boldsymbol{k})$ by a similar amount resulting in essentially no net shift in the two-magnon frequency.

The exchange coupling theory, including magnon-magnon interactions, has been remarkably successful in predicting the two-magnon Raman scattering spectra of magnetic solids [4] and can even be used to determine accurately the antiferromagnetic exchange constant [11]. It thus can be applied with confidence to more complex situations such as those described here in this chapter on high T_c superconductors.

References

1. F. Bloch: Z. Phys. **61**, 206 (1930)
2. P.A. Fleury, S.P.S. Porto, L.E. Cheeseman, H.J. Guggenheim: Phys. Rev. Lett. **17**, 84 (1966)
3. Much of the Raman work of this period is summarized in the seminal conference on light scattering in solids held in New York, 3-6 September 1968: In *Light Scattering Spectra of Solids* ed. by G.B. Wright (Springer-Verlag, New York 1969)
4. M.G. Cottam, D.J. Lockwood: In *Light Scattering in Magnetic Solids* (Wiley, New York 1986)
5. F.G. Bass, M.I. Kaganov: Zh. Eksp. Tear. Fiz. **37**, 1390 (1959) [Sov. Phys. - JETP **37**, 986 (1960)]
6. R.J. Elliott, R. Loudon: Phys. Lett. **3**, 189 (1963)
7. T. Moriya: J. Phys. Soc. Japan **23**, 490 (1967)
8. R. Loudon: Adv. Phys. **17**, 243 (1968)
9. P.A. Fleury, R. Loudon: Phys. Rev. **166**, 514 (1968)
10. R.J. Elliott, M.F. Thorpe: J. Phys. C **2**, 1630 (1969)
11. D.J. Lockwood, G.J. Coombs: J. Phys. C **8**, 4062 (1975)

6 Raman Applications in Catalysts for Exhaust-Gas Treatment

Willes H. Weber

Abstract. Applications of Raman scattering in the characterization of materials used in the catalytic treatment of exhaust gases, both from vehicles and stationary sources, are reviewed. The primary materials used as supports in highly-dispersed precious-metal catalysts include alumina, silica, titania, zirconia, and ceria, and reference spectra are given for all of these as well as for the standard substrate material cordierite. Additional Raman spectra are given for most of the known oxides of the platinum-group metals, many of which are used in automotive catalysts. Raman scattering helps to elucidate the oxygen storage mechanism and in some cases the oxygen storage capacity of specific catalysts. Adsorbed oxides of both nitrogen and sulfur give characteristic Raman signatures, which are useful for identifying the interactions of various NO_x and SO_x species with different catalysts. Raman scattering is also used to quantify particle size, an important attribute for any surface-mediated process. Finally, three examples are cited that use the PdO Raman signal for quantitative analyses on Pd-containing catalysts.

Catalysis plays an important role in the production of many mass-produced chemicals and synthetic materials. For example, zeolite-based catalysts are widely used in the production of liquid petroleum-derived products, the most important of which on a volume basis is probably gasoline [1]. In addition, metal-oxide catalysts of Co and Mo dispersed on Al_2O_3 are used to remove sulfur from petroleum- or coal-derived feedstocks [2], and a variety of catalysts are central to the synthesis of polymeric materials [3]. Because of the enormous commercial importance of catalysis, the major chemical and petrochemical industries have devoted considerable effort toward understanding how catalysts function and toward inventing new and better catalysts. To this end a large number of analytical techniques have been applied to the characterization of catalytic materials and processes. The recent monograph by Wachs [4] lists thirty such techniques. Raman scattering is among the few techniques that can be done in situ under pressure and temperature conditions at which catalysts for gas-phase reactions typically operate, and it can be done in aqueous or transparent polymer-melt environments as well. Under ideal conditions Raman scattering can even identify adsorbed species. It also has good chemical specificity for identifying the phases of the catalyst materials, excepting, of course, the pure metallic phases, whose Raman spectra are generally unobservable. The inability of Raman scattering to detect these metallic phases is probably its greatest limitation with regard to catalyst characterization. Although many of the electron-spectroscopy based techniques have greater surface sensitivity, these generally operate only in ultrahigh vacuum and, thus, they are not suitable for in situ measurements under realistic conditions.

The importance of Raman scattering for catalytic studies is evidenced by the large number of excellent review articles, book chapters, and monographs devoted to the subject. These include the recent monograph by Stencel [5]; the review articles by Knözinger [6], Dixit et al. [7], and Wachs and co-workers [8–10]; and the book chapters by Bartlett and Cooney [11] and Mehicic and Grasselli [12]. All of these works have concentrated on catalytic applications directed toward chemical production, which has historically been the main use for catalysts. In the last two to three decades, because of increasing environmental concerns about air quality, there has been a new thrust to develop catalysts whose purpose is to remove noxious species from combustion processes occurring in fixed-source power-generating plants and in internal combustion engines used in vehicles. This catalytic process is referred to as *heterogeneous* catalysis, since the reactants are in the gas phase whereas the catalysts are in the solid phase. In order to limit the scope of this review to a manageable size, we will focus on these types of catalysts and mainly on those used in vehicles. A few other recent topics that have not been specifically addressed in previous reviews will also be included. In keeping with the main subject of the book, we will concentrate on the materials aspects of catalysis rather than the chemical reactions.

There are other optical techniques, besides Raman scattering, that have been used for in situ catalyst studies, including ellipsometry [13–18] and second-harmonic generation (SHG) [19–21]. These have greater surface sensitivity than Raman, but they lack its chemical specificity. In addition, both ellipsometry and SHG require a smooth specular-reflection surface, which means they cannot be used with most commercial catalysts.

The paper is organized in the following manner. Section 6.1 deals briefly with some typical support materials: alumina, silica, titania, ceria and zirconia. The supports are generally high-surface-area, porous phases throughout which the active precious-metal catalysts are dispersed. The primary substrate material, cordierite, is also discussed. Cordierite is a light-weight, crystalline ceramic onto which the active, supported-metal catalysts are applied. Section 6.2 catalogs Raman spectra of the oxides of the platinum-group metals, some of which are widely used in automotive exhaust catalysts for the simultaneous conversion of CO to CO_2 and the combustion of unburned hydrocarbons as well as for the reduction of NO to N_2 [22,23]. Although the pure metals themselves give no Raman signals, they are always used in a gas stream containing oxygen so that metal oxides can and often do form. Section 6.3 discusses the materials added to exhaust-gas catalysts to improve their oxygen storage capacity, which is an important attribute for emissions reduction from gasoline engines. Section 6.4 addresses the use of Raman scattering in identifying adsorbed species. The emphasis is on the oxides of nitrogen (NO_x) and sulfur (SO_x). There is considerable effort being devoted to the removal of NO_x in the presence of excess oxygen. Sulfur oxides are important, since sulfur is a trace impurity in most petroleum-based fuels, and it has a negative impact of several air-quality issues. Certain zeolite-based catalysts are being considered for NO_x removal, but these will be only briefly discussed, since the subject of Raman studies on zeolites has been included in the earlier general reviews [5,11,12] as well as in recent reviews focused specifically on zeolites [24,25]. Section 6.5 describes the use of Raman scattering to estimate particle size, which can have an important effect on the activity of the catalyst. Section 6.6 considers using Raman scattering to do quantitative analyses on catalysts. Although Raman scattering is generally considered to be a qualitative analysis technique (or semi-quantitative at best), when well-characterized reference samples are used for calibration, Raman can both identify and give good quantitative information about the concentration of a specific compound. Section 6.7 contains a brief summary.

6.1 Supports and Substrates

Support materials must satisfy several, sometimes conflicting, requirements. They must provide a porous medium through which the gas molecules can rapidly diffuse and a structure on which the precious metals and other active components can be dispersed, without undergoing unwanted chemical reactions. In addition, they must limit or prevent sintering of the active components at the elevated operating temperatures of the catalyst. The supports most compatible with Raman studies generally have weak and broad Raman spectra, but subtle phase changes can sometimes occur after high-temperature treatments that can substantially change their spectra.

A typical automotive catalyst employs a support material of which γ-Al_2O_3 is a major component. This phase of alumina is a sponge-like, porous material that has a spinel-type crystalline structure and a surface area, as measured by gas adsorption, as high as $100-200\,m^2/g$. The γ-Al_2O_3 is sometimes shaped into spherical beads, for a packed-bed reactor, but more often is coated onto a cordierite substrate, for a monolithic reactor. The coating, referred to as the washcoat, is first impregnated with the active catalytic components by way of a wet chemical process. Cordierite is an ideal substrate, since it is light weight, mechanically strong, chemically inert, and stable at high temperatures. It is processed by extrusion into a thin-walled honeycomb structure that gives good throughput for a flowing gaseous stream. Figure 6.1 shows an electron back-scattering image of the cross-section of a typical monolithic automotive catalyst that has been cut open and polished. The channel openings are square, $\approx 1\,mm$ on a side, the cordierite thickness is $\approx 140\,\mu m$,

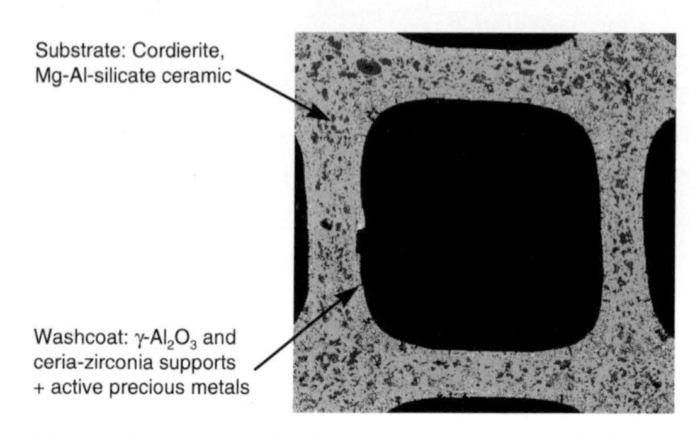

Fig. 6.1. Electron back-scattering image of the cross-section of a typical automobile catalyst. The square opening in the channel is roughly 1 mm on a side. The washcoat layer, which is $\approx 40\,\mu m$ thick, covers the inside of the channels and has a slightly different texture from the cordierite. This layer is partially peeled off on the left side. (Courtesy of A. Chen)

and the washcoat layer is $20-50\,\mu m$ thick. A single catalyst brick contains $\approx 1-2\times10^4$ channels, each 10–20 cm in length. Monolithic reactors are now used almost exclusively in automotive catalysts, because they produce lower back pressure and they are much less likely to degrade by attrition, a common failure mode for packed-bed reactors.

Figure 6.2 shows Raman spectra from several phases of alumina taken from the work of Deo et al. [26]. Of the materials shown in the figure, γ-Al_2O_3 is the one most compatible with Raman studies, since it does not contribute any spectral features that might interfere with those from adsorbates or catalytically active materials. Gamma-alumina is also widely used in catalytic applications, e.g., as a precious-metal support in exhaust-gas treatment and as a Ag support for the catalytic oxidation of ethylene to ethylene oxide, which is an important process in the chemical industry. The spinel structure of γ-Al_2O_3 is apparently stabilized by the presence of protonated cation vacancies and both surface and bulk hydroxyl groups [27,28]. Thermal dehydrogenation of γ-Al_2O_3 at temperatures in the 1100–1400 K range leads to the formation of transitional alumina phases, such as δ-Al_2O_3 and θ-Al_2O_3, which show characteristic Raman spectra. The presence of these phases can yield important information on the thermal history of the catalyst. On the other hand, their Raman spectra may interfere with those from other species. The high-temperature stable phase of alumina, α-Al_2O_3 (sapphire), produced by complete dehydrogenation, is a dense crystalline material with very strong and sharp lines. This material is not generally used as a support.

Figure 6.3 shows spectra from the substrate material cordierite and three other support oxides: amorphous silica (a-SiO_2) and the rutile and anatase phases of TiO_2. Cordierite is a magnesium-aluminum silicate. It has a strong and characteristic Raman spectrum that can sometimes interfere with in situ studies of production catalysts, unless the catalyst layer is sufficiently thick

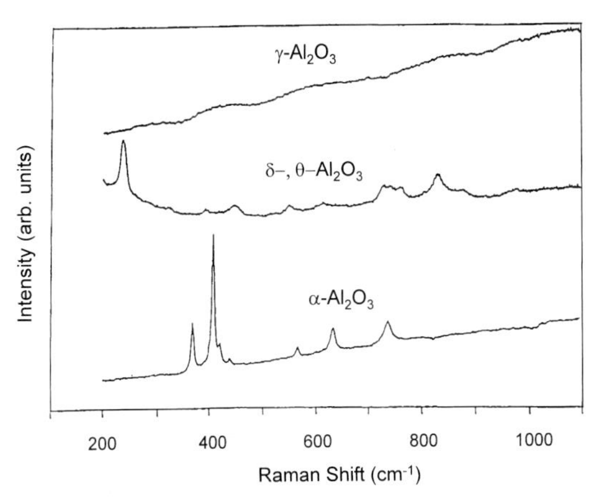

Fig. 6.2. Raman spectra of alumina phases. From Deo et al. [26]

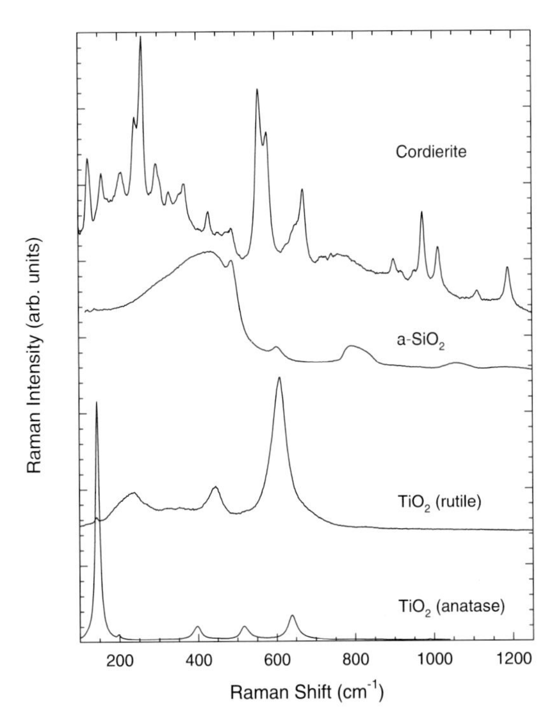

Fig. 6.3. Raman spectra of cordierite, amorphous silica, and two phases of titania

to prevent the laser beam from reaching the substrate. Amorphous silica gives only weak and broad lines, and there is little intensity above $600\,\mathrm{cm}^{-1}$. Both phases of titania give strong lines in the low frequency region but no interference above $\approx 700\,\mathrm{cm}^{-1}$.

During the last few years, the support materials in automotive catalysts have evolved into complicated mixtures of phases of which γ-alumina is only one component. Mixed oxides of $CeO_2 - ZrO_2$ are another important component. The addition of zirconia to ceria, which produces solid solutions of the form $Ce_{1-x}Zr_xO_2$, improves the oxygen storage capability of the material and stabilizes it against sintering at high temperatures. As the amount of Zr is increased, the material transforms from the fluorite-structured CeO_2 phase, with only a single Raman mode, into a tetragonal phase with six Raman-active modes. This phase transformation has been studied with Raman scattering by Yashima et al. [29] and by Vlaic et al. [30]. Results from the latter work are shown in Fig. 6.4.

Although it is clear from Figs. 6.2–6.4 that γ-Al_2O_3 is the support material most suitable for Raman studies, we note that trace impurities in γ-Al_2O_3 as well as in any of the other materials shown here can sometimes produce a large fluorescence background that obscures the Raman spectrum.

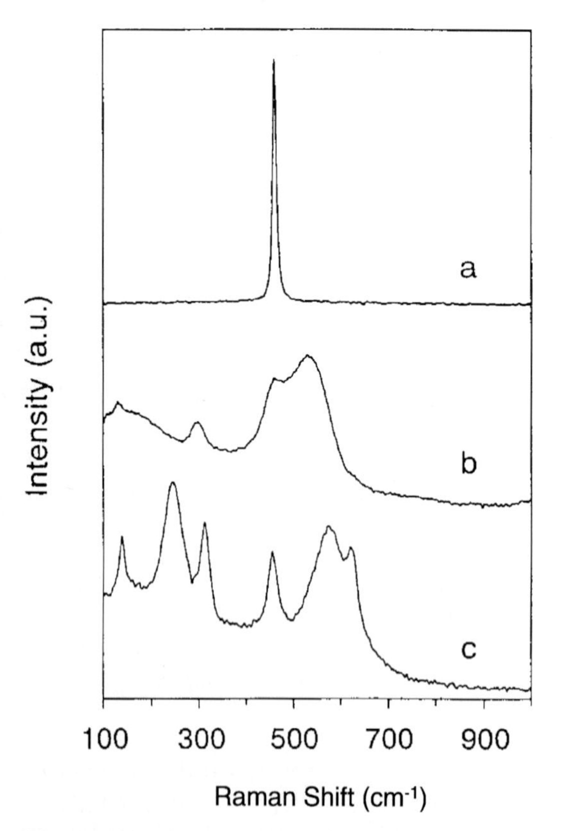

Fig. 6.4. Raman spectra of (*a*) CeO_2, (*b*) $Ce_{0.5}Zr_{0.5}O_2$, (*c*) $Ce_{0.2}Zr_{0.8}O_2$. From Vlaic et al. [30]

6.2 Oxides of the Pt-Group Metals

Automobile exhaust catalysts are designed to reduce the concentrations of carbon monoxide, unburned hydrocarbons, and oxides of nitrogen (NO_x). Catalysts that perform these functions are termed three-way-catalysts (TWC's), because of their threefold functionality [22,23]. The precious metals typically used in TWC's are Pt, Pd, and Rh. These are Group VIII metals, which have in common an electronic structure with a nearly filled *d* band. They are also called the Pt-group metals. Os, Ir, and Pt comprise the end of the 5*d* series of the Pt-group metals and Ru, Rh, and Pd the end of the 4*d* series. All of these metals and/or their oxides have important catalytic applications. In this section we catalog Raman spectra for the 5*d* and 4*d* series of these oxides. Oxides of the transition metals Fe, Co, and Ni, which constitute the end of the 3*d* series, also have important catalytic applications, particularly when combined with oxides of Mo. These materials will not be considered here, however, since most of them have been covered in earlier reviews [5,8–12] and they are not of primary importance in exhaust-gas treatment. Raman studies of supported molybdenum oxides [31] and supported vanadium oxides [32] have also been recently reviewed.

6.2.1 Platinum Oxides

Although Pt is considered a noble metal and thus resistant to oxidation, under extreme conditions such as during reactive sputtering in an O_2-containing plasma or high-temperature and high-O_2-pressure exposure, it can form a number of oxides [33]. Some of these oxides should be viewed as metastable phases, since they decompose under the typical temperature and oxygen partial pressure conditions of a catalytic reactor.

In a reactive sputtering environment, well-crystallized films of PtO and poorly crystallized samples of α-PtO_2 [34,35] have been grown as well as an amorphous phase, *a*-PtO_2 [36]. Raman spectra of these materials obtained by McBride et al. [36] are shown in Fig. 6.5. Symmetry species assignments for the α-PtO_2 lines are based on polarization measurements from the polycrystalline films and those for PtO on polarization measurements from single crystals of the isostructural compound PdO [37]. Raman studies of model catalysts containing a few per cent by weight of Pt dispersed on γ-Al_2O_3 [36,38] and on *a*-SiO_2 [39] indicated that the primary phase of platinum oxide present under oxidizing conditions was amorphous PtO_2.

A number of other oxides of Pt can be synthesized by reacting Pt in a high-pressure and high-temperature O_2 environment [40–42], and those for which Raman analyses have been done are shown in Fig. 6.6. The samples for this figure are thin oxide films grown on Pt foils by simply heating the foils to 800–1100 K in O_2 at pressures up to several hundred atmospheres [43]. Powdered samples can be synthesized in the same manner. The powders, however, are generally black, and it is often difficult to obtain good-quality

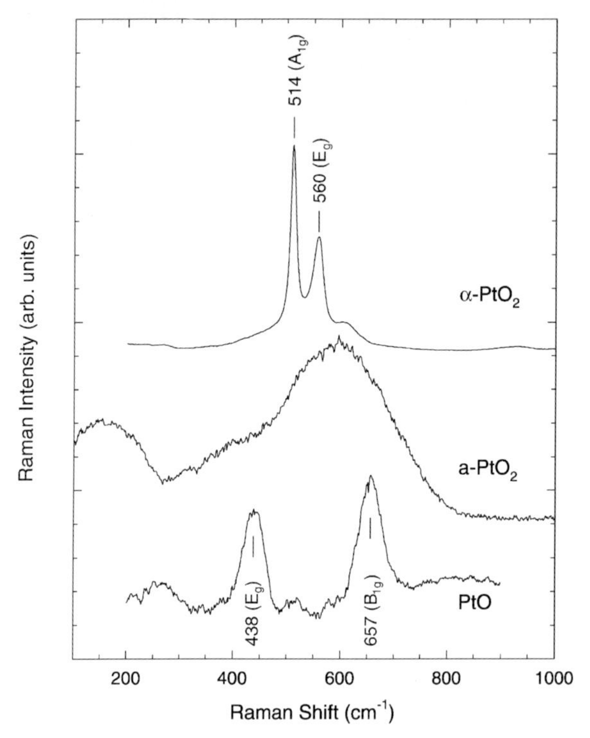

Fig. 6.5. Raman spectra of Pt oxides grown by reactive sputtering. From McBride et al. [36]

Raman spectra from them because of laser heating, which can lead to decomposition.

Cahen et al. [41] identified the oxide designated Pt_5O_6 in Fig. 6.6 primarily through its X-ray diffraction (XRD) pattern, which can be indexed to a small monoclinic cell or a slightly larger orthorhombic cell. The precise structure, however, is still unknown. The large number of Raman lines is consistent with a low symmetry and/or a large unit cell.

The oxide designated $Na_xPt_3O_4$ has been shown by Cahen and Ibers [44] to be the active component in the so-called "Adam's catalyst", which is commonly used in the reduction of organic compounds. This material is a member of a larger class of compounds referred to as the platinum bronzes, which have the general formula $M_xPt_3O_4$ (with $0 \leq x \leq 1$ and M = Li, Na, Mg, Ca, Zn, Cd, Co, Ni, or Pt) [41,45]. The platinum bronzes crystallize in a cubic structure that has four allowed Raman modes, an E_g, F_{2g} pair associated primarily with motion of the Pt ions and another E_g, F_{2g} pair associated primarily with motion of the O ions. The $Na_xPt_3O_4$ spectrum in Fig. 6.6, obtained from a polycrystalline thin film with $x \approx 0.4$, shows only two rather broad lines. However, Weber et al. [46] have recently identified all four

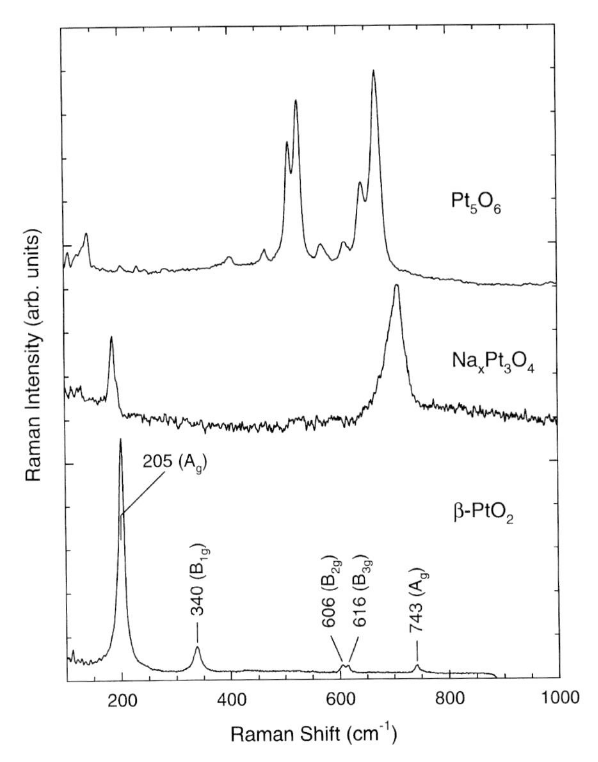

Fig. 6.6. Raman spectra of Pt oxides grown by high-temperature, high-oxygen-pressure reaction. From Graham et al. [43]

of the expected modes by using polarization measurements with a Raman microscope from small single crystals of $NaPt_3O_4$. Results from this study are shown in Fig. 6.7. These spectra provide a clear example of the use of group theory and selection rules, along with polarized Raman scattering measurements from an oriented single crystal, to identify the symmetries of the modes. The table inset at the top of the figure gives the expected intensity variations of the modes for the different scattering geometries. Since the O atoms are so much lighter than the Pt, we can immediately assign the pair of lines near $700\,cm^{-1}$ as modes in which primarily the O's move and the pair near $200\,cm^{-1}$ as modes in which primarily the Pt's move. The intensity variations with scattering geometry then identify the highest and lowest frequency lines as E_g modes and the intermediate lines as F_{2g}.

The remaining oxide in Fig. 6.6, β-PtO_2, is the most interesting one from a structural point of view. β-PtO_2 is the only known metal dioxide that has the $CaCl_2$ structure under ambient conditions. The $CaCl_2$ lattice is a slightly distorted form of the rutile structure [33,40], obtained from the latter by a rotation about the c axis of the oxygen octahedra surrounding each metal ion, followed by a small orthorhombic distortion. The symmetry-species assign-

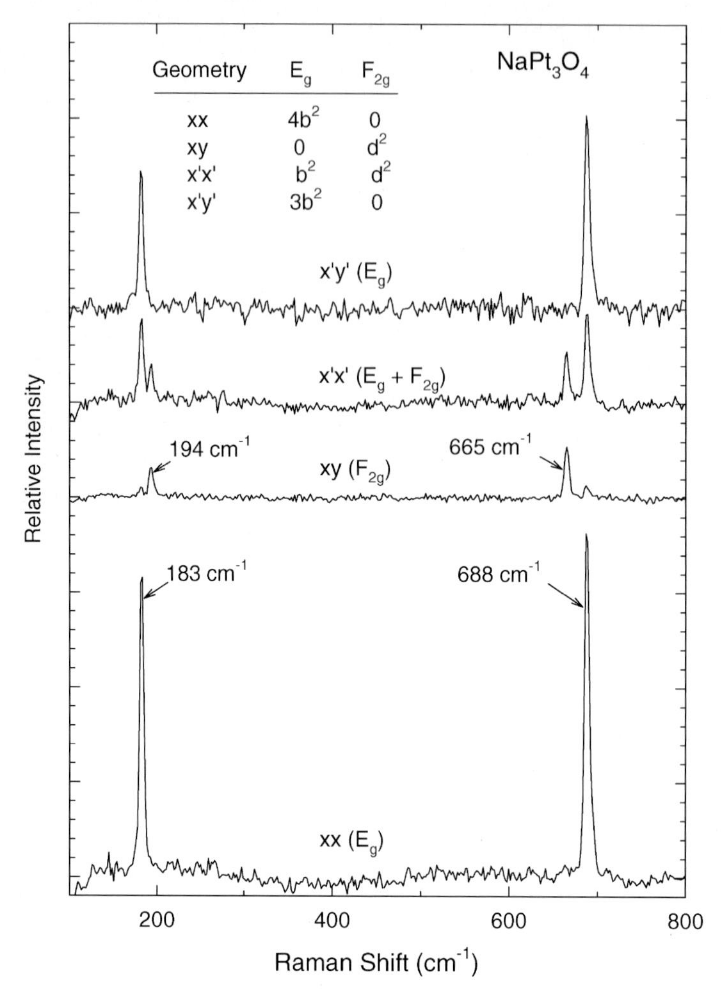

Fig. 6.7. Polarized Raman spectra from the (001) face of a $NaPt_3O_4$ single crystal. The table at the *top* gives the predicted relative line intensities for different scattering geometries. After Weber et al. [46]

ments given for the modes in the figure are based on the analogy with rutile, considering the changes in the modes expected from the deformation [47].

Metal dioxides from the other Pt-group metals tend to crystallize in the rutile structure, and the reason that β-PtO_2 deviates from this trend has long been a puzzle. However, recent first-principles density-functional calculations by Wu and Weber [48] have shown that indeed the $CaCl_2$ structure is the lowest energy structure for β-PtO_2 under ambient conditions. Moreover, they showed that the preferred structure depends on the lattice constant. If the lattice were expanded by about 10%, then the rutile structure would be favored, which is a result consistent with earlier temperature-dependent studies of the β-PtO_2 Raman lines [47].

6.2.2 Iridium and Osmium Oxides

Compared with Pt, there has been much less Raman work done on the oxides of Ir or Os. Both metals form a rutile-structured dioxide. Huang et al. [49] reported polarized Raman scattering from single crystals of IrO_2, results from which are shown in Fig. 6.8. Figure 6.9 shows a Raman spectrum from a small crystal of OsO_2, oriented so as to display all the lines. Mode assignments for OsO_2 are given based on polarization measurements from high-symmetry facets of several other small crystallites. The frequencies and relative intensities of these lines are very similar to those observed for IrO_2 and RuO_2.

Iridium is unusual in its ability to catalyze the reduction of NO to N_2 under oxidizing conditions [50]. Unfortunately, the high volatility of its oxide precludes the use of Ir in commercial TWC's.

6.2.3 Palladium Oxide

Palladium monoxide, PdO, is probably the most important material for automotive applications discussed in this section. It is also the material that has been most extensively studied. The first and second order Raman spectra

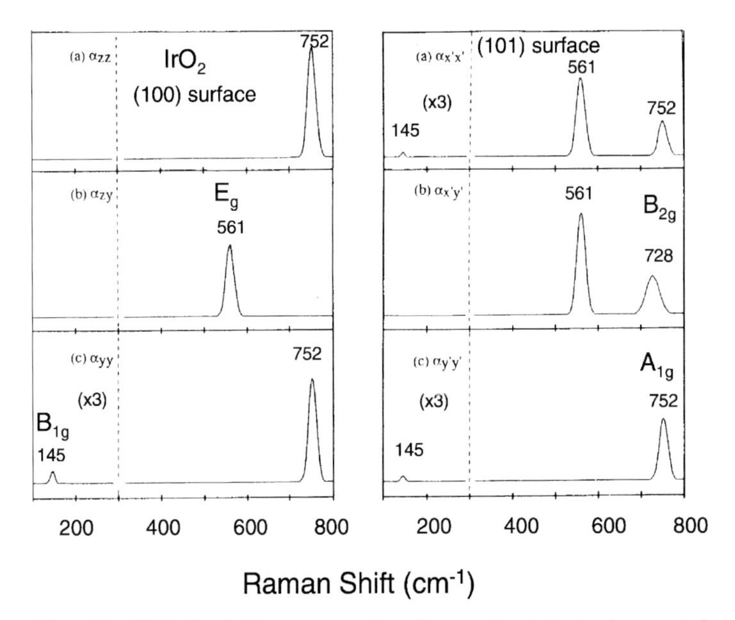

Fig. 6.8. Polarized Raman spectra from oriented single crystals of IrO_2. From Huang et al. [49]

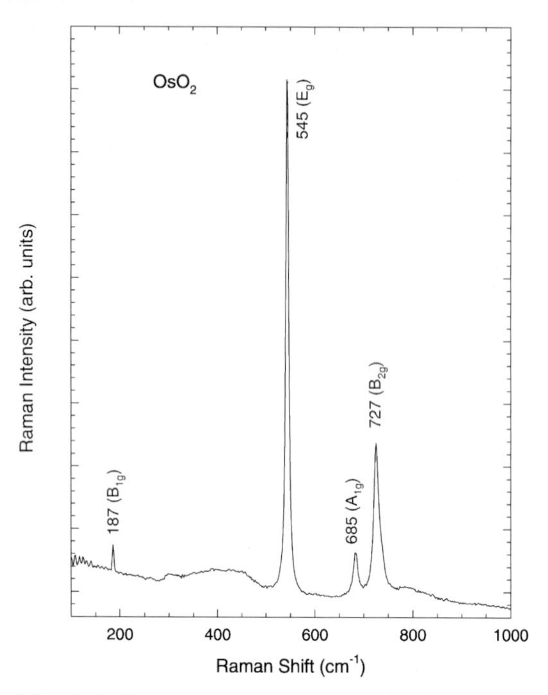

Fig. 6.9. Raman spectrum from small single crystal of OsO_2, recorded with 633-nm excitation

have been measured and assigned [37], as has the first order IR spectrum [51]; calculations have been done on the lattice dynamics [37], the electronic band structure [52,53] and the optical properties [54]; and the anisotropic optical properties have been measured [55].

PdO was first observed in Raman studies of catalysts by Chan and Bell in 1984 [56]. Not only is Pd a good oxidation catalyst for CO and hydrocarbons, but, as shown by McCabe and Usmen [57], its oxidation kinetics are sufficiently fast that the Pd $+1/2(O_2) \leftrightarrow$ PdO reaction can serve as an oxygen storage and release mechanism during the rich-lean excursions encountered by the catalyst. Such excursions are forced on the catalyst by the vehicle emissions control system.

The Raman spectrum of a Pd foil with an oxide coating a few hundred Angstroms thick, taken from the work of McBride et al. [37], is shown in Fig. 6.10. The two allowed Raman modes are identified from polarization measurements on oriented single crystals. These modes are resonantly enhanced, owing to a strong optical absorption feature in the PdO optical properties [55]. The resonance enhancement peaks near 2.4 eV, as shown in Fig. 6.11, which is conveniently close to the 514.5-nm line of the Ar-ion laser (2.41 eV). Remillard et al. have shown that an oxide layer $\approx 7\,\text{Å}$ thick (less

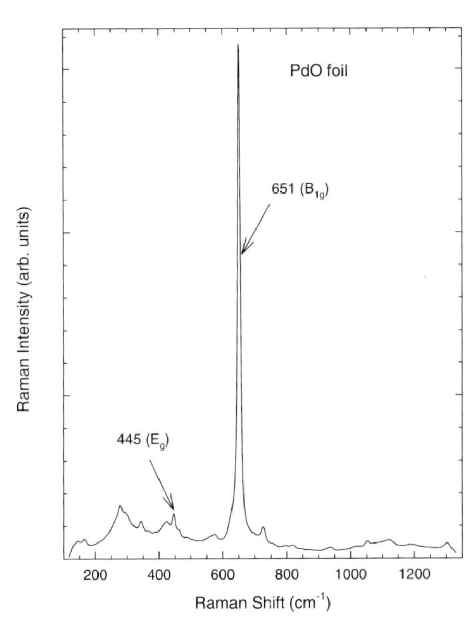

Fig. 6.10. Raman spectrum of PdO on an oxidized Pd foil. From McBride et al. [37]

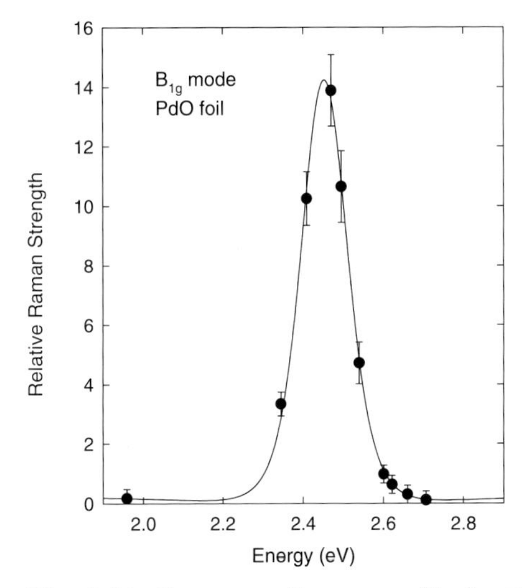

Fig. 6.11. Resonance Raman profile for the B_{1g} mode of PdO. From McBride et al. [37]

than two lattice constants) can be readily detected via Raman scattering on a Pd surface, if excitation with 514.5-nm light is used [58].

6.2.4 Rhodium Oxides

Rhodium is added to automotive three-way catalysts primarily to reduce oxides of nitrogen, NO_x [59]. It is one of the most important components and often the most expensive. Several oxides of Rh are known to exist, but there have been only limited Raman studies. Under moderate temperatures (900–1000 K) in air, the oxide forming on the surface of a Rh foil is a hexagonal phase of rhodium sesquioxide Rh_2O_3, which has the corundum structure [60,61]. Figure 6.12 shows Raman spectra from this phase along with assignments based on polarization results and comparisons with spectra from other oxides having the same structure [62]. With somewhat higher oxidation temperatures, above 1050 K, the surface oxide converts to an orthorhombic phase of Rh_2O_3. In a vapor transport method in excess O_2, a rutile phase RhO_2 can be produced [63,64]. Although no Raman spectrum of RhO_2 has been reported, it should be similar to that from IrO_2 (Fig. 6.8), OsO_2 (Fig. 6.9) and RuO_2 (discussed below).

6.2.5 Ruthenium Oxide

Ruthenium forms a rutile-structured dioxide RuO_2 that has a number of practical applications, including its use as a corrosion-resistant electrode in

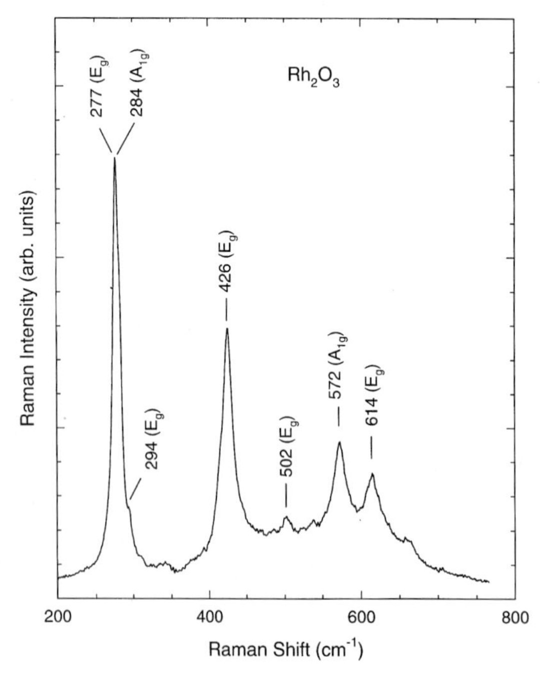

Fig. 6.12. Raman spectrum of the corundum phase of a thin film of Rh_2O_3 on a Rh foil. From Weber et al. [62]

electrochemistry and a catalyst for the photodecomposition of water [65,66]. Ruthenium metal is also considered a good NO_x reduction catalyst [67], but the high volatility of the oxide makes it generally unsuitable for automobile exhaust treatment [68]. In this regard, Ru behaves very much like Ir [50].

Raman studies of single crystals of RuO_2 were reported by Huang and Pollak [69], who gave line assignments from polarization measurements, and by Rosenblum et al. [70], who used a diamond-anvil cell to study the pressure-induced second-order phase transition from the rutile to the $CaCl_2$ structure. This transition was first seen in RuO_2 through X-ray diffraction studies of powder samples by Haines and Legér [71], and similar transitions are expected to occur in other Pt-group metal dioxides [48]. The transition pressure found in the X-ray work was 5 GPa, compared with the value of (11.8 ± 0.3) GPa from the Raman work. The discrepancy likely follows from differences in the pressure medium and/or the stoichiometry of the samples [70].

Figure 6.13 shows the RuO_2 Raman spectra from Huang and Pollak [69]. These results differ from those of Rosenblum et al. [70] only for the very weak, low-frequency B_{1g} mode, which the latter authors found at 165 cm^{-1} in contrast to the value 97 cm^{-1} in Fig. 6.13. This mode is the soft mode for the phase transition, and its frequency under ambient conditions may be sensitive to the O stoichiometry, since O vacancies may reduce the stability of the rutile structure.

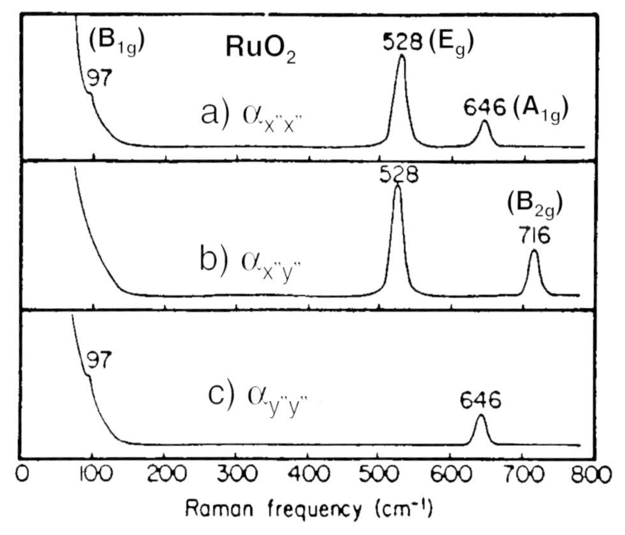

Fig. 6.13. Polarized Raman spectra from oriented single crystal of RuO_2. From Huang and Pollak [69]

6.2.6 Mixed Oxides

Raman spectra from three different ternary oxides of the Pt-group metals are shown in Fig. 6.14 [43,62]. The lower two spectra are from oxides with the delafossite structure, which have been synthesized and studied by Shannon and co-workers [72]. Such oxides can appear on the surfaces of alloys that have been heated in oxygen. For example, the bottom spectrum in Fig. 6.14 is from an alloy foil (56% Pd, 31% Pt, 13% Rh) oxidized at 1075 K. Alloys of Pd and Rh are particularly interesting. At low temperatures the first oxide to form is PdO, which is displaced by $PdRhO_2$ as the oxidation temperature is raised [73]. These studies address the possibility of surface enrichment effects in supported mixed-metal catalysts in which alloying might occur.

The remaining spectrum in Fig. 6.14 is from $CoPt_3O_6$, which is one of a class of oxides of the form MPt_3O_6, with M = Ni, Co, Fe, Mn, Mg, Zn, and Cd, that have an orthorhombic structure yielding nine Raman-active modes. Several groups have synthesized and studied these materials [74,75]. Table 6.1 gives a summary of the Raman assignments for the oxides of the Pt-group metals discussed in this section.

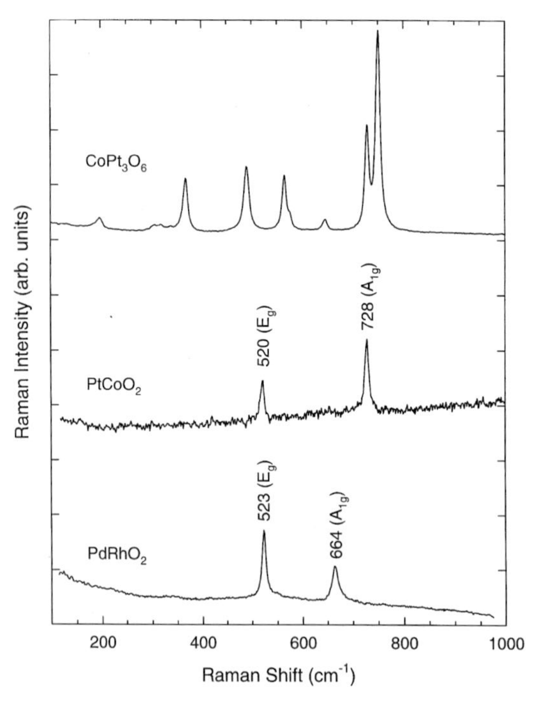

Fig. 6.14. Raman spectra of ternary oxides of Pt-group metals. $CoPt_3O_6$ and $PtCoO_2$ from Graham et al. [43]; $PdRhO_2$ from Weber et al. [62]

Table 6.1. Summary of Raman results on Pt-group metal oxides

Compound	Structure	Raman modes	Observed (cm^{-1})	Ref.
PtO	D_{4h}^9	$B_{1g} + E_g$	$438(E_g)$, $657(B_{1g})$	[36]
α-PtO_2	D_{3d}^3	$A_{1g} + E_g$	$514(A_{1g})$, $560(E_g)$	[36]
β-PtO_2	D_{2h}^{12}	$2A_g + 2B_{1g}$ $+B_{2g} + B_{3g}$	$205(A_g)$, $340(B_{1g})$, $606(B_{2g})$, $616(B_{3g})$, $743(A_g)$	[47]
$NaPt_3O_4$	O_h^3	$2E_g + 2F_{2g}$	$183(E_g)$, $194(F_{2g})$, $665(F_{2g})$, $688(E_g)$	[46]
IrO_2	D_{4h}^{14}	$A_{1g} + B_{1g}$ $+B_{2g} + E_g$	$145(B_{1g})$, $561(E_g)$, $728(B_{2g})$, $752(A_{1g})$	[49]
OsO_2	D_{4h}^{14}	$A_{1g} + B_{1g}$ $+B_{2g} + E_g$	$187(B_{1g})$, $545(E_g)$, $685(A_{1g})$, $727(B_{2g})$	This work
PdO	D_{4h}^9	$B_{1g} + E_g$	$445(E_g)$, $651(B_{1g})$	[37]
Rh_2O_3	D_{3d}^6	$2A_{1g} + 5E_g$	$277(E_g)$, $284(A_{1g})$, $294(E_g)$, $426(E_g)$, $502(E_g)$, $572(A_{1g})$, $614(E_g)$	[62]
RuO_2	D_{4h}^{14}	$A_{1g} + B_{1g}$ $+B_{2g} + E_g$	$97(B_{1g})$, $528(E_g)$, $646(A_{1g})$, $716(B_{2g})$,	[69]
			$165(B_{1g})$, $526(E_g)$, $646(A_{1g})$, $715(B_{2g})$	[70]
$PtCoO_2$	D_{3d}^5	$A_{1g} + E_g$	$520(E_g)$, $728(A_{1g})$	[43]
$PdRhO_2$	D_{3d}^5	$A_{1g} + E_g$	$523(E_g)$, $664(A_{1g})$	[62]

6.3 Oxygen Storage Materials

The closed-loop emission control system for most present-day gasoline engines requires that the air-to-fuel ratio of the charge injected into the engine be modulated in time between slightly rich and lean conditions. The average ratio corresponds to a stoichiometric mixture. During rich excursions there is insufficient O_2 in the gas stream to oxidize unburned hydrocarbons and CO, and thus a catalyst that can store excess oxygen during lean excursions and then release it during rich excursions will produce lower emissions of the unwanted species [23,76]. For this reason oxygen storage is important for optimum performance of automotive TWC's. Here we describe several instances in which Raman scattering can help to identify the mechanism and in some cases the oxygen storage capability (OSC) of specific catalysts.

Cerium dioxide, CeO_2, is added to γ-alumina-supported automotive catalysts to increase the oxygen storage capability and, in some cases, to improve the thermal stability of alumina [77,78]. In addition, CeO_2 has other potentially beneficial effects regarding interactions with the precious metals in a TWC [79,80]. These interactions, however, can be quite complicated. For example, Murrell et al. have shown using Raman scattering that many of the Pt-group metals form surface oxide complexes on CeO_2 with lines in the 700-cm^{-1} region [81]. The metal-oxide complexes are relatively stable under moderate temperatures ($\approx 750\,°C$) or continuously oxidizing conditions. However, after cyclic oxidation-reduction aging at $850\,°C$, the surface complexes disappear, severe sintering of the metals occurs, and the high surface area of the CeO_2 is lost, all of which are detrimental to the catalyst.

Oxygen storage occurs in cerium oxide when there are reversible reactions between Ce^{4+} and Ce^{3+}. In pure CeO_2 such reactions can involve either the formation of new compounds CeO_x (with $x < 2$) or the creation of O vacancies in the fluorite-structured lattice. Under mildly reducing conditions, the lattice loses O, forming substoichiometric CeO_{2-y} with $y \leq 0.28$ [82]. This material maintains the fluorite structure, although discrete phases associated with ordering of the O vacancies may occur. CeO_{2-y} readily converts back to stoichiometric CeO_2 under oxidizing conditions [83]. There is also evidence for the formation of hexagonal Ce_2O_3 under strongly reducing conditions [83], but no Raman spectrum of this phase has been reported.

Miki et al. have shown that the addition of lanthanum oxide improves the OSC of a typical precious metal/CeO_2 catalyst supported on alumina [84], and Cho has found the same result for the addition of gadolinium oxide [85]. In both cases $Ce_{1-x}RE_xO_{2-y}$ compounds are being formed, where RE is one of the other lanthanides (or actinides). These compounds maintain the same fluorite structure of CeO_2 up to quite large doping levels (as high as 55% for La) [86]. For a trivalent dopant such as La^{3+} or Gd^{3+}, one O vacancy is required for every two dopant ions in order to balance the charge. These vacancies apparently increase the diffusion rate of O, thereby increasing the ease with which the material can absorb and give off oxygen.

The Raman signature for dopant-induced O vacancies in CeO_2 is shown in Fig. 6.15. These spectra, obtained by McBride et al., are from a series of La-doped ceria samples [87]. Similar results were seen for other trivalent dopants Nd, Eu, and Gd. As the O-vacancy concentration increases, the $465\,cm^{-1}$ F_{2g} mode of the fluorite structure shifts to lower frequency, broadens, and becomes asymmetric. In addition, a broad feature near $570\,cm^{-1}$, attributed to the O vacancies, grows uniformly. The assignment of this new feature to O vacancies is supported by a simple model calculation using Green's function techniques [87].

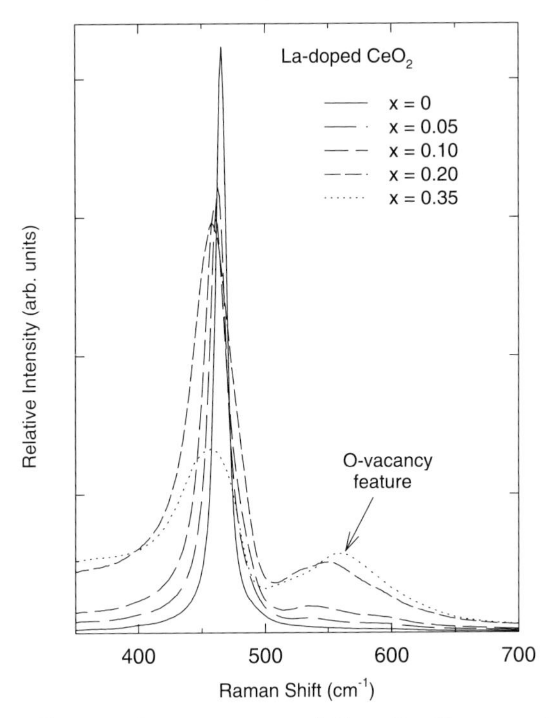

Fig. 6.15. Raman spectra of $Ce_{1-x}La_xO_{2-y}$ powders. From McBride et al. [87]

Ceria doped with tetravalent elements such as Pr and Zr and solid solutions of ceria and zirconia generally show better OSC than any of the dioxides by themselves. As shown by Vidmar et al. [88], the addition of trivalent dopants to the ceria-zirconia system further improves its OSC, in analogy with the effects of trivalent dopants in ceria alone.

In ceria supported on γ-alumina the formation of $CeAlO_3$ under reducing conditions can degrade the oxygen storage capacity of ceria, since the $CeAlO_3$ will not readily reoxidize. Shyu et al. gave results from X-ray photoelectron spectroscopy (XPS), Raman, and temperature-programmed reduction (TPR) that suggest the formation of a surface phase of $CeAlO_3$ [89].

In Pt/CeO_2 catalysts Brogan et al. gave evidence for the formation of a Pt-O-CeO_2 surface mixed-oxide species [90]. Figure 6.16 shows their in situ Raman results. The broad bands at 550 and 690 cm^{-1} in (a) occurred after calcination at 450 °C and were recorded at room temperature. After heating to 100 °C in flowing H_2-Argon, (b), the 550-cm^{-1} band became stronger and sharper, a new band appeared at 630 cm^{-1} and the 690-cm^{-1} band became weaker. The authors ascribed these changes to the decomposition of the Pt-O-CeO_2 species with movement of the bridging oxygen atom to the Pt site. After reduction at 200 °C, (c), all features due to the mixed oxide disappeared, indicating that oxygen was removed from the Pt/CeO_2 interface.

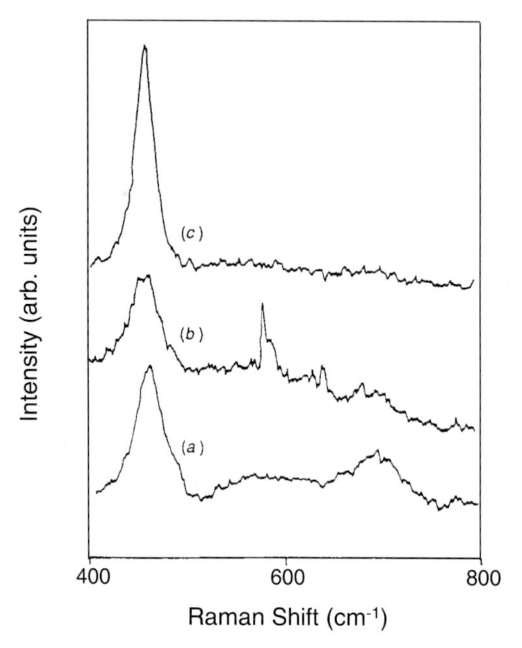

Fig. 6.16. Raman spectra of Pt/CeO_2 catalysts showing evidence for metal-support interaction. From Brogan et al. [90]

Otto et al. used Raman scattering to study the oxidation/reduction behavior of PdO in a Pd/γ-alumina catalyst [91]. Figure 6.17 shows their in situ measurements of the strength of the 651-cm^{-1} PdO line during reduction at room temperature in 50 Torr of hydrogen. Su et al. [92] performed similar time-resolved reduction measurements on a Pd/ZrO_2 catalyst, results from which are shown in Fig. 6.18. The oxygen content coordinates on the left-hand axes of this figure are the amounts of PdO remaining after a fixed reduction time, as determined by measuring the total amount of CO_2 released during TPR of the samples in CH_4. The Raman signal correlates well with the oxygen content, but appears to lag slightly behind it. The authors attributed the lag to the formation of a metallic Pd layer on the surface of the PdO particles, which would tend to scatter and attenuate the laser light and thereby reduce the Raman signal from the underlying PdO.

6.4 Adsorbed Species

The first step in a gas-phase catalytic process is the adsorption of a particular molecule on the surface of the active catalyst. The adsorbed species have their own characteristic vibrational modes, which can often be measured and identified with Raman scattering. Identifying the chemical state of the adsorbed molecule gives insight into the catalytic process in which it is

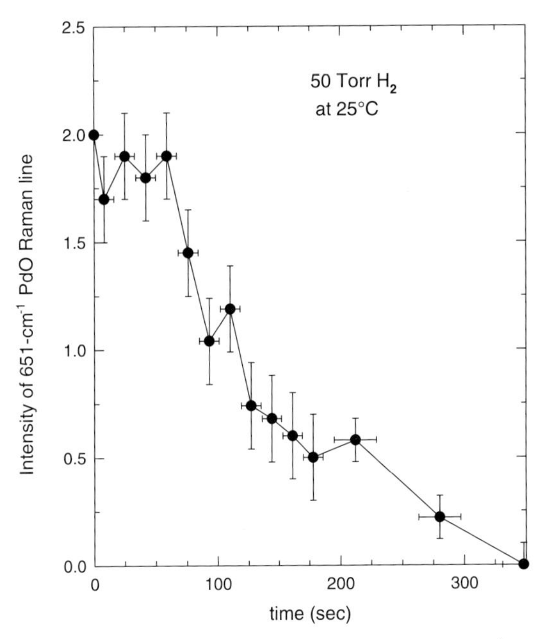

Fig. 6.17. Time dependence of the 651-cm^{-1} PdO Raman line in a previously oxidized Pd/γ-alumina catalyst after exposure to H_2 at room temperature. From Otto et al. [91]

participating. In this section we give examples of the use of Raman scattering to identify adsorbates containing oxides of nitrogen (NO_x) and sulfur (SO_x).

6.4.1 Oxides of Nitrogen

The removal of NO_x produced during high-temperature combustion has proven to be much more difficult than the removal of CO or unburned hydrocarbons [93–95]. This statement is particularly true for diesel engines and lean-burn gasoline engines, both of which offer better fuel economy than the conventional gasoline engine. For fixed power plants, supported vanadia catalysts are used for the selective catalytic reduction (SCR) of NO_x with the addition of NH_3 to the exhaust stream. Wachs and co-workers have studied these catalysts extensively with Raman scattering [96,97]. However, the addition of N-containing species to the exhaust is impractical for vehicles and has drawbacks even for stationary sources [93]. We will therefore focus in this section on catalysts that use more common reductants, such as CO and unburned hydrocarbons, that are already present in the exhaust stream.

Cooney and Tsai [98] identified the Raman peaks produced by some of the NO_x complexes adsorbed on alkali metal ion-exchanged zeolites and γ-alumina exposed to NO_2. An example of their spectra is shown in Fig. 6.19 for NO_2 on γ-alumina. Table 6.2 summarizes their results for all the materials.

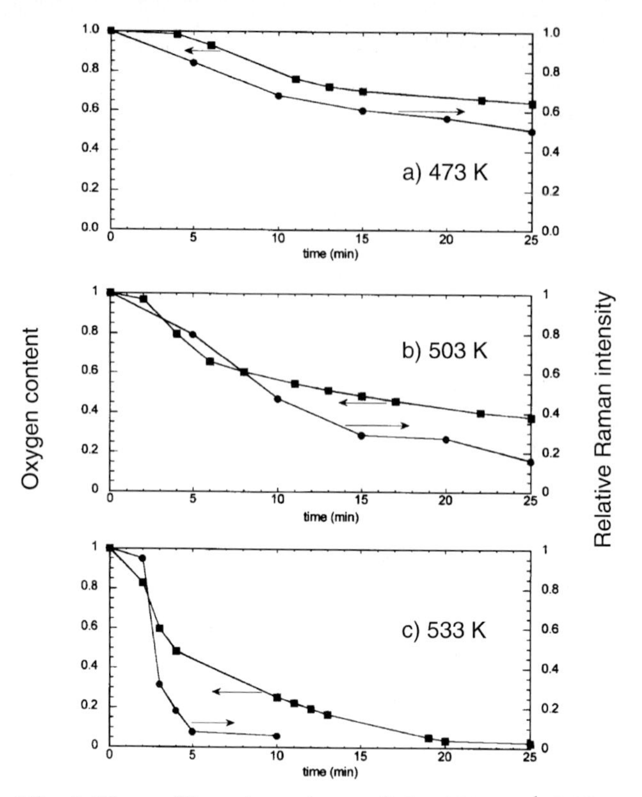

Fig. 6.18a–c. Time dependence of the 651-cm^{-1} PdO Raman line (*circles*) and of the oxygen content determined by TPR (*boxes*) in Pd/ZrO_2 during isothermal reduction in 1 atm of 3% CH_4/He at different temperatures. From Su et al. [92]

The zeolite spectra obtained by Cooney and Tsai suffered from the occurrence of a large fluorescence background. Fluorescence is a common problem with Raman studies of zeolites, since it can sometimes completely obscure the much weaker Raman features.

Saperstein and Rein [99] identified the Raman signatures for N_2 and O_2 adsorbed onto A-type zeolites. They also demonstrated a NaOH treatment for reducing the fluorescence problem.

Li and Stair recently described a UV Raman instrument that appears to eliminate the fluorescence background, at least for certain zeolites [100,101]. Ultra-violet (UV) excitation offers several advantages over conventional visible Raman systems: (1) the Raman signal is increased by the ν^4 factor because of the higher frequency of the light source; (2) there is an electronic enhancement of the Raman signal, since most catalytic species absorb in the near UV; (3) the Stokes shift of the fluorescence signal is often much greater than the Raman shifts, so the fluorescence does not interfere; and (4) the UV component of the thermal background from a heated catalyst being studied

Table 6.2. Raman lines[a] (cm^{-1}) of NO_x species adsorbed on zeolites and γ-alumina, after Cooney and Tsai [98]

NaX	CsX	NaY	NaA	γ-Al_2O_3	Assignment
1053 m	1048 m	1053	1051 m	1010 m	$\nu_1(NO_3^-, D_{3h})$
			1071 s	1050 s	
				1069 vs	
			1260 w		$\nu_1(NO_2, C_{2v})$
			1322 w		
				1384 w	$\nu_3(NO_3^-, D_{3h})$
2185 sh	2186 sh	2194 sh	2169 sh		$\nu(NO^+, C_{\infty,v})$
2257 s	2248 s	2257 s	2226 s		

[a] v = very, s = strong, m = medium, w = weak, sh = shoulder.

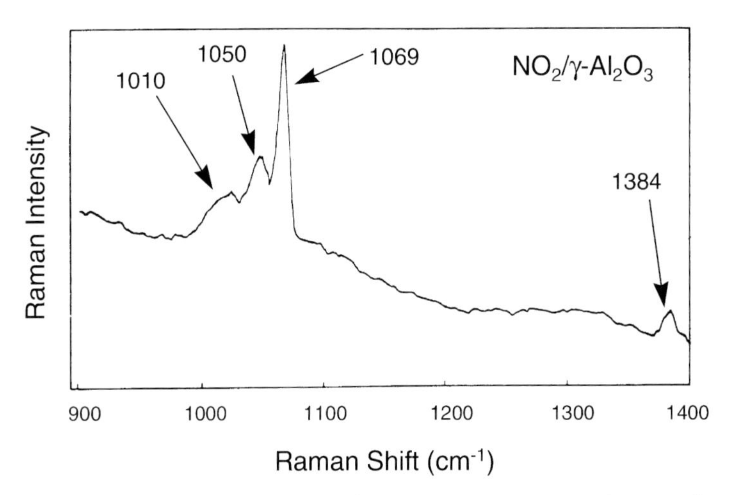

Fig. 6.19. Raman spectrum of NO_2 adsorbed on γ-alumina from Cooney and Tsai [98]

in situ is negligible. In spite of these advantages, the high-energy photons used in UV excitation can promote chemical reactions that would not normally occur, so the possibility that the spectra obtained result from exposure to UV must be ruled out.

An example of the results obtained by Li and Stair [100] is given in Fig. 6.20, which shows the adsorption of propene onto a ZSM-5 zeolite at room temperature (A), the subsequent coke formation that occurs when the sample is heated to 773 K (B and C), and the dehydrogenation that occurs after propene is removed from the gas stream at 773 K (D). Note that the

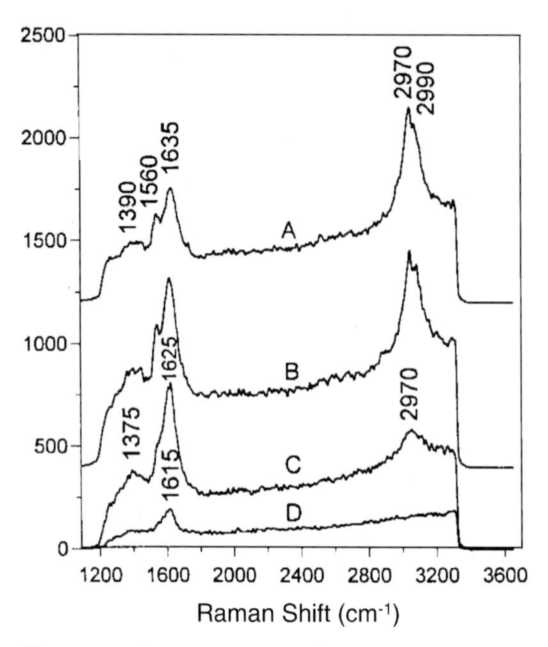

Fig. 6.20. Raman spectra obtained with UV excitation of ZSM-5 zeolite showing adsorbed propene at room temperature (*A*), coke formation after heating to 773 K (*B* and *C*), and dehydrogenation (*D*). From Li and Stair [100]

C–H vibrations near $3000\,\mathrm{cm}^{-1}$, which are often difficult to detect in adsorbed species with visible excitation, are quite prominent in Fig. 6.20. Although these results are not directly applicable to SCR of NO_x, they are relevant to zeolite studies in general, and one of the best NO_x reduction catalysts is Cu-ZSM-5 zeolite [93]. Attempts to follow the same reactions with visible Raman excitation yielded only a strong fluorescence background.

Recently, Lunsford and co-workers reported the decomposition of NO over barium oxide supported on magnesium oxide [102–105]. This system also lends itself well to Raman analyses. Figure 6.21 shows the usefulness of in situ measurements for following the complex reactions that occur when a fresh BaO_2/MgO catalyst is first exposed to a NO/He gas stream at different temperatures [104]. In (a) the characteristic lines of BaO_2 at 130, 200, and $830\,\mathrm{cm}^{-1}$ nearly disappear after only 5-min exposure to the NO/He stream, and new lines appear at 1059 and $718\,\mathrm{cm}^{-1}$, which the authors attribute to the formation of nitrate ions, NO_3^-. At the same time bands are observed at 160, 263, 420, 807, and $1239\,\mathrm{cm}^{-1}$, which are characteristic of the Ba-nitro complexes present under catalytic conditions. These features, however, disappear after further exposure, being replaced by bands and shoulders at 333, 456, 822, and $1335\,\mathrm{cm}^{-1}$, which are indicative of what the authors refer to as the "amorphous phase II″ containing NO_3^- and Ba-nitrito complexes".

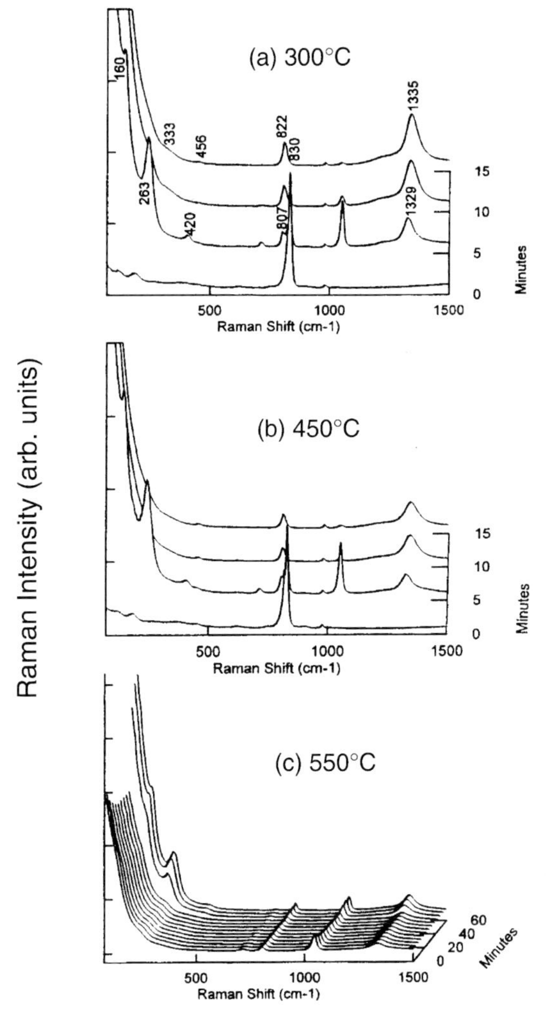

Fig. 6.21a–c. Time-resolved, in situ Raman spectra while a BaO_2/MgO catalyst is exposed to a NO/He gas stream at different temperatures. From Mestl et al. [104]

At somewhat higher temperature and lower NO partial pressure in Fig. 6.21b, the same transformations occur. At still higher temperatures in Fig. 6.21c, the catalytically active Ba-nitro complexes are again formed, but they now appear to remain stable indefinitely.

6.4.2 Oxides of Sulfur

Although sulfur oxides are not regulated pollutants from gasoline-powered engines, the presence of trace amounts of sulfur in most petroleum-based fuels leads to a variety of effects that are detrimental to air quality. The sulfur-containing hydrocarbon in the fuel is converted to SO_2 in the combustion process. The direct release of SO_2 is known to be a major contributor to the acid rain problem, but the interactions of this molecule with the catalyst can degrade its performance and increase emissions of other pollutant species as well [76,106]. Sulfur can accumulate on the catalyst support material during lean operation as surface sulfate compounds and then be released during rich excursions in the form of H_2S, producing a noxious odor. The sulfate species can also end up as hydrated sulfuric acid, which contributes to particulate emissions and is a serious problem for diesel engines. Sulfur can also poison the active components by forming surface sulfate compounds with them, an example of which is the formation of barium sulfate on the BaO_2-containing NO_x catalyst discussed in the previous subsection.

The primary Raman signatures for oxides of sulfur are lines near $1000\,cm^{-1}$ and $1400\,cm^{-1}$ associated with the stretching frequencies of the S$-$O and S$=$O bonds, respectively. Through analyses of these bands, Twu et al. [107] have identified a number of surface oxysulfur species and bulk cerium-oxygen-sulfur compounds that can be formed on ceria substrates. Spielbauer et al. [108,109] associated similar Raman bands with surface sulfate species on zirconia.

Figure 6.22 shows examples of the Raman spectra from surface species of SO_x adsorbed on two common support materials, CeO_2 and an 8 : 1

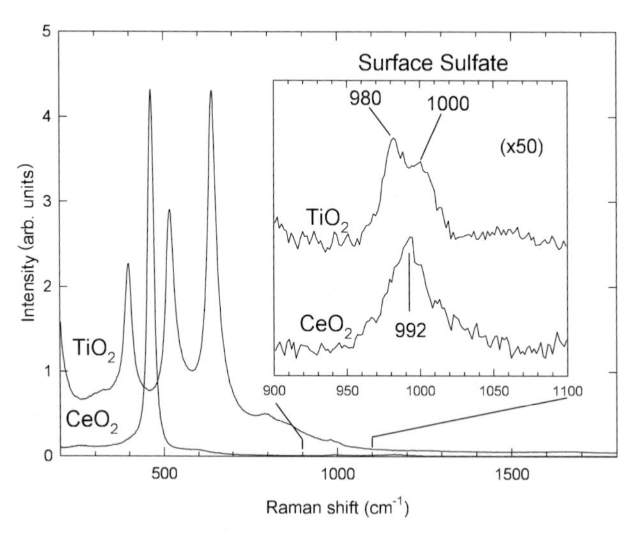

Fig. 6.22. Raman spectra obtained with a 633-nm laser of support materials CeO_2 and TiO_2 (anatase)/γ-Al_2O_3 after exposure to gas streams containing trace amounts of SO_2 as described in the text. The *inset* shows a $\times 50$ expanded plot of the surface sulfate spectral region in which the smooth background from the support materials has been subtracted. From Weber et al. [110]

mixture of the anatase phase of TiO_2 and γ-Al_2O_3 [110]. The ceria support was exposed at 500 °C to a simulated exhaust stream, containing 20 ppm SO_2, that is known to reproduce aging effects seen in field studies [111]. The titania/alumina support was exposed at 200 °C to a gas stream containing N_2, 10% O_2 and 800 ppm of SO_2. In both cases the spectra are dominated by the Raman lines from the supports. Expansion of the 900–1100 cm^{-1} region, however, yields clear signatures for adsorbed sulfur oxide species, as shown in the inset. In titania there is a doublet at 980 and 1000 cm^{-1}, whereas in ceria there is a single broader line centered at 992 cm^{-1}. These lines are weaker than the support Raman lines by at a least a factor of 100.

6.5 Particle-Size Effects

Gas-phase catalysis is fundamentally a surface-mediated process, and thus catalytic activity is generally increased when the surface areas of the active components are increased. For a fixed-volume catalyst, maximum surface area requires minimum particle size, which means that small particle sizes for the active components are desirable. One of the common degradation mechanisms for an automotive catalyst is that extended operation at extreme high temperatures leads to sintering of some of the active components, thus reducing their surface area and thereby the activity of the catalyst. Conventional methods for measuring particle size include transmission electron microscopy (TEM), in which the particles are directly imaged, and X-ray diffraction, in which the particle size can be inferred from the diffraction line widths using the Scherrer relation. Raman scattering can also be used in some cases to estimate particle size, but it must first be calibrated with one of the conventional methods.

Richter et al. [112], who studied microcrystalline Si, were the first authors to show a clear correlation between crystallite size and Raman spectra in well-characterized samples. Tiong et al. [113] obtained similar results on ion-implanted GaAs. Both groups interpreted their results using the spatial correlation model, also referred to as the phonon confinement model [112–115]. This model assumes that the lack of long-range order in the small particles relaxes the strict momentum conservation rule by allowing phonons with momentum vectors as large as the reciprocal particle size to contribute to the Raman scattering process. A Gaussian correlation is assumed, and the line shape can be written in the form

$$I(\omega, \xi) = \frac{\xi^3}{\pi^{3/2}} \int \mathrm{d}\boldsymbol{q} \frac{\exp(-\xi^2 q^2)}{[\omega - \omega(\boldsymbol{q})]^2 + \Gamma^2} , \tag{6.1}$$

where Γ is the half-width at half-maximum (HWHM) of the Raman line in a large crystal and ξ is the Gaussian correlation length [115]. Formally, the integral in (6.1) extends over all $\boldsymbol{q}$ space, but in practice only the region near the Brillouin zone center contributes because of the exponential factor.

The correlation length ξ is taken to be proportional to the crystallite size D, although different proportionality constants have been used by different authors, ranging from $\xi = D/2$ by Richter et al. [112] to $\xi = D/4\pi$ by Campbell and Fauchet [114]. In the limit $\xi \to \infty$ the integral in (6.1) contains $\delta(\boldsymbol{q})$, thus yielding a Lorentzian line centered at $\omega(0)$ with HWHM Γ, as expected for a large crystal. Since the dispersion of most Raman-active modes away from the zone center is towards lower frequency, this model generally predicts a Raman line that shifts down and broadens asymmetrically as the particle size decreases. A review of the spatial correlation model and its application to micro-crystalline semiconductors has been given by Pollak [116].

Figure 6.23 shows Raman spectra from Graham et al. [117] of a fresh CeO_2-containing TWC and a crystal of CeO_2. The Raman peak in the catalyst sample is shifted down by several cm^{-1} and broadened by about a factor of two compared with the single crystal, in qualitative agreement with the above model. Figure 6.24 shows that there is a linear relation, given by the solid line, between the Raman line-width and the inverse particle size, as determined by XRD, for a range of catalyst samples, some with CeO_2 particles as small as 30 Å [115].

As an empirical relation, the result in Fig. 6.24 is quite useful, since it allows the CeO_2 particle size in an unknown sample to be determined solely from the Raman spectrum. However, the spatial correlation model predicts a much smaller size effect than is observed, as shown by the dashed line in the figure. A similar behavior has been noted in BN [118], graphite [119], and

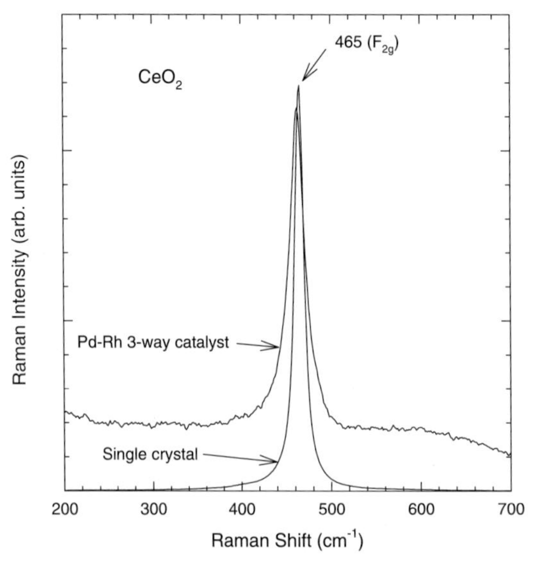

Fig. 6.23. Raman spectra of a single crystal of CeO_2 and a commercial three-way catalyst containing CeO_2. From Graham et al. [117]

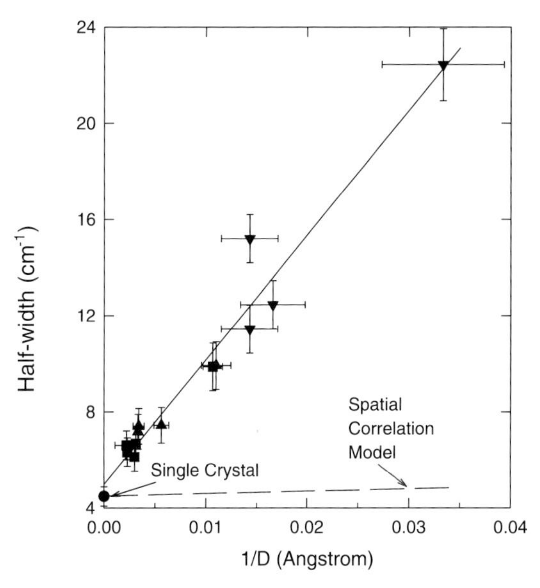

Fig. 6.24. Correlation between the half-width of the CeO_2 Raman line and the particle size as determined by X-ray diffraction from Weber et al. [115]. The *solid line* is a linear fit to the data; the *dashed line* a calculation based on (6.1)

diamond [120]. In each case the qualitative features of the size-dependence of the Raman peak are consistent with the model, but the magnitude of the observed broadening is much larger than predicted. Alternative explanations for the increased broadening, in terms of size-dependent changes in the phonon lifetime, have been offered by several authors [112,115,116].

6.6 Quantitative Analyses

In previous sections of this chapter we have dealt only with the relative intensities, frequencies, and widths of the lines in a Raman spectrum. Here we consider using the line intensities to infer information about the concentration of a particular material. Three examples, all involving PdO formation, will be considered below. The first measures the PdO Raman signal on highly dispersed Pd/γ-alumina catalysts and correlates it with the Pd loading. The second follows the oxidation of Pd in a Pd/ZrO_2 catalyst and correlates the Raman signal with the oxygen content. The third considers correlating the intensity of the Raman peak with the thickness of a thin PdO film on a Pd substrate. In all cases there are limited ranges over which Raman scattering can give useful quantitative information. In general, if reliable intensity data are to be obtained, there are a number of experimental details that must be accounted for, such as the laser power, the focusing and collecting optics, and the spectrometer slit widths.

Otto et al. [91] used Raman scattering to study a series of model Pd catalysts supported on γ-alumina, with the Pd concentration varying from 0 to 20 wt. %. The catalysts were oxidized by heating in air at 600 °C for 20 h, and the integrated intensity of the PdO Raman line at 651 cm^{-1} was measured versus the Pd concentration, giving the result shown in Fig. 6.25. The error bars in the figure represent the variations between similarly prepared samples. Note that for small Pd loadings, less than $\approx$ 2 wt. %, the Raman signal tracks the Pd concentration quite well, as shown by the solid line. For higher loadings the Raman intensity levels off and gives a signal independent of the loading and with considerable scatter. For Pd loadings less than 2 wt. % the catalyst maintains a light gray color. This means that the volume sampled by the Raman probe is constant, independent of loading. For higher Pd concentrations, the catalyst takes on a progressively darker appearance, which limits the depth probed by the laser. In addition, for the higher loadings some of the oxidized particles become larger than the attenuation length ($\approx$ 100 Å) of the light in PdO, which means that not all of the PdO contributes to the Raman signal. Using the result in Fig. 6.25 for calibration, Otto et al. found that the strength of the 651-cm^{-1} Raman line predicted the amount of Pd in various commercial catalysts to within only about a factor of two in the range between 0.2 and 2 wt. %. They concluded that different Pd-support interactions in the catalysts limited the accuracy.

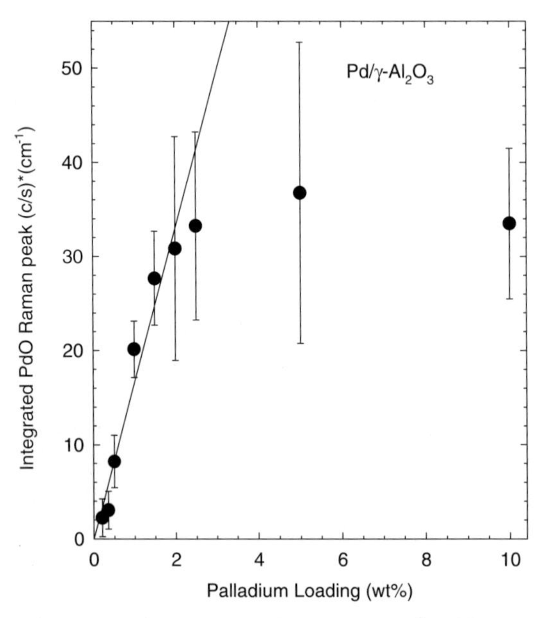

Fig. 6.25. Dependence of the 651-cm^{-1} PdO Raman line on Pd loading in oxidized Pd/γ-alumina catalyst. From Otto et al. [91]. The *solid line* is a linear least-squares fit to the points below 2 wt. % loading

The results of Su et al. [92], already discussed in Sect. 6.3, also pertain to quantitative analyses of the PdO concentration. Figure 6.26 shows their time-dependent PdO Raman signal in a 10 wt. % Pd/ZrO_2 catalyst during isothermal oxidation from a reduced state [92]. The oxidation was carried out in 1 atm of 5% O_2/He. For the same oxidation conditions they also determined the oxygen content, plotted on the left axes, from the CO_2 released during TPR of the sample in CH_4. As shown in the figure, the Raman signal generally follows the oxygen content but tends to lag slightly, particularly at the lower temperatures. The authors ascribe this effect to the formation of an amorphous layer of PdO that contributes to the oxygen content but not to the PdO Raman peak. At the higher temperatures the amorphous layer crystallizes more quickly, so there is less delay between the two signals.

Remillard et al. [58] measured the Raman intensity versus film thickness for PdO films on a Pd substrate over the range from 10 to 10^3 Å. The film thicknesses were measured by ellipsometry. They also derived a formula for the thickness dependence of the Raman signal that is sufficiently general and important that we will reproduce it here. For the case in which the sample probed by the laser is both absorbing and semi-infinite in extent,

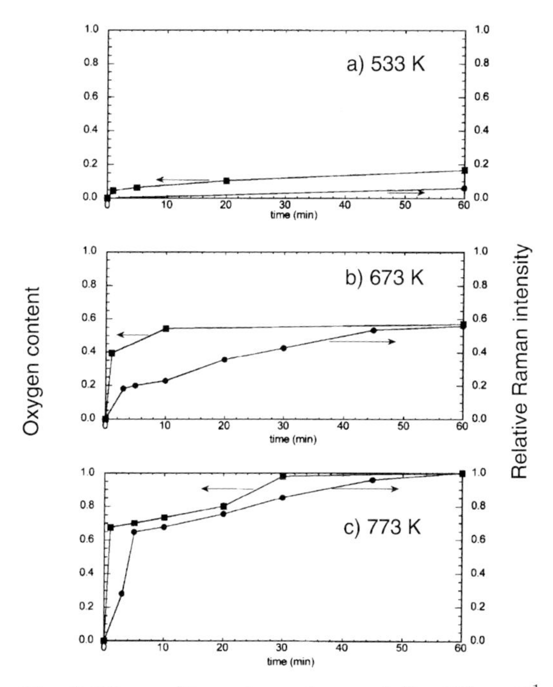

Fig. 6.26a–c. Time dependence of the 651-cm^{-1} PdO Raman line (*circles*) and of the oxygen content determined by TPR (*boxes*) in pre-reduced Pd/ZrO_2 during isothermal oxidation in 1 atm of 5% O_2/He at different temperatures. From Su et al. [92]

Cardona has given a simple formula for the Raman intensity in terms of the optical properties of the material [121]. Here we consider the modifications to that result when multiple reflections within the sample being probed become important.

Figure 6.27 defines the geometry for calculating the thickness dependence of the back-scattered Raman intensity. The first step is to calculate the electric field produced by the incident laser beam at an arbitrary point z within the film. This calculation can be accomplished by summing geometric series involving the Fresnel reflection and transmission coefficients. The result is

$$E_{\mathrm{i}}(z) = E_0 t_{12} \mathrm{e}^{\mathrm{i}k_2 z} \frac{1 + r_{23}\mathrm{e}^{2\mathrm{i}k_2(d-z)}}{1 + r_{12}r_{23}\mathrm{e}^{2\mathrm{i}k_2 d}} , \tag{6.2}$$

where E_0 is the electric field strength in the incident beam; t_{12}, r_{23}, and r_{12} are the amplitude transmission and reflection coefficients for the incident light beam (assumed to be s-polarized) at the interfaces defined in Fig. 6.27; k_2 is the z-component of the incident wave vector in medium 2; and we have suppressed a factor $\hat{\boldsymbol{y}} \exp[\mathrm{i}(k_x x - \omega t)]$, which modulates the fields in all media. The field in (6.2), when multiplied by the Raman tensor, induces a polarization at the Raman-shifted frequency. This polarization can now be treated as a distribution of radiating point dipoles, whose far-field radiation yields the Raman signal. The multiple reflections that modified the incident beam also affect the radiated signal, and the resulting expression contains the same Fresnel coefficients again, but evaluated at the Raman-shifted frequency [122,123]. We consider only the light emitted normal to the surface, which is generated by the components of the Raman-induced dipole parallel to the surface. The result for the far-field amplitude is ,

$$E_{\mathrm{s}}(z) = A E_{\mathrm{i}}(z) t'_{12} \mathrm{e}^{\mathrm{i}k'_2 z} \frac{1 + r'_{23}\mathrm{e}^{2\mathrm{i}k'_2(d-z)}}{1 + r'_{12}r'_{23}\mathrm{e}^{2\mathrm{i}k'_2 d}} , \tag{6.3}$$

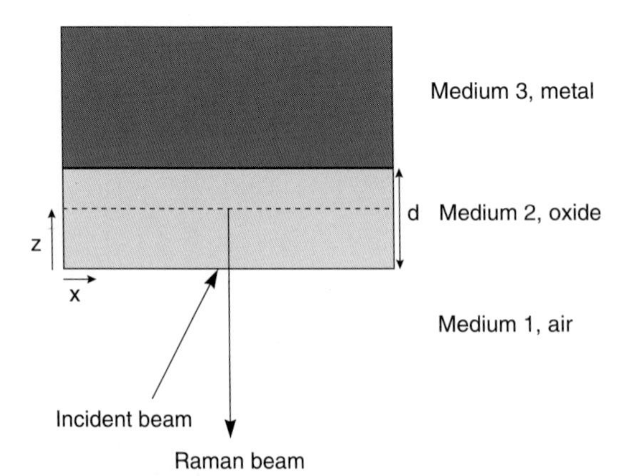

Fig. 6.27. Geometry for calculating the thickness dependence for Raman scattering from a thin film

where A is a constant that includes all factors independent of d and z, and the primed quantities are to be evaluated for normal incidence at the Raman-shifted frequency. Summing the scattered intensity associated with the field E_s across the thickness of the film yields an expression for the dependence of the Raman signal on thickness:

$$I(d) \propto \int_0^d \mathrm{d}z\, |E_s(z)|^2 \ . \tag{6.4}$$

Figure 6.28 shows the application of (6.4) to PdO films grown on a Pd substrate [58]. There are no adjustable parameters for the calculated curve (solid line) in this plot, other than an overall scale factor for the intensity. The data points are from two samples oxidized in air for increasing periods of time at 500 °C. Note that there is only a very limited range, below ≈ 100 Å, where there is a unique relation between the Raman intensity and the oxide thickness. There is also a peak in the intensity near 200 Å that is roughly three times greater than that for a bulk sample. Qualitatively similar results can be expected for any opaque oxide grown on a metal surface.

6.7 Summary and Outlook

Raman scattering is useful for characterizing catalyst materials incorporating any of the Pt-group metal oxides. This is particularly true for automotive

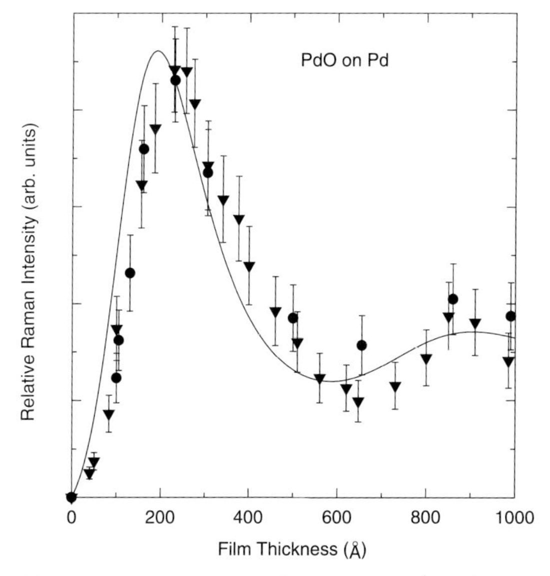

Fig. 6.28. Dependence of the 651-cm^{-1} PdO Raman line on oxide thickness (measured by ellipsometry) and the result calculated with (6.2)–(6.4). From Remillard et al. [58]

catalysts containing PdO. Palladium is the most heavily used of all the active precious metals, and PdO is easily detected by Raman scattering, even in commercial catalysts. Consequently, many of the examples cited in this chapter pertain to PdO. Other precious-metal additives (e.g., Pt and Rh) are generally present in such low concentrations and their oxides have such weak Raman signals that they can only be studied in model catalyst systems in which the concentrations are much higher. Ceria and doped-ceria form another class of catalyst materials in which Raman scattering has given useful information. Problems addressed for these materials include identifying the phases present and estimating particle sizes, both of which relate to the oxygen storage capacity of the material. Raman scattering has the potential for identifying adsorbed species, in situ, during catalytic reactions, but this potential has not yet been fully realized. This is the direction that future research is likely to take.

Acknowledgements. I am indebted to my colleague George Graham, one of the resident experts in automotive catalysts at the Ford Research Laboratory, for critically reading the manuscript and making many helpful suggestions; to my former student, James McBride, for providing some of the PdO figures; to Ann Chen for providing the electron backscattering image in Fig. 6.1; to Bert Chamberland for providing the samples used to obtain the Raman spectra in Figs. 6.7 and 6.9; and to Bob McCabe and Mordecai Shelef for many useful discussions on automotive catalysts.

References

1. G.T. Kerr: Sci. Am. **261**, 100 (1989)
2. R.R. Chianelli: In *Catalysis and Surface Science*, ed. by H. Heinemann, G.A. Somorjai (Dekker, New York 1984) p. 61
3. E.J. Vandenberg, J.C. Salamone (Eds.): *Catalysis in Polymer Synthesis*, ACS Symp. Ser. 496 (Am. Chem. Soc., Washington, D.C. 1992)
4. I.E. Wachs (Ed.): *Characterization of Catalytic Materials* (Butterworth-Heinemann, Manning, Greenwich, CT 1992)
5. J.M. Stencel: *Raman Spectroscopy for Catalysis* (Van Nostrand Reinhold, New York 1990)
6. H. Knözinger: Catalysis Today **32**, 71 (1996)
7. L. Dixit, D.L. Gerrard, H.J. Bowley: Appl. Spectrosc. Rev. **22**, 189 (1986)
8. I.E. Wachs, F.D. Hardcastle: *Catalysis*, Vol. 10 (The Royal Society of Chemistry, Cambridge 1993)
9. I.E. Wachs, F.D. Hardcastle, S.S. Chan: Spectrosc. **1**, 30 (1986)
10. I.E. Wachs: Catalysis Today **27**, 437 (1996)
11. J.R. Bartlett, R.P. Cooney: In *Spectroscopy of Inorganic-Based Materials*, ed. by R.J.H. Clark, R. E. Hester (Wiley, Chichester 1987) Chap. 3, p. 187
12. M. Mehicic, J.G. Grasselli: In *Analytical Raman Spectroscopy*, ed. by J.G. Grasselli, B.J. Bulkin (Wiley, New York 1991) Chap. 10, p. 325

13. G. Saidi, T.T. Tsotsis: Surf. Sci. **161**, L591 (1985)
14. L.T. Tsitsopoulos, T.T. Tsotsis: Surf. Sci. **187**, 165 (1987)
15. D. König, W.H. Weber, B.D. Poindexter, J.R. McBride, G.W. Graham, K. Otto: Catal. Lett. **29**, 329 (1994)
16. G. Haas, R.U. Franz, H.H. Rotermund, R.M. Tromp, G. Ertl: Surf. Sci. **352-354**, 1003 (1996)
17. H.H. Rotermund: Surf. Sci. **386**, 10 (1997)
18. G.W. Graham, D. König, B.D. Poindexter, J.T. Remillard, W.H. Weber: Top. Catal. **8**, 35 (1999)
19. F. Eisert, A. Rosén: Phys. Rev. B **54**, 14061 (1996)
20. W. Akemann, K.A. Friedrich, U. Linke, U. Stimming: Surf. Sci. **402-404**, 571 (1998)
21. M.A. Lovell, D. Roy: Appl. Surf. Sci. **135**, 46 (1998)
22. J.T. Kummer: J. Phys. Chem. **90**, 4747 (1986)
23. K.C. Taylor: In *Catalysis Science and Technology*, Vol. 5, ed. by J.R. Anderson, M. Boudart (Springer, Heidelberg, Berlin 1984) Chap. 2, p. 119
24. P.-P. Knops-Gerrits, D.E. De Vos, E.J.P. Feijen, P.A. Jacobs: Microporous Mater. **8**, 3 (1997)
25. C. Brémard, D. Bougeard: Adv. Mater. **7**, 10 (1997)
26. G. Deo, F.D. Hardcastle, M. Richards, A.M. Hirt, I.E. Wachs: In *Novel Materials in Heterogeneous Catalysis*, ed. by R.T.K. Baker, L.L. Murrell, ACS Symp. Ser. **437**, 317 (1990)
27. A.A. Tsyganenko, P.P. Mardilovich: J. Chem. Soc. Faraday Trans. **92**, 4843 (1996)
28. K. Sohlberg, S.J. Pennycook, S.T. Pantelides: Mater. Res. Soc. Symp. Proc., Advanced Catalytic Materials **1998**, 165 (1999)
29. M. Yashima, H. Arashi, M. Kakihana, M. Yoshimura: J. Am. Ceram. Soc. **77**, 1067 (1994)
30. G. Vlaic, P. Fornasiero, S. Geremia, J. Kašpar, M. Graziani: J. Catal. **168**, 386 (1997)
31. G. Mestl, T.K.K. Srinivasan: Catal. Rev. – Sci. Eng. **40**, 451 (1998)
32. I.E. Wachs, B.M. Weckhuysen: Appl. Catal. A **157**, 67 (1997)
33. K.B. Schwartz, C.T. Prewitt: J. Phys. Chem. Solids **45**, 1 (1984)
34. W.D. Westwood, C.D. Bennewitz: J. Appl. Phys. **46**, 2313 (1974)
35. M. Hecq, A. Hecq: J. Less-Common Met. **56**, 133 (1977)
36. J.R. McBride, G.W. Graham, C.R. Peters, W.H. Weber: J. Appl. Phys. **69**, 1596 (1991)
37. J.R. McBride, K.C. Hass, W.H. Weber: Phys. Rev. B **44**, 5016 (1991)
38. K. Otto, W.H. Weber, G.W. Graham, J.Z. Shyu: Appl. Surf. Sci. **37**, 250 (1989)
39. G.W. Graham, W.H. Weber, K. Otto: Appl. Surf. Sci. **59**, 87 (1992)
40. O. Muller, R. Roy: J. Less-Common Met. **16**, 129 (1968)
41. D. Cahen, J.A. Ibers, J.B. Wagner, Jr.: Inorg. Chem. **13**, 1377 (1974)
42. K.B. Schwartz, C.T. Prewitt, R.D. Shannon, L.M. Corliss, L.M. Hastings, B.L. Chamberland: Acta Cryst. B **38**, 363 (1982)
43. G.W. Graham, W.H. Weber, J.R. McBride, C.R. Peters: J. Raman Spectrosc. **22**, 1 (1991)
44. D. Cahen, J.A. Ibers: J. Catal. **31**, 369 (1973)
45. R.D. Shannon, T.E. Gier, P.F. Carcia, P.E. Bierstedt, R.B. Flippen, A.J. Vega: Inorg. Chem. **21**, 3372 (1982)

46. W.H. Weber, G.W. Graham, A.E. Chen, K.C. Hass, B.L. Chamberland: Solid State Commun. **106**, 95 (1998)
47. W.H. Weber, G.W. Graham, J.R. McBride: Phys. Rev. B **42**, 10969 (1990)
48. R.Q. Wu, W.H. Weber: J. Phys. C (to appear)
49. Y.S. Huang, S.S. Lin, C.R. Huang, M.C. Lee, T.E. Dann, F.Z. Chien: Solid State Commun. **70**, 517 (1989)
50. K.C. Taylor, J.C. Schlatter: J. Catal. **63**, 53 (1980)
51. G. Kliche: Z. Naturforsch. **44a**, 169 (1989)
52. K.C. Hass, A.E. Carlsson: Phys. Rev. B **46**, 4246 (1992)
53. K.-T. Park, D.L. Novikov, V.A. Gubanov, A.J. Freeman: Phys. Rev. B **49**, 4425 (1994)
54. R. Ahuja, S. Auluck, B. Johansson, M.A. Khan: Phys. Rev. B **50**, 2128 (1994)
55. W.H. Weber, J.T. Remillard, J.R. McBride, D.E. Aspnes: Phys. Rev. B **46**, 15085 (1992)
56. S.S. Chan, A.T. Bell: J. Catal. **89**, 433 (1984)
57. R.W. McCabe, R.K. Usmen: In Proc. 11th International Congress on Catalysis – 40th Anniversary, Studies in Surface Science and Catalysis, Vol. 101, ed. by J.W. Hightower, W.N. Delgass, E. Iglesia, A. T. Bell (Elsevier Science, Amsterdam 1996) p. 355
58. J.T. Remillard, W.H. Weber, J.R. McBride, R.E. Soltis: J. Appl. Phys. **71**, 4515 (1992)
59. M. Shelef, G.W. Graham: Catal. Rev. – Sci. Eng. **36**, 433 (1994)
60. J.M.D. Coey: Acta Crystallogr. B **26**, 1876 (1970)
61. A. Wold, R.J. Arnott, W.J. Croft: Inorg. Chem. **2**, 972 (1963)
62. W.H. Weber, R.J. Baird, G.W. Graham: J. Raman Spectrosc. **19**, 239 (1988)
63. D.B. Rogers, R.D. Shannon, A.W. Sleight, J.L. Gillson: Inorg. Chem. **8**, 841 (1969)
64. O. Muller, R. Roy: J. Less-Common Met. **19**, 209 (1969)
65. S. Trasatti: Electrochim. Acta **36**, 225 (1991)
66. T. Kawai, T. Sakata: Chem. Phys. Lett. **72**, 87 (1980)
67. M. Shelef, H.S. Gandhi: Platinum Mater. Rev. **18**, 2 (1974)
68. H.S. Gandhi, H.K. Stepien, M. Shelef: Mater. Res. Bull. **10**, 837 (1975)
69. Y.S. Huang, F.H. Pollak: Solid State Commun. **43**, 921 (1982)
70. S.S. Rosenblum, W.H. Weber, B.L. Chamberland: Phys. Rev. B **56**, 529 (1997)
71. J. Haines, J.M. Legér: Phys. Rev. B **48**, 13344 (1993)
72. R.D. Shannon, D.B. Rogers, C.T. Prewitt: Inorg. Chem. **10**, 713 (1971); C.T. Prewitt, R.D. Shannon, D.B. Rogers: Inorg. Chem. **10**, 719 (1971)
73. R.J. Baird, G.W. Graham, W.H. Weber: Oxid. Met. **29**, 435 (1988); G.W. Graham, T.J. Potter, W.H. Weber, H.S. Gandhi: Oxid. Met. **29**, 487 (1988)
74. H.R. Hoekstra, S. Siegel, F.X. Gallagher: Adv. Chem. Ser. **98**, 39 (1971)
75. K.B. Schwartz, J.B. Parise, C.T. Prewitt, R.D. Shannon: Acta Crystallogr. B **39**, 217 (1983)
76. R.W. McCabe, J.M. Kisenyi: Chemistry & Industry, 7 August 1995, p. 605
77. H.C. Yao, Y.F. Yao: J. Catal. **86**, 254 (1984)
78. Y.F. Yao, J.T. Kummer: J. Catal. **106**, 307 (1987)
79. J.C. Summers, S. Ausen: J. Catal. **58**, 131 (1979)
80. J.G. Nunan, H.G. Robota, M.J. Cohon, S.A. Bradley: J. Catal. **133**, 309 (1992)
81. L.L. Murrell, S.J. Tauster, D.R. Anderson: Studies in Surf. Sci. and Catal. **71**, 275 (1991)

82. M. Ricken, J. Nölting, I. Riess: J. Solid State Chem. **54**, 89 (1984)
83. V. Perrichon, A. Laachir, G. Bergeret, R. Fréty, L. Tournayan, O. Touret: J. Chem. Soc., Faraday Trans. **90**, 773 (1994)
84. T. Miki, T. Ogawa, M. Haneda, N. Kakuta, A. Ueno, S. Tatieshi, S. Matsuura, M. Sato: J. Phys. Chem. **94**, 6464 (1990)
85. B.K. Cho: J. Catal. **131**, 74 (1991)
86. E. Zintl, U. Croatto: Z. Anorg. Allg. Chem. **242**, 79 (1939)
87. J.R. McBride, K.C. Hass, B. D. Poindexter, W.H. Weber: J. Appl. Phys. **76**, 2435 (1994)
88. P. Vidmar, P. Fornasiero, J. Kašpar, G. Gubitosa, M. Graziani: J. Catal. **171**, 160 (1997)
89. J.Z. Shyu, W.H. Weber, H.S. Gandhi: J. Phys. Chem. **92**, 4964 (1988)
90. M.S. Brogan, T.J. Dines, J.A. Cairns: J. Chem. Soc. Faraday Trans. **90**, 1461 (1994)
91. K. Otto, C.P. Hubbard, W.H. Weber, G.W. Graham: Appl. Catal. B **1**, 317 (1992)
92. S.C. Su, J.N. Carstens, A.T. Bell: J. Catal. **176**, 125 (1998)
93. M. Shelef: Chem. Rev. **95**, 209 (1995)
94. K.C. Taylor: Catal. Rev. Sci. Eng. **35**, 457 (1993)
95. V.I. Pârvulescu, P. Grange, B. Delmon: Catal. Today **46**, 233 (1998)
96. I.E. Wachs, G. Deo, B.M. Weckhuysen, A. Andreini, M.A. Vuurman, M. De Boer, M.D. Amiridis: J. Catal. **161**, 211 (1996)
97. M.A. Vuurman, D.J. Stufkens, A. Oskam, G. Deo, I.E. Wachs: J. Chem. Soc. Faraday Trans. **92**, 3259 (1996)
98. R.P. Cooney, P. Tsai: J. Raman Spectrosc. **9**, 33 (1980)
99. D.D. Saperstein, A.J. Rein: J. Phys. Chem. **81**, 2134 (1977)
100. C. Li, P.C. Stair: Catalysis Today **33**, 353 (1997)
101. C. Li, P.C. Stair: In Proc. 11th International Congress on Catalysis – 40th Anniversary, Studies in Surface Science and Catalysis, Vol. 101, ed. by J.W. Hightower, W.N. Delgass, E. Iglesia, A.T. Bell (Elsevier Science, Amsterdam 1996) p. 881
102. S. Xie, G. Mestl, M.P. Rosynek, J.H. Lunsford: J. Am. Chem. Soc. **119**, 10186 (1997)
103. G. Mestl, M.P. Rosynek, J.H. Lunsford: J. Phys. Chem. B **101**, 9321 (1997)
104. G. Mestl, M.P. Rosynek, J.H. Lunsford: J. Phys. Chem. B **101**, 9329 (1997)
105. G. Mestl, M.P. Rosynek, J.H. Lunsford: J. Phys. Chem. B **102**, 154 (1998)
106. M. Shelef, R.W. McCabe: Catalysis Today, in press (1999)
107. J. Twu, C.J. Chuang, K.I. Chang, C.H. Yang, K.H. Chen: Appl. Catal. B **12**, 309 (1997)
108. D. Spielbauer, G.A.H. Mekhemer, E. Bosch, H. Knözinger: Catal. Lett. **36**, 59 (1996)
109. T. Riemer, D. Spielbauer, M. Hunger, G.A.H. Mekhemer, H. Knözinger: J. Chem. Soc., Chem. Commun. 1181 (1994)
110. W.H. Weber, D. Uy, G.W. Graham, A. Dubkov: Proc. XVII International Conf. on Raman Spectrosc., Beijing, August 2000 (to appear)
111. H.-W. Jen, G.W. Graham, W. Chun, R.W. McCabe, J.-P. Cuif, S.E. Deutsch, O. Touret: Catal. Today **50**, 309 (1999)
112. H. Richter, Z.P. Wang, L. Ley: Solid State Commun. **39**, 625 (1981)
113. K.K. Tiong, P.M. Amirtharaj, F.H. Pollak, D.E. Aspnes: Appl. Phys. Lett. **44**, 122 (1984)

114. I.H. Campbell, P.M. Fauchet: Solid State Commun. **58**, 739 (1986)
115. W.H. Weber, K.C. Hass, J.R. McBride: Phys. Rev. B **48**, 178 (1993)
116. F.H. Pollak: In *Analytical Raman Spectroscopy*, ed. by J.G. Grasselli, B.J. Bulkin (Wiley, New York 1991) Chap. 6, p. 137
117. G.W. Graham, W.H. Weber, C.R. Peters, R. Usmen: J. Catal. **130**, 310 (1991)
118. R.J. Nemanich, S.A. Solin, R.M. Martin: Phys. Rev. B **23**, 6348 (1981)
119. K. Nakamura, M. Fujitsuka, M. Kitajima: Phys. Rev. B **41**, 12260 (1990)
120. J.W. Ager, D.K. Veirs, G.M. Rosenblatt: Phys. Rev. B **43**, 6491 (1991)
121. M. Cardona: In *Light Scattering in Solids II*, Topics Appl. Phys. **50**, 38 (1982)
122. G.W. Ford, W.H. Weber: Phys. Rep. **113**, 197 (1984)
123. W. Richter: Springer Tracts Mod. Phys. **78** (Springer, Berlin, Heidelberg 1976) p. 121 (Note that the formulas given on p. 160 of this article for the incident field agree with our (6.2), but subsequent formulas differ, since this author neglects the multiple reflections of the Raman scattered field.)

VII Historical Perspective of Raman Spectroscopy in Catalysis

Israel E. Wachs

Raman spectroscopy is a very powerful catalyst characterization technique because it can provide fundamental and molecular-level information about catalyst structures, bulk as well as surface, and surface reaction intermediates. Furthermore, Raman spectroscopy is among a handful of characterization techniques that can provide such fundamental information about heterogeneous catalysis under in situ reaction conditions. Consequently Raman spectroscopy has been used to examine essentially every type of catalytic material: bulk metals, supported metals, bulk mixed metal oxides, supported metal oxides, bulk and supported metal sulfides, zeolites and molecular sieves, heteropoly anions and clays.

The combination of fundamental molecular structural information and in situ capabilities has resulted in an explosion of Raman spectroscopy characterization studies in the catalysis literature (see Fig. VII.1). The Raman instrumentation that was available in the 1970's, double monochromators with single-channel photo-multiplier detectors, primarily allowed only strong Raman signals to be detected. Thus, much of the early catalyst characterization studies focused on materials that gave strong Raman signals (e.g.,

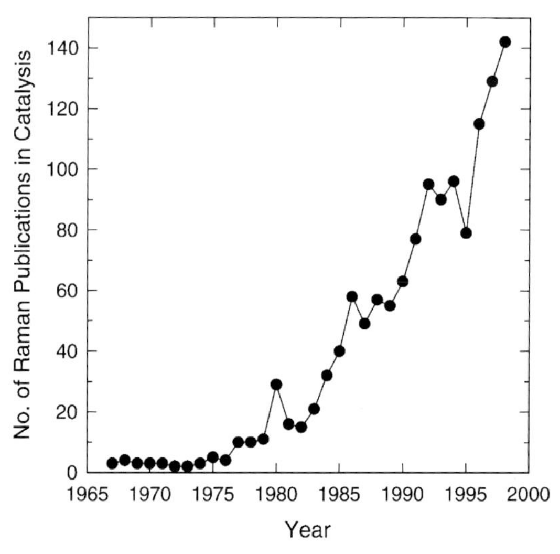

Fig. VII.1. Raman papers on catalysis versus publication year

crystalline mixed metal oxides). However, the Raman structural information was not unique, since the bulk structures of these materials could also be obtained with X-ray diffraction techniques. The first major growth in Raman characterization in catalysis occurred in 1977, when it was found that Raman spectroscopy can detect the very weak signals of materials that were X-ray amorphous (e.g., metal oxide crystallites smaller than ≈ 40 Å and 2D metal oxide overlayers) [1–3]. The weak Raman signals and fluorescence problems, however, dampened much of the initial enthusiasm. The development of optical multi-channel analyzers (OMA's) and triple monochromators in the early 1980's resulted in significantly stronger Raman signals and renewed interest in Raman spectroscopy. This renewed interest was also fueled by the appearance of several in situ Raman studies that demonstrated the dynamic nature of heterogeneous catalysis surfaces [4–8]. The introduction of commercial Raman systems with CCD detectors, notch filters, and single monochromators in the 1990's resulted in even stronger signals and the ability to perform real-time Raman analyses of heterogeneous catalysts under reaction conditions [9].

As catalysis science turns to a molecular-level understanding of the molecular structure – reactivity/selectivity relationships of heterogeneous catalysis, the unique information that can be provided by Raman spectroscopy becomes even more critical (especially with real-time analysis under in situ reaction conditions). Thus, the exponential growth of Raman spectroscopy in catalysis should continue for many more years. Eventually it may result in the molecular design and engineering of heterogeneous catalytic materials and their surfaces from fundamental principles.

References

1. F.R. Brown, L.E. Makovsky: Appl. Spectrosc. **31**, 44 (1977); F.R. Brown, L.E. Makovsky, K.H. Rhee: J. Catal. **50**, 162 and 385 (1977)
2. R. Thomas, J.A. Moulijn, F.P.J.M. Kerkhof: Recl. Trav. Chim. Pays-Bas **96**, M114 (1977)
3. S.L.K.F. de Vries, G.T. Pott: Recl. Trav. Chim. Pays-Bas **96**, M115 (1977)
4. L. Wang, W.K. Hall: J. Catal. **82**, 177 (1983)
5. G.L. Schrader, C.P. Cheng: J. Catal. **80**, 369 (1983)
6. S.S. Chan, I.E. Wachs, L.L. Murrell, L. Wang, W.K. Hall: J. Phys. Chem. **88**, 5831 (1984)
7. J.M. Stencel, L.E. Makovsky, T.A. Sarkus, J. de Vries, R. Thomas, J. Moulijn: J. Catal. **90**, 314 (1984)
8. J.M. Stencel, L.E. Makovsky, J.R. Diehl, T.A. Sarkus: J. Raman Spectrosc. **25**, 282 (1984)
9. G. Mestl, M.P. Rosynek, J.H. Lunsford: J. Phys. Chem. B **101**, 9321 (1997)

7 Raman Scattering Spectroscopy and Analyses of III-V Nitride-Based Materials

Leah Bergman, Mitra Dutta, and Robert J. Nemanich

Abstract. This paper reviews Raman studies of GaN, AlN, InN, and the alloy systems $Ga_xAl_{1-x}N$ and $In_xGa_{1-x}N$. The review focuses on the applications of Raman spectroscopy to material characterizations and phonon dynamics of the wurtzite (WZ) and the zincblende (ZB) polytypes. Among the topics addressed are structure determination and microstructure identification, stress and strain in film-substrate interfaces and in crystallites, phonon-plasmon coupling, and the determination of free carrier density. The issue of phonon mode-type in the alloys is also examined, as well as the topic of alloy order. The subject of the quasi-modes, their angular dispersion, and the dominance of the long-range electrostatic field are addressed. Also, the issues of the isotope effects on the various vibrational modes as well as the phonon lifetimes are discussed.

The potential of UV optoelectronic devices made of III-V nitrides has prompted numerous studies focusing on the properties of this family of materials [1–5]. One of the most attractive properties of the III-V nitrides is the nature of their wide band gaps, which in the wurtzite structure have the values of 1.95 eV for InN, 3.45 eV for GaN, and 6.4 eV for AlN [2,5]. In addition, the superior mechanical and thermal properties of the III-V nitrides, as well as the feasibility of alloying and the material polymorphism, have made this family of materials a unique candidate for synthesizing high-performance UV devices.

Raman spectroscopy has contributed a great deal to the advances in the III-V nitride field, and the key studies leading to these advances are reviewed here. This review focuses in particular on the application of Raman spectroscopy to topics that are of concern to material scientists including structure determination, stress analysis, determination of free carrier concentrations, and material quality characterization. Other issues such as alloy-disorder and alloy-mode, isotopic effects, and aspects of phonon dynamics and phonon lifetimes are also addressed. The implications of these issues for material characterizations are discussed as well.

In Sect. 7.1 a discussion of the underlying principles involved in Raman scattering experiments on wide band-gap semiconductors is presented. Sect. 7.2 focuses on Raman selection rules and studies concerning the structural properties of the III-V nitrides. In particular, Sect. 7.2.1 surveys polarized Raman analyses of the wurtzite and zincblende structures, and Sects. 7.2.2 and 7.2.3 review structure-related studies of GaN, and AlN and InN respectively. Stress effect studies on GaN and AlN films and crystallites are next reviewed in Sect. 7.3. The relevance of the hydrostatic and biaxial Raman stress coefficients to stress analysis of the nitride films and crystals is discussed. Specifically, Sects. 7.3.1 and 7.3.2 focus on stress analysis of GaN and AlN, respectively.

Section 7.4 is devoted to Raman analyses of the LO- and TO-quasi-modes and their frequency dispersion; the implications of the dispersion to film characterization are also presented. Another topic addressed in Sect. 7.5 is the LO phonon-plasmon coupling and the resulting spectral lineshape. The validity of using the lineshape as a measure of the free carrier concentration in GaN is discussed. In Sect. 7.6 studies of the isotopic effects on the phonon dynamics as well as studies concerning the mode lifetimes are summarized. Section 7.7 presents Raman studies of III-V nitride alloy systems. The one-mode and two-mode behavior of phonons in the alloy is described, and experimental and theoretical studies are presented. The topic of alloy disorder and its detectability via Raman spectroscopy is addressed as well. Section 7.8 concludes with comments about directions of future research.

7.1 Experimental Considerations for Raman Scattering of Wide Band-Gap Semiconductors

In its general form the Raman scattering intensity I can be expressed as

$$I(\omega_\mathrm{L}) \propto \omega_\mathrm{S}^4 \left|\hat{\mathbf{e}}_\mathrm{S} \cdot R \cdot \hat{\mathbf{e}}_\mathrm{L}\right|^2 \left|\sum_{\alpha\beta} \frac{1}{(E_\alpha - \hbar\omega_\mathrm{L})(E_\beta - \hbar\omega_\mathrm{S})}\right|^2 , \tag{7.1}$$

where ω_L, ω_S are the incoming and scattered laser frequencies, respectively; E_α and E_β are the energies of intermediate crystal states (to be defined); R is the Raman tensor; and $\hat{\mathbf{e}}_\mathrm{S}$and $\hat{\mathbf{e}}_\mathrm{L}$are the scattered and incident polarization vectors. In (7.1) the first term is due to the dipole transition radiation, the second represents the Raman selection rules, which come about from crystal symmetry considerations, and the last term leads to resonance effects [6].

Energy conservation relates ω_L and ω_S to the phonon frequency ω:

$$\omega = \omega_\mathrm{L} - \omega_\mathrm{S} . \tag{7.2}$$

As the incident laser frequency approaches the energies of the crystal intermediate states, the Raman intensity becomes larger and the signal is said to be resonance-enhanced; an additional enhancement comes from the dipole radiation term. In general, for semiconductors there may be three relevant types of intermediate states: Bloch states which are the conduction-valence bands, exciton states, and in-gap impurity states. Thus as the incident frequency approaches the band-gap frequency, an enhancement of the Raman intensity should be observed. Similarly, resonance may be achieved via the interaction of the incoming light with the exciton states as well as the impurity states.

The band gaps of InN, GaN, and AlN are listed in Table 7.1 [2]. It is evident that if enhancement of the Raman signal of pure GaN is to be achieved, a laser line in the lower energy range of the UV ($\approx$ 2.7–4 eV) should be utilized [7], unless impurity resonance is present [8]. Resonance Raman scattering of AlN and AlGaN alloys, on the other hand, requires laser excitation lines in the deep UV range ($\approx$ 4–7 eV) depending on the alloy composition [9]. The resonance of the Raman signal of InN, due to its narrower band gap, can be achieved via excitation in the visible optical range. For laser excitation energy much above the fundamental band-gap energy of the semiconductor, the light is absorbed and thus only a small volume of the material is probed. In the case of strong absorption the Raman signal would be significantly weakened.

Table 7.1. Band-gap energies in units of eV of GaN, AlN, and InN [2]

Structure	GaN	AlN	InN
Wurtzite	3.39 (200 K)	6.2 (300 K)	1.89 (300 K)
	3.50 (1.6 K)	6.28 (5 K)	
Zincblende	3.2–3.3 (300 K)	5.11 (300 K)	2.2 (300 K)

7.2 Raman Scattering of GaN, AlN, and InN Films and Crystallites

GaN, AlN, and InN based materials can be grown in the wurtzite as well as the zincblende structure, depending mainly on the choice of substrates and growth conditions [10–12]. Unintentional coexistence of both structures in a film thus may occur. This section focuses on the identification of the structure in the films via polarized Raman spectroscopy and analysis.

7.2.1 Raman Tensors and Structure Identification of GaN, AlN, and InN

The III-V nitrides are highly stable in the hexagonal WZ structure and high quality material in that structure has been achieved [13]. However, it has been demonstrated that the growth of films in the less stable cubic ZB structure is feasible as well [14–17]. Due to the polymorph nature of the III-V nitrides, the identification of material structure as well as its purity, i.e., the possibility of polytype material, is an issue. The following section addresses the topic of Raman spectroscopy for structure determination of GaN, AlN, and InN films and crystallites.

The WZ crystal structure belongs to the space group $C_{6\nu}^{4}$, and the group theory analysis predicts the zone-center optical modes $A_1 + 2B_1 + E_1 + 2E_2$ [18]. The A_1, E_1, and the two E_2 modes are Raman active, while the B_1 modes are silent, i.e., forbidden in Raman scattering. Furthermore, the A_1 and E_1 modes are polar: their vibrations polarize the unit cell, which results in the creation of a long-range electrostatic field. The effect of this field manifests itself in the splitting of the A_1, E_1 modes into longitudinal optical (LO) and transverse optical (TO) components, thus creating the A_1(LO,TO) and E_1(LO,TO) modes. Figure 7.1 depicts the scheme of the vibrational modes in the wurtzite structure [19]. The ZB structure belongs to the space group T_d^2 and group theory predicts one Raman active mode of F_2 representation: it is a polar mode which splits into the TO and LO components [18,20].

Tables 7.2 and 7.3 list the experimental Raman frequencies that have been reported for WZ- and ZB-GaN, respectively [14,16,17,21–28]. In Table 7.4 the Raman frequencies of AlN and InN are listed as well. There are small differences in the Raman frequencies presented in Tables 7.2, 7.3 and 7.4, which in most cases are due to differences in material quality. Moreover, as can be observed in the Tables, the Raman frequency of the TO mode of the ZB lies between the frequencies of the E_1(TO) and the A_1(TO) of the WZ; similarly, the frequencies of the LO, E_1(LO), and the A_1(TO) lie

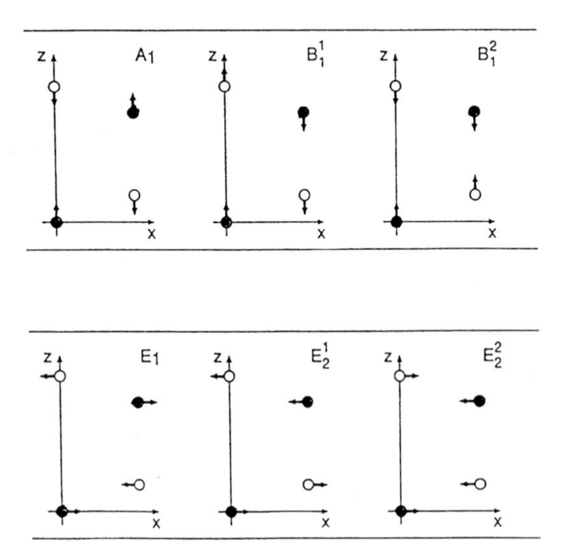

Fig. 7.1. Optical phonon modes in the wurtzite structure. The Raman active modes are the polar A_1 and E_1, and the two non-polar E_2 modes. The B_1 modes are silent [19]

Table 7.2. Raman frequencies (cm^{-1}) of the WZ-GaN modes

E_2^1	145		144	143		144	142
E_2^2	568	568	569	568	570	569	567
A_1(TO)	533	533	532	533		533	530
A_1(LO)					738	735	
E_1(TO)	559	559	560	559		561	558
E_1(LO)				726		743	
Ref.	[21]	[22]	[23]	[25]	[24]	[26]	[27]

Table 7.3. Raman frequencies (cm^{-1}) of the ZB-GaN modes

TO		554	555	555
LO	730	740	740	737
Ref.	[16]	[17]	[28]	[14]

Table 7.4. The Raman frequencies (cm^{-1}) of WZ-AlN and WZ- and ZB-InN

	WZ-AlN				WZ-InN		ZB-AlN
E_2^1	241	252	246	249			655(TO)
E_2^2	660	660	655	657	495	491	
A_1(TO)	607	614	608	610			
A_1(LO)		893	890		596	590	902(LO)
E_1(TO)		673	668	670			
E_1(LO)	924	916		913			
Ref.	[30]	[31]	[27]	[32]	[33]	[34]	[15]

in the same range. The close proximity of the ZB and the WZ phonons has been attributed to the close relation between these two structures [29], and is explained as follows. The WZ structure, consisting of four atoms per primitive cell, can be obtained from the ZB structure, consisting of two atoms per cell, by a rearrangement of the atom-planes that are perpendicular to the (111) axis [29]. Due to this relation the WZ phonon dispersion curve along [0001] may be obtained via folding of the ZB phonon dispersion along the [111] direction, thus doubling the phonon modes in the WZ Brillouin zone as required by the increased number of the atoms per primitive cell.

Figure 7.2 presents a description of the phonon dispersion of the two structures. In the figure, the LOs and the TOs of the ZB structure are plotted

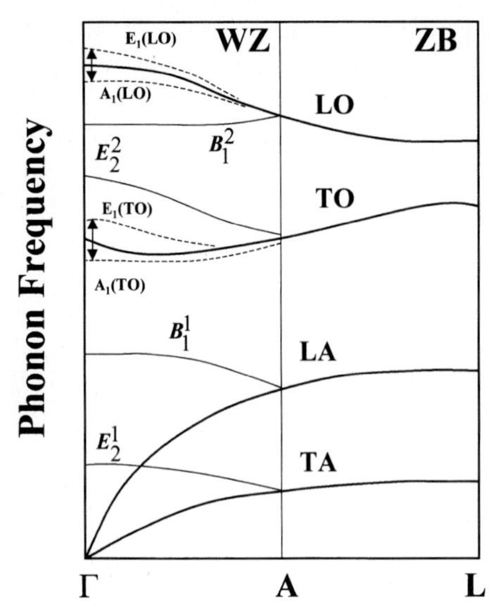

Fig. 7.2. Schematic phonon dispersions for the ZB and the WZ structures. The phonon curves: TA, LA, TO, and LO of the ZB are along the Γ–L [111] direction (*thick lines*). The folding (*thin lines*) of these curves into the WZ Brillouin zone Γ–A [0001] direction creates the: E_2^1 from the TA, B_1^1 from the LA, E_2^2 from TO and B_1^2 from LO. The splitting (*dashed lines*) creates the A_1(TO) and the E_1(TO) from the original TO as well as the A_1(LO) and the E_1(LO) from the original LO

along the Γ – L [111] direction (thick lines); the folding creates the two E_2 branches and the two B_1 branches (thin lines) along the Γ – A [0001] direction of the WZ structure. Moreover, due to the WZ crystal potential, a splitting occurs in which the E_1(TO) and A_1(TO), and the E_1(LO) and A_1(LO) branches (dashed lines) are created from the original TO and LO branches, respectively [29]. Thus, as can be seen in Fig. 7.2, the folding and splitting mechanisms cause the Raman frequencies of the A_1(TO) and E_1(TO) modes of the WZ structure to be in the same range as that of the TO mode of the ZB structure; the same holds for the LO modes.

The identification of the Raman mode may be achieved using the Raman tensors. Equations (7.3) and (7.4) present the Raman tensors for the ZB and WZ structures respectively [18]. The x, y, and z in the equations represent the phonon-polarization direction for each of the polar modes. Each non-zero entry in the matrices represents a coupling of the polarization of the incoming and scattered light to the lattice vibrations that results in an allowed Raman spectral line via the second term of (7.1). Throughout this review Porto notation $a(bc)d$ will be used to describe a Raman geometry: a and d represent the wavevector direction of the incoming and scattered light respectively, while b and c represent their polarization state [35]. Additionally, for each of the polar modes the direction of the phonon propagation with respect to its polarization direction determines whether the observed spectral line is of a longitudinal or a transverse mode. The phonon propagation direction can be ascertained via the wavevector conservation law: $\boldsymbol{k}_{\mathrm{L}} = \boldsymbol{k}_{\mathrm{S}} + \boldsymbol{q}$, where $\boldsymbol{k}_{\mathrm{L}}$ and $\boldsymbol{k}_{\mathrm{S}}$ are the wavevectors of the incoming and the scattered light, respectively, and $\boldsymbol{q}$ is the phonon wavevector.

Thus, for example, the Raman backscattering configuration $z(xy)\bar{z}$ would give rise to the LO mode in a case of ZB-(001) oriented crystallographic plane or to the two E_2 modes in a case of WZ-(0001) plane. The other modes are said to be symmetry forbidden at that configuration and will not appear in the spectra unless a mechanism exists, such as internal reflection due to defects and interfaces, that may cause the forbidden scattering.

$$u_x : \begin{pmatrix} 0 & 0 & 0 \\ 0 & 0 & d \\ 0 & d & 0 \end{pmatrix} \quad u_y : \begin{pmatrix} 0 & 0 & d \\ 0 & 0 & 0 \\ d & 0 & 0 \end{pmatrix} \quad u_z : \begin{pmatrix} 0 & d & 0 \\ d & 0 & 0 \\ 0 & 0 & 0 \end{pmatrix} \tag{7.3}$$

$$A_1(z) : \begin{pmatrix} a & 0 & 0 \\ 0 & a & 0 \\ 0 & 0 & b \end{pmatrix} \quad E_1(x) : \begin{pmatrix} 0 & 0 & c \\ 0 & 0 & 0 \\ c & 0 & 0 \end{pmatrix} \quad E_1(y) : \begin{pmatrix} 0 & 0 & 0 \\ 0 & 0 & c \\ 0 & c & 0 \end{pmatrix}$$

$$E_2 : \begin{pmatrix} f & 0 & 0 \\ 0 & -f & 0 \\ 0 & 0 & 0 \end{pmatrix} \begin{pmatrix} 0 & -f & 0 \\ -f & 0 & 0 \\ 0 & 0 & 0 \end{pmatrix} . \tag{7.4}$$

7.2.2 Wurtzite and Zincblende Phases of GaN

GaN in the WZ structure has been grown in the crystalline form of needles and platelets as well as heteroepitaxial films [1,2,5]. The most widely used substrates are sapphire(0001), 6H – SiC(0001), and ZnO(0001). Their thermal expansion coefficients and lattice constants are approximately matched to GaN. Among the first to identify the Raman modes in WZ-GaN were Manchon et al. [21], Lemos et al. [22], Burns et al. [23] and Cingolani et

Table 7.5. Scattering configurations and observable Raman modes of WZ-GaN(0001) [26]

Raman geometry	Raman mode
$y(xx)\bar{y}$	A_1(TO), E_2^1, E_2^2
$y(zz)\bar{y}$	A_1(TO)
$y(zx)\bar{y}$	E_1(TO)
$z(xy)\bar{z}$	E_2^1, E_2^2
$z(xx)\bar{z}$	A_1(LO), E_2^1, E_2^2

al. [25], followed by Murugkar et al. [24] and Azuhata et al. [26]. Table 7.5 lists the scattering configurations and the observable modes for the WZ-GaN film studied by Azuhata et al. [26]. The 2 μm GaN film in that study was grown on a sapphire(0001) substrate via MOCVD and was capped with a thin layer of InGaN. The sample quality was reported to be comparable to that of a double-heterostructure light-emitting diode [26]. Figures 7.3 and 7.4 show the Raman spectra of that film.

The values of the LO- and the TO-phonon may be used to determine the dielectric constants of a material via the Lyddane–Sachs–Teller relation [36]

$$\omega_{\mathrm{L}}^2 = \frac{\varepsilon_0}{\varepsilon_\infty}\omega_{\mathrm{T}}^2 . \tag{7.5}$$

In this relation ε_0 is the static dielectric constant of the crystal, while ε_∞ is the dielectric constant at optical frequencies, and ω_{L} and ω_{T} are the LO- and the TO-frequencies, respectively. Since GaN is an uniaxial crystal, there are two values for each of the dielectric constants defined as $\varepsilon_\perp$ and $\varepsilon_\parallel$, respectively; the directions given are relative to the c-axis. In that respect, the phonons of E_1 symmetry correspond to the perpendicular Lyddane–Sachs–Teller relation whereas the A_1 phonons to the parallel relation. Moreover, infrared reflectivity measurements have indicated that for GaN $\varepsilon_{\perp\infty} \cong \varepsilon_{\parallel\infty}$, i.e., in the optical frequency range GaN is approximately an isotropic material [21]. Using the Raman frequencies of the LO and TO, Azuhata et al. calculated the perpendicular and the parallel static dielectric constants of the GaN film to be $\varepsilon_{\perp 0} = 9.28$ and $\varepsilon_{\parallel 0} = 10.1$, respectively (for $\varepsilon_{\perp\infty} = \varepsilon_{\parallel\infty} = 5.29$) [26]. These values are in accord with the ones obtained via infrared reflectivity and Kramers–Kronig analyses [37].

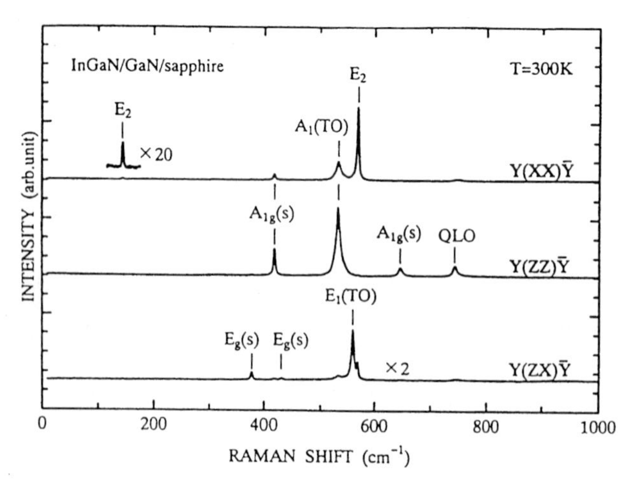

Fig. 7.3. Polarized Raman spectra of GaN film in the y-direction backscattering geometry; the GaN modes are present as well as the sapphire substrate modes [26]

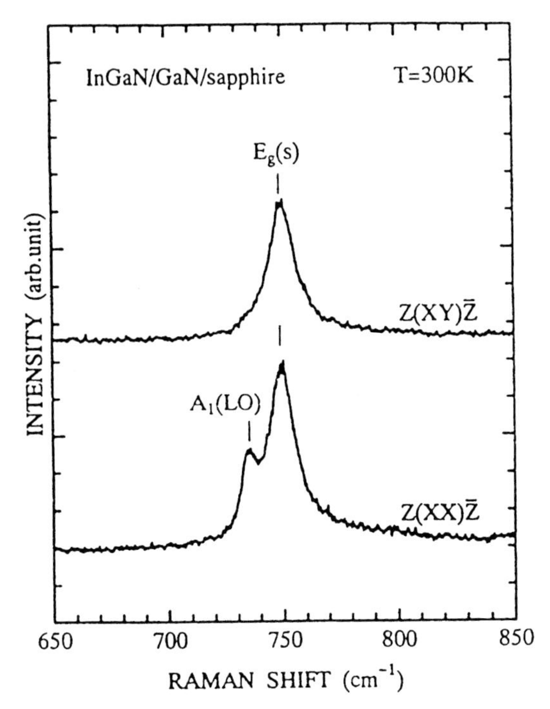

Fig. 7.4. Polarized Raman spectra of GaN for the z-direction back-scattering Raman configuration [26]

In contrast to the extensive studies that have been conducted on WZ-GaN material, the cubic phase is as yet a new and largely unexplored area. Raman and X-ray scattering studies of ZB-GaN films, grown on GaAs(001) substrates by metalorganic vapor phase epitaxy (MOVPE), have been reported by Miyoshi et al. [16]. In their study, the structure of the films was first identified via X-ray diffraction and then confirmed by Raman analysis. The X-ray spectra exhibit the (200) diffraction peak of the cubic face whereas the (0002) diffraction of the WZ symmetry is absent. In order to verify the X-ray findings, Raman spectra were acquired in the $z(xy)\bar{z}$ configuration for which the LO mode for cubic symmetry is allowed from the (001) face. The spectra of that study show a line at $\approx 730\,\mathrm{cm}^{-1}$ which has been attributed by Miyoshi et al. to the LO mode of ZB-GaN [16]. Moreover, from the X-ray data the authors calculated the lattice constant of ZB-GaN to be 4.5 Å, which agrees well with their theoretical calculation. In light of that, Miyoshi et al. concluded that despite the 20% lattice mismatch with the GaAs substrate, the ZB-GaN films ($\approx 0.4\,\mu\mathrm{m}$ thick) are almost strain-free due to a relaxation mechanism occurring at the early stages of the growth [16].

Phonon characteristics of cubic, hexagonal, and mixed phase GaN films grown via molecular beam epitaxy (MBE) on GaAs(001) substrates were studied by Giehler et al. [17]. Each film in this study was grown under different conditions to invoke the growth of the different structure types. The structural properties were analyzed by high-energy electron diffraction and X-ray scattering, and the phonon characteristics were investigated utilizing infrared transmission and Raman spectroscopy. The structural analysis conducted by Giehler et al. revealed that the sample grown under Ga-rich conditions (sample A) consists of nonepitaxial hexagonal columns whose c-axes are aligned parallel to the [001] direction of the substrate. The film grown under N-rich conditions (sample B) exhibits a phase mixture, while the film grown under near-stoichiometric conditions (sample C) was mainly cubic.

The infrared transmission spectra of sample A consist of lines at 558 and 735 cm^{-1}, which have been attributed by Giehler et al. to the E_1(TO) and A_1(LO) phonons, respectively, of the WZ-GaN. The spectra of sample C reveals transmittance features at 552 and 739 cm^{-1} that were assigned to the TO and LO phonons, respectively, of the ZB-GaN [17]. The spectra in the TO frequency range of the mixed sample exhibit a broad band centered approximately between the E_1(TO) and the TO modes; similarly, the band in the LO range is broad and centered between the A_1(LO) and the LO modes. The Raman spectra of the three samples are presented in Fig. 7.5; the spectra were acquired in the $z(xy)\bar{z}$ configuration. The spectra of sample A exhibit the E_2^2 mode at 569 cm^{-1} and a broad band near 680 cm^{-1}, which has been

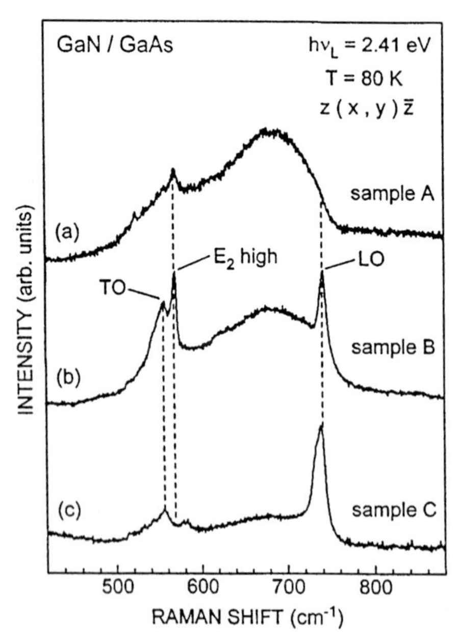

Fig. 7.5. Depolarized Raman spectra of GaN films grown on GaAs substrates. Sample A is mainly WZ, sample C is mainly ZB, and sample B is a mixed phase structure [17]

assigned by the authors to a disordered state. The spectra of sample C consist of the allowed LO mode at 740 cm^{-1} and the symmetry forbidden TO mode at 554 cm^{-1}. The forbidden scattering has been explained by Giehler et al. to occur via multiple reflections inside the film. The spectra of the mixed-phase film exhibit the E_2^2 of the WZ structure, the LO and the TO of the ZB, and a broad band due to a disorder-activated scattering [17].

Comparative Raman studies of cubic and hexagonal GaN have been reported also by Tabata et al [14]. In that study, both WZ and ZB films were grown on GaAs(001) substrates via MBE. The nucleation of the two structures was achieved by varying the Ga-to-N flux ratio in the initial stage of the growth: the growth of the ZB film was initiated using N-rich conditions, and that of the WZ phase was obtained under a slight excess of Ga [14]. The X-ray spectra of the sample grown under Ga-rich condition (sample A) exhibit the (0002) and the (0004) diffraction peaks associated with the WZ-GaN, whereas the spectra of the sample grown under N-rich condition (sample B) exhibit the (002) peak of the ZB-GaN. From these findings, the authors concluded that sample A has a WZ structure with the surface-normal parallel to the c-axis, and sample B consists of ZB structure with the surface-normal parallel to the cubic axis [14].

The Raman spectra of samples A and B are presented in Figs. 7.6 and 7.7, respectively. The spectra were acquired in nearly backscattering geometry

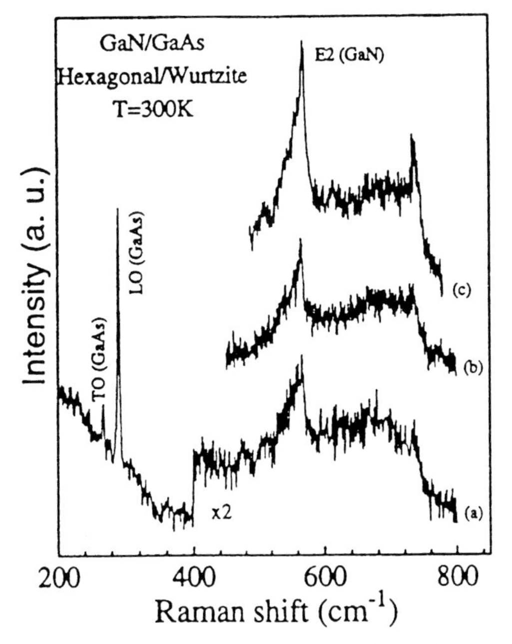

Fig. 7.6. Raman spectra of WZ-GaN film at three incident laser wavelengths: (a) 514.5 nm, (b) 488 nm, and (c) 457.9 nm [14]

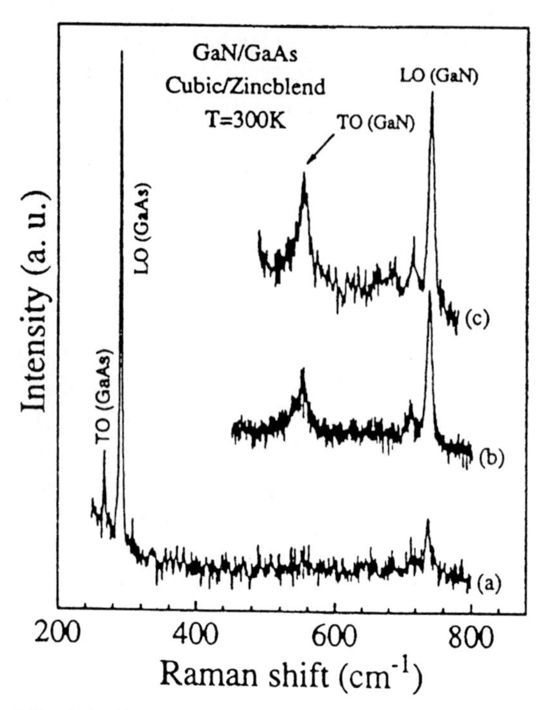

Fig. 7.7. Raman spectra of the ZB-GaN film at the three incident laser wavelengths as in Fig. 7.6 [14]

with polarized incident light and unanalyzed scattered light. Moreover, the spectra in Tabata's study were acquired at three different incident wavelengths in order to enhance the Raman signal [14]. Figure 7.6 exhibits the allowed E_2^2 mode at $571\,\text{cm}^{-1}$ as well as the A_1(LO) mode at $737\,\text{cm}^{-1}$, which is also allowed in that configuration. The Raman spectra of the ZB-GaN in Fig. 7.7 exhibit the allowed LO mode at $741\,\text{cm}^{-1}$, and the forbidden TO mode appears as well at $555\,\text{cm}^{-1}$. The appearance of the TO mode was attributed by Tabata et al. to the presence of a short-range perturbation in the film, which relaxes the Raman selection rules [14].

7.2.3 Wurtzite and Zincblende Structure of AlN and InN

In the following, a review of Raman scattering studies in AlN and InN is presented. AlN and InN, like GaN, in the stable state have a wurtzite crystal structure and belong to the space group $C_{6\nu}^4$. The cubic zincblende phases of AlN and InN can be grown as well; however, these materials have been studied much less than the wurtzite polytype. The Raman selection rules leading to mode assignments and polytype identification are the same as those for GaN, as described in Sect. 7.2.1.

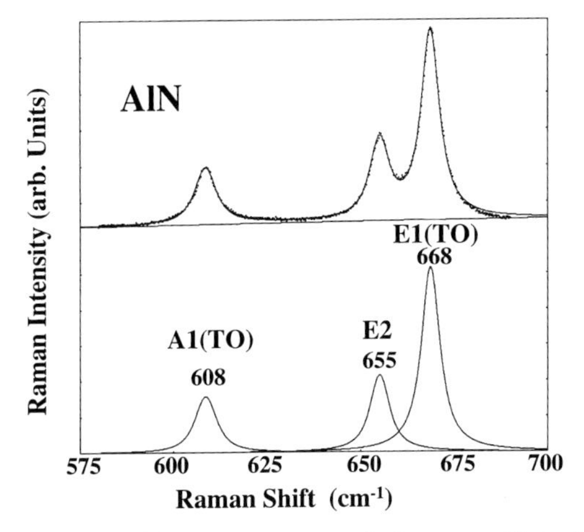

Fig. 7.8. Raman spectra and the curve fit to the spectra of AlN crystallite [27]

Figure 7.8 shows typical Raman spectra of a WZ-AlN crystallite grown via the sublimation method [27]. The spectra were acquired in a backscattering geometry from the *a*-face of the crystallite, and the Raman frequencies are listed in Table 7.4. Figures 7.9 and 7.10 present the Raman spectra of a ZB-AlN film along with that of ZB-AlGaN films; the symmetry-allowed LO and the forbidden TO modes of the cubic ZB are both present in the spectra [15].

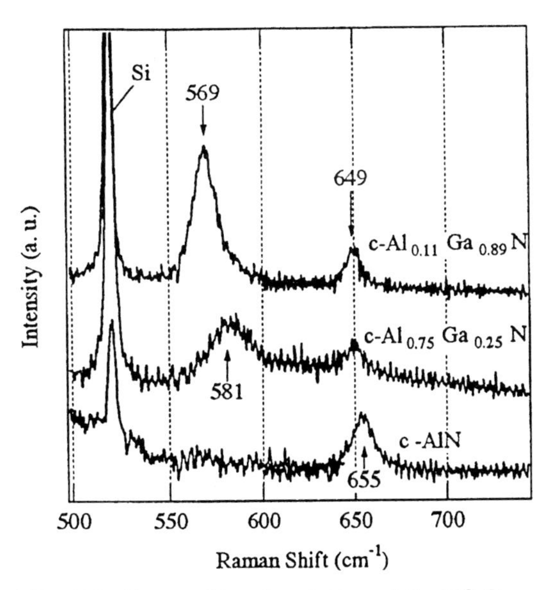

Fig. 7.9. Compositional variation of the TO-Raman mode of ZB-AlGaN alloy. The phonon frequency of the ZB-AlN is $655\,\mathrm{cm}^{-1}$ [15]

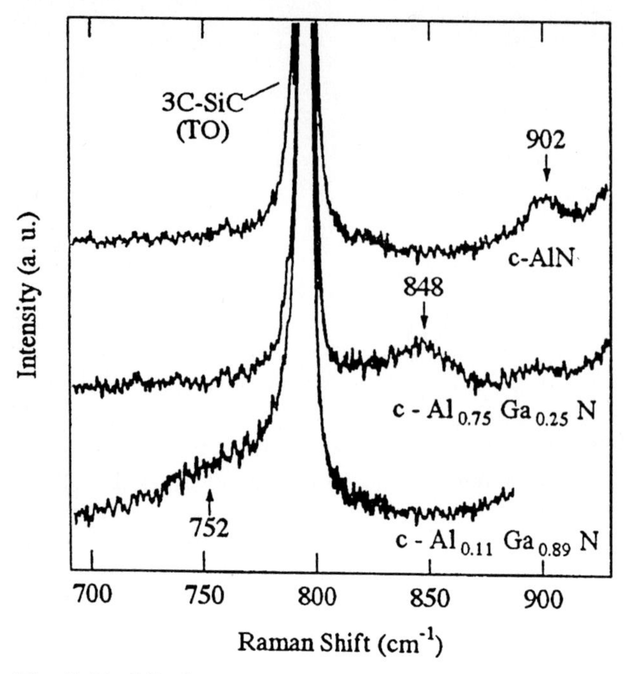

Fig. 7.10. LO phonon spectra of ZB-AlGaN alloy. The LO mode of the ZB-AlN is observed at $902\,\mathrm{cm}^{-1}$ [15]

These cubic films were grown by Harima et al. via the MBE method on cubic SiC/Si substrates, and the cubic phase was confirmed by X-ray diffraction as well to be the major structure [15].

Due to the difficulties in achieving reasonably good quality InN films, information on its material properties is as yet scarce. Raman scattering studies of InN films grown on (0001) sapphire substrates via metalorganic vapor phase epitaxy were first reported by Kwon et al. [33]. X-ray diffraction indicated that the structure of the InN films was wurtzite. Identification of the Raman modes was carried out via the Raman selection rules in a backscattering geometry; the modes are listed in Table 7.4. The authors reported a linewidth of $20\,\mathrm{cm}^{-1}$ for the E_2^2 mode, a value that is larger compared to the E_2^2 linewidths of WZ-GaN and AlN, which usually range from 3 to $8\,\mathrm{cm}^{-1}$. Lee et al. investigated the unusual line broadening of the InN Raman modes as a function of growth temperature [34]. These films were grown at temperature ranges of 325–600 °C via the same method used by Kwon et al. It was found that in the elevated temperature range (450 °C and above), the E_2^2 linewidth is $\approx 6\,\mathrm{cm}^{-1}$ while in the lower range the linewidth is $\approx 70\,\mathrm{cm}^{-1}$; the broadening of the spectral lines was accompanied by a frequency shift as well. Lee et al. suggested that this significant line broadening reflects the coexistent of cubic and amorphous phases in the WZ structure of the InN films [34].

7.3 Stress Analysis and Substrate Issues for Epitaxial Growth

Among the important factors leading to the synthesis of high quality films is the availability of matching substrates. In order to minimize the internal stress, the substrate under consideration should have lattice parameters and thermal expansion coefficients similar to these of the film. Table 7.6 lists some of the substrates most commonly used in the growth of III-V nitride films, as well as their lattice parameters and thermal expansion coefficients [2].

One of the most informative methods of measuring stress in materials is Raman spectroscopy [19,38,39]; however, it requires an a-priori knowledge of the Raman pressure coefficients that relate the peak position to the stress. The following reviews some of the Raman studies concerning the determination of the pressure coefficients and the stress state in GaN and AlN.

Table 7.6. Material properties for the GaN- and AlN-substrate system [2]

Material	Lattice parameters (Å)	Coefficients of thermal expansion (10^{-6} /K)
6H – SiC	$a = 3.080$	4.2
	$c = 15.12$	4.68
Sapphire	$a = 4.758$	7.5
	$c = 12.99$	8.5
ZnO	$a = 3.252$	2.9
	$c = 5.213$	4.75
WZ-AlN	$a = 3.112$	4.2
	$c = 4.982$	5.3
WZ-GaN	$a = 3.189$	5.59
	$c = 5.185$	3.17

7.3.1 Stress Analysis of GaN Films

A detailed study of the effect of pressure on the crystalline WZ-GaN Raman spectral lines was reported by Perlin et al. [38]. The applied pressure was hydrostatic in the range 0 to 50 GPa. The pressure dependence of the Raman frequencies of the A_1(TO), E_1(TO), E_2^2, and E_2^1 modes was shown to follow the quadratic relation [38]:

$$\omega(\mathrm{cm}^{-1}) = \omega_0 + \sigma_1 P + \sigma_2 P^2 \ , \tag{7.6}$$

where P is the applied pressure in units of GPa, ω_0 is the Raman frequency at zero applied pressure, and σ_1 and σ_2 are the first and second order pressure coefficients, respectively. The hydrostatic pressure coefficients derived from first principal density-functional calculations by Gorczyca et al. [19] were found to agree well with the experimental results [38]. Table 7.7 summarizes the set of parameters for the various Raman modes.

Table 7.7. The hydrostatic pressure coefficients of WZ-GaN, σ_1 and σ_2, in units of cm^{-1}/GPa and $\mathrm{cm}^{-1}/(\mathrm{GPa})^2$, respectively, as obtained from theory [19] and experiment [38]

Raman mode	Theory		Experiment	
	σ_1	σ_2	σ_1	σ_2
E_1(TO)	4.10	0.013	3.68	−0.0078
E_2^1	−0.15	0.006	−0.25	−0.0017
E_2^2	4.46	0.018	4.17	−0.0136
A_1(TO)	4.08	0.024	4.06	−0.0127

The stress-strain relation for an isotropic material under hydrostatic pressure is [40,41]:

$$\varepsilon = E^{-1}\sigma(1 - 2\nu) \ , \tag{7.7}$$

where ε is the strain, E the Young's modulus, σ the stress, and ν the Poisson ratio. However, for thin epitaxial films the stress is more likely to be two dimensional due to film-substrate lattice and thermal mismatches (see Table 7.6). For the case of biaxial stress, σ_a, a strain is induced in the basal plane, ε_a, and along the c-direction, ε_c; the isotropic constituent equations are [41]:

$$\varepsilon_a = E^{-1}\sigma_a(1 - 2\nu) \ , \tag{7.8a}$$

$$\varepsilon_c = -E^{-1}\sigma_a 2\nu \ . \tag{7.8b}$$

For thin epitaxial films, the biaxial stress approximations in (7.8a,b) may be used to describe the stress state of a thin film if the stress contribution from point defects is negligible [40].

Kisielowski et al. have analyzed the stress state of GaN films grown on the c-face of SiC and sapphire [40]. The authors analyzed the stress in the films in terms of the biaxial isotropic approach described in (7.8). It was shown that, due to point defects, an additional hydrostatic component has to be superimposed to the biaxial component in order to fully describe the stress state of the films. Moreover, Kisielowski et al. determined that a biaxial stress of 1 GPa would shift the Raman peak position of the E_2^2 mode by $\approx 4.2\,\mathrm{cm}^{-1}$, a result consistent with the values listed in Table 7.7.

Strain effect studies in epitaxial GaN grown on AlN-buffered Si(111) have been reported by Meng et al. [42]. One of the objectives of the study was the determination of the Raman stress coefficient for the biaxial stress state, taking into account the hexagonal anisotropy. An outline of their method is presented here. The films were grown via RF-glow discharge reactive magnetron sputtering at various RF input powers. Raman and X-ray spectroscopy established that the film structure is WZ with an (0001) orientation. In their analysis, Meng et al. assumed a biaxial stress state and took into consideration the hexagonal anisotropy; the constituent relation in that case is [42]:

$$\sigma_a = \frac{2C_{13}^2 - C_{33}(C_{11} + C_{12})}{2C_{13}}\varepsilon_c \,. \tag{7.9}$$

Using the elastic constants C_{ij} for WZ-GaN in (7.9), the authors obtained $\sigma_a = -202\varepsilon_c$. The values of the vertical strain ε_c determined via X-ray measurements were correlated to the RF input power and to the E_2^2 Raman peak position; in both cases a direct correlation was found. The vertical strain-Raman frequency relation found for the E_2^2 mode was $\omega(\varepsilon_c) = 561 + 701\varepsilon_c$, which in conjunction with (7.9) yields the relation between the biaxial stress and the Raman frequency to be $\omega(\sigma_a) = 561 + 4.4\sigma_a$. Thus the biaxial pressure coefficient found in the study of Meng et al. ($4.4\,\mathrm{cm}^{-1}/\mathrm{GPa}$) [42] is comparable to the hydrostatic pressure coefficient [38].

Other studies of stress-related phenomena have been reported [43–45]; a brief summary is given below. Thermal stress effects on GaN films of differing thickness grown on the c-plane of sapphire substrates were investigated by Kozawa et al [45]. As can be seen in Table 7.6, a significant difference exists between the thermal expansion coefficients of GaN and those of sapphire, which produces a compressive stress in the film. In the studies of Kozawa et al., the magnitude of the biaxial compressive stress in each of the GaN films was obtained via the curvature wafer bending method; the obtained values were correlated to the peak positions of the E_2^2 Raman mode. It was found that the Raman pressure coefficient is $6.2\,\mathrm{cm}^{-1}/\mathrm{GPa}$, a value somewhat greater than the one listed in Table 7.7. Additionally, the authors demonstrated that the compressive stress decreases as the thickness of the film increases: for $\approx 5\,\mu\mathrm{m}$ and $50\,\mu\mathrm{m}$ films the Raman peak position was found to be at 569 and $567.5\,\mathrm{cm}^{-1}$, respectively [45].

Rieger et al. investigated the influence of the AlN buffer layer, when deposited on the c-plane of sapphire substrates, on the stress state of the GaN films [43]. In order to estimate the stress magnitude, they analyzed the Raman peak position of the E_2^2 mode as a function of the AlN layer thickness.

Rieger et al. observed a pronounced reduction of the compressive stress in the GaN films with increasing AlN buffer layer thickness. According to the authors, an AlN layer thicker than 200 nm eliminates the compressive stress completely [43]. Depth profiles of the strain in a 220-µm thick GaN film grown on sapphire were investigated by Siegle et al. utilizing Raman spectroscopy PL and CL imaging [44]. The depth profile of the stress was determined from the peak position of the E_2^2 mode. The authors found that the stress decreases nearly exponentially with increasing distance from the substrate; the film was found to be relaxed at a distance $\approx 35\,\mu$m away from the substrate.

7.3.2 Stress Analysis in WZ-AlN

The influence of hydrostatic pressure on the Raman frequencies of WZ-AlN has been investigated by various groups [19,30,46]. The Raman pressure coefficients for the AlN modes, defined in (7.6), are listed in Table 7.8.

Table 7.8. The hydrostatic pressure coefficients of WZ-AlN, σ_1 and σ_2, in units of cm^{-1}/GPa and cm^{-1}/(GPa)^2 respectively as obtained from theory [19] and experiment [30]

Raman mode	Theory		Experiment	
	σ_1	σ_2	σ_1	σ_2
E_1(TO)	4.36	0.059		
E_2^1	−0.29	0.022		
E_2^2	4.79	0.063	3.99	0.035
A_1(TO)	4.29	0.019	4.63	−0.01
E_1(LO)			1.67	0.27

Figures 7.11 and 7.12 [19] show the functional behavior of the Raman modes under pressure; the dots represent the experimental data of [30] while the lines represent the model calculation [19]. One of the most intriguing aspects of these results is the anomalous behavior of the E_2^1 mode of AlN as well as of GaN; unlike the other modes it decreases to lower frequencies with increasing pressure. This tendency is better expressed in terms of the Grüneisen parameterγ_i [47]:

$$\gamma_i = \frac{B_0}{\omega_i}\left(\frac{\mathrm{d}\omega_i}{\mathrm{d}P}\right) , \tag{7.10}$$

where the ω_i's are the given phonon modes, B_0 is the bulk modulus, and P is the pressure. Thus all the AlN and GaN modes have positive γ values, excepting the E_2^1 mode, which has a negative γ value. Similar behavior has

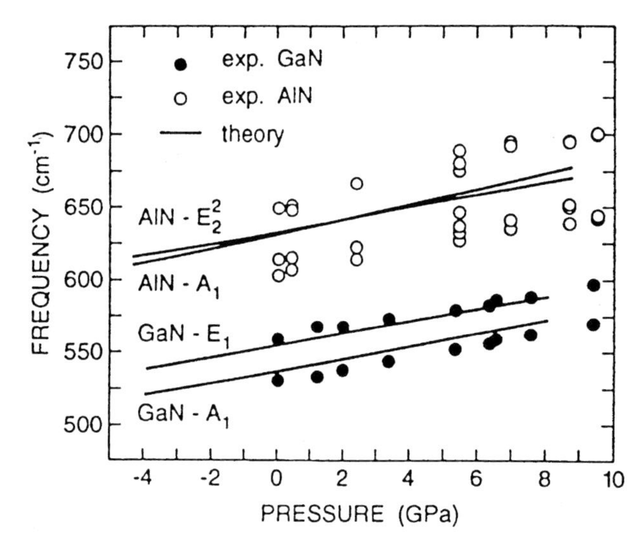

Fig. 7.11. Model calculations (*lines*) of the high frequency phonon of WZ-AlN and GaN under hydrostatic pressure. The *dots* represent experimental data from [30,38] [19]

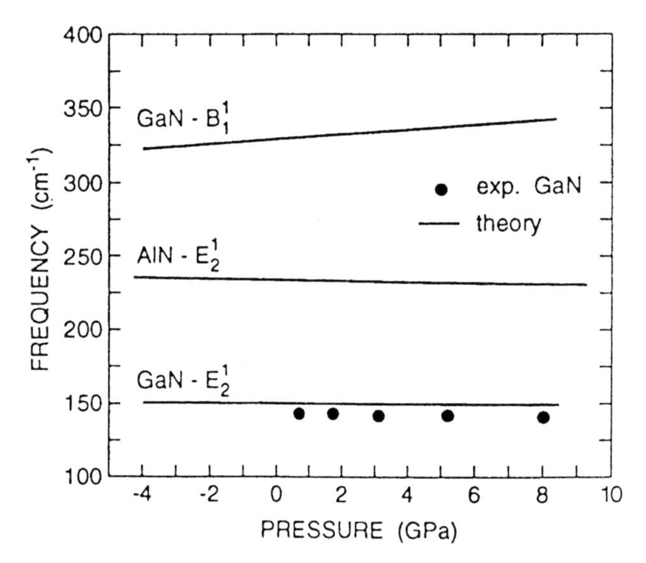

Fig. 7.12. Model calculations (*lines*) of the low frequency phonons of WZ-AlN and GaN under hydrostatic pressure. The *dots* represent experimental data from [30,38] [19]

been found in CdS and ZnO, both of which have a WZ structure [47]. It has been suggested that the negative γ value may indicate a softening of the lattice to a particular mode under compression, preceding a first-order transition [19,47]. Moreover, it has been established that WZ-ZnO and -CdS undergo a phase transition to the NaCl structure [47]; the same phase transition has been suggested to occur in GaN crystallites at $\approx 45\,\mathrm{GPa}$ [38]. However, the reason for the softening of the E_2^1 mode in the wurtzite structure materials needs further investigation.

7.4 Raman Analysis of the Quasi-Modes in AlN

The theory developed by Loudon formulates that in uniaxial materials the polar phonon characteristics may be affected via two interaction mechanisms: one due to the long range electrostatic field, and the other due to the short range field which exhibits the anisotropy of the vibrational force constants [18,48]. The phonon dynamics and thus the Raman spectra depend on which of the two mechanisms is the dominant interaction. Figure 7.13 depicts the Raman frequency scheme and the direction of phonon vibrations for both interactions [49].

For the case where the long range electrostatic field is the dominant mechanism, the interaction of the polar phonons with the long range electrostatic field may result in a significant frequency separation between the group of the TO phonons relative to that of the LO phonons. Moreover, the TO phonons belonging to different symmetry are grouped together in a relatively narrow frequency range; the same holds for the LO phonons. One consequence of the dominant electrostatic field interaction is that under certain propagation and polarization conditions, phonons of mixed A_1 and E_1 symmetry character exist and can be observed in the Raman spectra [48]. These mixed symmetry modes are termed quasi-LO and quasi-TO modes. The frequencies

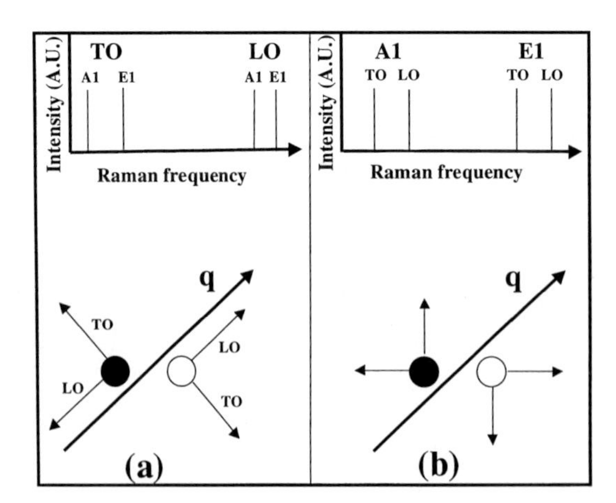

Fig. 7.13. Raman frequency scheme for (**a**) the case when the long range electrostatic field is the dominant interaction, and (**b**) when the anisotropy of the short range interaction dominates [49]

of such quasi modes are predicted by the theory to be between the values of the pure A_1 and E_1 modes for each of the LO and the TO bands. Alternatively, for the case where the short range interatomic forces are dominant, the LO-TO splitting will be small and in this case the TO and LO mode of each symmetry group will occur in a relatively narrow band.

In AlN the E_1(TO) and A_1(TO) Raman frequencies are grouped together within about 60 cm^{-1}, the E_1(LO) and A_1(LO) within 20 cm^{-1}, and the LO-TO group-splitting is ≈ 220 cm^{-1}. The frequency scheme of AlN thus implies the dominance of the long-range electrostatic force interaction. The mode-mixing in AlN may occur if the propagation direction ($\boldsymbol{q}$-vector) or polarization of the quasi-polar phonons lie in the plane that spans between the c and the a_1 (or a_2) crystallographic axes [50]. This plane is referred as to the mixing plane. For example, the pure A_1 phonon has a c-direction polarization while the pure E_1 phonon is polarized in the basal plane; thus a quasi-phonon with the $\boldsymbol{q}$-vector between the c and the a_1 axes would exhibit a mixed polarization of A_1-E_1 symmetry. When the $\boldsymbol{q}$-vector lies along the crystallographic axes, only pure phonons are observed in the spectra. A detailed analysis of the Raman configurations that enable the observation of the quasi modes in WZ materials is given in [50] and [49].

Raman studies of quasi modes in AlN have been reported by Filippidis et al. [32] and Bergman et al. [49]. In these studies the $\boldsymbol{q}$-vector direction

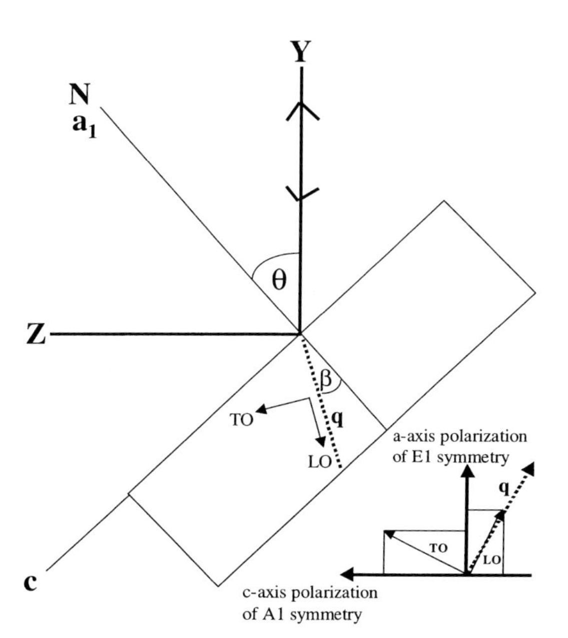

Fig. 7.14. Geometry of the Raman experiment setup for the observation of the quasi-modes. The spectra were acquired in a backscattering geometry from the Y axis, θ is the rotation angle, and β is the angle of the phonon propagation [49]

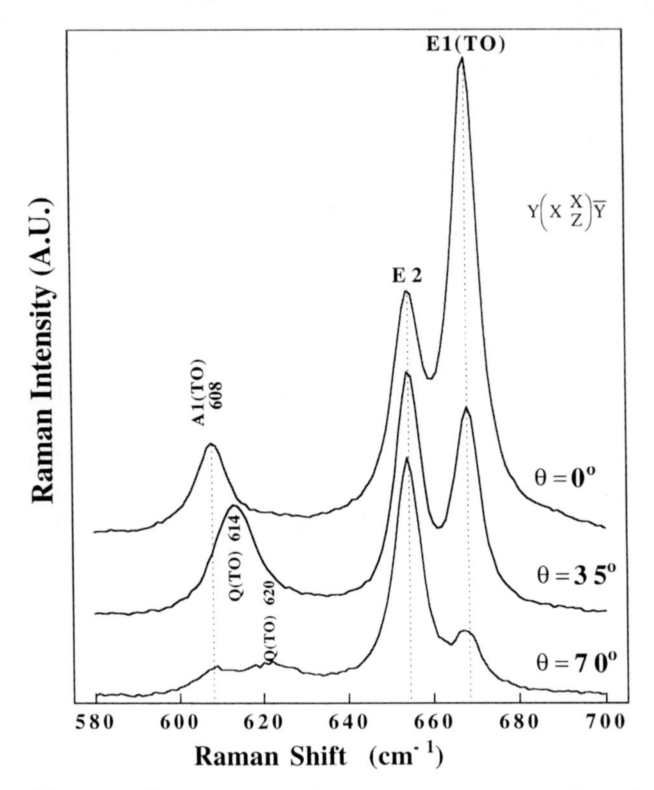

Fig. 7.15. Raman spectra in various scattering configurations that enable the observation of the quasi-TO modes in AlN. The pure-modes: E_2^2, and E_1(TO) are present as well [49]

was achieved by rotating the crystallite by an angle θ relative to a fixed laboratory coordinate system, as depicted in Fig. 7.14. Figures 7.15 and 7.16 present Raman spectra for various rotations in a scattering configuration that enables the quasi-TO and quasi-LO to be observable [49]. Figure 7.17 shows the behavior of the quasi-modes as a function of the phonon propagation angle β, where β is related to θ via Snell's law and a small anisotropy is assumed. In Fig. 7.17 the points represent the experimental results while the lines represent Loudon's model [48,50]:

$$\omega^2_{\mathrm{Q(TO)}} = \omega^2_{\mathrm{E1(TO)}} \cos^2(90 - \beta) + \omega^2_{\mathrm{A1(TO)}} \sin^2(90 - \beta) \,, \qquad (7.11\mathrm{a})$$

$$\omega^2_{\mathrm{Q(LO)}} = \omega^2_{\mathrm{A1(LO)}} \cos^2(90 - \beta) + \omega^2_{\mathrm{E1(LO)}} \sin^2(90 - \beta) \,. \qquad (7.11\mathrm{b})$$

As was discussed in [49], one important implication of these results involves the accurate determination of the Raman frequency acquired from AlN thin films. In such films, due to their µm-size dimension along one of the crystallographic axes, the determination of the various Raman frequencies requires the data to be acquired at grazing angle geometry. Such experimental setup,

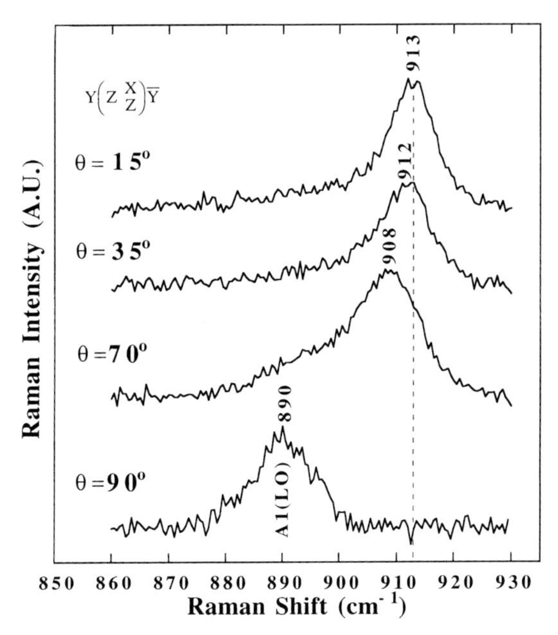

Fig. 7.16. Raman spectra of the quasi-LO modes of an AlN crystallite [49]

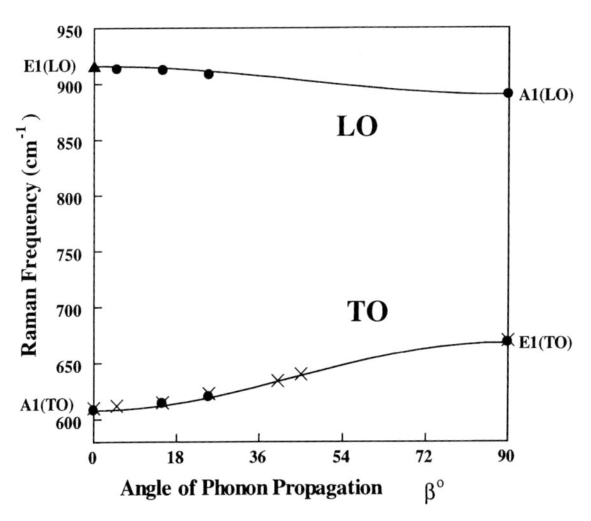

Fig. 7.17. Angular dispersion of the quasi-LO and quasi-TO modes. The *lines* represent Loudon's model; the *dots* and *X's* are the experimental results from [49] and [32], respectively [49]

which deviates by an angle from the crystallographic coordinate system, may result in a shifting of the observed Raman frequency. Due to the strong angular dispersion of the quasi-TO mode frequency (see Fig. 7.17), a disparity in the TO Raman frequency values of thin films is expected to be observed; less disparity is expected for the LO mode. A similar effect in which the frequency is shifted from its pure value may also occur in films grown at off-crystallographic directions. In measuring the stress state of the films, the possible additional shift due to the electrostatic coupling needs to be considered.

7.5 Phonon–Plasmon Interaction in GaN Films and Crystallites

Raman spectroscopy has proven to be a useful tool in analyzing the effect of free electrons on the lattice dynamics of GaN, which tends to be an n-type material as grown [4]. A free carrier concentration in GaN may be achieved via intentional n-type doping or via the growth method [5,51]. In general, when an appreciable free carrier concentration is present in a polar-semiconductor, a coupling of the LO-phonons to the plasma oscillations of the free carriers (plasmons) may occur [52]. The phonon–plasmon interaction results in a characteristic Raman scattering that may yield information on the free carrier density of a given sample [51,53–58]. This section examines the topic of Raman phonon–plasmon coupled modes in GaN.

The determination of the free carrier concentration via Raman spectroscopy is often more advantageous than that via electrical measurements since no contacts are required. The plasma frequency ω_p at wavevector $\boldsymbol{q} = 0$ is related to the free carrier concentration n through the relation [36]:

$$\omega_\mathrm{p}^2 = \frac{4\pi n e^2}{m^* \varepsilon_\infty}\,, \tag{7.12}$$

where ε_∞ is the high frequency dielectric constant, e is the charge, and m^* is the effective mass of the free carriers. Moreover, in the case when the plasma oscillations are not overdamped the frequencies of the coupled plasmon–LO-phonon may be approximated by [59]:

$$\omega_\pm^2 = \frac{1}{2}\left\{\left(\omega_\mathrm{LO}^2 + \omega_\mathrm{p}^2\right) \pm \left[\left(\omega_\mathrm{LO}^2 + \omega_\mathrm{p}^2\right)^2 - 4\omega_\mathrm{p}^2\omega_\mathrm{TO}^2\right]^{1/2}\right\}\,. \tag{7.13}$$

Therefore the Raman spectra in principle should exhibit two spectral lines corresponding to an upper branch coupling ω_+ and a lower branch ω_- (also known as L_+ and L_-) whose frequencies depend on the free carrier concentration. No LO-mode is expected in the spectra. The two coupled frequencies have been observed in the Raman spectra of n-type GaAs as well as other III-V semiconductors of the ZB structure [20].

In the case of overdamped plasmons (7.13) is not applicable; the spectra cannot be analyzed in terms of the ω_+ and ω_- coupled modes. However, the case of large damping of the plasma oscillations in 6H–SiC (doped with nitrogen $\sim 10^{19}\,\mathrm{cm}^{-3}$) has been analyzed and modeled by Klein et al. [60]. In their study the Raman spectra of the SiC do not exhibit the ω_+ and ω_- branches. Instead, the influence of the coupling manifests itself in the lineshape of the LO-mode: it is asymmetrically broadened and slightly shifted toward the frequency range where the ω_+ was supposed to be (in the case of small damping). Moreover, Klein et al. demonstrated that the line broadening as well as intensity reduction is directly correlated to the increase of carrier concentrations. The influence of damping effects on the Raman scattering efficiency of the coupled modes has also been investigated by Hon et al. [61] and by Irmer et al. [62] in the case of n-type GaP. The Raman spectra were found to exhibit characteristics similar to those of n-type SiC. The model developed to explain the effect of the large damped plasmons on the coupled modes can be expressed as [60–62]:

$$I(\omega) \approx [1, C, C^2]\mathrm{Im}(-1/\varepsilon) \tag{7.14a}$$

$$\varepsilon(\omega) = \varepsilon(\infty)\left[1 + \frac{\omega_\mathrm{L}^2 - \omega_\mathrm{T}^2}{\omega_\mathrm{T}^2 - \omega^2 - \mathrm{i}\gamma\omega} - \frac{\omega_\mathrm{p}^2}{\omega^2 + \mathrm{i}\gamma_\mathrm{p}\omega}\right], \tag{7.14b}$$

where $I(\omega)$ is the Raman cross section, $\varepsilon(\omega)$ is the dielectric function, and $[1, C, C^2]$ is the interference factor expressed in term of the Faust–Henry coefficient C [61]. In (7.14b) ω_L, ω_T and ω_p are the frequencies of the longitudinal phonon, the transverse phonon, and the plasmons, respectively. Additionally, γ and γ_p are the damping constants of the phonons and the plasmons, respectively. Equation (7.14a), when fully expanded, predicts the Raman lineshape of the phonon–plasmon coupled modes.

In a manner similar to SiC the plasma oscillations in GaN are considered to be overdamped. Kozawa et al. investigated the A_1(LO) phonon–plasmon coupled mode in WZ-GaN film as a function of a relatively low level of Si-dopant concentration [51]. The dopant concentration in these samples ranged from about 1×10^{16} to $2 \times 10^{18}\,\mathrm{cm}^{-3}$. Figure 7.18 depicts the Raman spectra of the A_1(LO) mode at several carrier densities; here the Raman band shifts towards the high frequency side as well as broadens and weakens as the carrier concentration increases. The spectra of the A_1(LO) coupled mode as well as the model predicted by the lineshape (7.14) are presented in Fig. 7.19. The inset to the figure lists the values of the fitting parameters, among which are the plasma damping constant $\gamma_\mathrm{p} = 400\,\mathrm{cm}^{-1}$ and the plasma frequency $\omega_\mathrm{p} = 119\,\mathrm{cm}^{-1}$, the latter from which the carrier concentration may be evaluated via (7.12). Figure 7.20 shows the carrier concentration n calculated from the Raman data versus the one obtained from the Hall measurements. As shown in the figure, the values of the concentrations obtained in both techniques agree [51].

Kirillov et al. studied the effect of free carriers on phonon–plasmon interactions in WZ-GaN films in the high concentration regime [54]. The concentrations in their study were determined via Hall measurements to be $4 \times 10^{19}\,\mathrm{cm}^{-3}$ for sample A and $8 \times 10^{19}\,\mathrm{cm}^{-3}$ for sample B. From those meas-

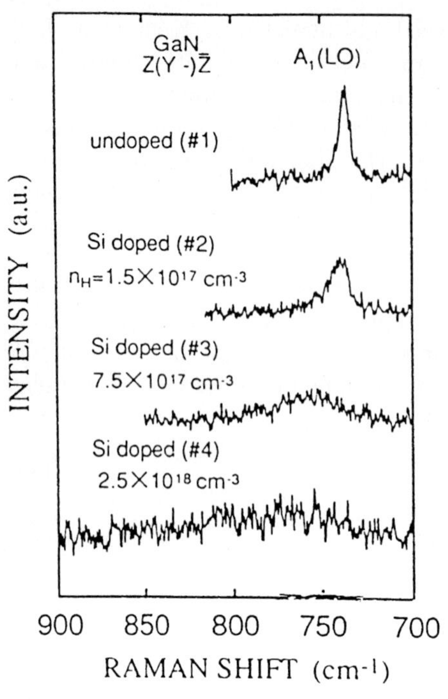

Fig. 7.18. Raman spectra of the A_1(LO) mode of GaN for several Si doping concentrations n_H obtained via Hall measurements [51]

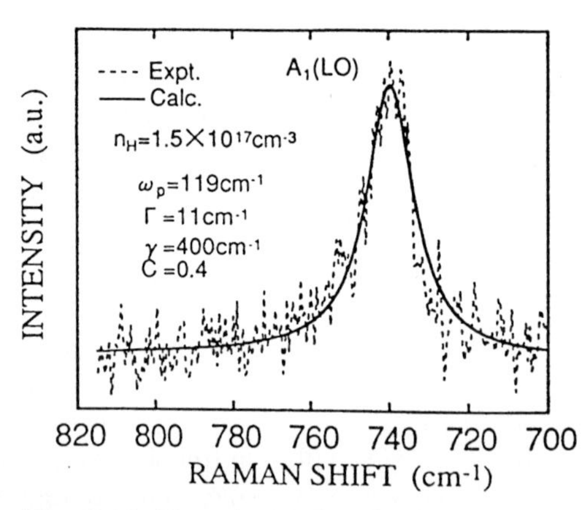

Fig. 7.19. Experimental and calculated Raman lineshape of the A_1(LO)-plasmon coupled mode [51]

urements the plasma damping constant for sample A was been determined to be 1116 cm^{-1}. Figure 7.21 presents the Raman spectra acquired in a backscattering geometry from the c-face of the two films: neither spectra exhibits the allowed A_1(LO) mode at $\approx 735\,\mathrm{cm}^{-1}$. In order to gain further insight into

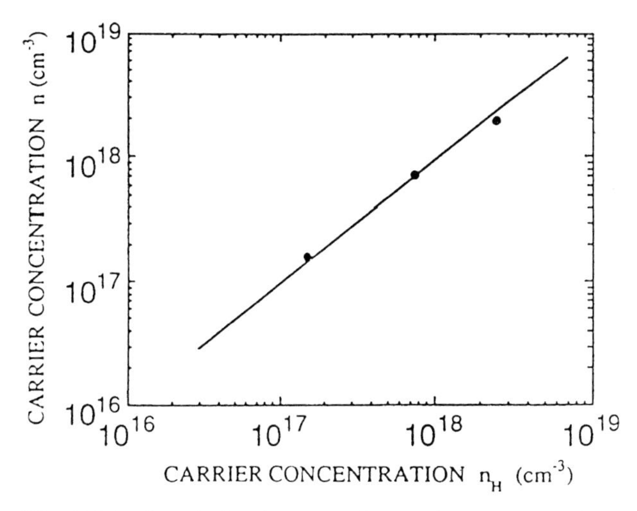

Fig. 7.20. Correlation between the carrier concentration n found from the Raman analysis and the concentration n_H obtained via Hall measurements [51]

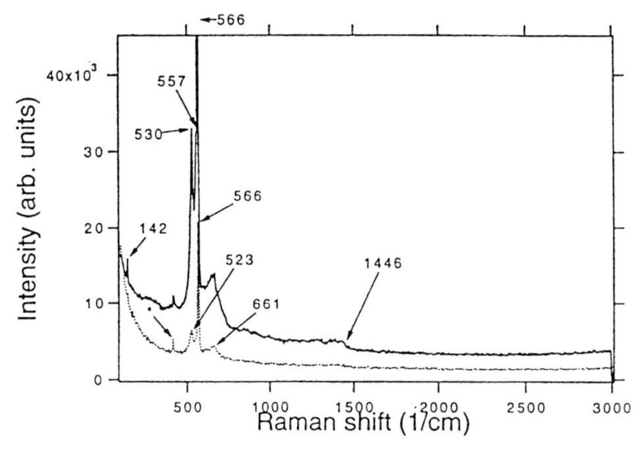

Fig. 7.21. Raman spectra of highly conductive GaN films. *Dotted line*: spectra from sample with free carrier density $n = 4 \times 10^{19}\,\mathrm{cm}^{-3}$; *solid line*: from sample with $n = 8 \times 10^{19}\,\mathrm{cm}^{-3}$ [54]

the dynamics of the phonon-free carrier interactions, Kirillov et al. calculated the expected spectra using Klein's model expressed in (7.14a,b). Figure 7.22 shows the calculated spectra for a free carrier concentration of $4 \times 10^{19}\,\mathrm{cm}^{-3}$ for two values of the plasma damping constant. The dotted line in Fig. 7.22 is a calculated spectrum for a hypothetical sample with a low damping constant of $100\,\mathrm{cm}^{-1}$. This value is typical for plasmons in GaAs films; in that material the upper branch coupling ω_+ and a lower branch ω_- have been clearly identified. The solid line in Fig. 7.22 is the calculated spectrum where

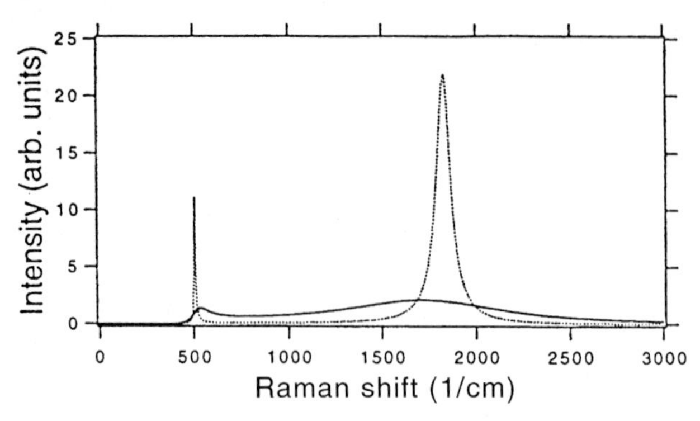

Fig. 7.22. Calculated spectra of the coupled plasmon-phonon modes for a sample with $n = 4 \times 10^{19}\,\mathrm{cm}^{-3}$. The *dotted line* corresponds to a plasma damping constant of $100\,\mathrm{cm}^{-1}$; the *solid line* to $1116\,\mathrm{cm}^{-1}$, which is a realistic value for the GaN film [54]

the damping constant is taken to be that of sample A ($1116\,\mathrm{cm}^{-1}$). As shown in the figure the lines become much broader as the plasma damping constant increases, with the upper branch being the most affected.

In contrast to the study of Kozawa et al. in the low dopant regime, the study of Kirillov et al. indicates that for high free carrier concentrations Raman spectroscopy may not be a useful tool to extract qualitative information on the concentration [54]. This is because the line of the upper branch of the phonon–plasmon coupling is too broad to allow a meaningful analysis. However, the model calculations presented in Fig. 7.22 reveal that the line of the lower branch at $\approx 530\,\mathrm{cm}^{-1}$ is not too broad and thus potentially may be used in the concentration analysis. This line corresponds to the 523 and $530\,\mathrm{cm}^{-1}$ spectral lines of samples A and B, respectively (see Fig. 7.21). However, due to defects and inhomogeneities in the film, which relax the Raman selection rules, the A_1(TO) mode near $530\,\mathrm{cm}^{-1}$ may be present in the spectra as well. Thus the identification of the lower phonon–plasmon coupling mode has to be dealt with cautiously.

A detailed investigation into the issue of the lower branch of the phonon–plasmon coupled modes in the case of high carrier concentration was carried out by Demangeot et al. [55]. The GaN sample was found via infrared reflectivity measurements to contain a high level of free carrier $\approx 10^{20}\,\mathrm{cm}^{-3}$, and the plasma damping constant was determined to be $1480\,\mathrm{cm}^{-1}$ [55]. Figure 7.23 shows Raman spectra of the GaN film that were acquired at two locations on the sample. As can be seen in the figure, a relatively broad Raman peak at $525\,\mathrm{cm}^{-1}$ is present in the spectra. Demangeot et al. used the model of (7.14a,b) and found a good fit (dotted line in the figure) to the Raman data of the low branch coupled mode. In that study the Raman spectra in the high frequency range do not exhibit the A_1(LO) mode nor a pronounced high-branch coupled mode, results which are consistent with the model cal-

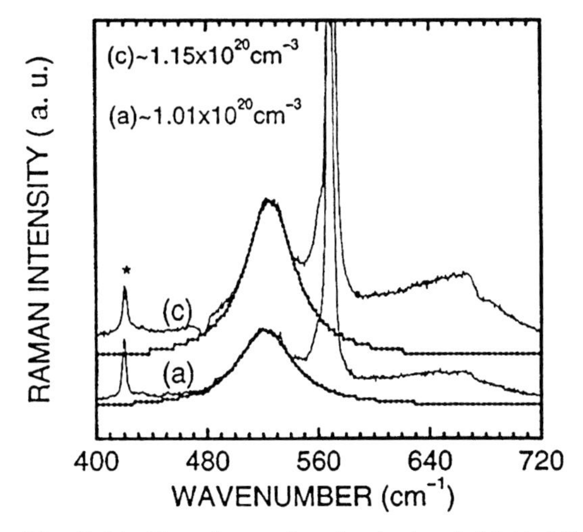

Fig. 7.23. Experimental and calculated (*dotted line*) lineshape of the low-branch A_1(LO) phonon–plasmon coupled mode in GaN films [55]

culation. Additionally, these results indicate that the distribution of the free carriers is not uniform across the GaN sample.

The fact that in the limit of low free carrier concentration the modified A_1(LO) is present in the Raman spectra led Wetzel et al. to quantitatively determine the concentrations [57]. In this study they investigated bulk GaN crystallites and correlated the free carrier concentration to the Raman peak position of the coupled A_1(LO)-plasmon mode. The authors suggested that the following correlation holds in the concentration range of $10^{17}\,\mathrm{cm}^{-3} < n < 10^{19}\,\mathrm{cm}^{-3}$ [57]:

$$n(\omega) = 1.1 \times 10^{17}\,\mathrm{cm}^{-3}(\omega - 736)^{0.764}\,. \tag{7.15}$$

In the above approximation ω is the A_1(LO) phonon–plasmon coupled mode in cm^{-1}, and the value 736 is the frequency of the A_1(LO) Raman mode. At that low level of free carriers, Ponce et al. investigated the spatial distribution in GaN crystallites via a Raman imaging technique [56]. The underlying principle of the technique involves the recording and digitization of the A_1(LO)-plasmon mode intensity across the sample. The images presented by Ponce et al. show a variation of brightness across the crystallite, with the areas of brightness corresponding to low doping concentration [56].

7.6 Isotopic Effects and Phonon Lifetimes in the Wurtzite Materials

Since the isotope mass affects the phonon frequency, Raman analysis of isotopic films may convey information on the elements controlling the mode-vibrations. Such studies of WZ-GaN films made from natural Ga and N as well as the isotope ^{15}N have been reported by Zhang et al. [63]. In the film containing the ^{15}N all of the observed Raman modes, excluding the E_2^1, were found to exhibit prominent downward frequency shifts. Additionally, the Raman frequencies of the polar modes A_1(LO,TO) and E_1(LO,TO) were found by Zhang et al. to be shifted according to the inverse square root of the reduced masses, as would be expected from the first principal calculations of crystal dynamics [63]. Since the reduced mass is $\mu^{-1} = 1/m_{\mathrm{Ga}} + 1/m_{\mathrm{N}}$, the vibrations of the polar modes are mainly due to the nitrogen oscillations. However, the isotope shifts of the non-polar modes E_2^1 and E_2^2 were found to deviate from the expected reduced mass behavior, with significant deviation found for the E_2^1 which seems not to be affected by the isotopic mass. The model calculations used to explain the vibrational dynamics of the non-polar modes led the authors to conclude that although the E_2^2 mode is still dominated by nitrogen atom vibrations, it constitutes only 92% of the total vibrations. The rest of the 8% vibrations in the admixture are due to the movements of the Ga atoms. A different scenario was inferred for the E_2^1 mode; its vibration mostly involves the motion of Ga, which explains the lack of frequency response to the isotope mass of the ^{15}N. Thus, Zhang's studies indicated that all of the Raman modes but the E_2^1 involve the vibrations of the nitrogen atoms, while that of the E_2^1 involves the vibrations of the heavier gallium atoms [63].

One crucial aspect impacting device performance is phonon interaction with free carriers. In general, the interaction can degrade the viability of the device; however, studies have also demonstrated that the phonon-electron interaction may be used to engineer certain laser devices [64–66]. The phonon lifetimes are important in both these aspects, and although the interaction involves only the LO phonons, knowledge of other mode-lifetimes may give insight into the characteristic dynamics of the material. One fundamental lifetime shortening mechanism in semiconductors has been established to occur via the anharmonic interaction [67–72]. In this mechanism, the Raman phonons decay into other normal modes in such a way that there is a conservation of momentum and energy in the process. More specifically, for a three-phonon decay process a phonon of frequency ω_1 and a wavevector $\boldsymbol{q}_1$ decays into two phonons of energies ω_2, ω_3, and wavevectors $\boldsymbol{q}_1$, $\boldsymbol{q}_2$, such that $\omega_1 = \omega_2 + \omega_3$ and $\boldsymbol{q}_1 = \boldsymbol{q}_2 + \boldsymbol{q}_3$. However, the additional lifetime shortening mechanism due to phonon scattering at point defects has to be considered as well [67].

Tsen et al. have reported phonon lifetime measurements via Raman spectroscopy [73]. In their investigation, the decay of the A_1(LO) mode in WZ-GaN film was studied via time-resolved Raman spectroscopy. The measured lifetime was found to be $\approx 3\,\mathrm{ps}$ at 300 K and $\approx 5\,\mathrm{ps}$ at 5 K. Tsen et al. hypothesized that the zone-center A_1(LO) phonons decay primarily into a large wavevector TO phonon and a large wavevector LA or TA phonon. [73].

Raman studies of phonon lifetimes in GaN, AlN, and ZnO crystallites have been reported by Bergman et al. [27]. The lifetimes were obtained from the Raman linewidth, after correcting for the instrument broadening, using the uncertainty relation $\Delta E/\hbar = l/\tau$, where ΔE is the linewidth and τ is the lifetime [74]. The lifetime analysis of Bergman et al. indicated that the phonon lifetimes in AlN, GaN, as well as ZnO crystallites fall into two main time regimes: a relatively long lifetime of the E_2^1 mode and much shorter lifetimes for the E_2^2, E_1(TO), A_1(TO), and A_1(LO) modes. The lifetime of the E_2^1 mode of high quality GaN crystallites was found to be ≈ 10 ps, whereas the lifetimes of the other modes were found to be approximately an order of magnitude shorter. A similar trend in the lifetimes was observed for phonons of high quality AlN, ZnO, and AlN that contain high levels of impurities. Moreover, the lifetimes of all the modes of both the high and low quality AlN crystallites were found to be correlated to their relative impurity concentrations [27].

A tentative explanation of the relative long lifetime of the E_2^1 in the WZ-crystallites was given by Bergman et al. in terms of the factors determining the anharmonic lifetimes, specifically the energy-conservation constraints, the density of the final states, as well as the anharmonic interaction coefficient. Unlike the other modes, the energy of the E_2^1 mode lies in the low energy regime of the wurtzite dispersion curve [75–77] and only the acoustic phonons are available as a channel of decay. At the zone edges the energies of the acoustic phonons are equal or larger than that of the E_2^1 mode; thus, in order for the energy conservation to hold the optical phonons have to decay into acoustic phonons at the zone-center for which their density is low. Although the contribution of the anharmonic coefficient, which has not yet been determined theoretically, has to be taken into account, the authors have suggested that the low density of states significantly reduces the scattering rate, thus increasing the phonon lifetime.

7.7 Wide Band-Gap Alloys

In the following, a review of the mode behavior of III-V nitride-based alloys is presented. Mixed crystals of the form $AB_{1-x}C_x$ are classified into two main groups according to the behavior of the $\boldsymbol{q} \approx 0$ optical phonons [78]. Figure 7.24 depicts the two classes for ZB material, referred to as the one-mode and two-mode material, respectively [79]. In general, if the frequencies of the AB and the AC components differ greatly, a two-mode behavior is expected; if the frequencies of both components have proximate values, a one-mode behavior results. In addition to the one-mode and two-mode classes of materials, an intermediate class exists that exhibits two-mode behavior over a certain composition range and one-mode behavior over the rest of the range. This intermediate type of behavior has been observed in some III-V crystals, including $AsGa_{1-x}In_x$, $GaAs_{1-x}Sb_x$, and $SbGa_{1-x}In_x$ [80]. Several criteria based on the elemental mass differences have been proposed to distinguish

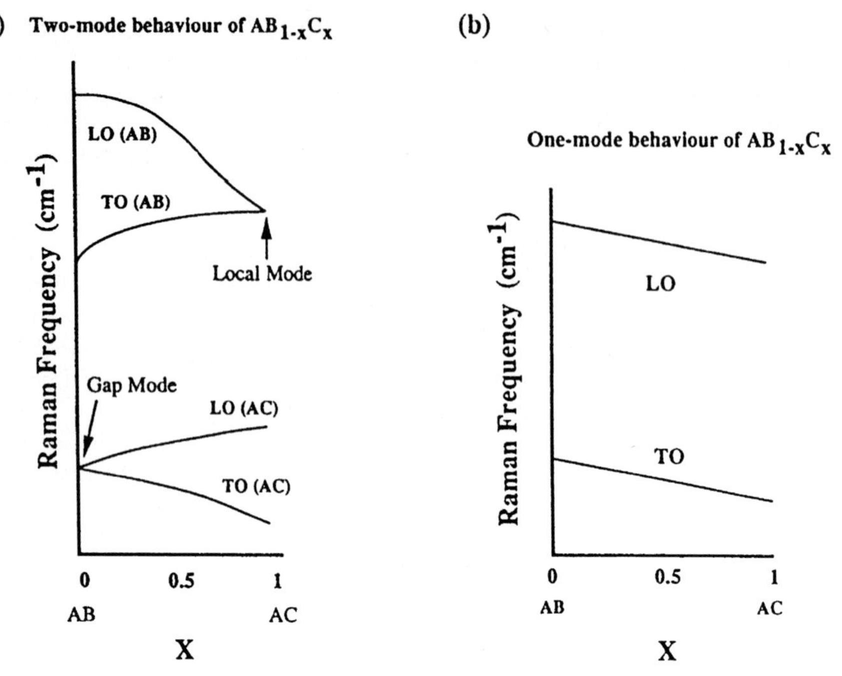

Fig. 7.24. The two classes of mixed ZB crystals: (**a**) two-mode type and (**b**) one-mode type material. For an intermediate composition x in the two-mode class of materials, two sets of frequencies would be observed in the spectra: one set is due to the LO and TO modes of the AB component and the other set is due to the LO and TO modes of the AC component. The one-mode exhibits only one set of LO and TO frequencies, which ideally are linear with composition [79]

between the one- and two-mode behavior; however, no consensus has been established [78].

Hayashi et al. conducted Raman studies of WZ-$Al_xGa_{1-x}N$ films in the composition range $0 < x < 0.15$ [81]. The investigation focused on the behavior of the E_2^2, $E_1(TO)$, $A_1(TO)$ and $E_1(LO)$ mode-frequencies as a function of the composition. They concluded that in the studied composition range, the $Al_xGa_{1-x}N$ alloy system exhibits, up to some deviation from linearity, a one-mode behavior [81]. Alloy studies in the same narrow composition range were conducted by Behr et al. [82]. These studies indicated that the $A_1(LO)$ phonon is most likely to be of a one-mode type in the AlGaN; in contrast, the alloying did not affect the E_2^2 mode. The alloy-type of $Al_xGa_{1-x}N$ in the whole composition range was investigated by Cros et al. [83], who studied the behavior of the $A_1(LO)$, E_2^2, and $A_1(TO)$ Raman modes. The experimental data led the authors to conclude that the $A_1(LO)$ phonon exhibits a one-mode behavior, whereas the E_2^2 phonon mode is a two-mode type; however, the type behavior of the $A_1(TO)$ mode was found to be inconclusive [83].

Similar Raman studies were conducted by Demangeot, where it was found that the A_1(LO), A_1(TO), and the E_1(TO) modes were all of a one-mode type [84]. Conflicting experimental results concerning the one-mode behavior of the E_1(TO) have been found utilizing infrared reflectance [85]. The infrared spectrum for each of the alloy compositions exhibits two features, which were attributed to the two-mode behavior of the E_1(TO) [85].

The polar phonons in III-V ternary nitride semiconductors of wurtzite structure were investigated theoretically by Yu et al. [86]. Within the modified-random-element isodisplacement model (MREI), they demonstrated that the polar modes of $Ga_xAl_{1-x}N$ and $In_xGa_{1-x}N$ follow a one-mode behavior [86]. The MREI model in [86] is based on the model developed previously for the ZB materials with the modifications that include the additional phonon modes and the anisotropy of the WZ structure. Figures 7.25 and 7.26 present the theoretical results and some of the compiled experimental results. As can be seen in Fig. 7.25, the theory concurs well with the experiment except for the E_1(TO) mode. This small deviation has been attributed to crystal quality. The one-mode behavior found by Yu et al. was explained in terms of the large mass difference between the nitrogen and the other alloy constituents. Since

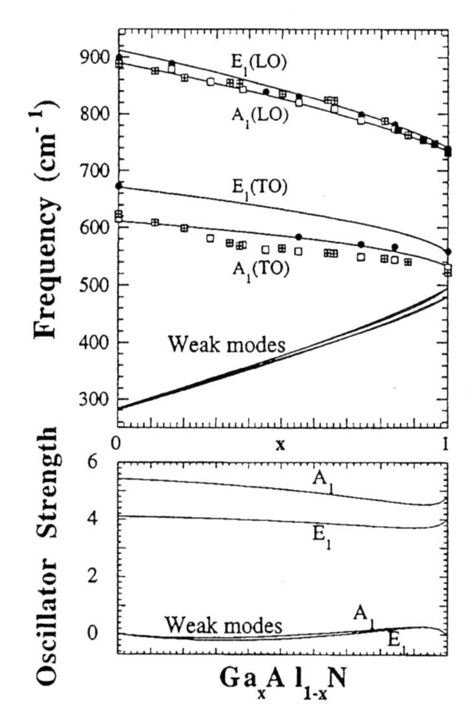

Fig. 7.25. Phonon mode behavior of WZ-GaAlN alloy. The *lines* represent the MREI model. The model calculations are for the A_1(TO), A_1(LO), E_1(TO), and E_1(LO) modes; the *points* represent the experimental data [86]

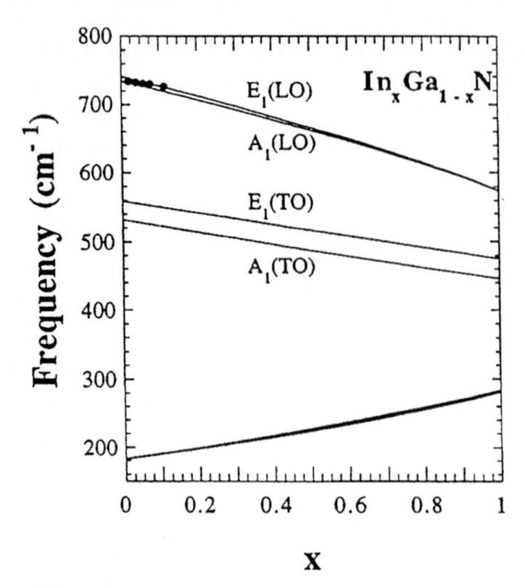

Fig. 7.26. Phonon mode behavior of WZ-InGaN alloy. Theory (*lines*) and experiment (*dots*). The polar modes are presented [86]

the atomic mass of nitrogen is much smaller that that of gallium, indium, and aluminum, the reduced masses of GaN, InN, and AlN are almost the same as that of nitrogen. As a result, no distinct GaN-like modes and AlN-like modes may exist in the $Ga_xAl_{1-x}N$ alloy system; the same situation holds for the $In_xGa_{1-x}N$ alloy [86]. It should be noted that in all of the aforementioned experimental results the other possible factors affecting the line position such as stress, phonon–plasmon coupling, and symmetry mixing were not taken into consideration.

Raman studies on phonon mode behavior in ZB-AlGaN films were reported by Harima et al. [15]. It was found that the LO mode is a one-mode type while the TO mode is a two-mode type. The authors calculated the mode behavior for the ZB-AlGaN alloy system and found it to be consistent with the experimental results. Figure 7.27 presents the phonon frequency versus the composition for the experimental as well as the calculated results [15].

The topic of ordering in the AlGaN alloy system was addressed by Korakakis et al., utilizing X-ray diffraction [87], and by Bergman et al., utilizing Raman and X-ray spectroscopy [88]. The films in the former study were grown by MBE at a temperature of $\approx 750\,°C$ on SiC and sapphire substrates. The ordering in the films was inferred from the superlattice (SL) lines of the (0001), (0003), and (0005) diffraction [87]. These lines are forbidden reflections in WZ structure and appear as SL lines only when lattice ordering exists. The intensity ratio of the SL line to the allowed (0002) line of the WZ structure is thus a measure of the extent of the order. The X-ray analysis of Korakakis indicated a long-range order; the relative intensity of SL lines was found to be strong. In contrast to these results the AlGaN films, which were grown

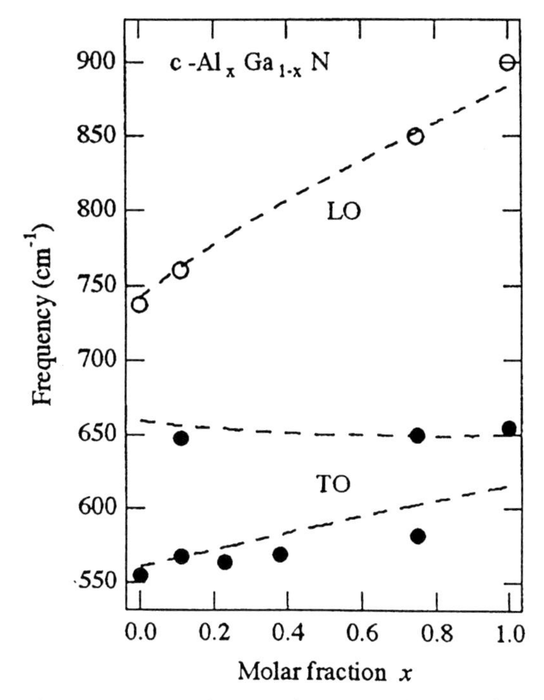

Fig. 7.27. The LO and TO mode behavior of ZB-AlGaN [15]

via MOCVD at the elevated temperature range $\approx 1100\,°C$ and which have been investigated by Bergman et al., were found to exhibit a much smaller intensity ratio thus implying a disordered state of the alloy [88]. The different order states found in both studies may be attributed to the temperature employed in growth, since it has been demonstrated that the achievement of an ordered alloy in some families of ternary tetrahedral semiconductors is a function of the growth temperature [89].

The influence of the disorder on the E_2^2 Raman mode of the $Al_xGa_{1-x}N$ MOCVD films in the composition range $0 < x < 1$ has been investigated as well by Bergman et al. [88]. Figure 7.28 presents the Raman spectra of the E_2^2 mode at various alloy compositions; it is shown that the spectral lines exhibit asymmetric broadening toward the higher frequency range. The asymmetry of the E_2^2 lines was explained in light of the spatial correlation model [90–92]. The foundation of the model lies in the wave vector uncertainty: $\Delta q = 2\pi/L$. In that relation Δq is the phonon wavevector range; this range is due to the relaxation of the $q = 0$ Raman selection rules caused by the disorder. The parameter L may be regarded as the size of the embedded ordered domain in the disordered matrix: thus the larger the disorder, the smaller the ordered

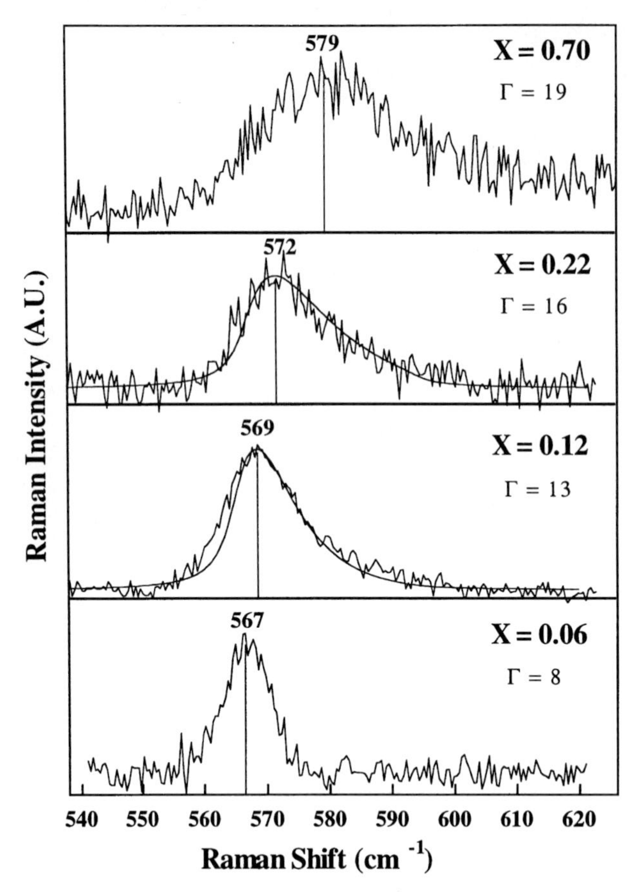

Fig. 7.28. Raman spectra of the E_2^2 mode of WZ-AlGaN films at various compositions. The *solid lines* represent the spatial correlation model [88]

domain. The model may be expressed as [90–92]

$$I(\omega) \propto \int \exp(-q^2L^2/4)\frac{\mathrm{d}^3q}{[\omega-\omega(\boldsymbol{q})]^2+[\Gamma_0/2]^2}\,. \tag{7.16}$$

In this relation Γ_0 is the linewidth of the bulk alloy of composition $x = 0$ and $\omega(\boldsymbol{q})$ is the phonon dispersion curve. Few and to some extent conflicting theoretical predictions of the dispersion relation in the III-V nitrides semiconductors have been reported [75,76,93–99]. Moreover, due to the lack of experimental results no consensus has been reached regarding the phonon dispersion relations in these materials. In order to fit the Raman data to the model, the authors in [88] assumed an averaged value of $\omega(\boldsymbol{q})$ of the form $A + Bq^2$ (in units of cm^{-1} and with $A = 568$ and $B = 100$). As can be seen in Fig. 7.28, the Raman data for the $x = 0.12$ and $x = 0.22$ can be fit with this model (represented by the solid lines), albeit with different values

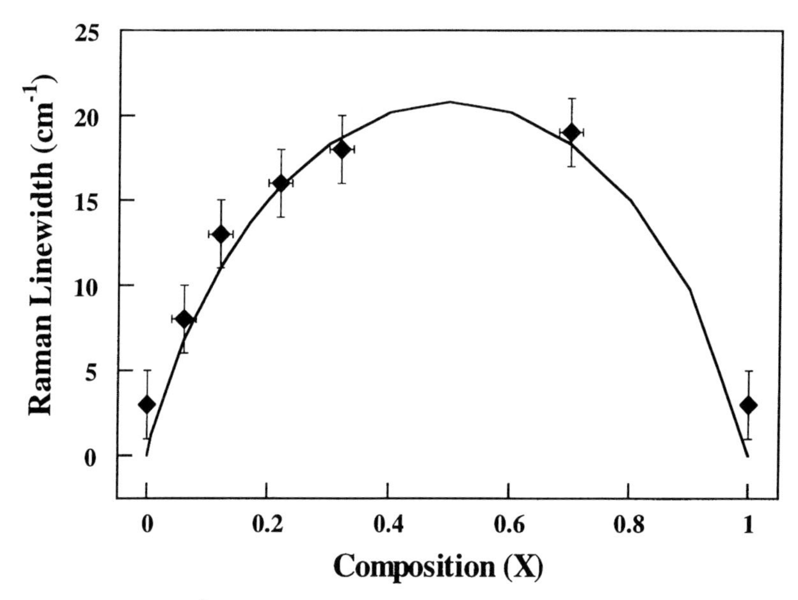

Fig. 7.29. The E_2^2 Raman linewidth as a function of composition (*dots*). The *solid line* is obtained from calculations of the entropy of mixing of an alloy system (in arbitrary units) [88]

for Γ. These fits suggest that the extent of the $q = 0$ relaxation is correlated with the Al composition [88]. Figure 7.29 shows the E_2^2 Raman linewidth as a function of composition; a maximum is observed at composition $x \approx 0.5$, a value at which maximum disorder would be expected in a random alloy.

7.8 Concluding Remarks

The Raman effect arises from the interaction of light with matter; as such, Raman spectroscopy is a nondestructive and powerful tool in the study of lattice dynamics. The various investigations reviewed here demonstrate the utility of Raman spectroscopy in characterizing the material properties of the III-V nitrides. Although significant progress has been made many issues are still in the initial stages of research. The cubic phases, the InN and its alloy system, and the deconvolution of the combined phonon dynamics effect on the alloy Raman spectra merit closer investigation. Moreover, the topic of Raman spectroscopy of phonon-interfaces and phonon-superlattices needs to be addressed. The successful investigation of many of the above issues is contingent on advances in material quality of the wide band-gap semiconductors.

Acknowledgements. Leah Bergman acknowledges the US Army Research Office and the National Research Council for supporting this work.

References

1. J.W. Orton, C.T. Foxon: Rep. Prog. Phys. **61**, 1 (1998)
2. H. Morkoc, S. Strite, G.B. Gao, M.E. Lin, B. Sverdlov, M. Burns: J. Appl. Phys. **76**, 1363 (1994)
3. J.H. Edgar: J. Mater. Res. **7**, 235 (1992)
4. J.H. Edgar (ed.): *Properties of Group III Nitrides* (INSPEC, London 1994)
5. R.F. Davis: Proc. IEEE **79**, 702 (1991)
6. M. Cardona: In: *Light Scattering in Solids I*, ed. by M. Cardona (Springer-Verlag, New York 1983) p. 1
7. D. Behr, J. Wagener, J. Schneider, H. Amano, I. Akasaki: Appl. Phys. Lett. **68**, 2404 (1996)
8. M. Ramsteiner, J. Menniger, O. Brandt, H. Yang, K.H. Ploog: Appl. Phys. Lett. **69**, 1276 (1996)
9. M. Kuball, F. Demangeot, J. Frandon, M.A. Renucci, H. Sands, D.N. Batchelder, S. Clur, O. Briot: Appl. Phys. Lett. **74**, 549 (1999)
10. H. Okumura: In: *Gallium Nitride and Related Semiconductors*, ed. by J.H. Edgar, S. Strite, I. Akasaki, H. Amano, C. Wetzel (INSPEC, London 1999) p. 402
11. Y. Takeda, M. Tabuchi: In: *Gallium Nitride and Related Semiconductors*, ed. by J.H. Edgar, S. Strite, I. Akasaki, H. Amano, C. Wetzel (INSPEC, London 1999) p. 381
12. A.D. Hanser, R.F. Davis: In: *Gallium Nitride and Related Semiconductors*, ed. by J.H. Edgar, S. Strite, I. Akasaki, H. Amano, C. Wetzel (INSPEC, London 1999) p. 386
13. P. Lawaetz: Phys. Rev. B **5**, 4039 (1972)
14. A. Tabata, R. Enderlein, J.R. Leite, S.W. da Silva, J.C. Galzerani, D. Schikora, M. Kloidt, K. Lischka: J. Appl. Phys. **79**, 4137 (1996)
15. H. Harima, T. Inoue, S. Nakashima, H. Okumura, Y. Ishida, S. Yoshida, T. Koizumi, H. Grille, F. Bechstedt: Appl. Phys. Lett **74**, 191 (1999)
16. S. Miyoshi, K. Onabe, N. Ohkouchi, H. Yaguchi, R. Ito, S. Fukatsu, Y. Shiraki: J. Cryst. Growth **124**, 439 (1992)
17. M. Giehler, M. Ramsteiner, O. Brandt, H. Yang, K.H. Ploog: Appl. Phys. Lett. **67**, 733 (1995)
18. W. Hayes, R. Loudon: *Scattering of Light by Crystals* (Wiley, New York 1978)
19. I. Gorczyca, N.E. Christensen, E.L. Peltzer y Blanca, C.O. Rodriguez: Phys. Rev. B **51**, 11936 (1995)
20. F.H. Pollak: In: *Analytical Raman Spectroscopy*, ed. by J.G. Grasselli, B.J. Bulkin (Wiley, New York 1991) pp. 137–221
21. D.D. Manchon, A.S. Barker, P.J. Dean, R.B. Zetterstrom: Solid State Commun. **8**, 1227 (1970)
22. V. Lemos, C.A. Arguello, R.C.C. Leite: Solid State Commun. **11**, 1352 (1972)
23. G. Burns, F. Dacol, J.C. Marinace, B.A. Scott: Appl. Phys. Lett. **22**, 356 (1973)
24. S. Murugkar, R. Merlin, A. Botchkarev, A. Salvador, H. Morkoc: J. Appl. Phys. **77**, 6042 (1995)
25. A. Cingolani, M. Ferrara, M. Lugara, G. Scamarcio: Solid State Commun. **58**, 823 (1986)
26. T. Azuhata, T. Sota, K. Suzuki, S. Nakamura: J. Phys.: Condens. Matter **7**, L129 (1995)

27. L. Bergman, D. Alexson, P.L. Murphy, R.J. Nemanich, M. Dutta, M.A. Stroscio, C. Balkas, H. Shin, R.F. Davis: Phys. Rev. B **59**, 12977 (1999)
28. H. Siegle, I. Eckey, A. Hofmann, C. Thomsen: Solid State Commun. **96**, 943 (1995)
29. J.L. Birman: Phys. Rev. **115**, 1493 (1959)
30. P. Perlin, A. Polian, T. Suski: Phys. Rev. B **47**, 2874 (1993)
31. L.E. McNeil, M. Grimsditch, R.H. French: J. Am. Ceram. Soc. **76**, 1132 (1993)
32. L. Filippidis, H. Siegle, A. Hoffmann, C. Thomsen, K. Karch, F. Bechstedt: Phys. Stat. Sol. B **198**, 621 (1996)
33. H.J. Kwon, Y.H. Lee, O. Miki, H. Yamano, A. Yoshida: Appl. Phys. Lett. **69**, 937 (1996)
34. M.C. Lee, H.C. Lin, Y.C. Pan, C.K. Shu, J. Ou, W.H. Chen, W.K. Chen: Appl. Phys. Lett. **73**, 2606 (1998)
35. S.P.S. Porto: In: *Light Scattering Spectra of Solids*, ed. by G.B. Wright (New York 1969) p. 1
36. N.W. Ashcroft, N.D. Mermin: Solid State Physics (Holt, Rinehart and Winston, New York 1976)
37. A.S. Barker Jr., M. Ilegems: Phys. Rev. B **7**, 743 (1973)
38. P. Perlin, C.J. Carillon, J.P. Itie, A.S. Miguel, I. Grzegory, A. Polian: Phys. Rev. B **45**, 83 (1992)
39. V.Y. Davydov, N.S. Averkiev, I.N. Goncharuk, D.K. Nelson, I.P. Nikitina, A.S. Polkovnikov, A.N. Smirnov, M.A. Jacobson: J. Appl. Phys. **82**, 5097 (1997)
40. C. Kisielowski, J. Krüger, S. Ruvimov, T. Suski, J.W. Ager III, E. Jones, Z. Liliental-Weber, M. Rubin, E.R. Weber, M.D. Bremser, R.F. Davis: Phys. Rev. B **54**, 17745 (1996)
41. J.F. Nye: *Physical Properties of Crystals* (Clarendon Press, Oxford 1984)
42. W.J. Meng, T.A. Perry: J. Appl. Phys. **76**, 7824 (1994)
43. W. Rieger, T. Metzger, H. Angerer, R. Dimitrov, O. Ambacher, M. Stutzmann: Appl. Phys. Lett. **68**, 970 (1996)
44. H. Siegle, A. Hoffmann, L. Eckey, C. Thomsen, J. Christen, F. Bertram, D. Schmidt, D. Rudloff, K. Hiramatsu: Appl. Phys. Lett. **71**, 2490 (1997)
45. T. Kozawa, T. Kachi, H. Kano, H. Nagase, N. Koide, K. Manabe: J. Appl. Phys. **77**, 4389 (1995)
46. J.A. Sanjurjo, E.L. Cruz, P. Vogl, M. Cardona: Phys. Rev. B **28**, 4579 (1983)
47. S.S. Mitra, O. Brafman, W.B. Daniels, R.K. Crawford: Phys. Rev. **186**, 942 (1969)
48. R. Loudon: Advances in Physics **13**, 423 (1964)
49. L. Bergman, M. Dutta, C. Balkas, R.F. Davis, J.A. Christman, D. Alexson, R.J. Nemanich: J. Appl. Phys. **85**, 3535 (1999)
50. C.A. Arguello, D.L. Rousseau, S.P.S. Porto: Phys. Rev. **181**, 1351 (1969)
51. T. Kozawa, T. Kachi, H. Kano, Y. Taga, M. Hashimoto, N. Koide, K. Manabe: J. Appl. Phys. **75**, 1098 (1994)
52. B.B. Varga: Phys. Rev. **137**, 1896 (1965)
53. M. Ramsteiner, O. Brandt, K.H. Ploog: Phys. Rev. B **58**, 1118 (1998)
54. D. Kirillov, H. Lee, J.S. Harris: J. Appl. Phys. **80**, 4058 (1996)
55. F. Demangeot, J. Frandon, M.A. Renucci, C. Meny, O. Briot, R.L. Aulombard: J. Appl. Phys. **82**, 1305 (1997)
56. F.A. Ponce, J.W. Steeds, C.D. Dyer, G.D. Pitt: Appl. Phys. Lett. **69**, 2650 (1996)

57. C. Wetzel, W. Walukiewicz, E.E. Haller, J. Ager III, I. Grzegory, S. Porowski, T. Suski: Phys. Rev. B **53**, 1322 (1996)
58. G. Popovici, G.Y. Xu, A. Botchkarev, W. Kim, H. Tang, A. Salvador, H. Morkoc, R. Strange, J.O. White: J. Appl. Phys. **82**, 4020 (1997)
59. E. Burstein, A. Pinczuk, S. Iwasa: Phys. Rev. **157**, 611 (1967)
60. M.V. Klein, B.N. Ganguly, P.J. Colwell: Phys. Rev. B **6**, 2380 (1972)
61. D.T. Hon, W.L. Faust: Appl. Phys. **1**, 241 (1973)
62. G. Irmer, V.V. Toporov, B.H. Bairamov, J. Monecke: Phys. Stat. Sol. B **119**, 595 (1983)
63. J.M. Zhang, T. Ruf, M. Cardona, O. Ambacher, M. Stutzmann, J.M. Wagner, F. Bechstedt: Phys. Rev. B. **56**, 14399 (1997)
64. M.V. Kisin, V.B. Gorfinkel, M.A. Stroscio, G. Belenky, S. Luryi: J. Appl. Phys. **82**, 2031 (1997)
65. H.B. Teng, J.P. Sun, G.I. Haddad, M.A. Stroscio, S.G. Yo, K.W. Kim: J. Appl. Phys. **84**, 2155 (1998)
66. M.A. Stroscio: J. Appl. Phys. **80**, 6864 (1996)
67. P.G. Klemens: In: *Solid State Physics; Advances in Research and Applications*, ed. by F. Seitz, D. Turnbull (Academic, New York 1958) Vol. 7, pp. 1–98
68. P.G. Klemens: Phys. Rev. **148**, 845 (1966)
69. W.J. Borer, S.S. Mitra, K.V. Namjoshi: Solid State Commun. **9**, 1377 (1971)
70. A. Debernardi: Phys. Rev. B **57**, 12847 (1998)
71. J. Menéndez, M. Cardona: Phys. Rev. B **29**, 2051 (1984)
72. B.K. Ridley: J. Phys.: Condens. Matter **8**, L511 (1996)
73. K.T. Tsen, D.K. Ferry, A. Botchkarev, B. Sverdlov, A. Salvador, H. Morkoc: Appl. Phys. Lett. **72**, 2132 (1998)
74. B. Di Bartolo: *Optical Interactions in Solids* (Wiley, New York 1969)
75. J.C. Nipko, C.K. Loong, C.M. Balkas, R.F. Davis: Appl. Phys. Lett. **73**, 34 (1998)
76. J.C. Nipko, C.K. Loong: Phys. Rev. B **57**, 10550 (1998)
77. A.W. Hewat: Solid State Commun. **8**, 187 (1970)
78. I.F. Chang, S.S. Mitra: Phys. Rev. **172**, 924 (1968)
79. L. Bergman, R.J. Nemanich: Ann. Rev. Mater. Sci. **26**, 551 (1996)
80. G. Lucovsky, M.F. Chen: Solid State Commun. **8**, 1397 (1970)
81. K. Hayashi, K. Itoh, N. Sawaki, I. Akasaki: Solid State Commun. **77**, 115 (1991)
82. D. Behr, R. Niebuhr, J. Wagner, K.H. Bachem, U. Kaufmann: Appl. Phys. Lett. **70**, 363 (1997)
83. A. Cros, H. Angerer, O. Ambacher, M. Stutzmann, R. Hopler, T. Metzger: Solid State Commun. **104**, 35 (1997)
84. F. Demangeot, J. Groenen, J. Frandon, M.A. Renucci, O. Briot, S. Clur, R.L. Aulombard: Appl. Phys. Lett. **72**, 2674 (1998)
85. P. Wisniewski, W. Knap, J.P. Malzac, J. Camassel, M.D. Bremser, R.F. Davis, T. Suski: Appl. Phys. Lett. **73**, 1760 (1998)
86. S. Yu, K.W. Kim, L. Bergman, M. Dutta, M.A. Stroscio, J.M. Zavada: Phys. Rev. B **58**, 15283 (1998)
87. D. Korakakis, K.F. Ludwig, T.D. Moustakas: Appl. Phys. Lett. **71**, 72 (1997)
88. L. Bergman, M.D. Bremser, W.G. Perry, R.F. Davis, M. Dutta, R.J. Nemanich: Appl. Phys. Lett. **71**, 2157 (1997)
89. A. Zunger: Appl. Phys. Lett. **50**, 164 (1987)
90. P. Parayanthal, F.H. Pollak: Phys. Rev. Lett. **52**, 1822 (1984)

91. R.J. Nemanich, S.A. Solin, R.M. Martin: Phys. Rev. B **23**, 6348 (1981)
92. P.M. Fauchet, I.H. Campbell: Critical Reviews in Solid State and Materials Sciences **14**, S79 (1988)
93. T. Azuhata, T. Matsunaga, K. Shimada, K. Yoshida, T. Sota, K. Suzuki, S. Nakamura: Physica B **219–220**, 493 (1996)
94. V.Y. Davydov, Y.E. Kitaev, I.N. Goncharuk, A.N. Smirnov, J. Graul, O. Semchinova, D. Uffmann, M.B. Smirnov, A.P. Mirgorodsky, R.A. Evarestov: Phys. Rev. B **58**, 12899 (1998)
95. K. Karch, F. Bechstedt, P. Pavone, D. Strauch: Physica B **219–220**, 445 (1996)
96. K. Karch, F. Bechstedt: Phys. Rev. B **56**, 7404 (1997)
97. K. Karch, J.M. Wagner, F. Bechstedt: Phys. Rev. B **57**, 7043 (1998)
98. K. Miwa, A. Fukumoto: Phys. Rev. B **48**, 7897 (1993)
99. H. Siegle, G. Kaczmarczyk, L. Filippidis, A.P. Litvinchuk, A. Hoffmann, C. Thomsen: Phys. Rev. B **55**, 7000 (1997)

8 Raman Scattering in Fullerenes and Related Carbon-Based Materials

M.S. Dresselhaus, M.A. Pimenta, P.C. Eklund, and G. Dresselhaus

Abstract. The application of Raman spectroscopy to the characterization and study of the physical properties of fullerenes and related carbon-based materials is reviewed. Carbon can exhibit different arrangements in the solid state, such as the well known graphite and diamond structures, as well as the more recently discovered structures based on fullerenes and carbon nanotubes. Each one of these structural arrangements exhibits a characteristic Raman spectrum. The classification of the vibrational modes and their characteristic Raman spectra are presented for these different forms of carbon. The effect on the characteristic Raman spectra of various perturbations, such as disorder, doping, and variation of temperature and pressure, is reviewed. Since some of these structures exhibit electronic interband separations close to the energy of the exciting photons, the resonant behavior of the various Raman spectra is also discussed.

Raman scattering is a very useful spectroscopic technique for the study and identification of the various forms of carbons. [1–4] The many ways in which carbon-carbon bonding may take place to form molecular and crystalline structures is remarkable, especially in comparison to other elements in the periodic table. In the case of a single crystal diamond, the carbon atoms are bonded to their neighbors by strong covalent sp^3 bonds, forming a cubic structure belonging to the O_h^7 ($Fd3m$) space group. The diamond crystal has only one triply-degenerate optical mode at the center of the Brillouin zone (T_{2g} symmetry), which appears in the Raman spectrum as a sharp line at 1332 cm^{-1} [5] (see Fig. 8.1). Increasing the laser excitation energy to above 3.0 eV reduces the luminescent background, allowing clear observation of the second-order diamond Raman lines [9]. No resonant enhancement of the diamond lines nor shift in frequency has been observed for laser excitation energies from 2.4 to 4.8 eV, consistent with the bandgap energy of diamond (5.5 eV) [9].

Under ambient conditions, the graphite structure with strong in-plane sp^2 bonding is the most stable phase, and the crystalline structure belongs to the D_{6h}^4 ($P6_3/mmc$) hexagonal space group. Graphite is a semimetal with an overlap of the valence and conduction bands of about 40 meV at 300 K. The graphite crystal exhibits two Raman-active modes. The most prominent feature in the Raman spectrum is the E_{2g_2} mode at $1582\,\mathrm{cm}^{-1}$, and under special conditions, the E_{2g_1} mode at $42\,\mathrm{cm}^{-1}$ can be observed.

In addition, carbon can bond in linear carbyne chains, forming either an alternating polyyne sequence of single and triple bonds [10], or a cumulene chain of double bonds, with the polyyne sequence having lower energy, being stabilized by the Peierls distortion [7]. The characteristic Raman feature associated with the polyyne structure is a simple fully symmetric breathing mode at $\sim 2150\,\mathrm{cm}^{-1}$ in the first-order Raman spectrum [7,11,12]. The polyyne line at $\sim 2150\,\mathrm{cm}^{-1}$ is highly dispersive, showing an upshift with laser excitation energy of $60\,\mathrm{cm}^{-1}/\mathrm{eV}$ [11]. Unlike the sharp Raman features for the sp^2 and sp^3 bonded carbon, the Raman polyyne line is quite broad, and this large linewidth is attributed to the large distribution of chain lengths (8–14 *sp* bonds) in a typical polyyne sample. Since the bonding for each type of carbon is different, so are the force constants and vibrational frequencies. The strongest bond is the *sp* bond, which corresponds to the highest vibrational frequency (see Fig. 8.1).

Fullerenes and nanotubes are carbon molecular solids based on a rolled-up single sheet of graphite, with approximately spherical and cylindrical shapes, respectively. Fullerenes form a molecular crystalline solid with an optical ab-

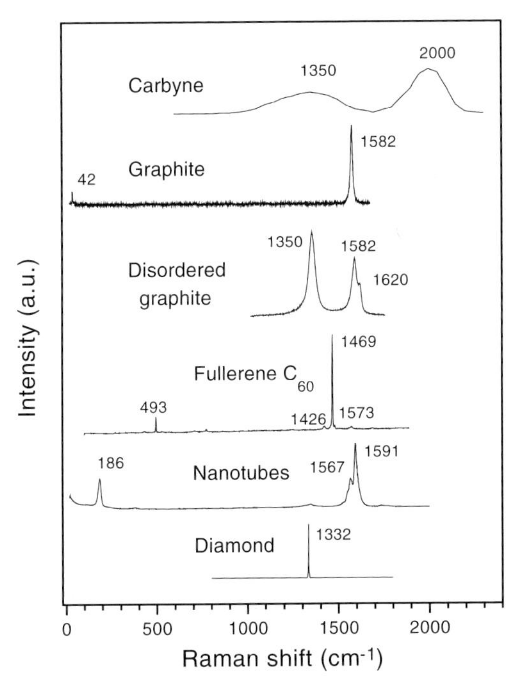

Fig. 8.1. Characteristic Raman spectra for the various carbon-based materials: carbyne (*sp* bonded carbon), graphite (sp^2 bonded carbon), disordered sp^2 bonded graphite, fullerene C_{60}, carbon nanotubes, and diamond (sp^3 bonded carbon) [6–8]

sorption edge at $\sim 1.7\,\mathrm{eV}$, while single-wall carbon nanotubes can be either semiconducting or metallic depending on their symmetry [3]. The Raman spectra of these two new forms of carbons are very rich compared to diamond and graphite (as shown in Fig. 8.1), and exhibit a number of modes associated with the intra-molecular vibrations of the various carbon atoms in these structures. Inter-molecular vibrations are also observed in some cases. The intra-molecular vibrations with predominantly tangential character are derived from the E_{2g_2} graphitic mode and occur at higher frequencies, whereas the intra-molecular vibrations of radial character and the inter-molecular modes appear at lower frequencies.

Raman spectroscopy is also a particularly useful technique for the characterization of disorder in carbon-based materials. The analysis of specific features in the Raman spectra provides a way to estimate the crystallite sizes in disordered carbons [2]. Disordered sp^2 carbon gives rise to a strong Raman band in the vicinity of $\sim 1350\,\mathrm{cm}^{-1}$ at a laser excitation wavelength of 488 nm, an example of which is shown in Fig. 8.1 for the case of activated charcoal. The technique is a very sensitive probe for small amounts of sp^2 bonding in diamond films because of the much larger Raman cross section for the sp^2 graphite vibrations relative to the sp^3 diamond-like vibrations [13]. Disordered diamond shows a disorder-induced peak in the $1200-1300\,\mathrm{cm}^{-1}$ range, as expected from the phonon density of states [14]. Typical sp bonded carbon samples also show an asymmetrical broadened feature, extending from $1000-1600\,\mathrm{cm}^{-1}$ and identified with additional highly disordered sp^2 and sp^3 bonded carbon atoms.

This review of Raman spectroscopy in carbon-based systems focuses on graphite-related materials (Sect. 8.1), carbon fullerenes (Sect. 8.2) and nanotubes (Sect. 8.3) because of the large research activity on these topics in recent years. Special attention is given to some of the newer topics that have not been previously reviewed [7,15,16], such as the dispersion of the disorder-induced D-bands of sp^2 carbon, resonant Raman effects in carbon nanotubes, and charge transfer effects in doped carbon nanotubes.

8.1 Graphite Related Materials

The crystal structure of graphite consists of layers in which the carbon atoms are arranged in a honeycomb network. Figure 8.2 shows the structure of the two-dimensional (2D) graphene sheet (which is a single atomic layer of 3D crystalline graphite normal to the hexagonal axis), together with the corresponding 2D Brillouin zone. There are two distinct atoms A and B per unit cell on a two-dimensional (2D) graphene sheet. In 3D graphite, the graphene layers are stacked with two inequivalent layers per primitive unit cell, so that the 3D unit cell contains 4 distinct carbon atoms as shown in Fig. 8.3 [18], and the crystal structure belongs to the D_{6h}^4 ($P6_3/mmc$) hexagonal space group. The small in-plane nearest-neighbor separation of 1.421 Å gives rise to

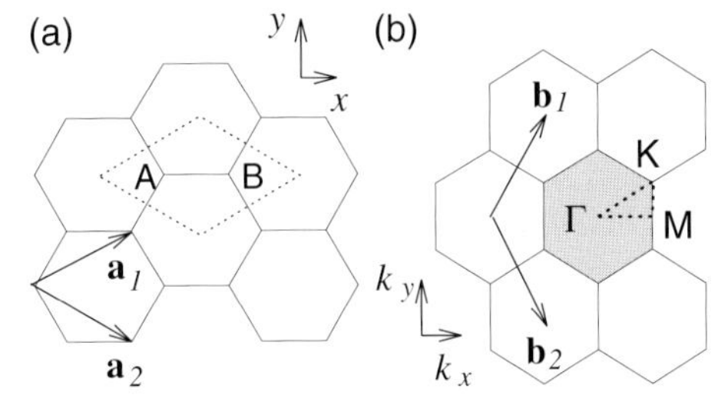

Fig. 8.2. (**a**) The unit cell and (**b**) Brillouin zone of a two-dimensional graphene layer are shown as the dotted rhombus and the shaded hexagon, respectively. $\mathbf{a}_i$, and $\boldsymbol{b}_i$, $(i = 1, 2)$ are, respectively, unit vectors in real and reciprocal space. Phonon dispersion relations are obtained along the perimeter of the dotted triangle connecting the high symmetry points, Γ, K and M. At the K point in the Brillouin zone, the electronic energy bands of a 2D graphene layer are degenerate at the Fermi level because of symmetry, thereby forming a zero-gap semiconductor [4]

very strong in-plane bonding, and the large interlayer separation of 3.35 Å is consistent with very weak binding between the adjacent graphene layers. The high crystalline anisotropy of graphite accounts for many of its interesting physical properties.

8.1.1 Single Crystal Graphite and 2D Graphene Layers

Using the crystal symmetry of the graphite lattice, numerous calculations of the phonon dispersion curves for graphite have been carried out [17], both from first principles and from a phenomenological standpoint using a Born–von Kárman model [1,17]. Because of the weak coupling between graphene layers, the phonon dispersion relations in 2D graphene layers and in 3D graphite are quite similar. An updated version of the phonon dispersion relations and the phonon density of states derived for 3D graphite, but relevant to the high symmetry directions for a 2D graphene layer, are given in Fig. 8.4. A listing of the graphite force constants up through fourth nearest neighbors, used in calculations of the phonon dispersion relations and phonon density of states (DOS), is given in [19].

From the symmetry properties of the point group D_{6h}, the zone-center optic phonon modes (at $k = 0$) for 3D graphite can be decomposed into the following irreducible representations

$$\Gamma_{\text{opt}} \rightarrow 2E_{2g} + E_{1u} + A_{2u} + 2B_{1g}. \quad (8.1)$$

Of these irreducible representations, only the E_{2g} modes are Raman active, the E_{1u} and A_{2u} being infrared active, and the B_{1g} modes being optically inactive. The atomic displacements corresponding to these normal modes are shown in Fig. 8.3. The E_{2g_1} mode corresponds to in-plane rigid layer shear

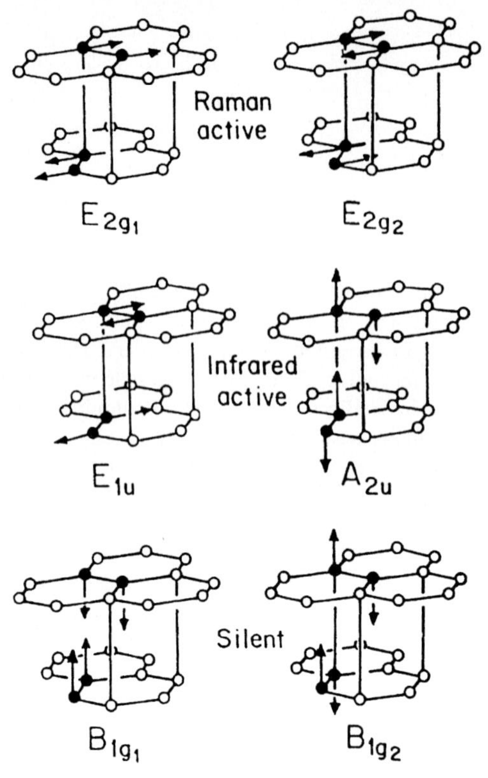

Fig. 8.3. Normal mode atomic displacements for the zone-center ($k = 0$) optical modes for 3D graphite. For the in-plane modes [E_{1u} (1587 cm^{-1}), E_{2g_1} (42 cm^{-1}), and E_{2g_2}, (1582 cm^{-1})] only one of the degenerate pair of modes is shown. The c-axis modes [A_{2u} (868 cm^{-1}), B_{1g_1} (127 cm^{-1}), B_{1g_2} ($\sim$ 870 cm^{-1})] are non-degenerate. The zero frequency acoustic modes (E_{1u}, A_{2u}) correspond to rigid translations of the lattice and are not shown [17]

displacements which occur at a very low frequency (42 cm^{-1}) because adjacent layers are rigidly displaced with respect to each other against the weak interlayer restoring force. The E_{2g_1} mode feature is not usually displayed in typical Raman spectra for graphite samples because of its weak intensity and low frequency. On the other hand, the E_{2g_2} mode represents in-plane displacements that occur at a high frequency (1582 cm^{-1}) because the neighboring atoms in each of the layer planes are displaced with respect to each other against a strong in-plane restoring force.

The first-order Raman spectrum of a large single crystal of graphite is thus usually displayed as a single high-frequency band at 1582 cm^{-1}, the so-called G-band corresponding to the Raman-allowed E_{2g_2} mode [21,22], as shown in Fig. 8.1 for highly oriented pyrolytic graphite (HOPG), the synthetic carbon material [23] commonly used for approximating the properties

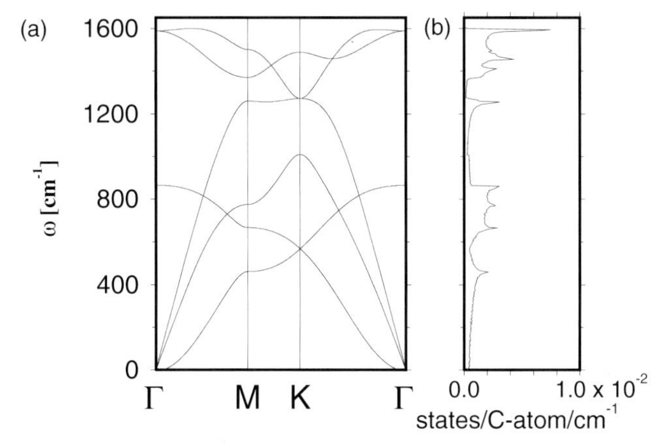

Fig. 8.4. (**a**) The phonon dispersion curves for a graphene layer, plotted along high symmetry directions [19]. (**b**) The corresponding DOS vs energy for phonon modes in units of states/C-atom/cm^{-1} $\times$ 10^{-2} [20]

of single-crystal graphite. The Raman-active E_{2g_2} mode can be observed in the (x, x), (y, y), and (x, y) back-scattering configurations, where the polarization directions x and y are in the layer planes.

8.1.2 Raman Spectra of Disordered sp^2 Carbons

Raman spectroscopy provides a very powerful tool for investigating perturbations that break the translational symmetry in crystals. The introduction of lattice disorder (or the reduction in crystallite size) relaxes the selection rule regarding the conservation of crystal momentum, so that phonons throughout the Brillouin zone may contribute to Raman scattering in accordance with the magnitude of the symmetry-breaking perturbation. In the case of graphitic materials, a modest amount of disorder gives rise to a new Raman band, which is not present in the Raman spectrum of single crystal graphite [21]. This so-called disorder-induced band, or D-band, occurs at about 1350 cm^{-1} for a laser excitation wavelength (λ_L) of 488 nm, and the ratio of the integrated intensities of the D-band to the symmetry-allowed G-band (E_{2g_2} mode) at $\approx$ 1582 cm^{-1} is commonly used to characterize different kinds of disordered sp^2 carbon [17], such as carbon fibers, glassy carbon carbon blacks pregraphitic carbons and ion-implanted carbons

As an example of the D-band and the G-band, we show in Fig. 8.5 the first-order Raman spectra for benzene-derived carbon fibers heat treated to various temperatures T_{HT} [24]. By varying the heat treatment temperature T_{HT}, vapor grown carbon fibers can be used as a model system for studying the Raman spectra as a function of disorder. Here the dominant features are the Raman-allowed line at 1582 cm^{-1} (G-band) and the disorder-induced band at $\approx$ 1350 cm^{-1} (D-band). In addition, a small disorder-induced feature

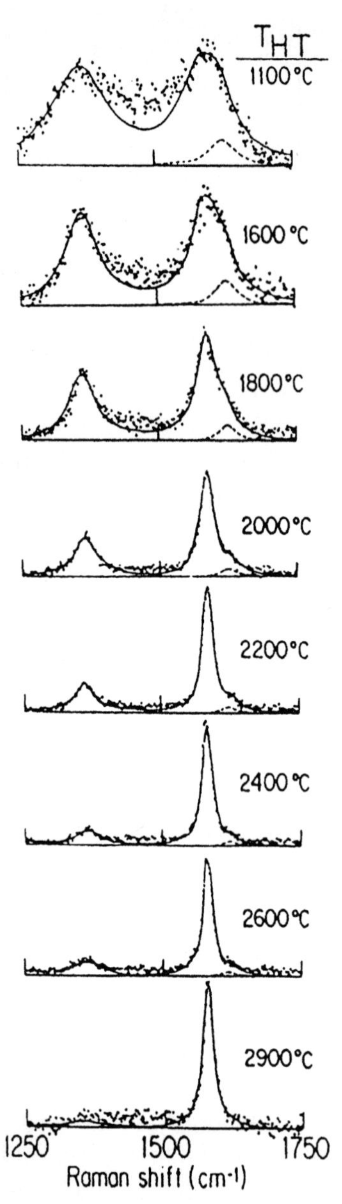

Fig. 8.5. First-order Raman spectra for Benzene-derived carbon fibers heat treated at various temperatures (T_{HT}). As T_{HT} is increased, the intensity of the disorder-induced line at $\sim$ 1350 cm^{-1} decreases, and the linewidth of the Raman-allowed line decreases. At the highest T_{HT} of 2900 °C the line at $\sim$ 1350 cm^{-1} can barely be detected. *Solid lines* represent a Lorentzian fit to the experimental points. *Dashed lines* represent a Lorentzian fit to a line at $\sim$ 1620 cm^{-1} [24]

near $\approx 1620\,\text{cm}^{-1}$ is seen in some of the traces. This feature is identified with the high phonon density of states for mid-zone phonons near the maximum optic phonon frequency (Fig. 8.4).

As the heat treatment temperature T_{HT} is increased, the intensity and linewidth of the disorder-induced D-band decreases, and the linewidth of the G-band also decreases, typical of the increasing structural order associated with heat treatment in the graphitization process [13,21,24–26]. Figure 8.6 illustrates a plot of the the ratio of the integrated Raman intensities of the disorder-induced peak at $\sim 1350\,\text{cm}^{-1}$ to the Raman-allowed peak at $\sim 1582\,\text{cm}^{-1}$ ($R = I_D/I_G$) vs. T_{HT} for the benzene-derived fibers [24], showing the decrease in the intensity ratio R with increasing T_{HT}.

Tuinstra and Koenig [21] were the first to relate the intensity ratio $R = I_D/I_G$ to the in-plane crystallite size L_a which is typically determined from X-ray diffraction measurements [21], so that L_a vs. T_{HT} is also plotted in Fig. 8.6 (see right hand scale). Knight and White [2] have investigated several different carbon materials and have shown the linear dependence of L_a on R^{-1}

$$L_a(\text{Å}) = 44R^{-1} \tag{8.2}$$

to hold approximately over the extended range $25 < L_a < 3000\,\text{Å}$, and the relation is valid for laser excitation wavelengths near 514.5 nm.

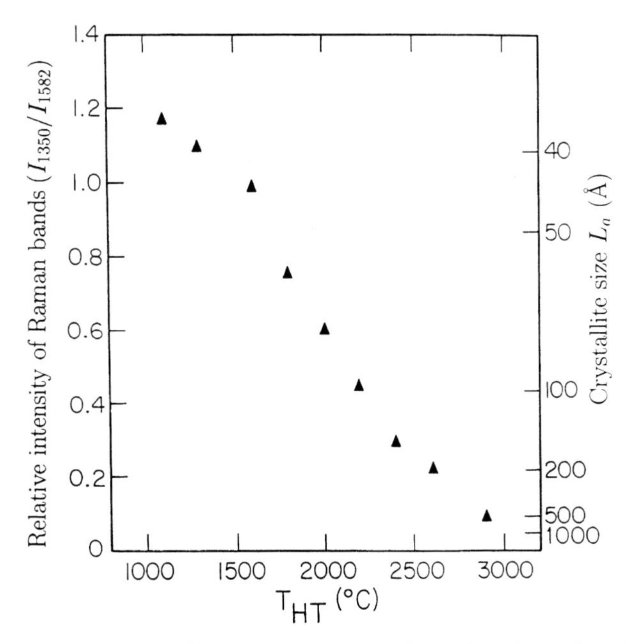

Fig. 8.6. Plot of the intensity ratio of the disorder-induced D band at $\sim 1350\,\text{cm}^{-1}$ to the symmetry-allowed G band (E_{2g_2} mode) at $\sim 1582\,\text{cm}^{-1}$ vs. heat-treatment temperature (T_{HT}) for benzene-derived carbon fibers. On the right scale, the intensity ratio $R = I_{1350}/I_{1582}$ is related to the crystallite size L_a [21,24]

The correlation between R and L_a makes Raman spectroscopy a powerful tool for characterizing the amount of structural order in small graphite crystallites, in polycrystalline graphite and in graphites with random disorder, and (8.2) has been commonly (and often incorrectly) used to estimate the crystallite size in disordered graphites. However, it must be emphasized that the ratio $R = I_D/I_G$ is very sensitive to the resonant Raman effect, and therefore (8.2) can only be used over a narrow range of laser excitation energies E_{L}. As an example, Fig. 8.7 shows the Raman spectra of polyparaphenylene (PPP) heat treated to a temperature of 2400 °C [28] for three different laser excitation energies. This material consists of short graphene ribbons that are stacked with interplanar correlation lengths of only a few layers, based on X-ray diffraction data [29]. Note in Fig. 8.7 that for the same sample the ratio of the intensities of the D and G-bands ($R = I_D/I_G$) depends significantly on the value of E_{L} used to measure the Raman spectrum. In particular, R increases for decreasing laser excitation energies (or for increasing laser wavelengths), due to the resonant enhancement of the disorder-induced D-band (Fig. 8.8).

The laser wavelength (energy) dependence of R has led to a misinterpretation [30] of the empirical formula, $L_a = C(I_D/I_G)^{-1}$ where $C = 44\,\text{Å}$, as in (8.2). This value of the prefactor C is only applicable near $\lambda_{\mathrm{L}} = 514.5\,\mathrm{nm}$, since C must depend on λ_{L} to accommodate for a λ_{L}-dependent intensity ratio $R(\lambda_{\mathrm{L}}) = I_D/I_G$. To estimate the appropriate $C(\lambda_{\mathrm{L}})$ for a given wavelength λ_{L}, we use the observation that for PPP-2400 and for all other available data on glassy carbons [31,32] and carbon blacks [33] heat-treated above

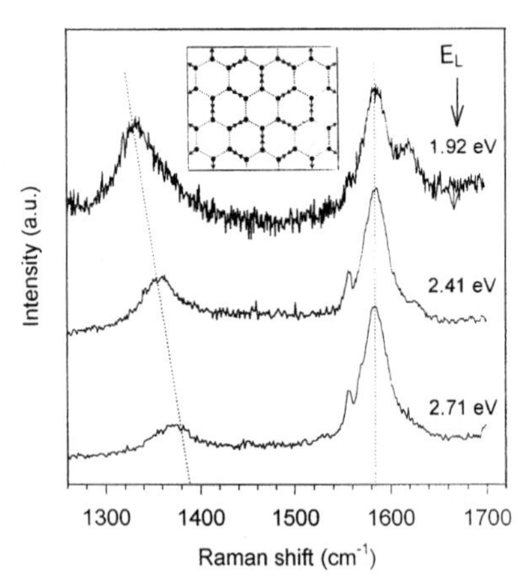

Fig. 8.7. Raman spectrum of PPP-2400 (poly(paraphenylene) heat-treated to 2400 °C) obtained with laser excitation energies 1.92 eV (647 nm), 2.41 eV (514.5 nm) and 2.71 eV (457.9 nm). The *inset* shows the displacement of the carbon atoms corresponding to the vibrational mode at the K point of the optic phonon branch, shown as heavy lines in Fig. 8.8 [27]

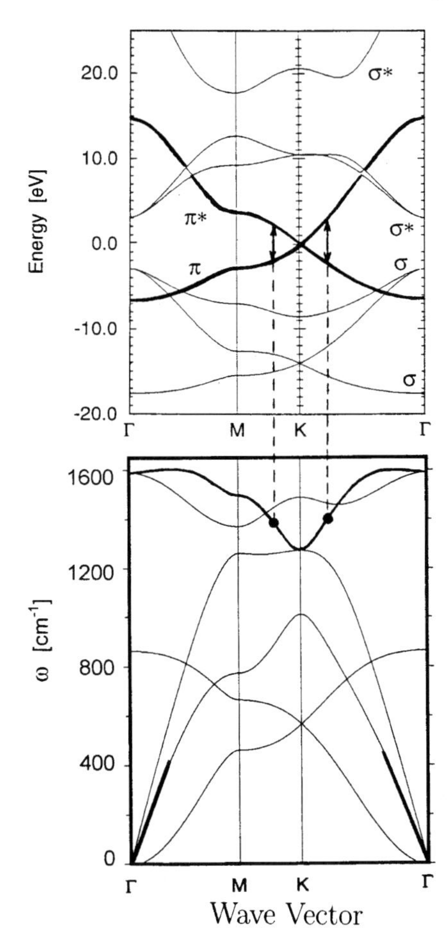

Fig. 8.8. Electronic energy bands of 2D graphite (*top*) [4,19]. Phonon dispersion curves of 2D graphite (*bottom*) [4,19]. Both the phonon branch that is strongly coupled to electronic bands in the optical excitation, and the electronic bands near the Fermi level ($E = 0$) that are linear in k are denoted by heavy lines. The slope for the low frequency TA phonon branch is also denoted by heavy lines

2000 °C, R is linear in λ_{L} in the range $400 < \lambda_{\mathrm{L}} < 700\,\mathrm{nm}$, so that we can write $R(\lambda_{\mathrm{L}}) \approx R_0 + \lambda_{\mathrm{L}} R_1$. We can then evaluate $C(\lambda_{\mathrm{L}})$ in the linear regime as $C(\lambda_{\mathrm{L}}) \approx C_0 + \lambda_{\mathrm{L}} C_1$, where C_0 and C_1 were, respectively, estimated to be $C_0 = -126\,\mathrm{\AA}$ and $C_1 = 0.033$, for λ_{L} given in Å, using $R(\lambda_{\mathrm{L}})$ data for PPP-2400. This empirical expression is accurate to about 10%. In the general case, a calibration of I_D/I_G vs. crystallite size L_a (from X-ray data) is needed when using other conventional laser lines, such as the 633 nm-line of the He-Ne laser or the 647 nm-line of the Kr ion laser.

We note in Fig. 8.7 that the peak frequency of the D-band also depends on the laser wavelength λ_{L} (or on the laser energy E_{L}). This is another characteristic feature of the resonant behavior of the Raman D-band in carbon materials. Vidano et al. [34] were the first to perform a systematic investigation of the Raman spectra of different kinds of carbon materials by varying the laser excitation energy E_{L}, and they observed that the frequency of this

D-band upshifts rapidly with increasing E_L. The same kind of dispersion was observed for the second-order G'-band around $2700\,\text{cm}^{-1}$, which is the overtone or second harmonic of the D-band frequency. This phenomenon occurs, in fact, in different kinds of carbon materials, such as graphon carbon black [33], hydrogenated amorphous carbon [35], glassy carbon and disordered crystalline graphite [31,32,34], multi-component carbon films [36], and carbon nanotubes [37,38].

In Fig. 8.9 we plot the laser energy (E_L) dependence of the frequencies of the D-band for PPP-2400 [27] and for glassy carbon [32]. The D-band frequency ω vs. E_L data can be well fit by straight lines, and the slopes $\partial\omega/\partial E_L$ are the same, $\partial\omega/\partial E_L \approx 50\,\text{cm}^{-1}/\text{eV}$, for the two sp^2 carbon materials shown in Fig. 8.9. The corresponding results for the second-order G'-band around $2700\,\text{cm}^{-1}$ are also plotted in Fig. 8.9 for PPP-2400, glassy carbon [32] and graphite. The fitting parameters of the linear regression for the second-order G'-band show that the slopes for all three samples are the same ($\partial\omega/\partial E_L \approx 100\,\text{cm}^{-1}/\text{eV}$), and are approximately two times the slope $\partial\omega/\partial E_L$ for the D-band, as would be expected since this second-order G'-band is the second-harmonic of the D-band. It is interesting to note that although single crystal graphite has no D-band, it does have a G'-band very

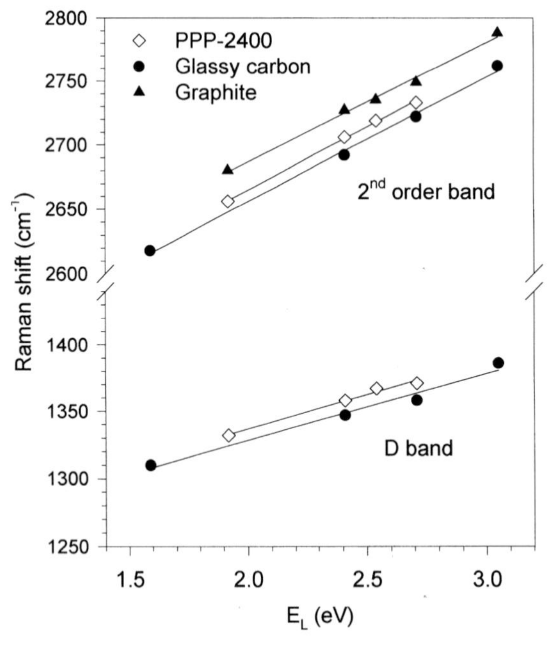

Fig. 8.9. Dependence of the frequencies of the D-band and the second-order G'-band on laser energy (E_L) for PPP-2400 and glassy carbon [32]. Measurements on crystalline graphite (HOPG) are included for the second-order Raman band $\omega_{G'} = 2\omega_D$ [28]

similar to that of the disordered carbons, and this G'-band is due to second-order scattering. This result shows that the dispersion of the D and G'-bands is an intrinsic property of the crystalline graphite lattice. The variation in the D and G'-band frequencies obtained with the same laser energy for the different carbon materials is ascribed to the slightly greater force constants for more well-ordered carbons. Therefore, the frequency of the second-order G' band for crystalline graphite is the highest among all carbon materials, as confirmed experimentally in Fig. 8.9.

Some authors [9] had previously attributed the dispersion effect to the selective sampling of different sized carbon clusters. However, this selective sampling model cannot explain, in the limit of large crystallites ($L_a \to \infty$), the dispersion of the second-order G'-band observed experimentally for ideal single crystal graphite, which shows no disorder-induced scattering. The D-band dispersion effect was recently explained by considering a selective coupling between phonons and electronic states having the same magnitude of the wave vector [27,31,39].

Figure 8.8 shows dispersion curves for the electronic energy bands (top) [4,19] and for the phonons (bottom) of 2D graphite [4,19]. In the upper part of Fig. 8.8 we see that electronic transitions between the π and π^* electronic states with energies corresponding to visible photons only occur in the vicinity of the K point in the Brillouin zone (BZ). Because of the linear k dispersion of the π and π^* bands near the K point, the energy difference ΔE between the valence and conduction bands can be written as

$$\Delta E(\Delta k) = E^c(\Delta k) - E^v(\Delta k) = \sqrt{3} a_0 \gamma_0 \Delta k, \tag{8.3}$$

where Δk is the distance between a given $\boldsymbol{k}$ point within the BZ and the K point ($\Delta \boldsymbol{k} = \boldsymbol{k}_K - \boldsymbol{k}$), while $\gamma_0 \sim 3.0\,\text{eV}$ is the nearest-neighbor C-C overlap energy and a_0 is the in-plane graphite lattice constant (2.46 Å). The phonons that are associated with the D-band are those that have wavevectors Δq with the same magnitude as the wave-vectors Δk of the electronic transitions that are in resonance with the laser energy (see dashed line in Fig. 8.8). Therefore, the resonance Raman condition can be written as $E_\text{L} = \Delta E(\Delta k) = \sqrt{3} a_0 \gamma_0 \Delta q$, and we can convert the experimental value for the slope ($\partial\omega/\partial E_\text{L}$) of the ω vs. E_L plot of Fig. 8.9 ($\partial\omega/\partial E_\text{L} \approx 50\,\text{cm}^{-1}/\text{eV}$) to relate $\partial\omega/\partial E_\text{L}$ to the slope of the D-band phonon dispersion in a ω vs. Δq plot as

$$\frac{\partial\omega}{\partial \Delta q} = \sqrt{3} a_0 \gamma_0 \frac{\partial\omega}{\partial E_\text{L}} \approx 650\,\text{Å}\,\text{cm}^{-1}. \tag{8.4}$$

The strong coupling of the light to the phonon branch associated with the D-band (shown as a heavy line in the phonon dispersion curves in Fig. 8.8) is due to the fact that the displacements of the atoms corresponding to the vibrational mode at the K point of this particular branch have a breathing-mode type behavior (see the inset of Fig. 8.7). In general, the modes which most strongly modulate the polarizability, and therefore have the largest Raman cross-sections, are of a symmetric breathing type.

The transverse acoustic (TA) phonon branch near $\omega = 0$ and $q = 0$ in Fig. 8.8 is also represented by a heavy line. The group velocity v_{TA} of this acoustic phonon in the long wavelength limit is $v_{\mathrm{TA}} = 1.23 \times 10^4\,\mathrm{m/s} = 650\,\mathrm{Å\,cm^{-1}}$ [40], which is very close to the value of $\partial\omega/\partial\Delta q$ shown in (8.4). This important result shows that the highly dispersive nature of the D and G'-bands is due to the coupling at the K point between the optic branch associated with these Raman bands and the transverse acoustic branch. The contributions from all the phonons around the K-point explain why the D and G' Raman bands tend to be broader than the Γ-point E_{2g_2} graphitic phonon band around $1582\,\mathrm{cm}^{-1}$.

8.2 Introduction to Fullerene Materials

This section reviews vibrational modes of fullerene molecules and related materials, emphasizing the unusual aspects of Raman scattering in these remarkable materials. The fullerene molecule is the fundamental building block of the crystalline phase [3], which is held together by weaker van der Waals forces that couple the fullerene molecules to each other. As a result, many of the properties of fullerene-related materials, including their Raman spectra, are closely connected with the molecular properties of isolated fullerenes.

The name of "fullerene" was given by Kroto and Smalley to the family of closed shell cage carbon molecules, consisting of 12 pentagonal rings and a variable number of hexagonal rings, (see Fig. 8.10) [41]. The more stable fullerenes obey the isolated pentagon rule, which states that pentagons should not be adjacent to each other, and this rule is based on stability arguments. The most stable and abundant fullerene is C_{60} and the next most common (or abundant) is C_{70} which can be visualized by inserting five hexagons around the equatorial belt of C_{60} normal to a 5-fold axis [see Figs. 8.10(a) and (b)]. Because of their different structures and symmetries, C_{60}, C_{70} and each of the higher mass fullerene [such as the two isomers of C_{80} shown in Figs. 8.10(c) and (d)] would be expected to show a unique and distinct Raman spectrum. The early demonstration of the 10-line first-order Raman spectrum of C_{60} [42] provided verification for the structural identification of the C_{60} molecule with icosahedral symmetry. Various dopants can be intercalated into crystalline C_{60} to a number of possible stoichiometries [3]. These dopants greatly modify the electronic properties of the host crystal depending on the dopant species and doping concentration. For the case of alkali metal dopants M_3C_{60} (where M = K, Rb) a conducting material with a relatively high superconducting transition temperature T_c is obtained [43], the highest being $T_c \sim 40\,\mathrm{K}$ in Cs_3C_{60} under a pressure of 12 kbar [44]. Other ordered solid state structures based on these dopants are possible (e.g., M_4C_{60} and M_6C_{60}).

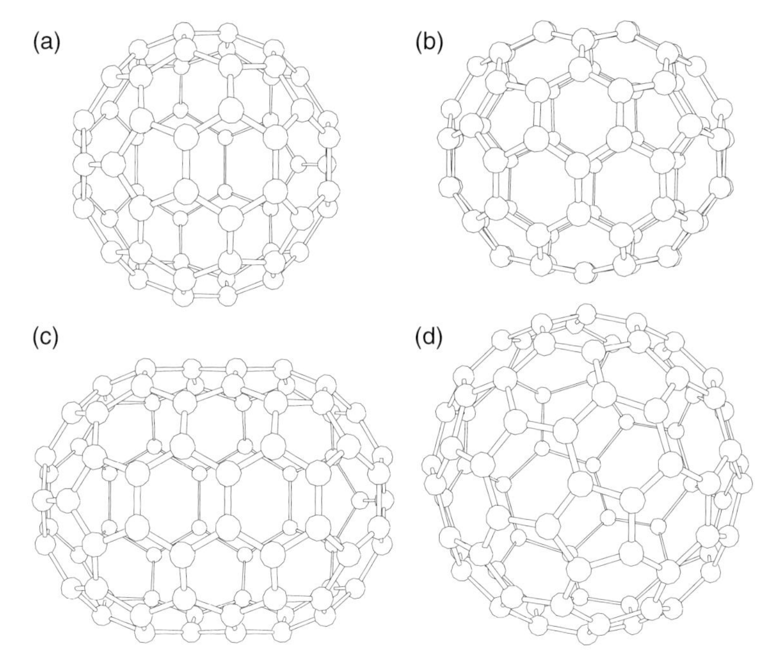

Fig. 8.10. (**a**) The icosahedral soccer-ball shaped C_{60} molecule. (**b**) The rugby ball–shaped C_{70} molecule (D_{5h} symmetry). (**c**) a C_{80} isomer, which is an extended rugby ball (D_{5d} symmetry). (**d**) a C_{80} isomer, which is also a truncated icosahedron (I_h symmetry) [3]

8.2.1 Mode Classification in Fullerene Molecules

The weak inter-molecular force leads to a molecular orientational ordering temperature ($T_{01} \sim 260\,\mathrm{K}$), above which the C_{60} molecules are freely rotating about their lattice positions. Consistent with the disparity between the strength of the inter- and intra-molecular bonds, the vibrational modes of solid C_{60} are divided into inter-molecular vibrations (or lattice modes) and intra-molecular vibrations (or simply "molecular" modes). The inter-molecular modes can be further divided into three subclasses: acoustic, optical, and librational modes, and occur at low frequencies (see Fig. 8.11). For doped C_{60} a new subclass of inter-molecular modes is introduced involving the relative motion of the cation (M^+) sublattice with respect to the fullerene molecular anion sublattice, shown in Fig. 8.11(c). Although several measurements [46–48] and calculations [42,49–52] of the inter-molecular fullerene modes have been made, most of the emphasis has been given to the intra-molecular vibrational modes.

The schematic picture in Fig. 8.11 is generally valid for all doped and undoped crystalline fullerene solids. The librational (a) modes lie lowest in frequency (10–$30\,\mathrm{cm}^{-1}$), followed by the inter-molecular optic (b) modes involving relative motion between neutral C_{60} molecules or C_{60}^{n-} anions

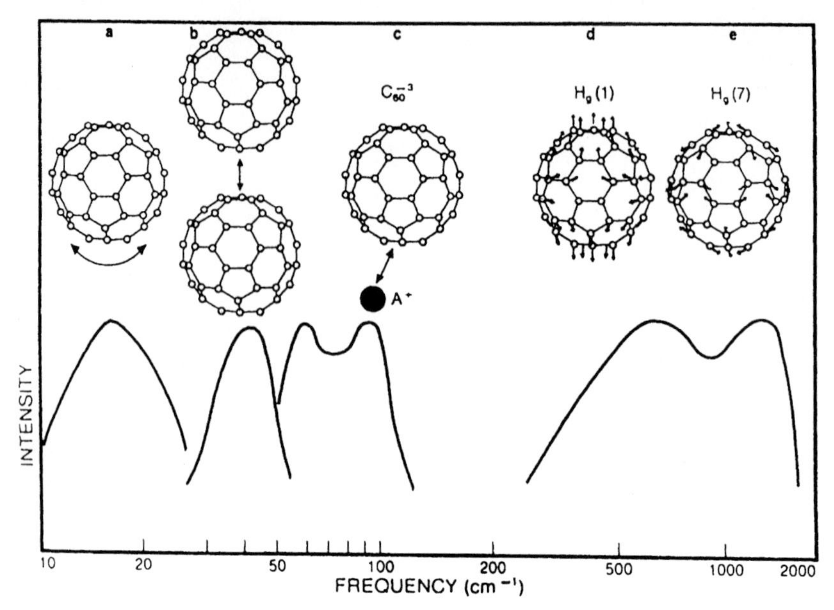

Fig. 8.11. Schematic phonon DOS (*solid curve*) for the various classes of vibrations in M_3C_{60} compounds (shown at the *top*). At lower frequencies, the compounds exhibit librational modes of individual C_{60} molecules (**a**), inter-molecular modes (**b**), and optic modes between C_{60} anions and dopant species (**c**). At higher frequencies (above $\sim 270\,\mathrm{cm}^{-1}$), the intra-molecular modes are dominant and follow the schematic vibrational density of states shown in the figure. These "intra-molecular" modes have predominantly radial character (**d**) at lower frequencies ($\omega < 700\,\mathrm{cm}^{-1}$) [for example the $H_g(1)$ mode] and at higher frequencies ($\omega > 700\,\mathrm{cm}^{-1}$) the mode displacements have predominantly a tangential character (**e**) [for example, the $H_g(7)$ mode] [45]

($30-60\,\mathrm{cm}^{-1}$), then the optic modes (c) involving relative motion between the C_{60}^{n-} anions and the M^+ cations ($50-120\,\mathrm{cm}^{-1}$), and finally the C_{60} intra-molecular or "molecular" (d, e) modes ($270-1700\,\mathrm{cm}^{-1}$), which involve relative C atom displacements on a single molecule or anion. The two broad bands schematically represent the lower frequency modes with predominantly radial displacements and the higher frequency modes with predominantly tangential displacements. As discussed below, these two broad density of states bands actually refer to 46 distinct molecular vibration frequencies. Some of the first-order intra-molecular modes have vibrational quanta as large as 0.1 to 0.2 eV ($800-1600\,\mathrm{cm}^{-1}$), which is significant when compared to the very small electronic bandwidths (~ 0.5 eV) for energy bands of crystalline C_{60} located near the Fermi level [3].

8.2.2 C_{60} Intra-Molecular Modes

Despite the large number of degrees of freedom ($60 \times 3 = 180$) for an isolated C_{60} molecule, its high icosahedral (I_h) symmetry results in a relatively simple Raman spectrum. After subtracting the six degrees of freedom corresponding to three translations and three rotations, we get 174 vibrational degrees of freedom for the C_{60} molecule. Standard group theoretical methods [53–55] yield 46 distinct intra-molecular mode frequencies with the following symmetries for an isolated C_{60} molecule

$$\begin{aligned}\Gamma_{\mathrm{mol}} = {} & 2A_g + 3F_{1g} + 4F_{2g} + 6G_g + 8H_g + A_u + 4F_{1u} \\ & + 5F_{2u} + 6G_u + 7H_u , \end{aligned} \tag{8.5}$$

where the subscripts g (gerade) and u (ungerade) refer to the symmetry of the eigenvector under the action of the inversion operator, and the symmetry labels (e.g., F_1, H) refer to irreducible representations of the icosahedral symmetry group (I_h) [3]. The degeneracy for each mode symmetry (given in parentheses) also follows from group theory: $A_g(1)$ $A_u(1)$; $F_{1g}(3)$, $F_{2g}(3)$, $F_{1u}(3)$, $F_{2u}(3)$; $G_g(4)$, $G_u(4)$; and $H_g(5)$, $H_u(5)$. Group theory, furthermore, indicates that 10 of the 46 mode frequencies are Raman-active ($2A_g + 8H_g$) in first-order, 4 are infrared (IR)-active ($4F_{1u}$), and the remaining 32 are optically silent. Experimental values for all of the 46 distinct mode frequencies, determined by first- and second-order Raman [56] and IR [57] spectroscopic features, are displayed in Table 8.1 [57], where the mode frequencies are identified with their appropriate symmetries and are listed in order of increasing vibrational frequency for each symmetry type.

The carbon atom displacements for the two nondegenerate A_g mode are shown in Fig. 8.12 and the corresponding eigenvectors for the 8 Raman-active mode of H_g symmetry [61] are found in [3]. The $A_g(1)$ "breathing"

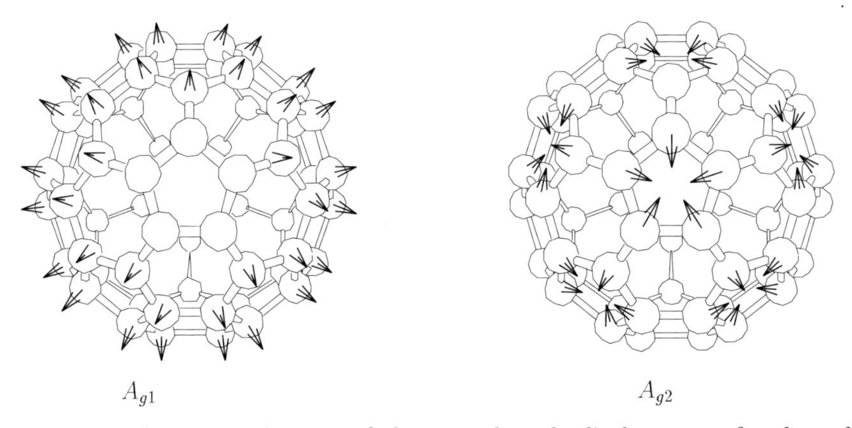

Fig. 8.12. Schematic diagram of the normal mode displacements for the radial breathing mode A_{g1} shown on the *left* and the pentagonal pinch mode A_{g2} shown on the *right* [3]

Table 8.1. Experimental determination of intra-molecular vibrational frequencies of the C_{60} molecule and their symmetries[a]

$\omega_i(\mathcal{R})$	Even-parity Frequency (cm^{-1}) optical[b]	tunneling[c]	$\omega_i(\mathcal{R})$	Odd-parity Frequency (cm^{-1}) optical[b]	tunneling[c]
$\omega_1(A_g)$	497.5	484	$\omega_1(A_u)$	1143.0	1145
$\omega_2(A_g)$	1470.0	1476			
			$\omega_1(F_{1u})$	526.5	532
$\omega_1(F_{1g})$	502.0	508	$\omega_2(F_{1u})$	575.8	589
$\omega_2(F_{1g})$	975.5	984	$\omega_3(F_{1u})$	1182.9	1186
$\omega_3(F_{1g})$	1357.5	–	$\omega_4(F_{1u})$	1429.2	1420
$\omega_1(F_{2g})$	566.5	573	$\omega_1(F_{2u})$	355.5	339
$\omega_2(F_{2g})$	865.0	871	$\omega_2(F_{2u})$	680.0	653
$\omega_3(F_{2g})$	914.0	895	$\omega_3(F_{2u})$	1026.0	1024
$\omega_4(F_{2g})$	1360.0	–	$\omega_4(F_{2u})$	1201.0	1202
			$\omega_5(F_{2u})$	1576.5	–
$\omega_1(G_g)$	486.0	460			
$\omega_2(G_g)$	621.0	637	$\omega_1(G_u)$	399.5	387
$\omega_3(G_g)$	806.0	–	$\omega_2(G_u)$	760.0	750
$\omega_4(G_g)$	1075.5	1065	$\omega_3(G_u)$	924.0	936
$\omega_5(G_g)$	1356.0	1355	$\omega_4(G_u)$	970.0	960
$\omega_6(G_g)$	1524.5[d]	1525	$\omega_5(G_u)$	1310.0	1299
			$\omega_6(G_u)$	1446.0	–
$\omega_1(H_g)$	273.0	266			
$\omega_2(H_g)$	432.5	428	$\omega_1(H_u)$	342.5	290
$\omega_3(H_g)$	711.0	710	$\omega_2(H_u)$	563.0	557
$\omega_4(H_g)$	775.0	774	$\omega_3(H_u)$	696.0	678
$\omega_5(H_g)$	1101.0	1089	$\omega_4(H_u)$	801.0	799
$\omega_6(H_g)$	1251.0	1250	$\omega_5(H_u)$	1117.0	1129
$\omega_7(H_g)$	1426.5	1420	$\omega_6(H_u)$	1385.0	1396
$\omega_8(H_g)$	1577.5	1589	$\omega_7(H_u)$	1559.0	1557

[a]Present status of mode frequency determination.
[b]Optical mode frequencies are derived from a fit to first-order and higher-order Raman [56] and IR spectra [57]. The Raman-active modes have A_g and H_g symmetry and the IR-active modes have F_{1u} symmetries. The remaining 32 modes are optically silent.
[c]Tunneling mode frequencies are from analysis of the inelastic electron tunneling spectroscopy experiments [58,59].
[d]Interpretation of an isotopically-induced line in the Raman spectra suggests that this mode should be at 1490 cm^{-1} [60].

mode ($492\,\mathrm{cm}^{-1}$) involves identical radial displacements for all 60 carbon atoms, whereas the higher-frequency $A_g(2)$ mode, or "pentagonal pinch" mode ($1469\,\mathrm{cm}^{-1}$), involves tangential displacements, with a contraction of the pentagonal rings and a corresponding expansion of the hexagonal rings operative for one set of displacements. The eigenvector displacements for the 8 fivefold-degenerate H_g modes are much more complex and span the frequency range from 273 [$H_g(1)$] to $1578\,\mathrm{cm}^{-1}$ [$H_g(8)$]. The ten Raman-active mode frequencies can be seen in the experimental Raman spectrum shown in the top trace of Fig. 8.13 for a vacuum-deposited film of polycrystalline C_{60}

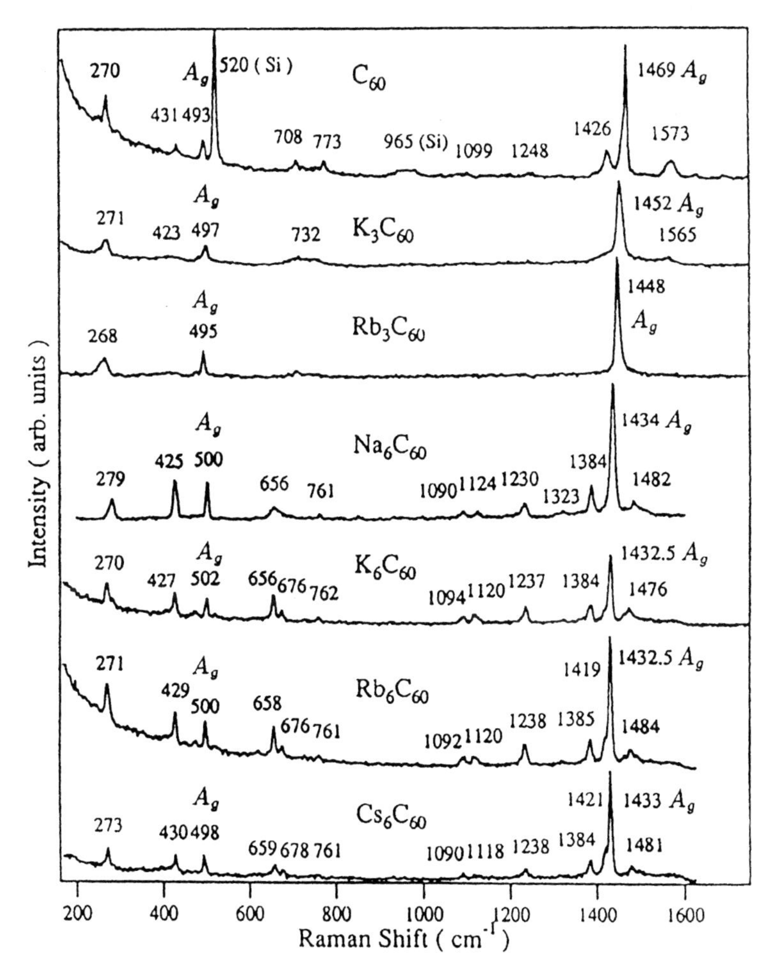

Fig. 8.13. Unpolarized Raman spectra ($T = 300\,\mathrm{K}$) for solid C_{60}, and alkali metal doped K_3C_{60}, Rb_3C_{60}, Na_6C_{60}, K_6C_{60}, Rb_6C_{60} and Cs_6C_{60} [47,62]. The tangential and radial modes of A_g symmetry are identified, as are the features associated with the Si substrate

[42,47,48,53,62–64]. In Table 8.1 we list the frequencies of the Raman-active A_g and H_g modes as measured experimentally.

Since the interactions between C_{60} molecules in the solid state are weak, the vibrational spectra in gas [65], solid [53,66], and solution phases [66], are very similar and crystal field effects are small. At room temperature, the C_{60} molecules are rotating rapidly, at a frequency which is low compared with the intra-molecular vibrational frequencies ($10-40\,\mathrm{cm}^{-1}$). Therefore, the C_{60} molecules in the crystalline phase at room temperature are orientationally disordered.

8.2.3 Higher-Order Raman Modes in C_{60}

Because of the highly molecular nature of fullerene crystals, Raman scattering measurements on a C_{60} film show a well-resolved second-order Raman spectrum (see Fig. 8.14) [56], which is very unusual for a crystalline solid. From group theoretical considerations, the expected number of second-order Raman lines with A_g and H_g symmetry is very large for the icosahedral C_{60} molecule [56], consistent with the experimental results in Fig. 8.14. The second-order Raman spectra include both overtones (i.e., integral multiples of the modes in Table 8.1) and combination modes (i.e., sums and differences of the modes in this table). Using group theoretical methods [56,57], study

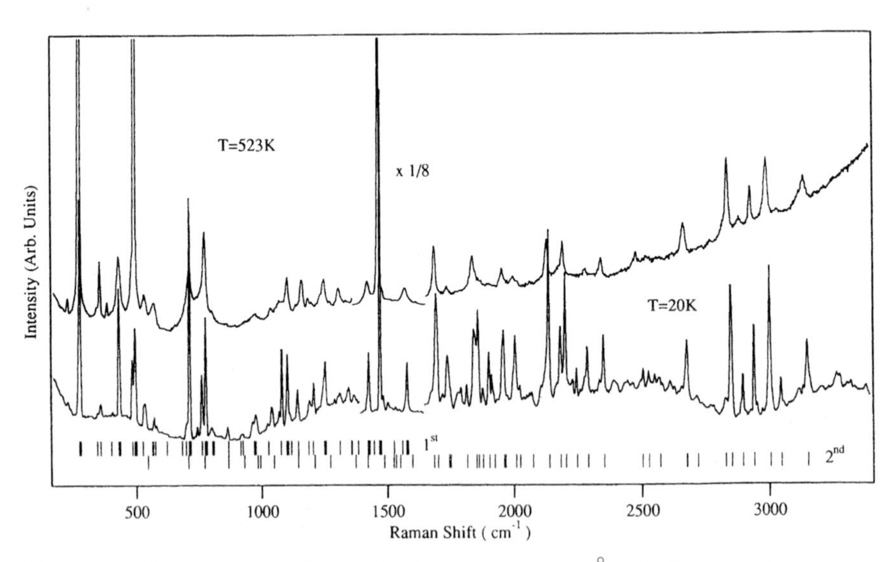

Fig. 8.14. Raman spectra for solid C_{60} films (~ 7000 Å thick) taken at temperatures $T = 523$ and $20\,\mathrm{K}$, showing overtones, combination modes, and modes arising from isotopic symmetry-breaking effects. The data were taken using 488.0 nm Ar laser radiation and the scale of the 523 K spectrum has been multiplied by a factor of 8 relative to the low temperature spectrum. From the second-order spectra the frequencies of many of the silent modes are obtained [56]

of the second-order overtones and combination modes of the Raman spectra along with the combination modes seen in the infrared spectra of C_{60} provides a powerful technique for the determination of the frequencies of the 32 silent modes (see Table 8.1) [57,66]. The table also shows the good agreement obtained for mode frequencies found by fitting the second-order Raman spectrum and the frequencies obtained from inelastic electron tunneling data [58,59], which are not subject to symmetry selection rules.

8.2.4 Perturbations to the Raman Spectra

Perturbations to the Raman spectra have been reported from a variety of origins, including temperature-dependent effects, and phase transitions, pressure-dependent effects, excited state effects, isotope effects, and surface related effects [3]. Except for the onset of phase transitions associated with inter-molecular ordering, the temperature dependence of the Raman spectra of fullerenes is weak. The abrupt changes in the Raman spectrum associated with structural phase transitions [15,67] provide complementary information about the phase transitions, consistent with that provided by many other techniques [3]. Raman scattering provided much useful information regarding pressure-dependent effects [68,69], because of the compatibility with diamond anvil pressure cells.

Because of the meta-stability (long lifetime of $\sim$ 50 ms at low temperature) of the lowest-lying triplet electronic state in C_{60} above the Fermi level, it is possible to observe vibrational features [70] in the Raman spectra associated with this excited state [71,72]. The mode frequency of the $A_g(2)$ pentagonal pinch mode in the excited triplet state is found to be downshifted relative to that for the vibration in the electronic ground state, because of the weaker binding and consequently reduced force constants for the excited state [70]. The presence of the ^{13}C isotope (with natural abundance of 1.1 %, the remaining 98.9 % being ^{12}C), influences the Raman spectrum of C_{60} by lowering the molecular symmetry, thereby turning on symmetry-forbidden modes [3,60].

8.2.5 Vibrational Spectra for Phototransformed Fullerenes

Phototransformed C_{60} has been identified as a perturbed polymeric structure with photo-induced covalent bonds formed between adjacent molecules in the crystal structure. Because of the symmetry lowering of the phototransformed fullerene phases, Raman and infrared spectroscopies provide powerful tools for the characterization of the phototransformation process. The overall effect of phototransformation on the Raman spectrum of solid C_{60} is the appearance of many more features due to symmetry lowering effects [73]. Phototransformation occurs at a moderate optical flux ($> 5\,W/cm^2$) in the visible and in the ultraviolet (UV) regions of the spectrum when light is incident on thin solid films [73]. One of the most important effects of phototransformation on the Raman spectrum of C_{60} is the quenching of the intensity of the $A_g(2)$ pentagonal pinch mode at 1469 cm^{-1} in pristine C_{60}, with a new mode appearing close by at 1458 cm^{-1} in the phototransformed material. The shift of the pentagonal pinch mode to lower frequencies has been modeled using molecular dynamics [74] and the calculated frequency shifts are in good agreement with experiment.

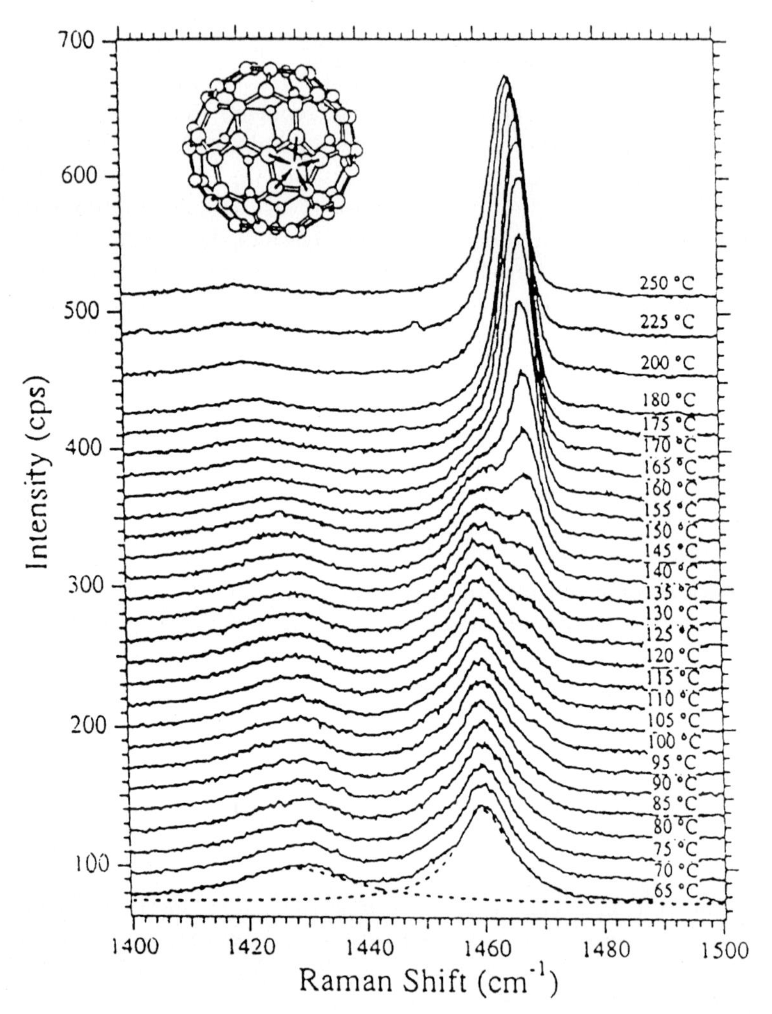

Fig. 8.15. Raman spectra in the vicinity of the pentagonal pinch mode for a photochemically polymerized C_{60} film ($d \approx 4500$ Å) on a Suprasil (fused silica) substrate for various temperatures. The *dashed curves* in the bottom spectrum represent a Lorentzian lineshape fit to each of the Raman lines in the spectrum at 65 °C. As the temperature is increased, the features associated with the polymerized phase are quenched and the spectrum of the freely rotating molecule is obtained [75]

In Fig. 8.15 the Raman spectra for polymerized C_{60} are shown in the vicinity of the $A_g(2)$ pentagonal pinch mode for various temperatures. The competition between the photo-induced attachment of C_{60} monomers to form oligomers and the thermal detachment process to form monomers has been monitored by plotting the temperature dependence of the integrated intensity of the 1458 cm^{-1} feature associated with the polymerized phase and the corresponding intensity of the 1469 cm^{-1} sharp Raman line associated with the C_{60} monomer [75]. The relative intensity of these features can be explained in terms of rate equations for oligomer formation by the phototransformation process which competes with the thermal dissociation process for monomer formation. Detailed fits of the model to the experimental data show that the rate constant for polymerization is dependent on the photon flux, the temperature, and the wavelength of the light. A new Raman-active mode, identified with a stretching of the cross-linking bonds *between* adjacent C_{60} molecules [75], is observed at 118 cm^{-1} in phototransformed C_{60} [73], in good agreement with theoretical predictions for this mode [76]. The various characteristic features of the Raman spectra are often used to characterize phototransformed C_{60} material.

8.2.6 Inter-Molecular Modes

Since the C_{60} molecules rotate rapidly about their equilibrium positions on an fcc lattice above the orientational ordering temperature $T_{01} \sim 261$ K [77], the orientational restoring force and libron effects are probed by Raman studies below T_{01}. The strongest Raman-active libron has a frequency of 17.6 cm^{-1} and upshifts by 0.52 cm^{-1}/kbar, while the next strongest spectral feature is at ~ 20.7 cm^{-1} and upshifts under pressure by 0.37 cm^{-1}/kbar [78].

8.2.7 Vibrational Modes in Doped C_{60}-based Solids

Since the discovery of moderately high-temperature superconductivity in the alkali metal (M)–doped C_{60} solids M_3C_{60} (M = K, Rb), considerable activity has been expended to document the doping-induced changes in the vibrational modes of these materials and to investigate whether or not the superconducting pairing interaction is mediated by vibrational modes and, if so, by which modes, optical or acoustic. The Raman spectra for insulating compositions of M_xC_{60}, such as M_6C_{60} (see Fig. 8.13), have also been studied for comparison to the behavior of C_{60} itself. Also, the Raman spectra of M_3C_{60} phases have been studied to determine the strength of the electron-phonon coupling for their A_g and H_g modes.

The Raman spectra for alkali metal doped C_{60} has been carried out over a range of stoichiometries in order to study the charge transfer between the M_x and C_{60} sublattices. In fact, measurements of the alkali metal induced shift have been widely used to characterize the charge transfer to the fullerene

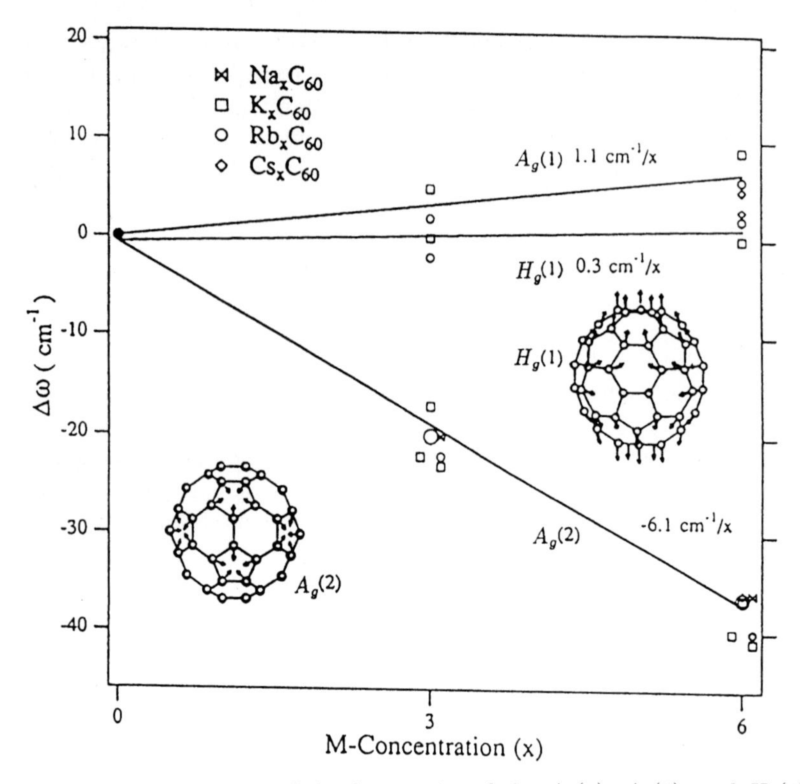

Fig. 8.16. Dependence of the frequencies of the $A_g(1)$, $A_g(2)$, and $H_g(1)$ modes on the alkali metal concentration x in M_xC_{60}, where M = Na, K, Rb, Cs. The frequency shifts of these Raman-active modes are plotted relative to the frequencies $\omega[A_g(1)] = 493\,\text{cm}^{-1}$, $\omega[A_g(2)] = 1469\,\text{cm}^{-1}$, and $\omega[H_g(1)] = 270\,\text{cm}^{-1}$ for C_{60} at $T = 300\,\text{K}$. A schematic of the displacements for the eigenvectors for the $A_g(2)$ pentagonal pinch and $H_g(1)$ modes are shown. All the C atom displacements for the $A_g(1)$ modes are radial (see Fig. 8.12) and of equal magnitude [53,79,80]

anion, which can then be used to estimate the alkali metal uptake in the C_{60} lattice (see Fig. 8.16) [53,79,80].

The introduction of alkali metal dopants into the lattice also gives rise to low-frequency vibrational modes, whereby the alkali metal ions vibrate relative to the large fullerene molecules (see Fig. 8.11). The presence of alkali metal dopants in the lattice also influences the characteristics of the intermolecular modes in crystalline C_{60}. Because of the very heavy damping of the vibration between the alkali metal cation and the C_{60}^{3-} anion, this overdamped mode is not directly observed by Raman scattering, but can be seen by pump-probe femtosecond spectroscopic techniques [81] at $\sim 150\,\text{cm}^{-1}$ in K_3C_{60} and Rb_3C_{60}.

Mode broadening in M_xC_{60} occurs with increasing x, and this broadening is most pronounced near $x \sim 3$ in the metallic phases of M_3C_{60}. However, the Raman lines for $x = 6$ are quite narrow because it is relatively easy to make these samples in single phase form. The line broadening for the $H_g(2)$, $H_g(7)$, and $H_g(8)$ Raman-active modes for the M_3C_{60} phases has been attributed to electron-phonon coupling predominantly, while for other H_g modes this broadening is so great that the lines cannot be observed in the M_3C_{60} spectrum [82,83]. Temperature-dependent studies of the features in the Raman spectra for Rb_3C_{60} show that the mode frequencies and linewidths remain unchanged across the normal–superconducting transition [83,84].

8.2.8 Vibrational Spectra for C_{70} and Higher Fullerenes

Except for C_{70}, relatively little is known about the detailed vibrational spectra of the higher mass fullerenes [3]. Because of their generally lower symmetry, their larger number of degrees of freedom, and the many possible isomers, little systematic experimental or theoretical study has been undertaken on the Raman spectra of high-mass fullerene molecules or crystalline solids [85]. The lack of adequate quantities of purified and characterized samples of higher mass fullerenes has curtailed experimental studies.

The Raman spectrum for C_{70} is much more complicated than for C_{60} because of the lower symmetry and the large number of Raman-active modes (53), out of a total of 122 possible vibrational mode frequencies for C_{70} [69,86–92]. The D_{5h} symmetry for the C_{70} molecule implies that there are five inequivalent atomic sites [3], which, in turn, allows for the existence of eight different bond lengths as well as 12 different angles between the bonds connecting nearest-neighbor atoms [93]. For this reason simplified models are used to explain the large number of Raman lines observed in the first-order spectrum [87]. As for the case of C_{60} films, the dependence of the Raman spectra on temperature, pressure and doping has been studied for C_{70} films. However, a detailed interpretation of these spectra has been more difficult.

8.3 Raman Scattering in Carbon Nanotubes

Raman scattering has provided an important technique for characterizing carbon nanotubes and for studying their unique features which are connected with their unusual 1D electronic structure and their unusual mechanical strength and compliance. Soon after the announcement of their experimental observation [94], carbon nanotubes attracted the attention of the scientific community because of the theoretical prediction [95] that about 1/3 of the nanotubes are metallic and 2/3 are semiconducting depending on their diameter d_t and chiral angle θ [3,4].

The early observations of carbon nanotubes involved multi-wall nanotubes [94,96]. The synthesis of single-wall carbon nanotubes in 1993 [97,98] further stimulated work in the field, since most of the theoretical predictions were made for single-wall carbon nanotubes [4]. The discovery in 1996 of a much more efficient synthesis route for single-wall carbon nanotubes, based on the laser vaporization of graphite, in a catalyzed reaction [99] greatly stimulated systematic experimental studies of carbon nanotubes and facilitated comparisons to theoretical calculations.

After a brief review of the structure (Sect. 8.3.1) and phonon dispersion relations for carbon nanotubes (Sect. 8.3.2), we review progress in the currently active area of resonant Raman spectroscopy studies of carbon nanotubes (Sect. 8.3.3), and the effect of charge transfer in doped nanotubes (Sect. 8.3.5).

8.3.1 Structure of Carbon Nanotubes

The basic carbon nanotube is a cylinder of carbon atoms arranged on a honeycomb lattice, as in a single layer of graphite, and with almost the same nearest-neighbor C–C spacing (1.421 Å in graphite and $\sim$ 1.44 Å in carbon nanotubes) [4,100]. Carbon nanotubes can have any one of the three basic geometries shown in Fig. 8.17. Carbon nanotubes are usually capped at either end by half of a fullerene, so that the smallest diameter fullerene obeying the isolated pentagon rule, C_{60}, corresponds to the smallest diameter nanotube (7.1 Å).

The structure of the nanotube can be understood by referring to Fig. 8.18, which demonstrates the rolling of a segment of a single graphite layer (called a graphene sheet) into a seamless cylinder. In this figure we see that points O and A are crystallographically equivalent on the two dimensional (2D) graphene sheet. The points O and A can be connected by a chiral vector $\boldsymbol{C}_h = n\hat{a}_1 + m\hat{a}_2$, where $\hat{a}_1$ and $\hat{a}_2$ are unit vectors for the honeycomb lattice of the graphene sheet. Lines OB and AB' are perpendicular to $\boldsymbol{C}_h$ at points O and A. If we now roll the graphene sheet and superimpose OB onto AB', then we obtain a cylinder of carbon atoms which constitutes a carbon nanotube. A single-wall carbon nanotube is thus uniquely determined by the integers (n, m) which specify the chiral vector $\boldsymbol{C}_h$ in terms of the number n of $\hat{a}_1$ units and the number m of $\hat{a}_2$ units contained in $\boldsymbol{C}_h$. The nanotube can also be uniquely specified by its diameter d_t and chiral angle θ (see Fig. 8.18), where

$$d_t = C_h/\pi = \sqrt{3}a_{\mathrm{C-C}}(m^2 + mn + n^2)^{1/2}/\pi \tag{8.6}$$

in which $a_{\mathrm{C-C}}$ is the nearest-neighbor C–C distance, C_h is the length of the chiral vector $\boldsymbol{C}_h$. The chiral angle θ is defined as the angle that $\boldsymbol{C}_h$ makes with the zigzag direction ($\hat{a}_1$), and is given by

$$\theta = \tan^{-1}[\sqrt{3}m/(m + 2n)] \tag{8.7}$$

as shown in Fig. 8.18. The unit cell of the 1D carbon nanotube is, therefore, a rectangle formed by the vectors $\boldsymbol{C}_h$ and $\boldsymbol{T}$ shown in Fig. 8.18, where $\boldsymbol{T}$ is

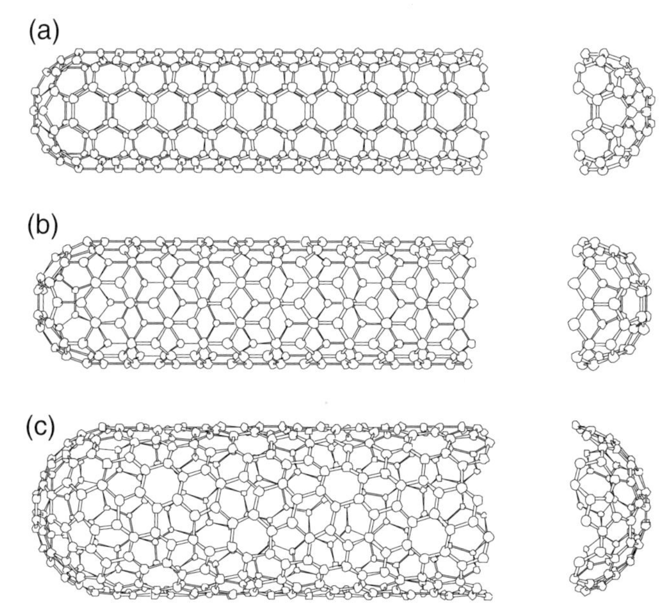

Fig. 8.17. Schematic models for single-wall carbon nanotubes with different symmetries using the notation of Fig. 8.18: (**a**) an "armchair" (n, n) nanotube with $\theta = 30°$, (**b**) a "zigzag" $(n, 0)$ nanotube with $\theta = 0°$, and (**c**) a "chiral" (n, m) nanotube with a general direction, such as $\overrightarrow{OB}$ in Fig. 8.18, where $0 < \theta < 30°$. The actual nanotubes shown here correspond to (n, m) values of: (**a**) (5, 5), (**b**) (9, 0), and (**c**) (10, 5) [101]. The nanotubes in this figure are capped by one half of a fullerene molecule. Single-wall nanotubes currently being synthesized have a typical aspect ratio (length/diameter) of $10^3 – 10^4$

the smallest lattice vector from O in the direction normal to $\boldsymbol{C}_h$. Since the basis vectors $\boldsymbol{C}_h$ and $\boldsymbol{T}$ of the 1D unit cell are large compared with the basis vectors $\hat{a}_1$ and $\hat{a}_2$ of the unit cell for the graphene sheet (see Fig. 8.18), the reciprocal space 1D unit cell (Brillouin zone) for carbon nanotubes is small compared to that for the graphene sheet.

Ropes (or bundles) of single-wall nanotubes can now be prepared by a number of methods, including laser vaporization [99] and carbon arc methods [102] with a narrow distribution of diameters close to that of the (10,10) armchair nanotube (1.38 nm). Variations in the most probable diameter and in the width of the diameter distribution are sensitively controlled by the composition of the catalyst, the growth temperature and other growth conditions [103]. Characterization of nanotubes with regard to their diameters and chiral angles can be carried out either by transmission electron microscopy (TEM) or by scanning tunneling microscopy (STM) [4].

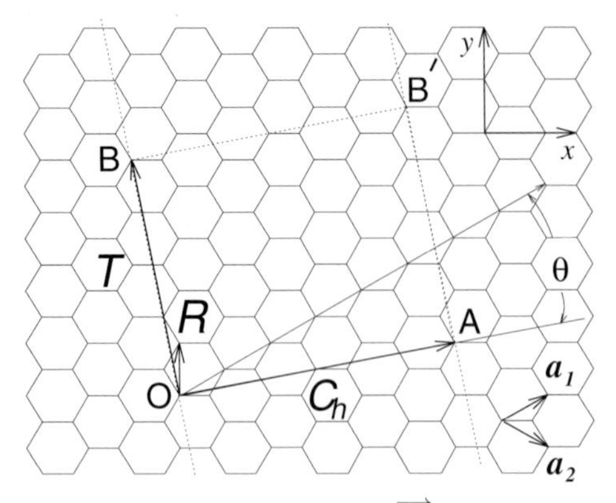

Fig. 8.18. The chiral vector $\overrightarrow{OA}$ or $\boldsymbol{C}_h = n\hat{a}_1 + m\hat{a}_2$ is defined on the flat honeycomb lattice of carbon atoms by the graphene unit vectors $\hat{a}_1$ and $\hat{a}_2$ and (n, m) are integers. The chiral angle θ is defined as the angle between $\boldsymbol{C}_h$ and $\hat{a}_1$. Along the zigzag axis $\theta = 0°$. Also shown is the lattice vector $\overrightarrow{OB} = \boldsymbol{T}$ of the 1D nanotube unit cell. The vector $\boldsymbol{R}$ denotes the basic symmetry operation for the carbon nanotube. The diagram is constructed for $(n, m) = (4, 2)$. The area defined by the rectangle $(OAB'B)$ is the area of the 1D unit cell of the nanotube [3]

8.3.2 Nanotube Phonon Modes

The phonon dispersion relations for nanotubes can be obtained from those of the 2D graphene sheet (see Fig. 8.4) by using a zone folding approach [19,104],

$$\omega_{1D}(k) = \omega_{2D}\left(k\,\hat{K}_2 + \mu\,\hat{K}_1\right) ,$$
$$\mu = 0, 1, 2, ..., N-1, \tag{8.8}$$

where the subscripts 1D and 2D refer, respectively, to the one-dimensional nanotube and the two-dimensional graphene sheet, k is a wave vector along $\hat{K}_2$ the nanotube axis direction, μ is a non-negative integer used to label the wave vectors or the states along the $\hat{K}_1$ reciprocal space direction, normal to the nanotube axis, and N denotes the number of hexagons in the 1D unit cell of the nanotube. Since there are 2 carbon atoms per hexagon, the number of vibrational degrees of freedom is $6N$.

The 1D dispersion relations assume that the lengths of the nanotubes are much larger than their diameters, so that the nanotubes can be described in the 1D limit where the nanotubes have infinite length, the k points are quasi-continuous, and the contributions from the carbon atoms in the caps can be neglected. The dispersion relations depend on the nanotube symmetry, and therefore the dispersion relations are different for armchair, zigzag and chiral nanotubes. There are three acoustic modes and one rotational mode (for an infinitely long axial system) which have $\omega \to 0$ as $k \to 0$. We show in Fig. 8.19

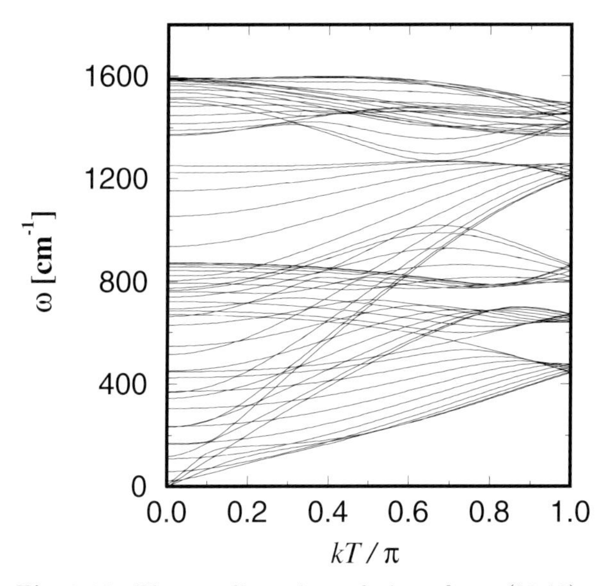

Fig. 8.19. Phonon dispersion relations for a (10,10) armchair carbon nanotube which is capped by a hemisphere of the icosahedral C_{240} fullerene. For the (10,10) nanotube, the number of hexagons per unit cell is $N = 20$, and there are 120 degrees of freedom, corresponding to 69 distinct non-zero mode frequencies, at the Γ-point ($k = 0$), of which 16 are Raman-active and 8 are infrared-active. In addition, 45 of the optical mode frequencies at $k = 0$ are optically silent, and there are also three acoustic (x, y, z) modes, and a rotational mode for which $\omega \to 0$ as $k \to 0$ [4]

an example of the phonon dispersion curves $\omega_{1D}(k)$ for a (10,10) single-wall armchair nanotube.

Group theory shows that the irreducible representations describing the vibrational modes for carbon nanotubes depend on the nanotube symmetry, which differs according to whether the nanotubes are of the armchair (n, n), zigzag $(n, 0)$, or chiral $[(n, m),\ m \neq n, 0]$ types (see Fig. 8.17) [3,4]. The results of the group theoretical analysis [3,4] are summarized in Table 8.2 for the various kinds of nanotubes.

Table 8.2. Symmetries of Raman-active and IR-active modes for carbon nanotubes

Nanotube structure	point group	Raman-active modes	IR-active modes
armchair (n, n) n even	D_{nh}	$4A_{1g} + 4E_{1g} + 8E_{2g}$	$A_{2u} + 7E_{1u}$
armchair (n, n) n odd	D_{nd}	$3A_{1g} + 6E_{1g} + 6E_{2g}$	$2A_{2u} + 5E_{1u}$
zigzag $(n, 0)$ n even	D_{nh}	$3A_{1g} + 6E_{1g} + 6E_{2g}$	$2A_{2u} + 5E_{1u}$
zigzag $(n, 0)$ n odd	D_{nd}	$3A_{1g} + 6E_{1g} + 6E_{2g}$	$2A_{2u} + 5E_{1u}$
chiral (n, m) $n \neq m \neq 0$	C_N	$4A + 5E_1 + 6E_2$	$4A + 5E_1$

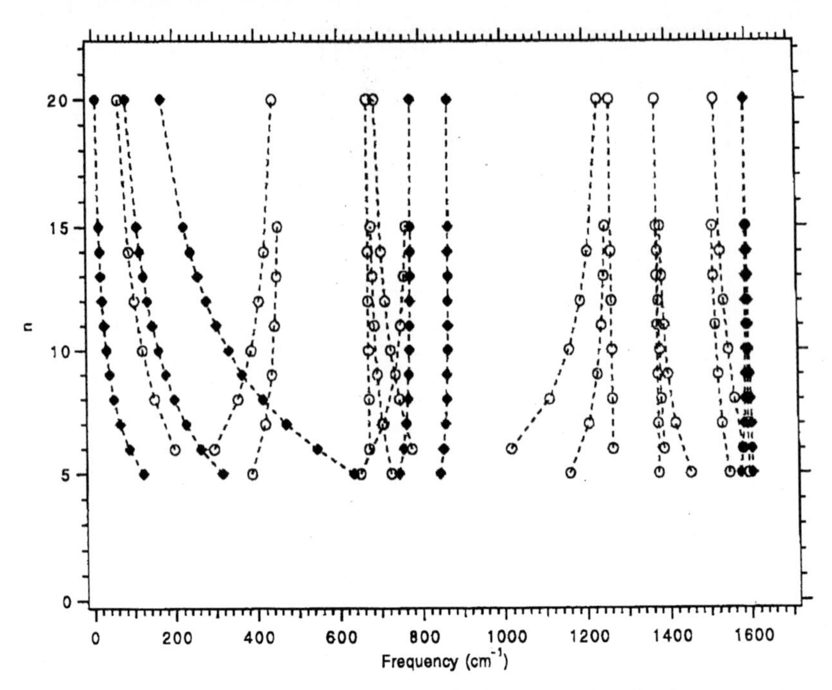

Fig. 8.20. The armchair index n vs. mode frequency for the Raman-active modes of single-wall armchair (n, n) carbon nanotubes [38]. For the armchair nanotubes the diameter is given in nanometers by $d_t = 0.138\,n$

Although the number of vibrational modes increases as the number of carbon atoms in the 1D unit cell increases, the number of Raman-active and infrared-active modes remains constant within each kind of nanotube (see Table 8.2). Thus, we can show the dependence of their Raman-active mode frequencies on the nanotube diameter for a given type of nanotube, such as in Fig. 8.20 for the armchair nanotubes. Here it is seen that certain Raman-active mode frequencies are strongly dependent of nanotube diameter and others are not. Shown in Fig. 8.21 are the normal mode displacements associated with the modes expected to have high Raman intensity (see Sect. 8.3.3) [4,38]; the displacements for modes with low intensity are not shown.

The most intense bands in the Raman spectra of single-wall carbon nanotubes (SWNT) are the radial breathing mode (RBM) near $180\,\mathrm{cm}^{-1}$ and the modes between 1500 and $1600\,\mathrm{cm}^{-1}$ associated with the tangential displacement C$-$C bond stretching motions of the nanotube (see Fig. 8.21). The RBM belongs to the identity representation (A_{1g} or A_1) and, according to theoretical predictions, its frequency is inversely proportional to the tube diameter, without any dependence on chiral angle θ [4,105]. Theoretical calculations show that the high frequency Raman-active modes associated with the tangential C$-$C bond stretching motions have A_{1g}, E_{1g} and E_{2g} (arm-

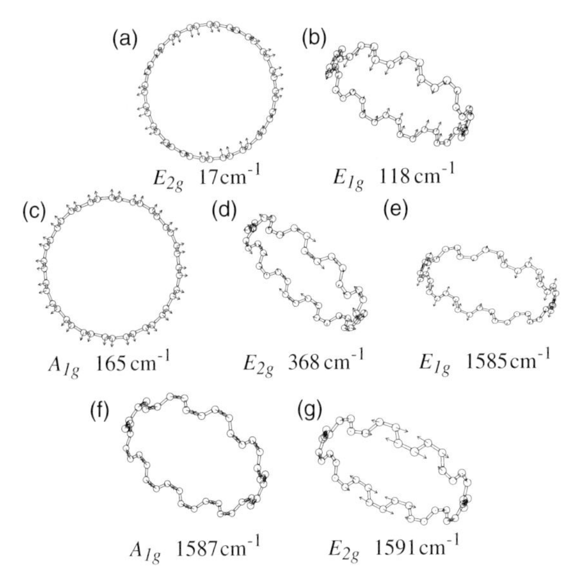

Fig. 8.21a–g. The calculated Raman mode atomic displacements, frequencies, and symmetries for those (10,10) nanotube modes with strong Raman intensity. The A_{1g} modes have no nodes, the E_{1g} modes have 2 nodes, and the E_{2g} modes have 4 nodes. For the A_{1g} mode at $1587\,\mathrm{cm}^{-1}$, all the A atoms move in the opposite direction to their nearest neighbor B atoms. We show the displacements for only one of the two partners of the doubly degenerate E_{1g} and E_{2g} modes [4]

chair and zigzag) [or A_1, E_1 and E_2 (chiral)] symmetries. The frequencies and number of Raman-active modes between 1500 and $1600\,\mathrm{cm}^{-1}$ depend on the diameter and chirality of the nanotube [3,4], and for the case of armchair (n,n) nanotubes, on whether n is even or odd. Therefore, several different modes are expected in a sample containing nanotubes of different types. However, within each type of nanotube, the modes in the $1580-1590\,\mathrm{cm}^{-1}$ range are expected to be very weakly dependent on the nanotube diameter, by perhaps an order of magnitude less than for the case of the diameter dependence of the radial breathing mode. In addition, the tangential modes near $1530\,\mathrm{cm}^{-1}$ with E_{1g} or E_{2g} (E_1 or E_2) symmetry are expected to show a diameter dependence of their mode frequencies that is intermediate between that for the modes near $180\,\mathrm{cm}^{-1}$ and near $1580\,\mathrm{cm}^{-1}$, as can, for example, be seen in Fig. 8.20 which shows the expected diameter dependence of the Raman-active modes for armchair nanotubes [4,106].

8.3.3 Raman Spectra of Single-Walled Carbon Nanotubes

Raman spectroscopy has provided a particularly valuable tool for examining the mode frequencies of carbon nanotubes with specific diameters, for evaluating the merits of theoretical models for the 1D phonon dispersion relations, for characterizing the diameter distribution in nanotube samples, and for studying the 1D electron DOS in resonant Raman experiments through the electron-phonon coupling mechanism [38].

Most of the early experiments on the vibrational spectra of carbon nanotubes were carried out on multi-wall carbon nanotubes which had too large a diameter to observe detailed quantum effects associated with the 1D dispersion relations discussed above [107,108]. A major advance in Raman spectroscopy studies of carbon nanotubes occurred through the availability of SWNT ropes [99] with a narrow diameter distribution [109]. Raman spectra taken on such samples (see Fig. 8.22) show a number of well-resolved Raman features, including strong lines near the frequency of the graphite E_{2g_2} mode at $1582\,\mathrm{cm}^{-1}$ [110,111], and a strong characteristic feature around $180\,\mathrm{cm}^{-1}$, which has A_{1g} symmetry and is associated with the radial breathing mode (RBM) of the nanotube.

Quantum effects are observed in the Raman spectra of single-wall carbon nanotubes through the resonant Raman enhancement effect, which is seen experimentally by measuring the Raman spectra at a number of laser excitation energies, as shown in Fig. 8.22. Resonant enhancement occurs when the laser excitation frequency corresponds to an electronic transition between the sharp features (i.e., $E^{-1/2}$ singularities) in the one-dimensional electronic DOS of the carbon nanotube, as shown in Fig. 8.23. Since the separation energies between these sharp features in the density of states are inversely proportional to the nanotube diameter, a change in the laser frequency may bring into optical resonance a carbon nanotube with a different diameter.

The presence of the RBM feature in the Raman spectrum is the signature for single-wall nanotubes in the sample [102,113]. By carrying out Raman experiments on nanotube samples with different diameter distributions, changes in the characteristics of the Raman spectra can be investigated. Variations in the shape of the RBM band can thus provide information on the nanotube diameters present in a given sample [114]. According to theoretical predictions [4,105,115], the frequency of the RBM for single-wall carbon nanotubes is inversely proportional to the nanotube diameter and is independent of chirality. In principle, the radial breathing mode frequency can provide the identity (n,m) of the individual nanotubes participating in the Raman scattering [102,113,114]. This follows from the observation that for the RBM mode $\omega \sim 1/d_{\mathrm{t}}$, and the value of d_{t} depends on the integers (n,m) [4,103,105,115]. However, several distinct (n,m) nanotubes can exhibit sufficiently similar diameters so that their RBM frequency differs only by $1-2\,\mathrm{cm}^{-1}$. This prevents using the RBM frequencies to determine (n,m) in a sample of single-wall nanotubes, since the natural linewidth of the RBM band is $\sim 6\,\mathrm{cm}^{-1}$ and the individual contributions to the band from nanotubes with similar diameters would be very difficult to resolve. A discussion of this issue is given in more detail in [103,116].

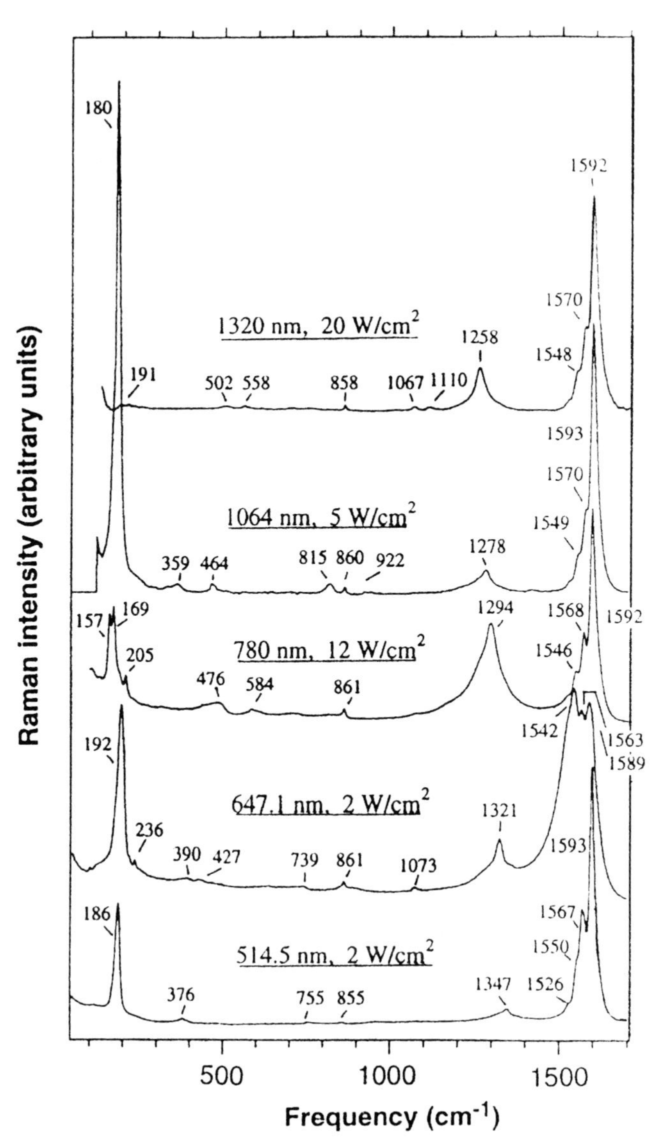

Fig. 8.22. Experimental room temperature Raman spectra for purified single-wall carbon nanotubes excited at five different laser excitation wavelengths. The laser wavelength and power density for each spectrum are indicated, as are the vibrational frequencies (in cm^{-1}) [38]. The equivalent photon energies for these laser wavelengths are: 1320 nm $\to$ 0.94 eV; 1064 nm $\to$ 1.17 eV; 780 nm $\to$ 1.58 eV; 647.1 nm $\to$ 1.92 eV; 514.5 nm $\to$ 2.41 eV

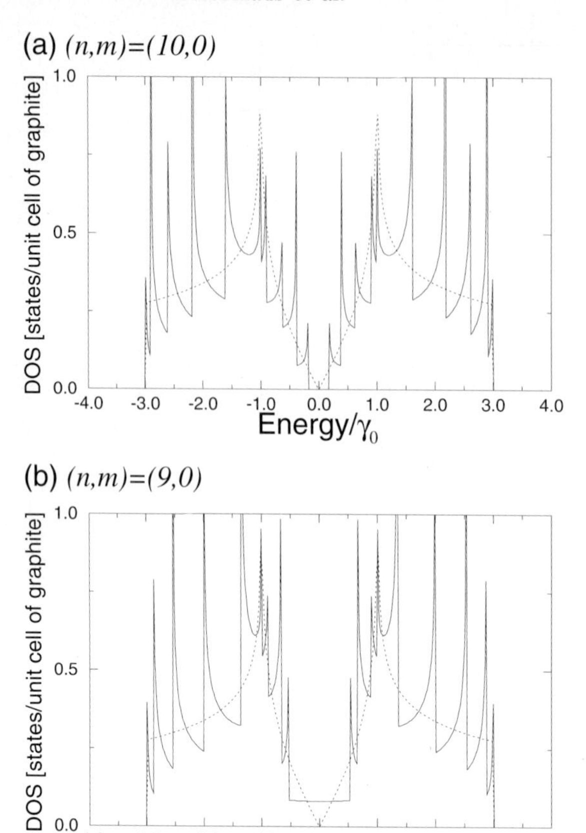

Fig. 8.23. Electronic 1D DOS per unit cell of a 2D graphene sheet for two $(n, 0)$ zigzag nanotubes: (**a**) the $(10, 0)$ nanotube which has semiconducting behavior, and (**b**) the $(9, 0)$ nanotube which has metallic behavior. Also shown in the figure is the density of states for the 2D graphene sheet (*dotted curve*). The value of γ_0 is taken to be 2.95 eV [112]

Figure 8.24 shows in more detail the Raman band associated with the RBM of a single-wall carbon nanotube sample using different laser lines. Here we note that the band-shapes are completely different from one laser line to another, as first pointed out by Rao et al. [38] This result indicates that the sample contains carbon nanotubes with different diameters, that the frequency of the RBM depends strongly on the nanotube diameter, and that the shape of the RBM depends on the relative populations of each nanotube type that is in resonance at the particular laser excitation energy. All the observed RBM bands have been fitted using a different set of Lorentzian peaks, as shown by the dotted lines in Fig. 8.24. Future measurements on

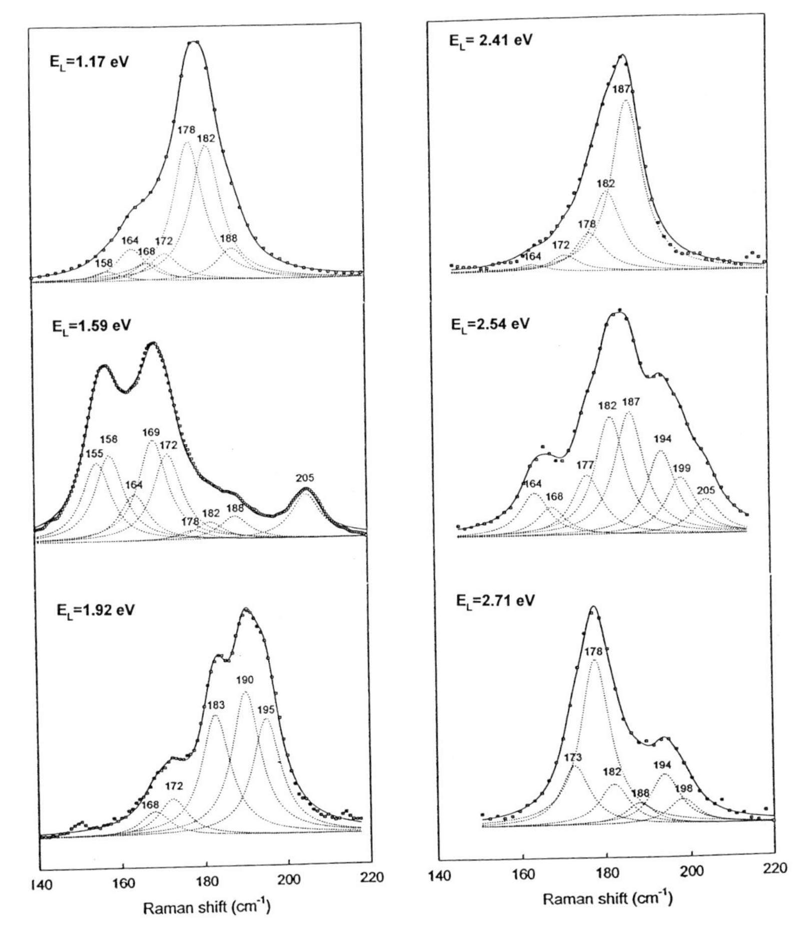

Fig. 8.24. Raman spectra of the radial breathing mode (RBM) of single-wall carbon nanotubes collected with six different laser lines: $E_{\mathrm{L}} = 1.17\,\mathrm{eV}$, $1.59\,\mathrm{eV}$, $1.92\,\mathrm{eV}$, $2.41\,\mathrm{eV}$, $2.54\,\mathrm{eV}$ and $2.71\,\mathrm{eV}$. The *dotted curves* represent the individual Lorentzian components, and the *solid curves* represent the fit of these Lorentzians to the experimental data. All Lorentzian curves are assumed to have the same linewidth (FWHM $= 8.4\,\mathrm{cm}^{-1}$), since the origin of the linewidth is expected to be the same [117]

individual single-wall carbon nanotubes will allow the association of each Lorentzian peak shown in Fig. 8.24 with a particular (n, m) nanotube.

The tangential Raman modes between 1500–$1600\,\mathrm{cm}^{-1}$ also exhibit an interesting resonant effect. It has been reported by different authors [38,117,118] that the shape of the Raman band associated with the tangential modes obtained with laser energies E_{L} around $2\,\mathrm{eV}$ is qualitatively different from those recorded with either higher or lower E_{L}. Figure 8.25 shows the Raman band associated with the tangential C–C stretching modes of the single-wall car-

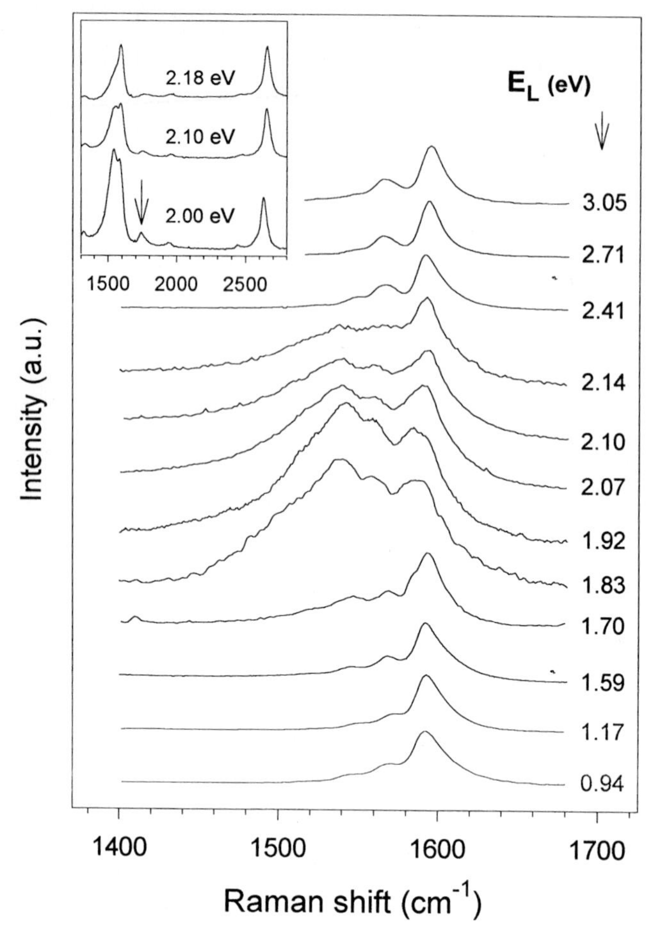

Fig. 8.25. Raman spectra of the tangential modes of carbon nanotubes in the range 1400–1680 cm^{-1} obtained with several different laser lines. The *inset* shows low resolution Raman spectra between 1300 and 2800 cm^{-1} in the critical range of laser energies 2.00–2.18 eV [119]

bon nanotubes, obtained with different laser lines over a wide energy range ($0.94 \leq E_{\mathrm{L}} \leq 3.05$ eV). Note that the spectra obtained for $E_{\mathrm{L}} < 1.7$ eV or $E_{\mathrm{laser}} > 2.2$ eV are quite similar, and each spectrum is dominated by a peak at 1593 cm^{-1}. In contrast, the spectra obtained in the narrow range $1.7 \leq E_{\mathrm{L}} \leq 2.2$ eV are qualitatively different; the bands are broader and are centered at lower frequencies.

Figure 8.26 taken at $E_{\mathrm{L}} = 2.41$ eV shows the fitting of one of the typical Raman bands in the ranges 0.94–1.7 eV and 2.2–3.05 eV by a sum of Lorentzian components. At least six different components are necessary to give a good fit to the experimental data, and their peak frequencies are: 1522,

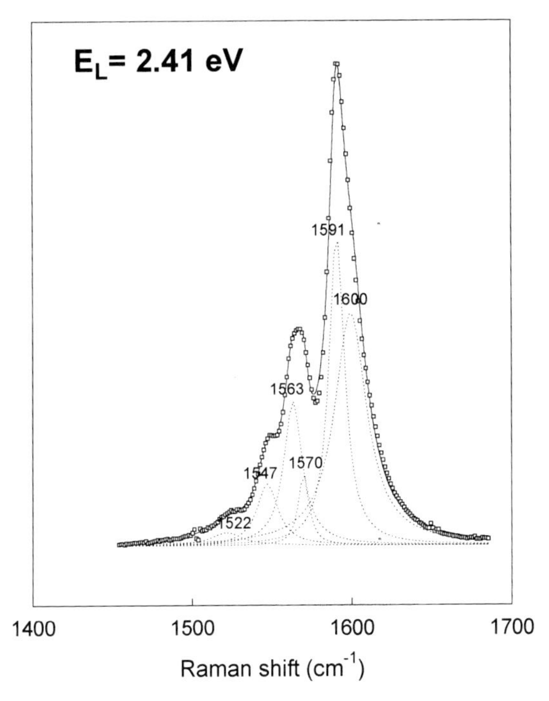

Fig. 8.26. Raman band associated with the tangential C–C stretching modes of single-wall carbon nanotubes obtained using the laser excitation energy 2.41 eV. The *dotted curves* represent the individual Lorentzian components and the *solid curve* represents the fit to the experimental data [117]

1547, 1563, 1570, 1591 and 1600 cm^{-1}. All the other high frequency Raman bands obtained in the laser energy ranges 0.94–1.7 eV and 2.2–3.05 eV are similar and have been fitted using the same set of frequencies (with small variations of $\pm 2\,cm^{-1}$) and linewidths. Tentative assignments of the individual Lorentzians that comprise these bands are given in [38,114,117].

To highlight the part of the tangential Raman band that is specially enhanced for $1.7\,eV < E_L < 2.2\,eV$, Fig. 8.27 shows the Raman band associated with the tangential modes of carbon nanotubes for $E_L = 1.92\,eV$, after subtracting the basic curve (Fig. 8.26) that fits the Raman bands obtained in the ranges $E_L < 1.7\,eV$ or $E_L > 2.2\,eV$. Note that the band in Fig. 8.27 can be fit by four Lorentzian peaks at 1515, 1540, 1558 and 1581 cm^{-1}, which are represented by the dotted curves. All Raman bands in this critical region of laser energy (1.7–2.2 eV) can be analyzed in a similar way, by using the appropriate normalization factor, and the subtracted data can always be fit using the same set of frequencies that fits Fig. 8.27. To represent the dependence of the intensity of the enhanced peaks (dotted peaks in Fig. 8.27) on E_L, we plot in Fig. 8.28 (solid circles) the E_L dependence of the ratio of the intensities of the peaks at 1540 cm^{-1} and 1593 cm^{-1}, I_{1540}/I_{1593}, which

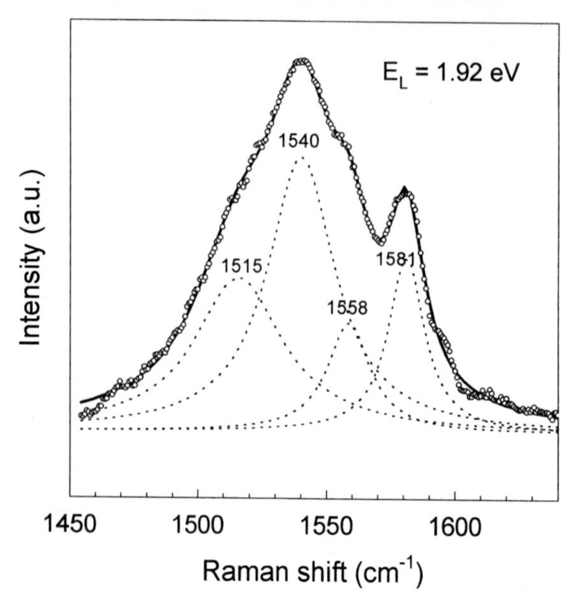

Fig. 8.27. Raman band associated with the tangential modes of carbon nanotubes for $E_{\rm L} = 1.92\,{\rm eV}$ after subtracting the basic curve that fits the bands obtained in the ranges $E_{\rm L} < 1.7\,{\rm eV}$ or $E_{\rm L} > 2.2\,{\rm eV}$. The *dotted curves* represent the four specially enhanced Lorentzian peaks, and the *solid curve* corresponds to the fit to the deconvolved experimental data (*open circles*) [119]

are the most prominent features inside and outside the critical range of laser energies 1.7–2.2 eV.

According to recent scanning tunneling microscopy and spectroscopy (STM and STS) experiments [120,121] done on single-wall carbon-nanotube samples [99] very similar to the sample used in the Raman experiment of Fig. 8.25, the metallic nanotubes exhibit an energy separation ($E_{11} = E_{c_1} - E_{v_1}$) between the first pair of DOS singularities in the valence (v) and conduction (c) bands in the energy range 1.7–2.0 eV, whereas the semiconducting carbon nanotubes exhibit a much lower electronic energy gap (E_{11}) of 0.5–0.65 eV. This result is consistent with the DOS plots shown in Fig. 8.23. Noting the similarity between the range of energy separations for the metallic nanotubes measured directly by STS (1.7–2.0 eV) and the range of laser energies associated with the specially enhanced tangential modes in the Raman spectra (1.7–2.2 eV) for similar single-wall carbon nanotube samples, we consider the resonant Raman effect for the metallic nanotubes. For a tube with a given diameter d, the enhancement of its Raman peaks will occur whenever the incident or scattered photon is in resonance with the energy separation between the DOS singularities $E_{11}(d) = E_{c_1}(d) - E_{v_1}(d)$ [115]. Considering that the sample contains a Gaussian distribution of nanotube diameters (see inset to Fig. 8.28), the Raman cross-section $I(E_{\rm L})$ is given by the sum of the

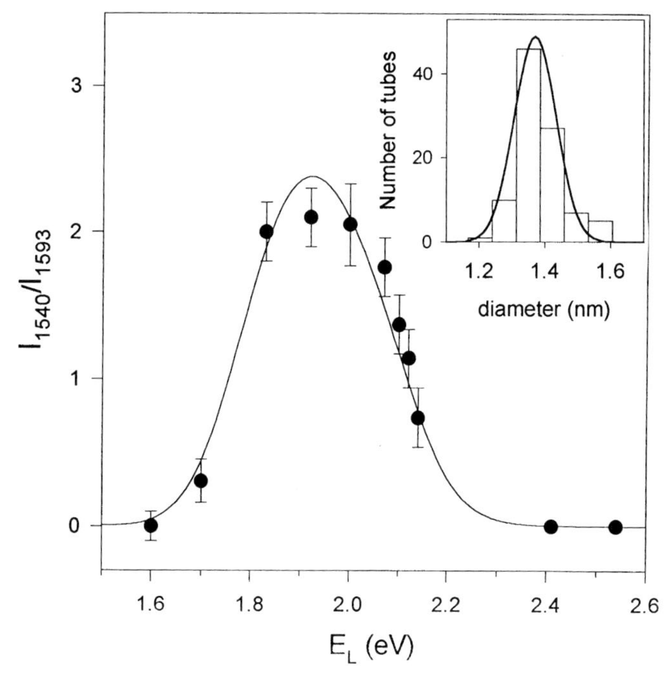

Fig. 8.28. The *solid circles* represent the intensity ratio of the Raman peaks at 1540 and 1593 cm^{-1}, and the *solid curve* represents the fit to the experimental data using (8.9). The *inset* shows the distribution of diameters measured by TEM [38] and the Gaussian fit to the data [119]

contributions of each individual nanotube with a given diameter d, weighted by the distribution of diameters:

$$\begin{aligned} I(E_\mathrm{L}) = & \sum_{d=1.1\,\mathrm{nm}}^{1.6\,\mathrm{nm}} A \exp\left[\frac{-(d-d_0)^2}{\Delta d^2/4}\right] \\ & \times \left[(E_{11}(d) - E_\mathrm{L})^2 + \Gamma_e^2/4\right]^{-1} \\ & \times \left[(E_{11}(d) - E_\mathrm{L} + E_\mathrm{phonon})^2 + \Gamma_e^2/4\right]^{-1}, \end{aligned} \tag{8.9}$$

where d_0 and Δd are the center and the width of the Gaussian distribution of the nanotube diameters (see TEM measurements in the inset to Fig. 8.28) and E_phonon is the average energy (0.197 eV) of the tangential phonons. The damping factor Γ_e avoids the divergence of the double resonance expression for the Raman cross-section [115] and accounts for the width of the singularities in the electronic DOS and the lifetime of the excited state. The diameter dependence of E_{11} is given by: [122,123]

$$E_{11}(d) = \frac{6a_{C-C}\Gamma_0}{d} \tag{8.10}$$

where a_{C-C} is the C–C distance and Γ_0 is the electronic overlap integral of the nearest neighbor carbon atoms.

The solid curve in Fig. 8.28 corresponds to the fit to the I_{1540}/I_{1593} vs. E_L data, and the relevant fitting parameters were found to be $\Gamma_0 = 2.95 \pm 0.05$ eV and $\Gamma_e = 0.04 \pm 0.02$ eV. The mean value for the energy separation $\langle E_{11} \rangle$ and the full width of the distribution ΔE_{11} obtained from (8.10) for ($\Gamma_0 = 2.95$ eV, $d_0 = 1.37$ nm and $\Delta d = 0.18$ nm) are $\langle E_{11} \rangle = 1.84$ eV and $\Delta E_{11} = 0.24$ eV, which are also in good agreement with the direct measurements of E_{11} by STS [120].

Therefore, the special tangential phonon modes observed in the range of laser energies 1.7–2.2 eV are enhanced by electronic transitions between the first singularities in the 1-D electronic DOS in the valence and conduction bands $v_1 \to c_1$ of the *metallic* carbon nanotubes. This result establishes the association of the specially enhanced modes with the *metallic* carbon nanotubes. The resonant Raman data are also in agreement with a recent Electron Energy Loss Spectroscopy (EELS) experiment [124], which shows a peak at 1.8 eV in the optical conductivity that originates from the metallic nanotubes. Thus, the optical quantum effects observed for the radial breathing mode and the tangential modes lend strong credibility to the 1D aspects of the electronic and phonon structure of single-wall carbon nanotubes and provide clear confirmation for the theoretical predictions about the singularities in the 1D electronic DOS.

8.3.4 Raman Scattering Studies at High Pressure

Raman scattering studies on bundles of "as-prepared" single-wall nanotubes have been carried out in a diamond anvil cell up to 5 GPa [125]. The data were collected under hydrostatic conditions at room temperature using 514.5 nm laser excitation. The Raman spectrum taken at a pressure of 1.0×10^5 Pa (1 bar) was found to closely resemble that shown in the bottom trace of Fig. 8.22. A 4 : 1 methanol: ethanol mixture was used as the pressure transmitting medium. The pressure dependence of the radial and tangential modes is summarized in Table 8.3 where comparisons are also made to the behavior of the Raman mode frequencies for solid C_{60} and graphite. Also summarized in the table is the agreement between experiment and theoretical calculations based on a Generalized Tight Binding Molecular Dynamics (GTBMD) approach using an optimized Lennard–Jones potential to describe the inter-nanotube C-atom interactions. Within this GTBMD approach, three models of differing complexity were considered in detail and were carried out on (9,9) tubes: Model I-uniform external radial compression of the triangular rope lattice, Model II-radial compression of a single tube (which ignores tube-tube interactions, and Model III-compression of the rope lattice (including inter-nanotube interactions). Model III provides for the penetration of the pressure medium into the interstitial channels in the rope lattice, whereas Model I does not. The model parameters were adjusted to produce a pressure dependence of the tangential mode frequency behavior that is consistent with experiment, and the performance of the model was assessed according to its ability to fit the observed radial mode pressure dependence. Model I was found to be in the best agreement with the Raman data and is the one used in Table 8.3 for comparison to the experimental data. An important theoretical result at 1 bar was the observation that the inter-tube interactions upshift the radial mode frequency of a bundle of (9,9) tubes by $\sim 14\,\text{cm}^{-1}$ over that calculated for an isolated tube. If this result is correct, this upshift must be taken into account when Raman scattering is used to estimate the nanotube diameters present in samples containing *bundles* of nanotubes.

Table 8.3. Experimental and calculated pressure dependence of the Raman-active radial and tangential band frequencies of single-wall carbon nanotubes at 300 K. The band frequency ω is fit to a second-order polynomial in pressure P, i.e., $\omega(P) = \omega(0) + a_1 P + a_2 P^2$ [125]

Material		Mode	$\omega(0)$ (cm^{-1})	a_1 (cm^{-1}/GPa)	a_2 ($\text{cm}^{-1}/\text{GPa}^2$)
SWNT	Experiment[a]	(R)	186±1	7 ± 1	–
SWNT	Theory[b]	(R)	186.2	9.6	−0.65
Solid C_{60}	Experiment[c]	$[A_g(1)]$	491	0.94	–
SWNT	Experiment[a]	(T_1)	1550±1	8 ± 1	–
SWNT	Experiment[a]	(T_2)	1564±1	10 ± 1	-0.9 ± 0.3
SWNT	Experiment[a]	(T_3)	1593±1	7 ± 1	-0.4 ± 0.2
SWNT	Theory[b]	(T_i)	1576.5	8.3	−0.31
graphite	Experiment[d]	(E_{2g_2})	1579±1	4.7	−0.08
graphite	Theory[d]	(E_{2g_2})	1593	3.3	−0.04
Solid C_{60}	Experiment[c]	$[A_g(2)]$	1465	1.7	–

[a] (R) denotes the radial breathing mode and (T_i; $i = 1, 2, 3$) denote three Raman-active tangential bands [125].
[b] Based on a generalized tight binding molecular dynamics calculation [125].
[c] Experimentally measured values for the $A_g(1)$ radial breathing mode and the $A_g(2)$ pentagonal pinch mode in solid C_{60} [3,69] are listed for comparison.
[d] Experimental [126] and theoretical [125] values for the E_{2g_2} mode in graphite are listed for comparison.

8.3.5 Charge Transfer Effects in Single-Wall Carbon Nanotubes

Extensive research was carried out on graphite intercalation compounds (GIC) in the 1970s and 80s to study the effects of charge transfer between the intercalant (dopant) and the graphene (host) [17,127], with donor dopants (e.g., alkali metals and rare earths) donating electrons, and acceptor dopants (e.g., Br_2, $FeCl_3$, AsF_5) removing electrons from the graphene layers. The charge transfer was found to involve primarily the graphene layers adjacent to the intercalate layers (the "bounding" layers), with much less charge transferred between the intercalate layers and the other, more distant graphene layers (i.e., the "interior layers"). In most cases, the doping shifted the Fermi energy substantially, i.e., by ~ 1 eV or more, upsetting the approximate balance between the electron and hole concentration in the semimetallic graphene host material.

The effect of charge transfer in GICs was measured by a variety of techniques, and especially useful were the Raman spectroscopy studies of the effect of charge transfer on the high frequency intralayer E_{2g_2} mode. The other Raman-active, low frequency E_{2g_1} (shear) mode, located at $42\,\mathrm{cm}^{-1}$ in pristine graphite, is more difficult to observe, and was therefore seldom studied as a function of charge transfer. Upon intercalation, the E_{2g_2} band was found to split by about $20\,\mathrm{cm}^{-1}$ into two bands, one associated with intralayer C-atom displacements in the bounding graphene layers adjacent to the intercalate, and the other associated with the interior graphene layers. Most of this splitting was attributed to the difference in the amount of charge transfer between these two distinct types of C-layers, with the "bounding layer" E_{2g_2} mode being perturbed the most, because most of the charge was transferred to this layer. The E_{2g_2} modes associated with the interior layers exhibited a frequency closer to that of the host, as expected. In general, the E_{2g_2} bounding layer modes upshifted (downshifted) in frequency by ~ 10–$30\,\mathrm{cm}^{-1}$ for acceptor- (donor-) type GICs relative to the value of $1582\,\mathrm{cm}^{-1}$ in pristine graphite. These shifts were attributed [128,129] to the contraction (acceptor-type GICs) and expansion (donor-type GICs) of the intralayer covalent C–C bonds induced by the charge transfer. These "single-layer" or graphene-based theories, however, neglect the contribution from perturbations due to c-axis interactions, whereas, 3D frozen phonon calculations on donor type GICs were found to be in good agreement with the stage dependence of the E_{2g_2} mode frequencies [130]. Consistent with this general view of contracting or expanding the C–C intralayer bonds were observations of the charge transfer-induced softening of the high frequency, tangential "pentagonal pinch" mode in alkali metal-doped solid C_{60} ($\sim 6\,\mathrm{cm}^{-1}$ per electron transferred to the C_{60} molecule, as found in Fig. 8.16). Since acceptor-type C_{60} solids are not generally formed, we cannot compare the charge transfer behavior of intercalated fullerene solids to GICs and to intercalated nanotubes.

Doping studies of the high frequency tangential modes in bundles of single-wall carbon nanotubes were carried out using synthesis methods similar to those used to form GICs from graphite [17], leading to charge transfer reactions with single-wall carbon nanotubes. The resulting shifts of the dominant Raman features associated with the tangential modes indicated the occurrence of both donor and acceptor compounds in the intercalated nanotubes [131–135], as were also confirmed by thermopower measurements [132]. To date, Raman scattering studies have only been carried out on as-prepared (and unpurified) material taken directly from the synthesis chamber and exposed to the following reactants: (donors) Li [134], K and Rb [131]; (acceptors) sulfuric acid (H_2SO_4) [135]; Br_2 [131], iodine (I_2) vapor [131], and molten iodine [132], though iodine does not form a GIC. To date, the presumption for all single-wall carbon nanotube charge transfer processes is that the dopant resides as ions (and also possibly as neutral atoms) in the interstitial channels between the tubes in the triangular nanotube lattice. The assumption, that the intercalant enters into these channels and then diffuses through them without opening the tube ends, has yet to be confirmed by X-ray diffraction, or other structural probes. Rare TEM images of tube ends show them to be closed in the as-prepared material. On the other hand, the most common nanotube purification process used to remove spent catalyst and carbon nanospheres has been proposed [136] to oxidatively open the tube ends.

Both the radial and tangential Raman-active nanotube modes upshift or downshift significantly with doping. Since the sign of the shift in nanotubes is usually consistent with earlier studies of intercalation in GICs and C_{60}, the shift has therefore been interpreted [131] in terms of C–C bond expansion or contraction. In Fig. 8.29 [131], the room temperature Raman spectra of several acceptor- and donor-type nanotube charge transfer compounds are shown. Here, spectra are shown for the pristine (as-prepared and unpurified) material, and for the same as-prepared material after exposure to various reactants, including K, Rb, Br_2 and I_2. The I_2 result is for vapor contact (not molten I_2, which is discussed below) with the bundles, and only a small charge transfer effect on the spectrum is found, while for the Br_2 vapor intercalant, the tangential modes are *upshifted* by $24\,\mathrm{cm}^{-1}$ upon doping. In the case of the donor dopants (K, Rb), the bands *downshift*, and the spectra are remarkably similar, suggesting that under the experimental conditions used, both reactions proceed to the same endpoint stoichiometry. The highest frequency tangential modes in the K and Rb intercalate nanotube Raman spectra were fit to a Breit–Wigner–Fano lineshape, similar to that found for the first stage MC_8 GICs (M = K, Rb, Cs), although the coupling constant ($1/q$) is a factor of three lower than observed in GICs [1], thereby leading to a narrower linewidth for the doped nanotubes. A variety of other Raman features in the region $900-1400\,\mathrm{cm}^{-1}$ are also evident in Fig. 8.29 for the donor intercalated nanotubes, and were assigned to charge transfer-induced softened modes, but no specific mode assignment was made, since the corresponding modes in the pristine sample were not observed.

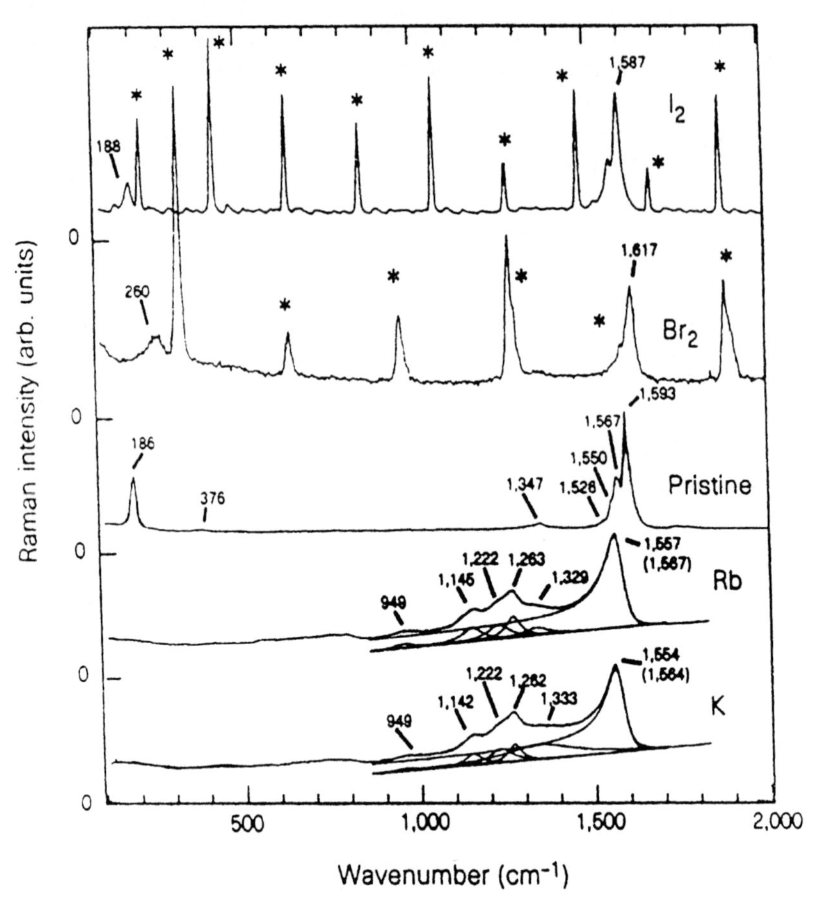

Fig. 8.29. Raman spectra for pristine as-prepared single-wall nanotube bundles reacted with various donor and acceptor reagents: from *top* to *bottom*: I_2 (vapor), Br_2, pristine (no intercalate) single-wall nanotube, Rb, and K. The backscattering spectra were taken at $T = 300\,K$ using 514.5 nm laser excitation, and the peak frequencies for the various samples are indicated. For both of the halogen-doped single-wall nanotube bundles, a harmonic series of peaks (indicated by an asterisk) is observed, and these peaks are identified with the fundamental stretching frequency ω_s: of $\sim 220\,cm^{-1}$ (I_2) and of $\sim 324\,cm^{-1}$ (Br_2). In the vicinity of the strongest high-frequency mode around $1550\,cm^{-1}$, the Raman spectra for single-wall nanotubes doped with K or Rb are fitted by a superposition of Lorentzian functions and an asymmetric Breit–Wigner–Fano lineshape, superimposed on a linear continuum. The peak frequency values in parentheses are renormalized phonon frequencies taking into account Breit–Wigner–Fano lineshape effects [131]

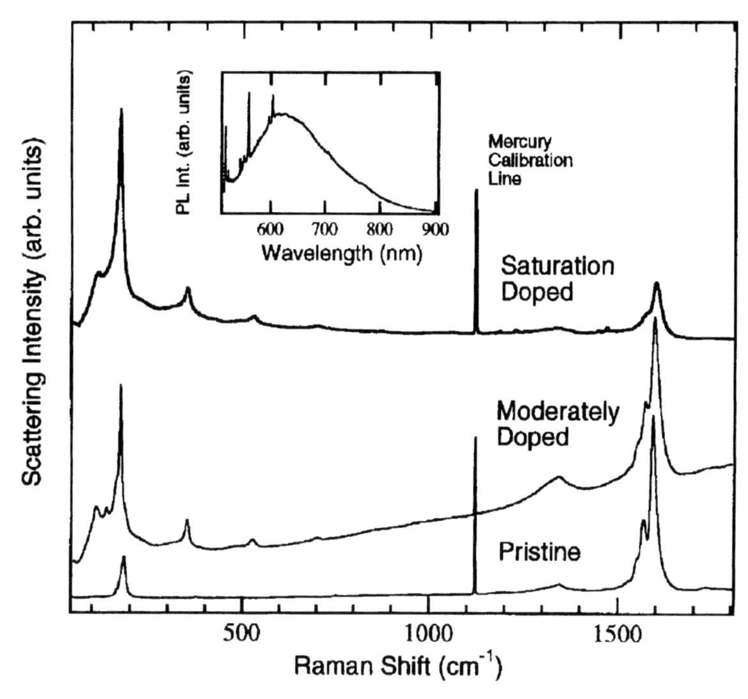

Fig. 8.30. Raman spectra of pristine, moderately I-doped, and saturation I-doped single-wall nanotube samples ($T = 300\,\mathrm{K}$, 514.5 nm laser excitation) where molten I_2 is the dopant. The *inset* shows the photoluminescence spectrum due to the intercalated polyiodide chains in the moderately doped sample where sharp Raman lines are superimposed on the broad PL spectrum [132]

Charge transfer reactions of as-prepared bundles of single-wall carbon nanotubes with *molten* iodine (see Fig. 8.30) produced a much more dramatic effect on the Raman spectrum than was observed when the reaction was carried out in iodine vapor (see Fig. 8.29) [132]. The Raman spectra clearly showed the intercalation of molten I_2 into the bundle to be reversible and uniform, producing an air stable compound of an approximate stoichiometry $C_{12}I$. The Raman spectra in Fig. 8.30 for pristine, moderately I-doped, and saturation I-doped single-wall nanotube bundles using molten iodine as an intercalant [132,133] show a low frequency region which is rich in structure, identified with the presence of intercalated charged polyiodide chains (I_3^- and I_5^-).

No Raman evidence for neutral I_2 ($215\,\mathrm{cm}^{-1}$) was found in the nanotube samples. Saturation doping was found to convert I_3^- observed in the moderately doped material into I_5^-. Analysis of these data showed that intercalation *downshifts* the radial band from 186 to $175\,\mathrm{cm}^{-1}$, while the tangential band is *upshifted* by $8\,\mathrm{cm}^{-1}$ from 1593 to $1601\,\mathrm{cm}^{-1}$. The radial mode downshift may be due to a coupling of the tube wall to the heavy iodine chains, while the small tangential mode upshift ($8\,cm^{-1}$) compared to the $21\,cm^{-1}$ upshift in Br_2 is attributed to the iodine chains being only singly ionized. If all the iodine was in the form of I_5^- in the $C_{12}I$ compound, we might write the formula as $C_{60}^+(I_5)^-$, which is equivalent to one hole per 60 C-atoms, and we therefore find an upshift of $8\,cm^{-1}$ per additional hole per 60 C-atoms in the single-wall nanotubes (or $480\,cm^{-1}$ per hole per C-atom). This shift can be compared to that of $6\,cm^{-1}$ per added electron in M-doped C_{60} (or $360\,cm^{-1}$ per electron per C-atom) as shown in Fig. 8.16.

In situ Raman scattering studies, performed during the electrochemical anodic oxidation of single-wall carbon nanotube bundles in sulfuric acid [135], are of special interest, and can be directly compared with similar studies on H_2SO_4 GICs [127]. In the case of the nanotubes, a rapid spontaneous shift of $\sim 15\,cm^{-1}$ in the tangential Raman modes was observed under open circuit conditions which was not observed in the GIC system. This observation may be due to the extremely high surface area per gram of the nanotube material ($> 1000\,m^2/g$). The analogs of the so-called electrochemical "over-charge" and "over-oxidation" regions found in the GIC system were also found in the carbon nanotube system. In these regions, the shift in the Raman frequency of the tangential modes could be identified with the electrochemical charge passing through the cell. From the overcharge region, a direct measure was obtained for the charge transfer effect on the tangential mode frequency, i.e., $320\,cm^{-1}$ per hole per C-atom, in reasonable agreement with values discussed above for M-doped C_{60}.

8.4 Summary

The Raman spectra for the various carbon materials reviewed here are each shown to have their own characteristic features, though many common and analogous behaviors can be identified. Raman spectroscopy is shown to be one of the most useful techniques for characterizing the perturbations to these materials associated with defects, doping, phase transitions and other perturbations.

Acknowledgments

The authors gratefully acknowledge the helpful discussions with Dr. A.M. Rao, Professors R. Nemanich, M. Endo and R. Saito. They are also thankful to many other colleagues for their assistance with the preparation of this article. The research at MIT is supported by NSF grant DMR 98-04734, and the work at UK by NSF grants OSR-9452895 and DMR 98-09686 and by the University of Kentucky Center for Applied Energy Research. One of us (M.A.P.) is thankful to the Brazilian agency CAPES for financial support during his visit to MIT.

References

1. M.S. Dresselhaus, G. Dresselhaus: Light Scattering in Solids III, Top. Appl. Phys. **51**, 3, (1982), ed. by M. Cardona, G. Güntherodt (Springer, Berlin) p. 3
2. D.S. Knight, W.B. White: J. Mater. Res. **4**, 385 (1989)
3. M.S. Dresselhaus, G. Dresselhaus, P.C. Eklund: *Science of Fullerenes and Carbon Nanotubes* (Academic Press, New York 1996)
4. R. Saito, G. Dresselhaus, M.S. Dresselhaus: *Physical Properties of Carbon Nanotubes* (Imperial College Press, London 1998)
5. S.A. Solin, A.K. Ramdas: Phys. Rev. B **1**, 1687 (1970)
6. A.R. Badzian, P.K. Backmann, T. Hartnett, T. Badzian, R. Messier: In *Amorphous Hydrogenated Carbon Films*, ed. by P. Koidl, P. Oelhafen, (Les Editions de Physique 1987). Vol. XVII, E-MRS Conference Proceedings, p. 67
7. J. Kastner, J. Winter, H. Kuzmany: Mater. Sci. Forum **191**, 161–170 (1995)
8. R.J. Nemanich, G. Lucovsky, S.A. Solin: In *Proceedings of the International Conference on Lattice Dynamics, Paris 1974*, ed. by M. Balkanski (Flamarion, Paris 1975) p. 619
9. J. Wagner, M. Ramsteiner, C. Wild, P. Koidl: Phys. Rev. B **40**, 1817 (1989)
10. T. Henning, F. Salama: Science **282**, 2204 (1998)
11. K. Akagi, Y. Furukawa, I. Harada, M. Nishiguchi, H. Shirakawa: Synth. Met. **17**, 557 (1987)
12. M. Nakamizo, R. Kammereck, P.L. Walker: Carbon **12**, 259 (1974)
13. N. Wada, P.J. Gaczi, S.A. Solin: J. Non-Cryst. Solids **35-36**, 543, (1980)
14. R.J. Nemanich, J.T. Glass, G. Lucovsky, R.E. Shroder: J. Vac. Sci. Technol. a **6**, 1783 (1988)
15. M.S. Dresselhaus, G. Dresselhaus, P.C. Eklund: J. Raman Spect. **27**, 351–371 (1995)
16. M.S. Dresselhaus, G. Dresselhaus, M.A. Pimenta, P.C. Eklund: In *Analytical Applications of Raman Spectroscopy* (Blackwell Science Ltd., Oxford, UK 1999) pp. 367–434
17. M.S. Dresselhaus, G. Dresselhaus: Adv. Phys. **30**, 139 (1981)
18. R.W.G. Wyckoff: In *Crystal Structures* Vol. 1. (Interscience, New York 1964)
19. R.A. Jishi, L. Venkataraman, M.S. Dresselhaus, G. Dresselhaus: Chem. Phys. Lett. **209**, 77 (1993)
20. R. Al-Jishi, G. Dresselhaus: Phys. Rev. B **26**, 4514 (1982)
21. F. Tuinstra, J.L. Koenig: J. Chem. Phys. **53**, 1126 (1970)
22. L.J. Brillson, E. Burstein, A.A. Maradudin, T. Stark: In Proceedings of the International Conference of the Applications of High Magnetic Fields to Semiconductor Physics, ed. by D.L. Carter, R.T. Bate, (Pergamon, New York 1971) p. 187
23. A.W. Moore: In *Chemistry and Physics of Carbon*, ed. by P.L. Walker, Jr., P.A. Thrower, M. Dekker (Inc., New York 1981) p. 233. Vol. 17
24. T.C. Chieu, M.S. Dresselhaus, M. Endo: Phys. Rev. B **26**, 5867 (1982)
25. R.P. Vidno, D.B. Fishbach: In *Extended Abstracts of the* 15th *Conf. on Carbon*, edited by F.L. Vogel, W.C. Forsman (American Carbon Society, University Park, PA, 1981) p. 468
26. R.O. Dillon, J.A. Woollam, V. Katkanant: Phys. Rev. B **29**, 3482 (1984)
27. M.J. Matthews, M.A. Pimenta, G. Dresselhaus, M.S. Dresselhaus, M. Endo: Phys. Rev. B **59**, R6585 (1999)

28. A. Marucci, S.D.M. Brown, M.A. Pimenta, M.J. Matthews, M.S. Dresselhaus, K. Nishimura, M. Endo: J. Mater. Res. **14**, 1124 (1999)
29. M. Endo, C. Kim, T. Hiraoka, T. Karaki, K. Nishimura, M.J. Matthews, S.D.M. Brown, M.S. Dresselhaus: J. Mater. Res. **13**, 2023 (1998)
30. L. Nikiel, P.W. Jagodzinski: Carbon **31**, 1313 (1993)
31. A.V. Baranov, A.N. Bekhterev, Y.S. Bobovich, V.I. Petrov: Opt. Spectrosc. USSR **62**, 612 (1987)
32. Y. Wang, D.C. Alsmeyer, R.L. McCreery: Chem. Mater. **2**, 557 (1990)
33. T.P. Mernagh, R.P. Cooney, R.A. Johnson: Carbon **22**, 39 (1984)
34. R.P. Vidano, D.B. Fishbach, L.J. Willis, T.M. Loehr: Solid State Commun. **39**, 341 (1981)
35. M. Ramsteiner, J. Wagner: Appl. Phys. Lett. **51**, 1355 (1987)
36. B. Marcus, L. Fayette, M. Mermoux, L. Abello, G. Lucazeau: J. Appl. Phys. **76**, 3463 (1994)
37. J. Kastner, T. Pichler, H. Kuzmany, S. Curran, W. Blau, D.N. Weldon, M. Dlamesiere, S. Draper, H. Zandbergen: Chem. Phys. Lett. **221**, 53 (1994)
38. A.M. Rao, E. Richter, S. Bandow, B. Chase, P.C. Eklund, K.W. Williams, M. Menon, K.R. Subbaswamy, A. Thess, R.E. Smalley, G. Dresselhaus, M.S. Dresselhaus: Science **275**, 187 (1997)
39. I. Pocsik, M. Hundhausen, M. Koos, L. Ley: J. Non-Cryst. Solids **227-230 B**, 1083 (1998)
40. B.T. Kelly: in *Physics of Graphite* (Appl. Sci. (London)1981)
41. H.W. Kroto, J.R. Heath, S.C. O'Brien, R.F. Curl, R.E. Smalley: Nature (London) **318**, 162 (1985)
42. D.S. Bethune, G. Meijer, W.C. Tang, H.J. Rosen: Chem. Phys. Lett. **174**, 219 (1990)
43. A.F. Hebard, M.J. Rosseinsky, R.C. Haddon, D.W. Murphy, S.H. Glarum, T.T.M. Palstra, A.P. Ramirez, A.R. Kortan: Nature (London) **350**, 600 (1991)
44. T.T.M. Palstra, O. Zhou, Y. Iwasa, P. Sulewski, R. Fleming, B. Zegarski: Solid State Commun. **93**, 327 (1995)
45. A.F. Hebard: Phys. Today **45**, 26 (1992). November issue
46. R.C. Haddon, A.F. Hebard, M.J. Rosseinsky, D.W. Murphy, S.J. Duclos, K.B. Lyons, B. Miller, J.M. Rosamilia, R.M. Fleming, A.R. Kortan, S.H. Glarum, A.V. Makhija, A.J. Muller, R.H. Eick, S.M. Zahurak, R. Tycko, G. Dabbagh, F.A. Thiel: Nature (London) **350**, 320 (1991)
47. K.A. Wang, Y. Wang, Ping Zhou, J.M. Holden, S.L. Ren, G.T. Hager, H.F. Ni, P.C. Eklund, G. Dresselhaus, M.S. Dresselhaus: Phys. Rev. B **45**, 1955 (1992)
48. G. Meijer, D.S. Bethune, W.C. Tang, H.J. Rosen, R.D. Johnson, R.J. Wilson, D.D. Chambliss, W.G. Golden, H. Seki, M.S. de Vries, C.A. Brown, J.R. Salem, H.E. Hunziker, H.R. Wendt: In *Clusters and Cluster-Assembled Materials, MRS Symposia Proceedings, Boston*, ed. by R.S. Averback, J. Bernholc, D.L. Nelson, (Materials Research Society Press, Pittsburgh, PA, 1991) p. 619
49. R.E. Stanton, M.D. Newton: J. Phys. Chem. **92**, 2141 (1988).
50. F. Negri, G. Orlandi, F. Zerbetto: Chem. Phys. Lett. **144**, 31 (1988)
51. D.E. Weeks, W.G. Harter: J. Chem. Phys. **90**, 4744 (1989)
52. Z.C. Wu, D.A. Jelski, T.F. George: Chem. Phys. Lett. **137**, 291, (1987)
53. P.C. Eklund, Ping Zhou, Kai-An Wang, G. Dresselhaus, M.S. Dresselhaus: J. Phys. Chem. Solids **53**, 1391 (1992)

54. M.S. Dresselhaus, G. Dresselhaus, P.C. Eklund: J. Mater. Res. **8**, 2054 (1993)
55. E. Brendsdal, J. Brunvoll, B.N. Cyvin, S.J. Cyvin: In *Quasicrystals, Network and Molecules of Five-fold Symmetry*, ed. by I. Harguttai, (VCH Publishers, New York, Weinheim, 1990) p. 277
56. Z.H. Dong, P. Zhou, J.M. Holden, P.C. Eklund, M.S. Dresselhaus, G. Dresselhaus: Phys. Rev. B **48**, 2862 (1993)
57. K.A. Wang, A.M. Rao, P.C. Eklund, M.S. Dresselhaus, G. Dresselhaus: Phys. Rev. B **48**, 11375 (1993)
58. S. Nolan, S.T. Ruggiero: Phys. Rev. B **58**, 10942 (1998)
59. S. Nolan, S.T. Ruggiero: Chem. Phys. Lett. **300**, 656 (1999)
60. S. Guha, J. Menéndez, J.B. Page, G.B. Adams, G.S. Spencer, J.P. Lehman, P. Giannozzi, S. Baroni: Phys. Rev. Lett. **72**, 3359 (1994)
61. M.A. Schlüter, M. Lannoo, M. Needels, G.A. Baraff, D. Tománek: Mater. Sci. Eng. **B19**, 129 (1993)
62. Ping Zhou, Kai-An Wang, Ying Wang, P.C. Eklund, M.S. Dresselhaus, G. Dresselhaus, R.A. Jishi: Phys. Rev. B **46**, 2595 (1992)
63. P. Zhou, A.M. Rao, K.A. Wang, J.D. Robertson, C. Eloi, M.S. Meier, S.L. Ren, X.X. Bi, P.C. Eklund, M.S. Dresselhaus: Appl. Phys. Lett. **60**, 2871 (1992)
64. S.J. Duclos, R.C. Haddon, S.H. Glarum, A.F. Hebard, K.B. Lyons: Science **254**, 1625 (1991)
65. C.I. Frum, R. Engleman Jr., H.G. Hedderich, P.F. Bernath, L.D. Lamb, D.R. Huffman: Chem. Phys. Lett. **176**, 504 (1991)
66. B. Chase, N. Herron, E. Holler: J. Phys. Chem. **96**, 4262 (1992)
67. M. Matus, H. Kuzmany: Appl. Phys. a **56**, 241 (1993)
68. S.H. Tolbert, A.P. Alivisatos, H.E. Lorenzana, M.B. Kruger, R. Jeanloz: Chem. Phys. Lett. **188**, 163 (1992)
69. D.W. Snoke, Y.S. Raptis, K. Syassen: Phys. Rev. B **45**, 14419 (1992)
70. P.H.M. van Loosdrecht, P.J.M. van Bentum, G. Meijer: Chem. Phys. Lett. **205**, 191 (1993)
71. H.J. Byrne, W.K. Maser, W.W. Rühle, A. Mittelbach, S. Roth: Appl. Phys. a **56**, 235 (1993)
72. H.J. Byrne, L. Akselrod, C. Thomsen, A. Mittelbach, S. Roth: Appl. Phys. A **57**, 299 (1993)
73. A.M. Rao, P. Zhou, K.-A. Wang, G.T. Hager, J.M. Holden, Ying Wang, W.T. Lee, Xiang-Xin Bi, P.C. Eklund, D.S. Cornett, M.A. Duncan, I.J. Amster: Science **259**, 955 (1993)
74. G.B. Adams, J.B. Page, O.F. Sankey, M.O'Keeffe: Phys. Rev. B **50**, 17471 (1994)
75. Y. Wang, J.M. Holden, Z.H. Dong, X.X. Bi, P.C. Eklund: Chem. Phys. Lett. **211**, 341 (1993)
76. M. Menon, K.R. Subbaswamy, M. Sawtarie: Phys. Rev. B **49**, 13966 (1994)
77. R. Tycko, G. Dabbagh, M.J. Rosseinsky, D.W. Murphy, A.P. Ramirez, R.M. Fleming: Phys. Rev. Lett. **68**, 1912 (1992)
78. P.J. Horoyski, J.A. Wolk, M.L.W. Thewalt: Solid State Commun. **93**, 575 (1995)
79. T. Pichler, M. Matus, J. Kürti, H. Kuzmany: Phys. Rev. B **45**, 13841, (1992)
80. S.H. Glarum, S.J. Duclos, R.C. Haddon: J. Am. Chem. Soc. **114**, 1996 (1992)
81. S.B. Fleischer, B. Pevzner, D.J. Dougherty, E.P. Ippen, M.S. Dresselhaus, A.F. Hebard: Appl. Phys. Lett. **71**, 2734 (1997)

82. M.G. Mitch, J.S. Lannin: J. Phys. Chem. Solids **54**, 1801 (1993)
83. P. Zhou, K.A. Wang, P.C. Eklund, G. Dresselhaus, M.S. Dresselhaus: Phys. Rev. B **48**, 8412 (1993)
84. G. Els, P. Lemmens, G. Güntherodt, H.P. Land, V. Thommen-Geiser, H.J. Güntherodt: Physica C **235–240**, 2475 (1994)
85. K. Kikuchi, N. Nakahara, T. Wakabayashi, S. Suzuki, H. Shiramaru, Y. Miyake, K. Saito, I. Ikemoto, M. Kainosho, Y. Achiba: Nature (London) **357**, 142 (1992)
86. P.H.M. van Loosdrecht, P.J.M. van Bentum, G. Meijer: Phys. Rev. Lett. **68**, 1176 (1992)
87. R.A. Jishi, M.S. Dresselhaus, G. Dresselhaus, K.A. Wang, Ping Zhou, A.M. Rao, P.C. Eklund: Chem. Phys. Lett. **206**, 187 (1993)
88. R.A. Jishi, R.M. Mirie, M.S. Dresselhaus, G. Dresselhaus, P.C. Eklund: Phys. Rev. B **48**, 5634 (1993)
89. D.S. Bethune, G. Meijer, W.C. Tang, H.J. Rosen, W.G. Golden, H. Seki, C.A. Brown, M.S. de Vries: Chem. Phys. Lett. **179**, 181 (1991)
90. Z.H. Wang, M.S. Dresselhaus, G. Dresselhaus, P.C. Eklund: Phys. Rev. B **48**, 16881 (1993)
91. P.H.M. van Loosdrecht, M.A. Verheijen, H. Meeks, P.J.M. van Bentum, G. Meijer: Phys. Rev. B **47**, 7610 (1993)
92. K.A. Wang, Ping Zhou, A.M. Rao, P.C. Eklund, M.S. Dresselhaus, R.A. Jishi: Phys. Rev. B **48**, 3501 (1993)
93. K. Raghavachari, C.M. Rohlfing: J. Phys. Chem. **95**, 5768 (1991)
94. S. Iijima: Nature (London) **354**, 56 (1991)
95. Riichiro Saito, Mitsutaka Fujita, G. Dresselhaus, M.S. Dresselhaus: Phys. Rev. B **46**, 1804 (1992)
96. M. Endo: *Mecanisme de croissance en phase vapeur de fibres de carbone (The growth mechanism of vapor-grown carbon fibers).* PhD thesis, University of Orleans, Orleans, France, 1975. (in French)
97. S. Iijima, T. Ichihashi: Nature (London) **363**, 603 (1993)
98. D.S. Bethune, C.H. Kiang, M.S. de Vries, G. Gorman, R. Savoy, J. Vazquez, R. Beyers: Nature (London) **363**, 605 (1993)
99. A. Thess, R. Lee, P. Nikolaev, H. Dai, P. Petit, J. Robert, C. Xu, Y.H. Lee, S.G. Kim, A.G. Rinzler, D.T. Colbert, G.E. Scuseria, D. Tománek, J.E. Fischer, R.E. Smalley: Science **273**, 483 (1996)
100. T. Ebbesen: In *Fullerenes and Nanotubes*, ed. by Pierre Delhaès, P.M. Ajayan (Gordon and Breach, Paris, France 1998). Series: World of Carbon, volume 2
101. M.S. Dresselhaus, G. Dresselhaus, R. Saito: Carbon **33**, 883 (1995)
102. C. Journet, W.K. Maser, P. Bernier, A. Loiseau, M. Lamy de la Chapelle, S. Lefrant, P. Deniard, R. Lee, J.E. Fischer: Nature (London) **388**, 756 (1997)
103. S. Bandow, S. Asaka, Y. Saito, A.M. Rao, L. Grigorian, E. Richter, P.C. Eklund: Phys. Rev. Lett. **80**, 3779 (1998)
104. R.A. Jishi, D. Inomata, K. Nakao, M.S. Dresselhaus, G. Dresselhaus: J. Phys. Soc. Jpn. **63**, 2252–2260 (1994)
105. R. Saito, T. Takeya, T. Kimura, G. Dresselhaus, M.S. Dresselhaus: Phys. Rev. B **57**, 4145 (1998)
106. M.S. Dresselhaus, G. Dresselhaus, P.C. Eklund, R. Saito: Physics World **11**(1), 33 (January 1998)
107. H. Hiura, T.W. Ebbesen, K. Tanigaki, H. Takahashi: Chem. Phys. Lett. **202**, 509 (1993)

108. N. Chandrabhas, A.K. Sood, D. Sundararaman, S. Raju, V.S. Raghunathan, G.V.N. Rao, V.S. Satry, T.S. Radhakrishnan, Y. Hariharan, A. Bharathi, C.S. Sundar: PRAMANA-J. Phys. **42**, 375 (1994)
109. J.M. Cowley, P. Nikolaev, A. Thess, R.E. Smalley: Chem. Phys. Lett. **265**, 379 (1997)
110. J.M. Holden, Ping Zhou, Xiang-Xin Bi, P.C. Eklund, Shunji Bandow, R.A. Jishi, K. Das Chowdhury, G. Dresselhaus, M.S. Dresselhaus: Chem. Phys. Lett. **220**, 186 (1994)
111. K. Tohji, T. Goto, H. Takahashi, Y. Shinoda, N. Shimizu, B. Jeyadevan, I. Matsuoka, Y. Saito, A. Kasuya, T. Ohsuna, K. Hiraga, Y. Nishina: Nature (London) **383**, 679 (1996)
112. R. Saito, G. Dresselhaus, M.S. Dresselhaus: J. Appl. Phys. **73**, 494 (1993)
113. D. Laplaze, P. Bernier, W.K. Maser, G. Flamant, T. Guillard, A. Loiseau: Carbon **36**, 685 (1998)
114. A. Kasuya, Y. Sasaki, Y. Saito, K. Tohji, Y. Nishina: Phys. Rev. Lett. **78**, 4434 (1997)
115. E. Richter, K.R. Subbaswamy: Phys. Rev. Lett. **79**, 2738 (1997)
116. A.M. Rao, S. Bandow, E. Richter, P.C. Eklund: Thin Solid Films **331**, 141 (1998)
117. M.A. Pimenta, A. Marucci, S.D.M. Brown, M.J. Matthews, A.M. Rao, P.C. Eklund, R.E. Smalley, G. Dresselhaus, M.S. Dresselhaus: J. Mater. Res. **13**, 2396 (1998); M.A. Pimenta, A. Marucci, S. Empedocles, M. Bawendi, E.B. Hanlon, A.M. Rao, P.C. Eklund, R.E. Smalley, G. Dresselhaus, M.S. Dresselhaus: Phys. Rev. B **58**, R16016 (1998)
118. A. Kasuya, M. Sugano, Y. Sasaki, T. Maeda, Y. Saito, K. Tohji, H. Takahashi, Y. Sasaki, M. Fukushima, Y. Nishina, C. Horie: Phys. Rev. B **57**, 4999 (1998)
119. M.A. Pimenta, A. Marucci, S. Empedocles, M. Bawendi, E.B. Hanlon, A.M. Rao, P.C. Eklund, R.E. Smalley, G. Dresselhaus, M.S. Dresselhaus: Phys. Rev. B **58**, R16012 (1998)
120. J.W.G. Wildöer, L.C. Venema, A.G. Rinzler, R.E. Smalley, C. Dekker: Nature (London) **391**, 59 (1998)
121. T.W. Odom, J.L. Huang, P. Kim, C.M. Lieber: Nature (London) **391**, 62 (1998)
122. C.T. White, T.N. Todorov: Nature (London) **393**, 240 (1998)
123. J.-C. Charlier, Ph. Lambin: Phys. Rev. B **57**, R15037 (1998)
124. T. Pichler, M. Knupfer, M.S. Golden, J. Fink, A. Rinzler, R.E. Smalley: Phys. Rev. Lett. **80**, 4729 (1998)
125. U.D. Venkateswaran, A.M. Rao, E. Richter, M. Menon, A. Rinzler, R.E. Smalley, P.C. Eklund: Phys. Rev. B **59**, 10928 (1999)
126. M. Hanfland, H. Beister, K. Syassen: Phys. Rev. B **39**, 12598 (1989)
127. P.C. Eklund, G.L. Doll: In *Graphite Intercalation Compounds II: Transport and Electronic Properties*, ed. by H. Zabel, S.A. Solin, (Springer-Verlag, Berlin 1992) pp. 105–162. Vol. 18, Springer Series in Materials Science
128. L. Pietronero, S. Strässler: Phys. Rev. Lett. **47**, 593 (1981)
129. M. Kertesz: Mol. Cryst. Liq. Cryst. **126**, 103 (1985)
130. C.T. Chan, W.A. Kamitakahara, K.M. Ho, P.C. Eklund: Phys. Rev. Lett. **58**, 1528 (1987)
131. A.M. Rao, P.C. Eklund, S. Bandow, A. Thess, R.E. Smalley: Nature (London), **388**, 257 (1997)

132. L. Grigorian, K.A. Williams, S. Fang, G.U. Sumanasekera, A.L. Loper, E.C. Dickey, S.J. Pennycook, P.C. Eklund: Phys. Rev. Lett. **80**, 5560, (1998)
133. L. Grigorian, G.U. Sumanasekera, A.L. Loper, S. Fang, J.L. Allen, P.C. Eklund: Phys. Rev. B **58**, R4195 (1998)
134. J.L. Allen, G. Sumanasekara, A.M. Rao, A. Loper, P.C. Eklund: (unpublished)
135. G. Sumanasekara, J.L. Allen, S. Fang, A.L. Loper, A.M. Rao, P.C. Eklund: J. Phys. Chem. B **103**, 4292 (1999)
136. J. Liu, A.G. Rinzler, H. Dai, J.H. Hafner, R.K. Bradley, P.J. Boul, A. Lu, T. Iverson, K. Shelimov, C.B. Huffman, F. Rodriguex-Macia, D.T. Colbert, R.E. Smalley: Science **280**, 1253 (1998)

VIII A Case History in Raman and Brillouin Scattering: Lattice Vibrations and Electronic Excitations in Diamond

A.K. Ramdas

Diamond occupies a special place in the history of modern physics and materials science. Raman and Brillouin scattering have proven to be spectroscopic techniques ideally suited to explore the fascinating properties of this unique material. Indeed, in the concluding remarks of his 1930 Nobel Prize address, Raman [1] stated: "The case of diamond, which has been investigated by Ramaswamy, Robertson and Fox, and with especial completeness by Bhagavantam, is of special interest. Very surprising results have been obtained with this substance, which may be the pathway to a fuller understanding of the crystalline state." Ramaswamy, cited above, was Raman's younger brother who studied the Raman spectrum of diamond at Raman's suggestion and observed for the first time that it consists of a single line with a Raman shift of 1332 cm^{-1}. This is the triply degenerate zone-center optical phonon of F_{2g} symmetry, in which the two sublattices of diamond rigidly vibrate against each other. The mode is Raman allowed and infrared forbidden. It is fascinating to read the paper by Nagendra Nath, one of Raman's students who collaborated with him in the famous "Raman-Nath theory of ultrasonic diffraction", and published a thirteen page article entitled "The dynamical theory of the diamond lattice." The group theoretical classification and the derivation of the F_{2g} frequency by Venkatarayudu in four pages underscore the power of group theory [2].

The observations of second-order Raman along with Brillouin components in the spectrum of light scattered by diamond were first reported by Krishnan [2], skillfully exploiting the Rasetti technique based on the 253.7 nm resonance line of Hg. In the post-laser period, Solin and Ramdas [3] used laser excitation and photon counting in the study of both first- and second-order spectra. Polarization measurements enabled them to interpret the first-order lines as well as the peaks, changes in slope and discontinuities in the second-order Raman spectrum in terms of the critical points of the full phonon dispersion curves as deduced from inelastic neutron scattering.

The calibration of the isotopic composition of diamond in terms of the Raman shift associated with the zone-center optical phonon, a critical assessment of the zero point motion and anharmonicity, and an improved set of critical point frequencies from a study of multiphonon Raman and infrared spectra are some of the significant outcomes of the study of isotopically controlled diamonds [4–7]. The Raman spectra of a natural and a ^{13}C diamond are shown in Fig. VIII.1.

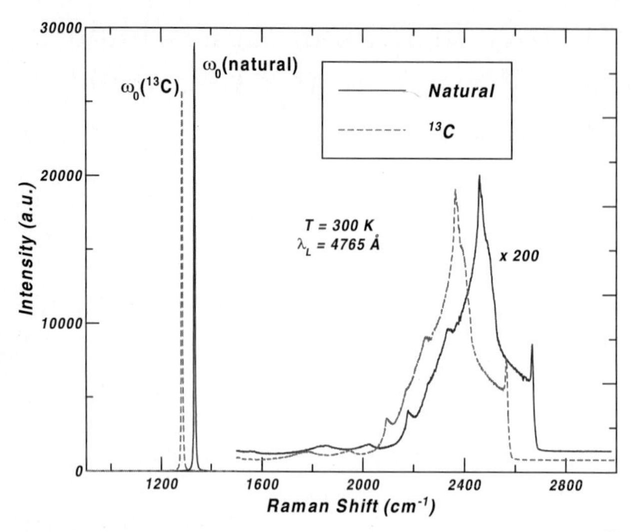

Fig. VIII.1. The Raman spectrum of a natural and a ^{13}C diamond. The spectra show the dominant first-order, Raman-active F_{2g} line and the significantly weaker, quasi-continuous, multi-phonon features [7]

Near the zone-center the vibrational frequency ω for each acoustic branch is linear with the magnitude of the wave vector $\boldsymbol{q}$. The energies of the phonons involved in Brillouin scattering studies with visible radiation are usually much smaller than the thermal energy, and the phonon wavelengths are much longer than the interatomic distances. The Brillouin shifts measured with high resolution Fabry-Perot interferometers have provided the elastic moduli of diamond with high precision [8], their isotopic [5] and temperature dependences [9], and the elasto-optic constants deduced from the Brillouin intensities [8]. The Brillouin shifts of diamond are sufficiently large with visible excitation to be observed with the conventional double grating Raman spectrometers (especially if equipped with holographic gratings). Thus the Brillouin components and the first-order Raman line can be recorded simultaneously in the same experiment under identical illumination and scattering (collection) optics. Such measurements have yielded an absolute cross-section of the first-order Raman line [8] and in turn, of the second-order Raman spectrum.

The ability to grow single crystals free of imperfections – be they lattice defects, or chemical impurities – followed by the controlled introduction of a desired imperfection is the prerequisite for a semiconductor to be significant in technology. While the high-pressure-high-temperature and chemical vapor deposition techniques of diamond synthesis are milestones in this respect, the incorporation of shallow impurities (group III acceptors or group V donors) has been successful to date only for boron acceptors. The study of the bound states of donors and acceptors can be performed with extraordinary detail using infrared spectroscopy. The Lyman spectrum and the associated Zeeman and piezo-spectroscopic effects of an acceptor or a donor yield the binding energies and the symmetries of the ground and excited states. To the extent that the bound states are described by the parameters of the band extrema with which they are associated, such a study is of clear value in the context of the properties of the host. Kim et al. [10–13] have investigated the Lyman spectrum of boron acceptors in isotopically enriched diamonds using Fourier transform infrared as well as Raman spectroscopy. They reported the electronic Raman transition Δ' between the spin-orbit-split ground states of boron acceptors (see Fig. VIII.2), its Zeeman effect and the Jahn–Teller splitting of the lower ground state, including the transition between the Jahn–Teller partners [13]. These studies illustrate the power of Raman spectroscopy in investigating electronic excitations.

From the above discussion, a good case can be made that the unique properties of diamond have been fruitfully addressed with Raman and Brillouin scattering. The phenomena discussed, however, do not exhaust the subject and it appears an enumeration, presented below, of topics/techniques not included will serve to underscore the continuing excitement in the scope and power of Raman spectroscopy with respect to diamond. They are: (a) electric field induced infrared absorption; (b) coherent Raman spectroscopy; (c) the diamond anvil cell and its impact on Raman spectroscopy; and (d) piezo-spectroscopy. It is clear Raman spectroscopy of diamond will continue to provide exciting scientific opportunities in the foreseeable future.

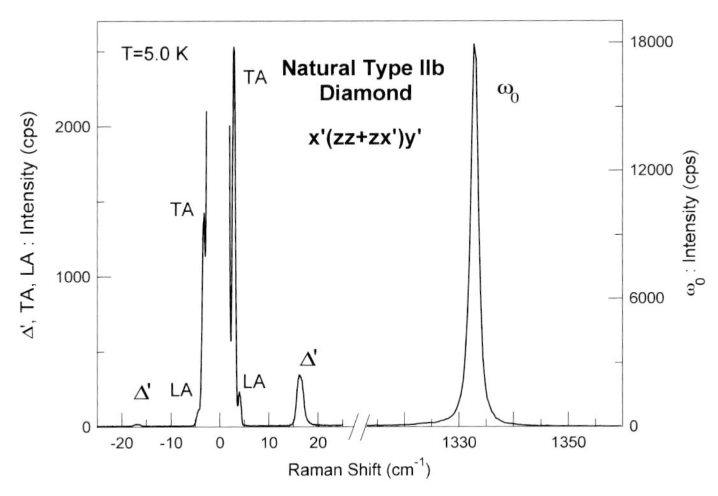

Fig. VIII.2. Comparison of the intensities of the Brillouin components (TA and LA), the Δ' line, and the zone center F_{2g} optical phonon at $\omega(0)$ in a natural Type IIb diamond, recorded in the right-angle scattering geometry $x'(zz+zx')y'$ [10,11]

References

1. S. Ramaseshan, ed.: Scientific Papers of C.V. Raman, *Scattering of Light*, Indian Academy of Science, Bangalore **1**, 434 (1988)
2. A.K. Ramdas: Raman and Brillouin Scattering in Diamond, *Proc. SPIE 1990 Int. Symp. Optical and Optoelectronic Appl. Sci. Eng., Diamond Optics III* **1325**, 17 (1990)
3. S.A. Solin, A.K. Ramdas: Phys. Rev. B **1**, 1687 (1970)
4. K.C. Hass, M.A. Tamor, T.R. Anthony, W.F. Banholzer: Phys. Rev. B **45**, 7171 (1992)
5. R. Vogelgesang, A.K. Ramdas, S. Rodriguez, M. Grimsditch, T.R. Anthony: Phys. Rev. B **54**, 3989 (1996)
6. A.K. Ramdas, S. Rodriguez, M. Grimsditch, T.R. Anthony, W.F. Banholzer: Phys. Rev. Lett. **71**, 189 (1993)
7. R. Vogelgesang, A.D. Alvarenga, H. Kim, A.K. Ramdas, S. Rodriguez, M. Grimsditch, T.R. Anthony: Phys. Rev. B **58**, 5408 (1998)
8. M.H. Grimsditch, A.K. Ramdas: Phys. Rev. B **11**, 3139 (1975)
9. E.S. Zouboulis, M. Grimsditch, A.K. Ramdas, S.Rodriguez: Phys. Rev. B **57**, 2889 (1998)
10. H. Kim, R. Vogelgesang, A.K. Ramdas, S. Rodriguez, M. Grimsditch, T.R. Anthony: Phys. Rev. Lett. **79**, 1706 (1997)
11. H. Kim, R. Vogelgesang, A.K. Ramdas, S. Rodriguez, M. Grimsditch, T.R. Anthony: Phys. Rev. B **57**, 15315 (1998)
12. H. Kim, A.K. Ramdas, S. Rodriguez, M. Grimsditch, T.R. Anthony: Phys. Rev. Lett. **83**, 3254 (1999)
13. H. Kim, A.K. Ramdas, S. Rodriguez, M. Grimsditch, T.R. Anthony: Phys. Rev. Lett. **83**, 4140 (1999)

9 Raman Spectroscopic Studies of Polymer Structure

Shaw Ling Hsu

Abstract. For polymer systems, Raman spectroscopy provides information concerning chemical composition, segmental orientation, conformational distribution, and phase identification. The technique is extremely versatile in that information can be obtained for liquid or solid (film, bulk or fiber) samples, ordered or disordered. The Raman intensity from vibrations associated with polymer backbone bonds is especially intense, and most structural information is contained in bands arising from these modes. Because of the availability of laser sources that produce highly polarized radiation capable of being focused to a volume of order 1 μm, sample anisotropy can be determined with accuracy, speed and high spatial resolution. Low lying vibrations ($< 200\,\mathrm{cm}^{-1}$) that originate from skeletal deformation modes or interchain interactions are easily accessible in Raman spectroscopy and yield much information regarding the microscopic structure of the 3-dimensional state. Increases in computation power and analysis capability have also greatly enhanced the utility of this technique.

One of the most valuable insights gained by a practicing polymer scientist dealing with structural characterization is that many different techniques are required to fully understand polymer structures. Gel permeation chromatography and mass spectrometry, particularly matrix assisted laser desorption mass spectrometry, are used to characterize overall molecular weight. Differential scanning calorimetry (DSC) measurements provide much insight regarding the structural changes as functions of temperature and processing history. Nuclear magnetic resonance and infrared (IR) spectroscopy are commonly used to elucidate the chemical compositions of polymers. Transmission electron microscopy is used to measure density fluctuations within the sample in order to characterize detailed morphological features such as crystallite form and degree of phase separation in compatible systems. For information regarding atomic placements, X-ray or electron diffraction methods are generally utilized. Raman spectroscopy can be used to characterize both crystalline and amorphous polymer structures. It can measure conformational order ranging from the shortest chemical repeat to several hundred Å. Although numerous techniques address polymer structure in the crystalline state, few provide information on disordered states. Raman spectroscopy provides structural information for polymers both in disordered states and in states containing chain segments with short-range order. Raman spectroscopy is routinely used to characterize molecular composition, segmental orientation, chain conformation, and intermolecular interactions.

Although the Raman effect was first reported in 1928 [1–3], its utility as a characterization tool was not realized until the advent of lasers in the early 1960s [4]. The earliest Raman studies on polystyrene and poly(methyl methacrylate) PMMA, soon after the discovery of the Raman effect, were possible only because the samples were totally clear and fluorescence free. Today Raman spectra can be readily obtained from most polymers, but there are still some that require a considerable degree of effort. Most Raman experiments on polymers deal with bulk samples, fibers, or films. Characterization of polymers in solution or melt can also be easily accomplished. Sample preparation is generally simple. Since scattering from water is extremely weak, biological samples can also be characterized. Raman spectroscopy can even be used to measure transient molecules with lifetimes of fractions of seconds. Confocal Raman microscopes are now available with spatial resolution approaching the diffraction limit of the focused laser beam. Because of the high spatial resolution achievable, various Raman techniques have been used on-line to elucidate effects of changing processing parameters on morphology development in uniaxial or biaxial systems. Raman spectra can also be used to characterize stress distribution in model composites.

Despite its capability for delivering detailed structural information, Raman scattering has still not achieved equal status with other characterization techniques and is generally *underutilized* in many polymer research laboratories. Raman spectroscopy has developed slowly in analytical applications because of the high initial investment and maintenance costs required for the instrumentation (high power lasers, sophisticated spectrometers, sensitive detectors, etc.). The Raman effect is extremely weak, e.g., less than 10^{-4} of the incident photons result in inelastic scattering, thus intense excitation sources are needed [5]. These sources may also cause fluorescence that overwhelms the Raman signals. Generally fluorescence is not an inherent problem with polymer samples. “Bleaching” is often an acceptable solution. Recrystallization or extraction has been effective in removing impurities that cause fluorescence. Another alternative is to use long wavelength excitation (1.06 μm) as in most Fourier transform instruments. Nevertheless, the Raman technique yields a tremendous amount of structural information and thus effectively complements other characterization techniques. The focus of this chapter is to cover the development of specific areas that illustrate the advantages of Raman spectroscopy in polymer structure research. The first section contains a general introduction to Raman spectroscopy. Subsequent sections deal with low-frequency vibrations, measurements of crystallite dimensions, structural transitions, orientation distribution functions, and recent studies on the characterization of disordered polymers.

Raman spectroscopy probes chemical composition and molecular structure on the microscopic scale, and these microscopic properties can affect macroscopic parameters such as tensile modulus or strength. For example, the mechanical properties of semi-crystalline polymers are dictated by the degree of crystallinity, crystallite size, orientation, and the interconnectivity between the crystalline units. In this case, the chemical composition of the polymer, such as configurational defects including racemic mixtures or the type of end groups, may affect the crystallization behavior and subsequently the development of higher morphological units [6]. In polymers containing multiple chemical repeats, the number of head-to-head versus head-to-tail sequences and their distribution along the chain may dictate the ultimate morphological units, and this can be measured by Raman spectroscopy. It is known that even the most ordered polymer structures possess some conformational defects; in most cases substantial conformational defects exist. These ordered and disordered chains are often described in terms of local parameters such as the relative occupation of the isomeric state along each backbone bond [7]. Aging phenomena in heterogeneous samples are strongly influenced by the processing history. How these structures change as a function of temperature and time is dictated by the changing conformational distribution along the chain. Typical Raman signatures are assigned to the ensemble of chain conformations. Their changes are characteristic of the physical states of polymers. Diffraction methods address only long-range ordered systems, whereas Raman spectroscopy deals with local order or disorder. In particular, considerable information can be obtained using low-frequency Raman modes. Lastly, the Raman intensity depends on the change in polarizability, which is large for carbon bonds that constitute the polymer backbone. Therefore, Raman scattering is particularly sensitive to polymer backbone conformations.

Vibrational spectroscopy is generally done by two methods: IR absorption and Raman scattering. A substantial literature exists regarding the use of vibrational spectroscopy for polymer characterization [8–15]. Because IR and Raman arise from different effects, with one depending on the dipole change and the other on changes in polarizability tensor [16], IR spectroscopy is more appropriate for studying chemical composition or side groups whereas Raman spectroscopy is more appropriate for studying chain conformation. Most IR studies determine the molecular composition of polymers by analyzing the characteristic vibrations of functional groups. In many studies Raman offers advantages over IR, since it can be implemented in a variety of configurations, including remote back scattering, for studying solids, liquids, and gases. Intractable samples can be studied without much preparation. Also, unlike IR, the source and detection systems operate in the visible and near-IR spectral regions, allowing the use of silica and glass lenses, low-cost fiber optics, and glass sampling windows and containers. Different vibrational modes are active in Raman and IR, and, generally speaking, the more symmetric

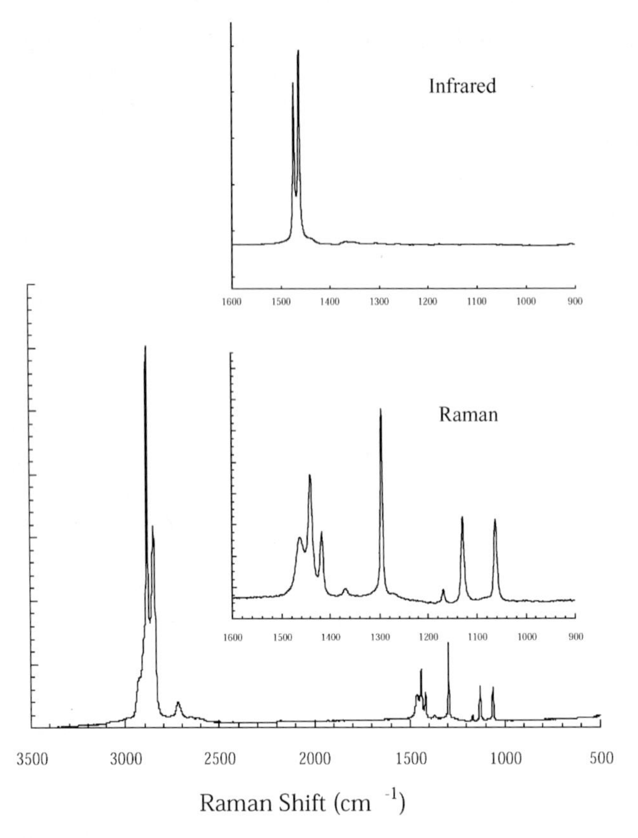

Fig. 9.1. Infrared and Raman spectra obtained for polyethylene

the molecule, the greater the difference between Raman and IR spectra [16]. Anyone using IR spectroscopy would benefit from the use of Raman as well, since the two techniques are truly complementary.

The IR and Raman spectra obtained for a highly crystalline polyethylene film are shown in Fig. 9.1. A large number of studies have been carried out to elucidate the vibrational features of polyethylene or n-alkanes [17–19]. For small molecules of N nuclei, there are $3N-6$ relative vibrations. There are 6 degrees of freedom associated with the overall molecular translations and rotations. However, the spectra obtained for most polymers of interest are not as complicated as might be imagined considering their extremely high molecular weight. In fact, for infinite polymers of well-defined conformation, there are only $3N-4$ optically active vibrations, where N now refers to the number of nuclei per translationally equivalent unit cell) [15,20]. For polyethylene, this unit is $-CH_2CH_2-$, and N is 6. For polypropylene N is 27, since there are nine atoms per chemical repeat, three of which form a translationally equivalent repeat unit. For highly ordered polymers or oligomers, group theory is a powerful tool for determining the number of vibrations to be expected and their optical activities [21]. The IR and Raman spectra shown in Fig. 9.1 contain unexpectedly few features, considering the high molecular weight of the polymer. Note also that none of the bands present in the IR are Raman active. This mutual exclusion simply reflects the fact that the polyethylene chain conformation has inversion symmetry [13,15,16,22–24]. The IR and Raman techniques are complementary, and often both are needed for structural characterization.

Polarized Raman scattering from polymers can yield much information concerning the orientation of the chain segments and assignment of particular bands. Unless the sample is an oriented macroscopic single crystal (almost never the case in polymer studies), some polarization information in the scattered light is lost. The molecular orientation with respect to the laboratory frame can be completely random as in liquids or melts. Partially oriented systems can be either uniaxial as in fibers or biaxial as in films. In these cases, the relationship between the molecular and laboratory frames is partially lost. Even under these circumstances, information about the scattering tensor elements associated with various modes can be obtained through measurements of the depolarization ratio ρ, defined as the ratio between the scattered light intensity obtained when the scattered and incident beams are linearly polarized perpendicular to each other and that obtained when they are polarized parallel to each other [4]. When the symmetry of the translational repeat unit is known, it is possible to obtain significant structural information about the chain segment orientation distribution and the parameters needed to achieve a specific degree of anisotropy [25–31].

Even for randomly oriented systems, the isotropic and anisotropic portions of the scattering tensor can still be measured [4]. In these cases, information can be deduced regarding the local chain conformation, interaction with solvents, and degree of chain extension [32–37]. Most polymer samples have carbon backbones, for which the differential polarizability tensor elements have large isotropic components. In many cases, only the isotropic spectrum is used for structural characterization, thus eliminating vibrations unrelated to conformational analysis [38–41].

The nature of intermolecular interactions can be studied by analyzing the frequency and intensity of the so-called external vibrations, or lattice modes, of the unit cell [42–44]. These intermolecular vibrations, which involve whole chain segment movements, are very low in frequency ($< 100\,\mathrm{cm}^{-1}$). If the elastic Rayleigh scattering can be removed, and there are many different ways to achieve that goal (for example the use of iodine gas filters [45] to selectively absorb the laser line), very low frequency vibrations, i.e. in the range of a few cm^{-1}, can be observed. These low-lying modes are extremely difficult to observe using other absorption or scattering techniques. These vibrations can be skeletal deformation modes or interchain lattice modes. For polymers with a well-defined crystalline state, the lattice modes are very sensitive to changes in the intermolecular interactions [42]. The effects of interchain interactions in the crystal are twofold: the splitting of low-frequency intrachain vibrations and the appearance of lattice modes. In trans-1,4-polybutadienes, the splitting, particularly at low temperatures, seen at $240\,\mathrm{cm}^{-1}$ is consistent with predictions from theoretical normal coordinate analysis [42]. The crystalline structure of polyethylene has been well characterized. There are two chains per crystalline unit cell [22]. Just as there is strong coupling between the vibrations along the polymer chain, the possibility exists that coupling can also occur for equivalent vibrations in the crystalline unit cell. A set of lattice vibrations has been observed for polyethylene and paraffins [46]. When these low lying vibrations are analyzed quantitatively, correlations between molecular and macroscopic properties such as heat capacity, thermal pressure, thermal expansion, P-V-T relations, and stress-strain behavior can be established.

Vibrational coupling (such as crystal field splitting in ordered polymers) is well understood. For each vibration with frequency ν_0 found for the individual chemical repeat units, a series of vibrations arising from backbone coupling may be found shifted from the unperturbed value. For a chain with N chemical repeat units or N oscillators, instead of observing a set of N degenerate vibrations of frequency ν_0, a series of vibrations or a progression may exist. Representative Raman spectra for $C_{20}H_{42}$ are shown in Fig. 9.2. The frequency separation of the progression modes is inversely proportional to the unperturbed frequency ν_0 [14]. Localized vibrations, such as the CH_2 stretching modes in the $3000\,\mathrm{cm}^{-1}$ region, are hardly perturbed by the interaction along the chain. In contrast, low-frequency components such as the CH_2 rocking vibration at $700\,\mathrm{cm}^{-1}$, can be significantly affected. As will be demonstrated, knowledge regarding these progression bands is crucial in con-

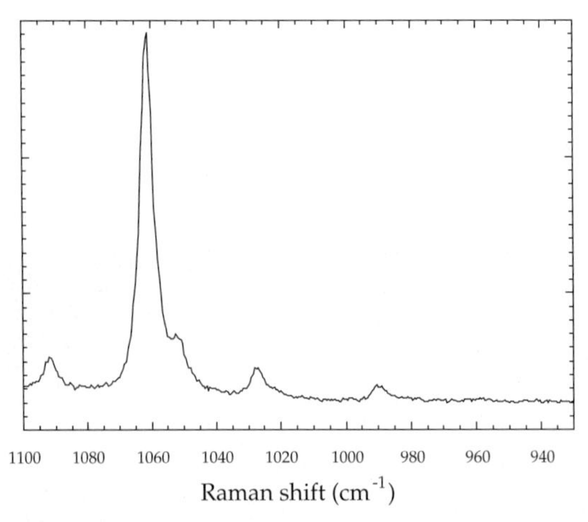

Fig. 9.2. Raman active progression modes observed for n-alkanes ($C_{20}H_{42}$)

sidering the validity of force fields and relative magnitudes of polarizability changes, thus elucidating polymer chain conformation.

9.1 Overview of Structural Characterization

Raman spectroscopy, unlike IR absorption, has not been used extensively for composition analysis, although it works equally well for many functional groups. For example, cis trans structures of polybutadienes, vinyl C=C stretching, compounds containing sulfurs, and perchlorates all exhibit strong, easily identified Raman bands. Aromatic units also usually exhibit strong Raman scattering. The use of Raman spectroscopy has been adequately summarized in previous publications [12]. Some applications combine Raman and IR spectroscopy for structural analysis. An example can be seen in the structure determination of 1:1 hexafluoroisobutylene and vinylidene fluoride copolymer. This problem is virtually impossible to solve with other techniques. The two monomers can be linked in two ways during polymerization: by formation of head-to-tail (the normal linking) or head-to-head linkages. By comparing IR and Raman data, we now know that the polymers formed are indeed alternating in nature [47], and that only head-to-tail linkages are present. The proposed chemical structure is shown below.

$$-(C(CF_3)_2CH_2CF_2CH_2)_n-$$

Here the CH_2 units are decoupled by the presence of a rather bulky CF_2 unit between them. If the copolymers were formed in a head-to-head linkage, considerable coupling of the stretching vibrations would be expected. In fact, in both IR and Raman spectra, the CH2 symmetric and asymmetric vibrations were observed at 3040 and 2985 cm^{-1}. This coincidence is proof that high frequency CH_2 vibrations are decoupled and thus degenerate. If the polymer were formed in a head-to-head linkage, mutual exclusion would be expected, contrary to observation. Similar configurational analyses can be extended to other polymers formed with substituted monomers.

In a head-tail alternating copolymer, the possibility exists that a head-head or tail-tail defect linkage can occur. Specific Raman features have been identified for analyzing chain-chain configuration in syndiotactic polypropylenes (sPP) containing varying racemic content [40]. Based on theoretical normal coordinate analysis, the syndiotacticity content of imperfect polypropylenes can be directly measured using the relative intensity of Raman-active bands in the 300 and 400 cm^{-1} region. The syndiotacticity index, shown in Fig. 9.3, provides a convenient way to quantify chain configuration. In addition, the relative intensity of the 875 and 830 cm^{-1} bands indicates the amount of helix or planar zigzag structure. Raman features obtained for various sPP's are shown in Fig. 9.4. The quantitative analysis associated with the interpretation of these features requires the use of chain statistics, which are presented in a later section.

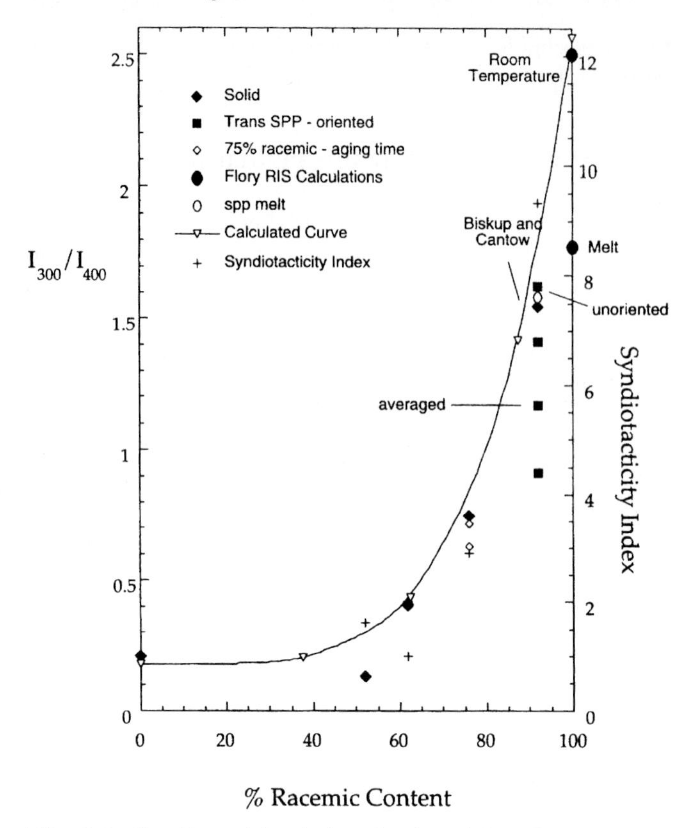

Fig. 9.3. Syndiotacticity index developed based on the relative intensity of the Raman-active 300 and 400 cm^{-1} bands

Diffraction methods are dependent on both local and long-range order; spectroscopy is dependent only on local order. Spectroscopy will obviously not displace diffraction for determining the atomic placement. It is, however, extremely useful for determining chain conformation and packing distribution. For example, polyoxymethylene (POM) exists in at least two crystalline forms [13,48–50]. It is generally known as a trigonal crystalline polymer consisting of chains of 9/5 or 29/16 helical conformations. Another modification is a metastable orthorhombic POM crystal consisting of 2/1 helical chains. Unusually well ordered needle-like crystals consisting of extended 9/5 helical chains have been prepared. Other typical trigonal single crystals are lamellae consisting of regularly folded molecular chains. The solution-grown crystals are easily changed by mechanical deformation. The Nujol mull method is therefore necessary to obtain IR spectra [49]. No significant differences for the two types of samples were obtained from Raman spectra. Wide angle X-ray diffraction yielded the same trigonal phase consisting of the 9/5 helices [49]. However, the IR spectra obtained are, significantly different. The

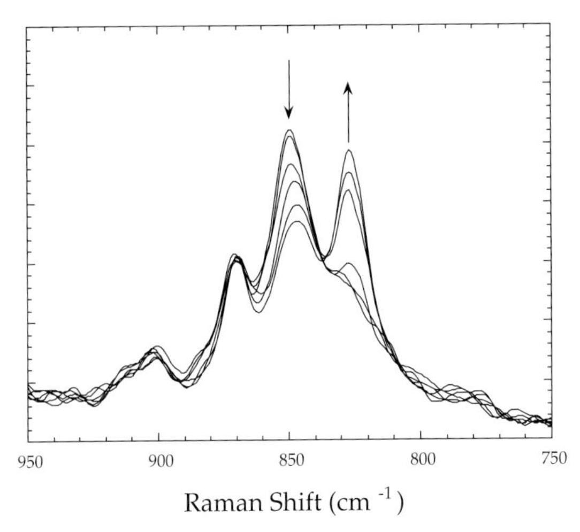

Fig. 9.4. Raman spectra obtained for sPP as a function of time at room temperature after initially quenching a molten sample into liquid nitrogen. The *arrows* indicate the trends of the peaks with increasing time

needle and solution-grown crystals, which give identical X-ray diffraction patterns, exhibit quite different vibrational spectra. In the spectra of folded chain crystals, only bands assigned to the A_2 block shift to higher frequencies. All bands of other species A_1 and E_1 are the same as those in extended chain crystals. Spectra of the POM specimen between these extremes appear as an overlap of the two spectra. This set of samples remains poorly understood. The differences demonstrate the necessity for use of different techniques to properly characterize polymer structure.

Raman spectroscopy can be a very useful supplementary technique when a fiber pattern cannot be obtained because the uniaxially oriented form is unstable or not easily synthesized. Polybutene is an example where Raman spectroscopy has been used to determine the conformation of the polymer chain in the solid state. Polybutene-1 is known to exist in at least three crystalline modifications [51]. Form I has a hexagonal unit cell with six 3_1 helices. Form II has a tetragonal unit cell containing 11_3 helices and is obtained by cooling from the melt. Polybutene will transform slowly and irreversibly from II to I at room temperature. Form III is prepared by casting a film from benzene, carbon tetrachloride, toluene, p-xylene, or decaline solutions. At elevated temperatures, form III will transform to form II and then spontaneously to form I. An X-ray fiber pattern of form III cannot be obtained due to its instability at elevated temperatures. Form III is thought to have an orthorhombic unit cell. Recent Raman results suggest a 10_3 helix. The Raman spectra for forms I, II, and III are shown in Fig. 9.5. Infrared and Raman data have been reported previously for all crystalline forms [51,52]. Because the bands at 774, 824, 875, and 982 cm^{-1} contain significant con-

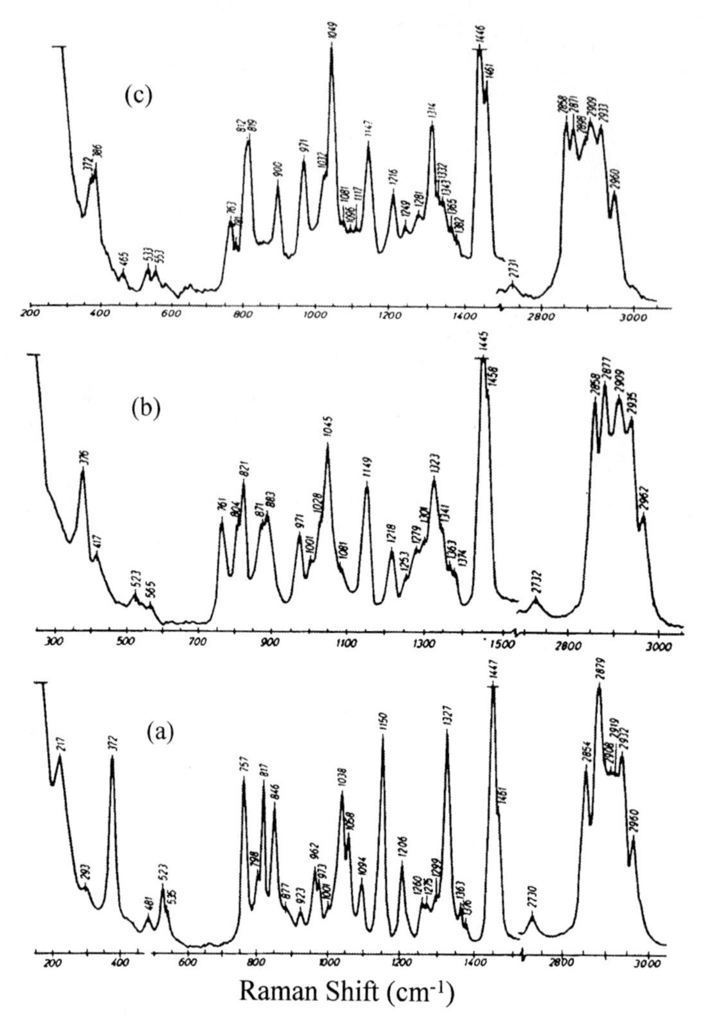

Fig. 9.5. Raman spectra of polybutene-1; (**a**) Form I; (**b**) Form II; (**c**) Form III [51]

tributions from backbone skeletal stretching [53], both their frequencies and intensities are sensitive to changes in chain conformation.

To determine the conformation of form III from the Raman spectrum, the functional relationship between these vibrational modes and changes in the chain conformation must be established. Normal coordinate analysis suggested a linear relationship between the calculated frequency and helix angle for the limited region between 98 and 120° [51]. In fact, this type of functional relationship between chain conformation and the frequency of skeletal deformation has also been observed in other polymers including polypeptides and proteins [54,55]. This frequency-structure correlation suggests that the observed frequencies for form I and form II may be plotted and that, from the observed frequencies of form III, an interpolation made to determine the helix angle for form III. The correlation would predict form III as having a 10_3 helix, consistent with X-ray diffraction data. In this case, Raman complements the diffraction techniques for elucidating polymer structure.

9.1.1 Amorphous Polymers: Low Frequency Observations

Many amorphous materials exhibit an extremely broad, low-frequency Raman band. This is true for polycarbonate, PMMA, and polystyrene [56]. These bands provide information on the density of states (directly related to the intensity distribution of the broad Raman-active band) as well as on the anomalous specific heat behavior [56]. Spectra from these amorphous polymers show many similarities. The separation of main-chain and side-chain bands is often important in modeling the specific heat, which makes band assignments important [57,58]. This separation is analogous to the inclusion of optic and acoustic vibrations in the specific heat calculations [59]. The low-frequency region is expected to contain both intrachain skeletal bending modes and interchain lattice modes. A strong, broad band assigned to the amorphous phase was also observed in PET [60] at $\sim 215\,\mathrm{cm}^{-1}$. Other bands in the 129 and $73\,\mathrm{cm}^{-1}$ region are delocalized skeletal vibrations associated with several monomer units. This interpretation is supported by normal coordinate analysis, which predicts a symmetric in-plane deformation mode at $280\,\mathrm{cm}^{-1}$ and torsional modes involving both ester and glycol units at 119 and $63\,\mathrm{cm}^{-1}$.

In the quasiharmonic approximation the specific heat at low temperature is given by

$$C_v = k_\mathrm{B} \int_{\omega_\mathrm{min}}^{\omega_\mathrm{max}} g(\omega) \left(\frac{\hbar\omega}{k_\mathrm{B}T} \right) \frac{\mathrm{e}^{-\hbar\omega/k_\mathrm{B}T}}{(\mathrm{e}^{-\hbar\omega/k_\mathrm{B}T} - 1)^2} \mathrm{d}\omega \,, \tag{9.1}$$

where $g(\omega)$ is the density of states in the low-frequency region, k_B is Boltzmann's constant, and T is temperature. The scattering intensity I is related to the density of states by

$$\frac{I}{\omega^2} \propto \sum_i g_i(\omega) \,. \tag{9.2}$$

Thus, the low-frequency Raman intensity is related to the specific heat. The Raman observations for amorphous polycarbonate are shown in Fig. 9.6. As shown in Fig. 9.7, the calculated specific heat [56] shows good agreement with experiments.

Low-lying vibrations can also be used to examine intermolecular interactions and associated steric hindrance to intrachain motions, thus providing information regarding free volume [61]. Nearly 100% syndiotactic polystyrene

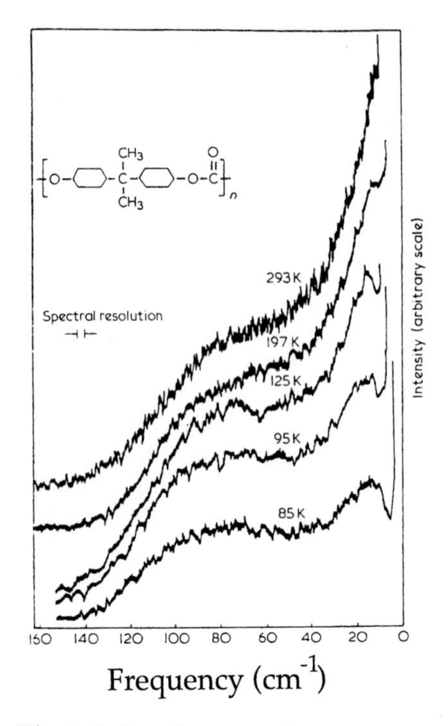

Fig. 9.6. Low frequency Raman spectra from amorphous polycarbonate at temperatures of 293, 197, 125, 95, and 85 K

(sPS) has been reported [62]. The solid-state structure includes several well-defined crystalline phases, which display at least 2 different chain conformations [63–69]. The all-trans planar zigzag structure is thermodynamically favored compared with the helical ttgg conformation [70]. Transformation from the helical to the all-trans phase can be accomplished by thermal annealing. The helical phase can be generated by exposing sPS to various solvents [63–66,69].

A striking spectroscopic feature associated with polystyrene-type polymers is the intense, broad, ($\sim 100\,\mathrm{cm}^{-1}$) polarized vibration observed in the $20–200\,\mathrm{cm}^{-1}$ region (Fig. 9.8) [61]. This band, first observed for sPS, is independent of molecular weight, degree of crystallinity, tacticity, or chain conformation [61]. As mentioned above, such low frequency bands are delocalized in most polymers, extremely sensitive to changes in chain conformation and packing, and usually difficult to assign unambiguously. Since this band is independent of the chemical structure of polystyrene, it is most likely associated with the torsional mode of an aromatic ring [61,71]. This band has also been associated with chain dynamics. Its usefulness for assessing intermolecular environments in various sPS crystals and complexes was explored.

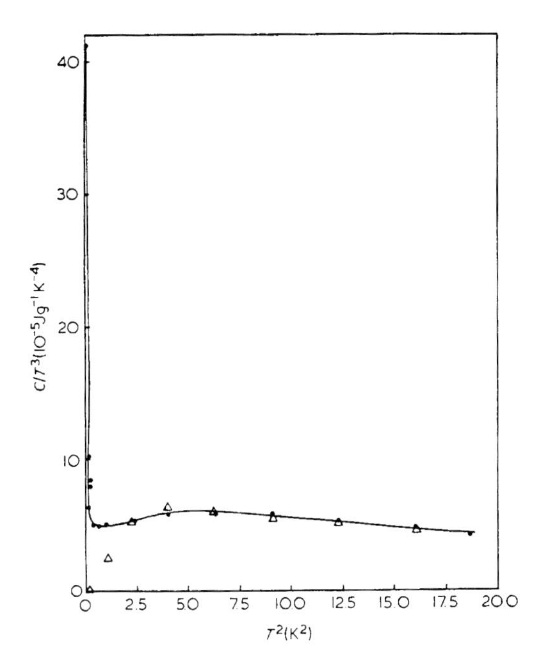

Fig. 9.7. Specific heat of amorphous polycarbonate as C/T^3 with T^2 variation. Δ Raman data; ∘ data from Cieloszyk et al. [242]; • data from Stephens [243]

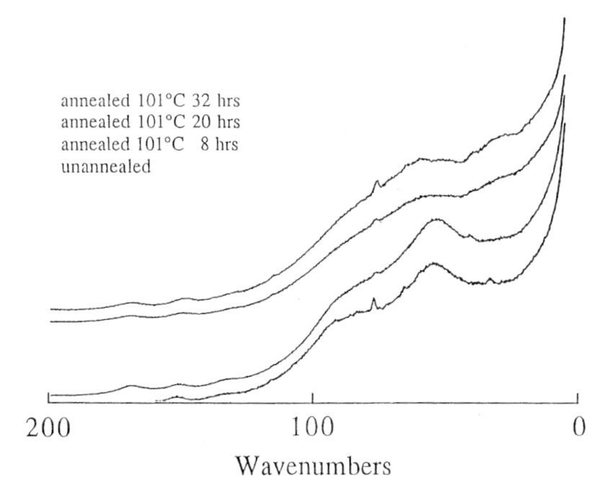

Fig. 9.8. Low-frequency band from annealed sPS, ordered top-to-bottom as in the legend

Several unusual features should be emphasized. The vibrational frequency depends on the exact location of substitution on the ring. A methyl group placed at the para-position leaves the frequency unperturbed. A substitution at the ortho-position changes the frequency. The band intensity is related

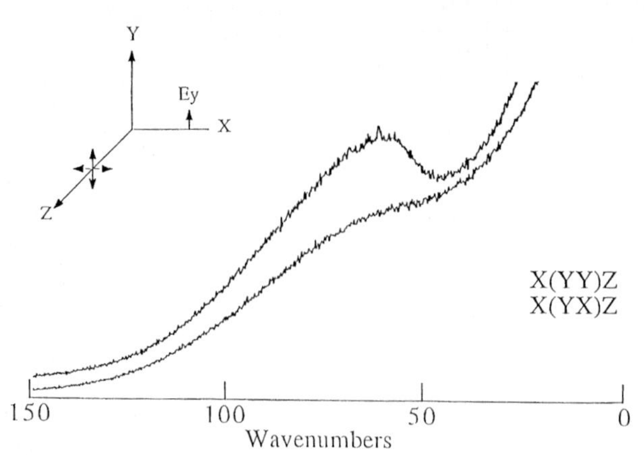

Fig. 9.9. Polarized Raman-active low-frequency band observed for sPS

to the mass of substitution [71]. The measured depolarization ratio ~ 0.22 (Fig. 9.9). Assigning this band to the torsional motion of the planar phenyl ring attached to the polymer backbone is consistent with its high intensity [61]. The optical activity of this vibration is such that the first overtone is responsible for the observed intensity, while the fundamental is forbidden [61]! The changes in the polarizability tensor elements have also been predicted using *ab initio* calculations performed with the GAUSSUN 86 package for the related compound *p*-difluorobenzene at the 3-21G level of theory. A quite strong 0–2 phenyl torsion transition has been reported in the 112–128 cm^{-1} region of the Raman spectrum of styrene. A torsional barrier of 623 cm^{-1} (1.78 kcal/mol) has been estimated from these data [72]. This mode has not been observed in the IR spectrum [73]. The reported Raman-active overtone band has approximately the same intensity as other strong bands of polystyrene. The torsional barrier height in styrene directly determines its torsional frequency and has been predicted with *ab initio* methods [74]. Although this vibration is intramolecular in origin, the bandwidth was found to be highly sensitive to the solvent-complexed crystalline state and, therefore, the intermolecular ring environment (Fig. 9.8). This observation suggests that such a vibrational band could be useful for examining the polymer-solvent specific interactions and, therefore, the formation mechanism of the helical crystalline phase in sPS. The similar behavior of this band in syndiotactic, atactic, and isotactic isomers suggests that this region of the Raman spectrum may be useful in determining the free volume associated with polymer chains in a variety of environments (Fig. 9.8) [61].

9.1.2 Solid State Properties

Raman spectroscopy can also determine chain packing and associated structural transitions in semi-crystalline polymer solids. Since the low-frequency spectrum is accessible in Raman spectroscopy, the magnitude and specificity of intermolecular interactions can be directly probed. Lattice modes representing interchain external vibrations or delocalized skeletal modes are found in the low-frequency region. In the crystalline portion of semi-crystalline polymers, the effects of intermolecular forces are usually quite small and difficult to analyze in a quantitative fashion. The frequencies associated with different species of the unit cell are not very different from the fundamental frequencies of an isolated chain. Furthermore, lattice modes have only been observed in a limited number of polymers [42,46]. Thus, intermolecular interactions in polymer crystals continue to be poorly understood.

For highly crystalline trans-1,4-polybutadiene (TPBD), the vibrations observed below 200 cm^{-1} in the IR and Raman spectra were all assignable to lattice modes (Fig. 9.10) [42]. This is indicated not only by the fact that the single-chain calculation predicts no modes below 200 cm^{-1} [42,75], but by their temperature dependence as well. When the temperature is lowered, the Raman bands at 48, 70, 104, and 118 cm^{-1}, all exhibit large shifts to higher frequencies. This is characteristic of interchain modes since the intermolecular interactions increase in magnitude as the distances diminish in the contracted unit cell. Crystal-field splittings should also increase under these circumstances, as seen for the bands near 240 cm^{-1}. Similar effects have been noted in polyethylene. In addition, if the TPBD sample is heated beyond the crystal phase transition, these four Raman bands disappear. This is as expected, since not only do distances between chains increase but there is an increase in torsional oscillations within the chain. Both features diminish the magnitude and specificity of interchain interactions. These characteris-

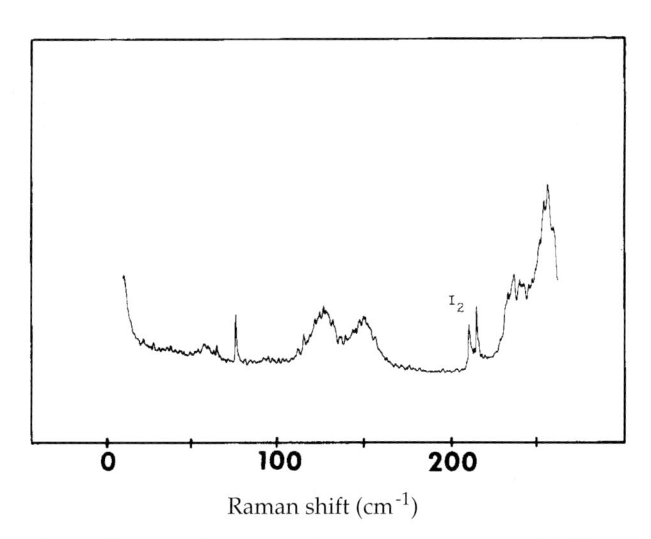

Fig. 9.10. Low-frequency Raman spectrum of TPBD at 110 K, using an iodine absorption cell. Bandpass, 3 cm^{-1}, Laser power, 150 mW at 5145.42 Å

tics unambiguously identify these bands as lattice modes, although specific assignments have to be determined from normal vibration calculations. Several studies have used TPBD as a model to investigate intermolecular forces and their influence on chain conformation. By incorporating intermolecular atom-atom interactions derived using other characterization techniques, these lattice modes and crystal-field effects have all been satisfactorily accounted for [42,75]. In the early studies, X-ray results from crystalline TPDB were used for a normal vibration analysis of the crystal [76–78]. The observed intermolecular modes and their crystal-field splittings were satisfactorily explained by a normal coordinate analysis involving addition of an interchain potential to an intrachain force field [42]. By studying TPBD in the highly crystalline state as well as in single crystals, the bands associated with noncrystalline conformations were also assigned.

TPBD has a first-order crystal-crystal phase transition at 76 °C [78,79]. Two crystalline forms are known to exist, and the transformation at 76 °C between the two forms is reversible. The crystal structure changes from a monoclinic unit cell to pseudo-hexagonal packing at high temperature. The conformational changes of the chain associated with the phase transition are: (1) a change in the $-CH-CH_2-$, ϕ angle, from $60-71°$ to $100°$, causing a shortening of the repeat length from 4.83 to 4.66 Å; (2) a considerable distortion in the molecular chain due to torsional disorder about C$-$C bonds; (3) an increase in the interchain distance in the unit cell from 4.54$-$4.60 to 4.95 Å. The distortion of the molecular chain may have the following consequences: (1) Chains may exist with irregular conformations, which have lost the inversion symmetry and cause breakdowns of the mutual exclusion principle; (2) In some chain segments, the CH_2-CH_2 groups may no longer be *trans* to each other, which may result in observation of *trans-gauche* amorphous bands.

The general similarity between the high-temperature and room temperature spectra shows that no additional bands can be attributed to *trans-gauche* conformations [75]. Even though considerable distortions must exist in the high-temperature form of TPBD, Raman spectra show that the molecular chain has a large degree of conformational regularity, consistent with X-ray results. The Raman-active lattice modes observed at 48, 70, 104, and 118 cm^{-1} disappear when the sample is heated beyond the phase transition. This is to be expected, since not only does the distance between chains increase but there is an increase of torsional disorder about the C$-$C bond as well as translational disorder along the chain axis. The intense CH_2-CH_2 torsional mode, observed at 239 cm^{-1} (room temperature), becomes diffuse and unobservable at high temperature. This is also expected as a result of the torsional disorder in the chain above the transition. Close to the transition temperature, the 549 (skeletal angle bend), 1305 (CH_2 twist), 1340 (CH_2 wag), and 1433 (CH_2 bend) cm^{-1} bands all show a doublet in the spectra. The two components correspond to the two structures above and below the phase transition. Only one band exists below 70 °C or above 80 °C. This is consistent with X-ray studies of single crystals, which indicate that the two forms coexist between 55.5 and 69.5 °C. This coexistence suggests that the energy minima for TPBD may not be as shallow as previously assumed [80]. The fact that there is only one component below and one above the phase transition temperature indicates that only one form of the chain exists with little interconversion between the different conformations. The interconversion takes place only at the onset of the crystal-crystal phase transition. At high temperatures the intramolecular conformational freedom may be large and will have the effect of significantly diminishing the magnitude and specificity broadening and cause the disappearance of some normal modes. Although a distribution of conformations exists at any given temperature, the spectra indicate that certain specific conformations are favored.

9.2 Polymer Anisotropy

9.2.1 Motivation

Material properties such as thermal conductivity, optical transmission, electrical conduction, piezoelectricity, shrinkage stability, and, particularly, mechanical performance are all dependent on the degree of anisotropy. For many polymer applications, high modulus and strength are especially important. These qualities can be obtained by deforming isotropic solids to achieve the necessary degree of anisotropy. In polymers secondary forces associated with interchain interactions are much weaker than the primary forces that form the polymer chain backbone. Therefore, the force necessary to deform or elongate the chain will always be higher than that required for chain separation. This is the motivation associated with processing isotropic to anisotropic states. The deformation process induces alignment of molecular chains along a specific direction that results in an increase in strength and modulus of the material in that direction. Uniaxial deformation is used to make samples of a fibrous nature. Biaxial orientation along two perpendicular axes has been successful in producing materials of interwoven fibers to yield films or sheets with desirable mechanical performance along the two plane axes. Double orientation has also been found in polymer samples such as rolled polyethylene and polypropylenes. In this case, double orientation occurs when the chain axis is aligned along one specific deformation direction and other crystal axes also possess preferred directions. For amorphous samples, it is important to understand the deformation behavior of individual chains. For crystalline polymers, the deformation behavior must be related to the relative motion of amorphous chains and crystalline regions. The deformation process must be correlated with the deformation mechanisms available to the crystalline portion. The crystalline degree, size, perfection, and interconnectivity are all important in the overall consideration of deformation mechanisms. The absence of a valid model to describe the deformation process makes experimental studies extremely important in relating the structure to the applied stress or strain.

A common measure of mechanical performance of a processed polymer is its modulus measured under extension. The experimental values generally fall substantially lower than theoretical estimates. This difference is undoubtedly due to both conformation and orientation disorder in the processed samples. The best illustration of the complex processing steps required and the resulting structure is polyethylene. Gel-spun polyethylene fibers and biaxial oriented blown films have better properties than the isotropic state [81]. Polyethylene has been manufactured for over 60 years and is widely viewed as a commodity plastic. Its primary material attributes are toughness, flexibility, and ease of processing. The modulus of as-polymerized material is typically in the range of 1 GPa. Pennings [81] showed that the modulus of polyethylene can be increased significantly to a value approaching theory, ~ 300 GPa [82,83]. The gel-spun method presented schematically in Fig. 9.11 is a complicated process, which stretches the chains from a "low entanglement" gel mass quench into a crystalline state before drawing into a highly oriented state [84]. In processed fibers virtually all chains are perfectly extended and aligned with respect to the deformation axis, thus producing an extremely high modulus. The force required to deform a backbone bond is approximately 10 times higher than that needed to deform a valence angle, which, in turn, is approximately 10 times higher than that needed to change the torsional angle of bonds along the backbone. If it is possible to orient and

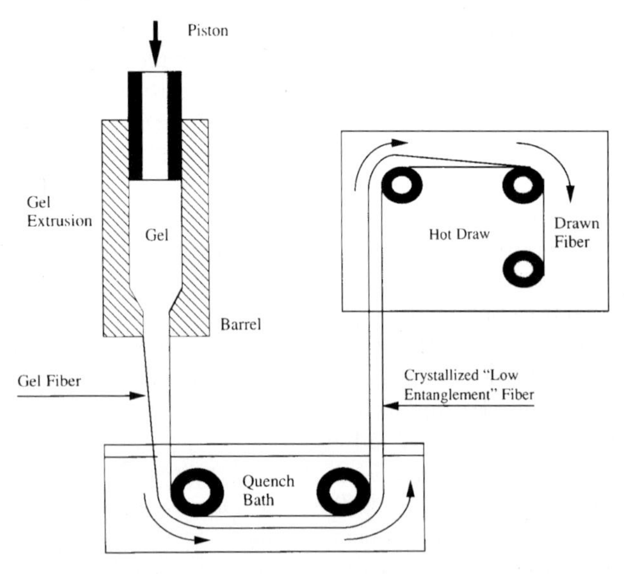

Fig. 9.11. Schematic diagram of the Smith and Lemstra gel-drawing experiment

extend individual polyethylene chains into a planar zigzag conformation as in these highly deformed fibers, thus eliminating the torsional deformation of disordered chains, then the modulus can be increased significantly [85]. The modulus for these gel-spun fibers exceeds that of most high-performance materials of comparable density.

Analysis of chain conformation distribution and segment orientation for both crystalline and amorphous regions is needed to assess the efficiency of various processing methods used to achieve anisotropy. Ultimately the mechanical properties obtained for highly deformed polymers are dependent on minute structural defects. The various postulated models are shown in Fig. 9.12. An estimate of the theoretical modulus is necessary for evaluating processing efficiency. A number of studies have been devoted to calculation of ultimate modulus of various polymers using force constants derived from vibrational spectroscopy [86]. Changes in the helical parameters can be related to changes in bond length, valence angles, and torsional angles. These relationships can be determined for virtually any helix, simple or complex [86]. For the simplest case, polyethylene, the calculated modulus is 340 GPa, which is close to the experimental value. Similarly the calculated

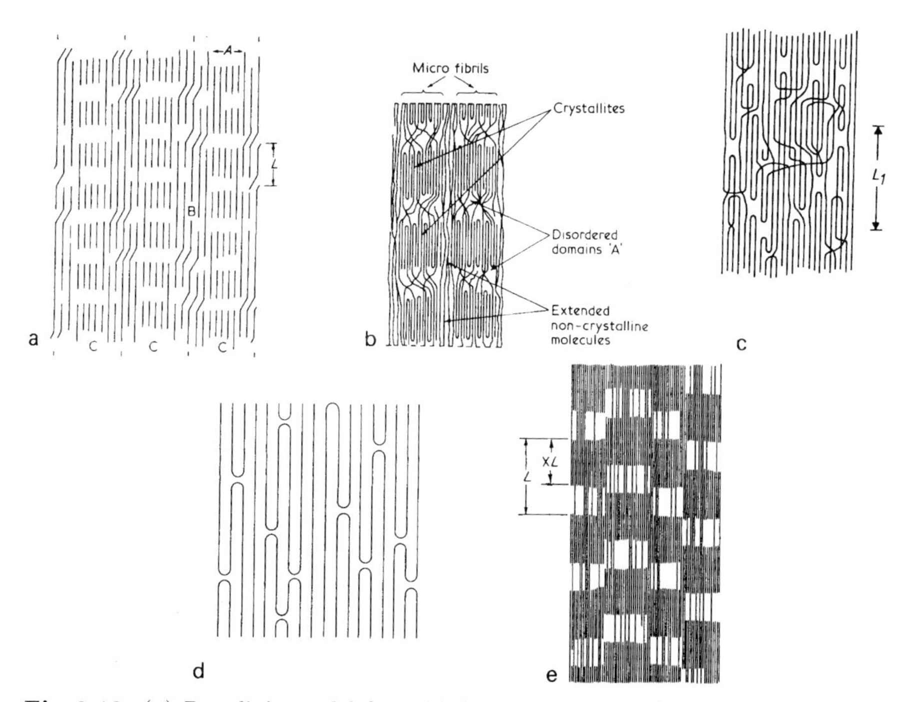

Fig. 9.12. (**a**) Peterlin's model for cold drawn microfibers [136]; (**b**) Prevorsek's model for nylon-6 [244]; (**c**) Fisher's model for drawn fibers [245]; (**d**) Clark's model for drawn fibers [246]; (**e**) Ward's model for hot drawn and solid state extruded fibers [247,248]

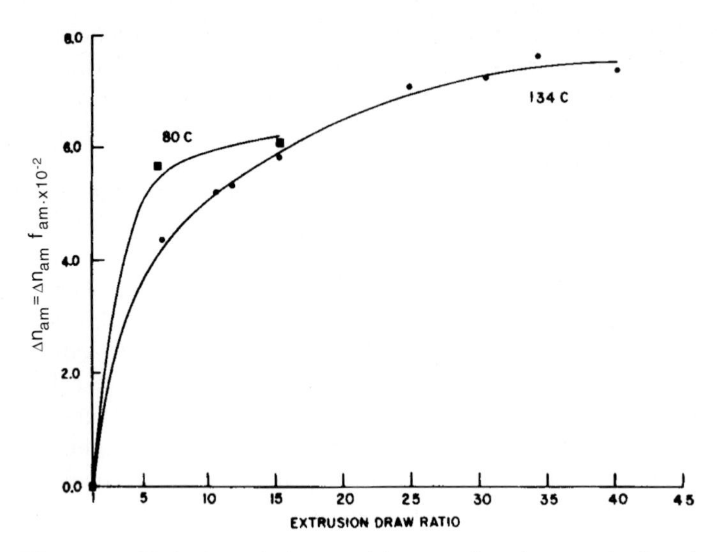

Fig. 9.13. Variation of Δn_{am} with extrusion draw ratio for ultra-oriented high density polyethylene extruded at 0.23 GPa and temperatures as shown [249]

modulus for polypropylene is ~ 42 GPa, also quite close to the experimental value [87].

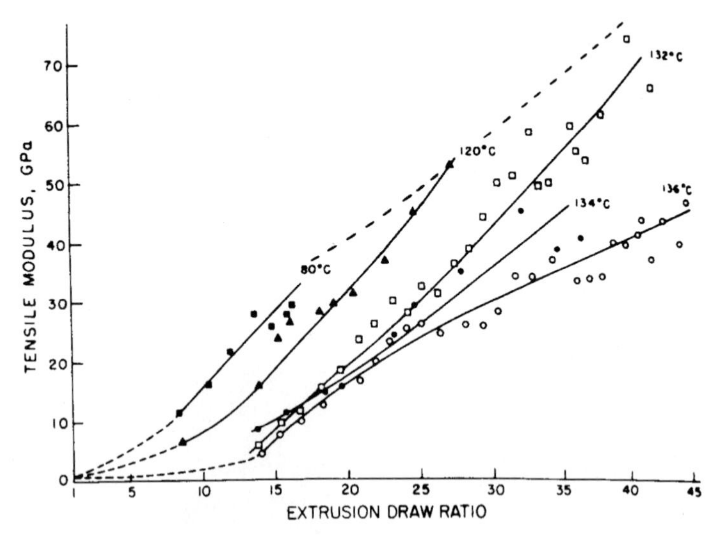

Fig. 9.14. Dependence of tensile modulus on fiber extrusion draw ratio at various temperatures. The high density polyethylene was crystallized at 0.23 GPa and 134 °C [249]

Polymers are processed to reduce structural defects and to achieve segmental orientation. Undoubtedly these two aspects are correlated, making the measurement of sample anisotropy a crucial one. Different characterization techniques measure segmental orientation of different morphological subunits. Some measure overall sample anisotropy. Others measure specific structural units, crystalline, amorphous, or interface. Different portions of samples experiencing different forces may yield different degrees of orientation. For example, the interior of an extrudate may yield a considerably different degree of orientation in comparison to its surface. Techniques for assessing the increase in sample anisotropy as a function of deformation include X-ray diffraction, IR dichroism, polarized Raman scattering, birefringence, NMR, fluorescence, and measurements of sample density. Yet, virtually all techniques lose sensitivity as draw ratio increases (Fig. 9.13). The one property that continues to change uniformly is sample modulus (Fig. 9.14).

9.2.2 Partially Oriented Systems

As demonstrated above, the modulus of a sample being deformed is directly related to the degree of deformation, draw ratio. The less than ideal modulus value is then related to the less than perfect segment orientation along the deformation axis. Both conformation and orientation changes become more difficult to measure as the sample deformation increases. Because of its sensitivity to short range order, however, Raman spectroscopy is capable of providing detailed information regarding orientation of chain segments, in the ordered or disordered states, and for a large variety of samples, including fibers and films.

The use of Raman spectroscopy to characterize segmental orientation is a well-developed subject. The Raman scattering tensor α for each vibration contains 6 independent elements. The forms of each tensor depend on the polymer and vibrational mode symmetries and are tabulated in most spectroscopy textbooks [21]. If the molecular axes are well defined with respect to the laboratory frame, it is possible to measure each individual element by varying the molecular orientation relative to the polarization of the incident radiation. The expression relating the scattering tensor in the molecular and laboratory frames is shown below:

$$\boldsymbol{P} = \alpha_{\mathrm{lab}}\boldsymbol{E} = T^{\mathrm{t}}\alpha_{\mathrm{mol}}T\boldsymbol{E} \,, \tag{9.3}$$

$$\alpha_{\mathrm{mol}} = \begin{pmatrix} \alpha_{11} & \alpha_{12} & \alpha_{13} \\ \alpha_{21} & \alpha_{22} & \alpha_{23} \\ \alpha_{31} & \alpha_{32} & \alpha_{33} \end{pmatrix} , \tag{9.4}$$

where $\boldsymbol{P}$ is the induced dipole due to the polarizability change, $\boldsymbol{E}$ is the electric field, and T is the matrix relating the molecular to the laboratory frame. It should be mentioned that care should be taken to account for the inherent polarization dependence of the spectrometer. In addition, sample

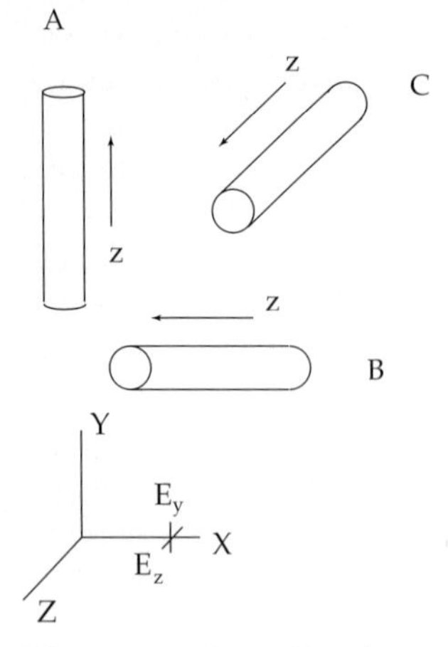

Fig. 9.15. Three fiber/uniaxial sample orientations relative to the incident radiation

inhomogeneity (for example large crystallite size) can scatter light and thus perturb the measured polarization characteristics.

For samples deformed to various stages of anisotropy, the relationship between the molecule and laboratory frames is partially lost. The macroscopic orientation of a sample is the average orientation of each of its microstructural units. For partially-oriented uniaxial systems, the 6 independent elements in the scattering tensor are mixed due to the 2 indistinguishable axes. In each case, only 4 elements can be measured. The polarized Raman scattering geometries for these partially oriented systems are shown in Figs. 9.15 and 9.16. For these cases, the average of all orientations will be necessary. For both uniaxial and biaxial systems, Snyder has tabulated the spatially averaged components of the differential polarizability tensor [27,28]. Spectra

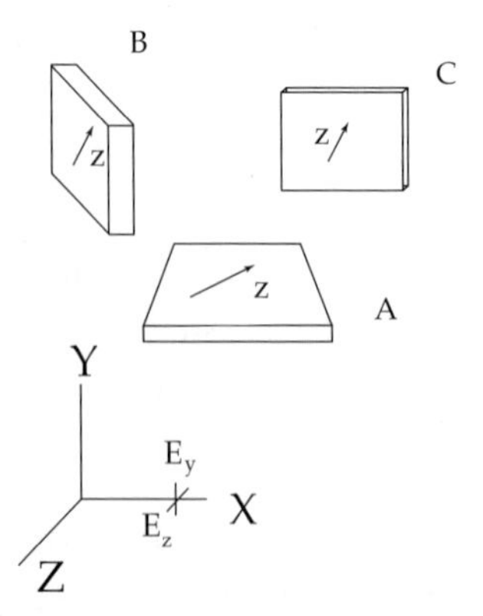

Fig. 9.16. Three film/uniplanar sample orientations relative to the incident radiation

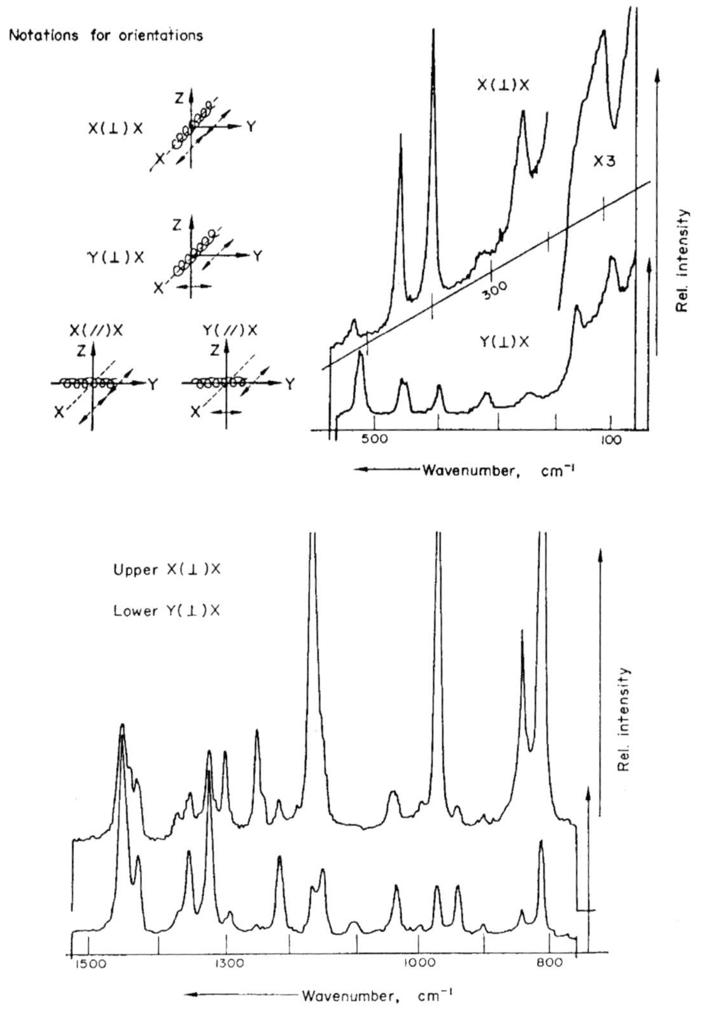

Fig. 9.17. Polarized Raman spectra of oriented isotactic polypropylene fibers [26]

obtained for a uniaxial isotactic polypropylene (iPP) sample are shown in Fig. 9.17.

Referring to Figs. 9.15 and 9.16, the tensors for partially oriented systems are

$$\text{Orientation A:} \quad \alpha = \begin{pmatrix} A_1 & A_3 & A_2 \\ A_3 & A_4 & A_3 \\ A_2 & A_3 & A_1 \end{pmatrix} \tag{9.5}$$

$$\text{Orientation B:} \quad \alpha = \begin{pmatrix} A_4 & A_3 & A_3 \\ A_3 & A_1 & A_2 \\ A_3 & A_2 & A_1 \end{pmatrix} \tag{9.6}$$

$$\text{Orientation C:} \quad \alpha = \begin{pmatrix} A_1 & A_2 & A_3 \\ A_2 & A_1 & A_3 \\ A_3 & A_3 & A_4 \end{pmatrix} . \tag{9.7}$$

For example in a uniaxial system the A's are

$$A_1 = \frac{1}{8}\left(2a^2 + b^2\right) , \; A_2 = \frac{1}{8}b^2 , \; A_3 = \frac{1}{2}c^2 , \; A_4 = d^2 , \tag{9.8}$$

where the coefficients a, b, c, and d are defined as

$$\begin{aligned} &a^2 = (\alpha_{xx} + \alpha_{yy})^2 , \; b^2 = (\alpha_{xx} - \alpha_{yy})^2 + 4\alpha_{xy}^2 , \\ &c^2 = \alpha_{yz}^2 + \alpha_{zx}^2 , \; d^2 = \alpha_{zz}^2 . \end{aligned} \tag{9.9}$$

The use of Raman spectroscopy in orientation studies is illustrated in the examples below. As shown in Fig. 9.18, the CH stretching vibration region changes significantly as a function of draw ratio. A completely symmetric A_g CH_2 stretching vibration should be found in this region at $2850\,\mathrm{cm}^{-1}$. The asymmetric CH_2 stretching vibration is found at $\sim 2900\,\mathrm{cm}^{-1}$. By using

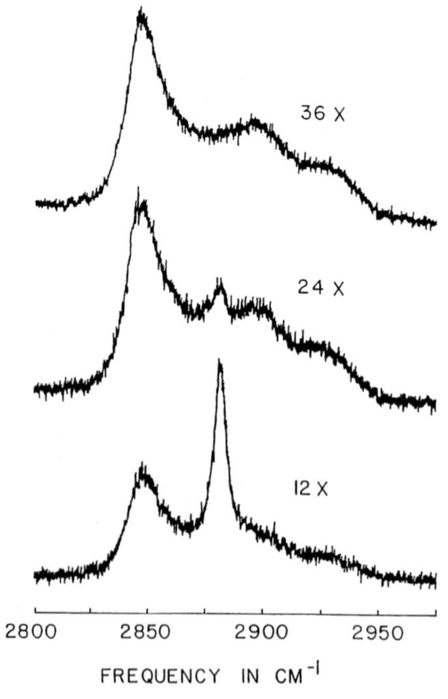

Fig. 9.18. CH stretching region change in polyethylene as a function of draw ratio X

the zz geometry, it is possible to eliminate contributions from the asymmetric stretching component. In fact, one can assign all 3 peaks to the same symmetric stretching vibration [29]. The additional components arise from Fermi resonance interactions between the fundamentals of the stretching and the second harmonic of the CH_2 bending vibrations. Similarly, in the CH_2 bending region, only two components are expected; four are observed. A detailed understanding of the optical activity of oriented polyethylene leads to assignments of the vibrational modes and allows their use in structural analysis [29].

Industrial processes include melt spinning, injection molding, blow molding and tubular film extrusion. When a polymer is extruded in the molten state, the major part of the deformation occurs prior to solidification. However, the degree of orientation depends on the relaxation process and stress at the time of solidification. Usually, the degree of orientation is not very high and may depend on the temperature profile in various portions of the sample. In order to obtain higher orientation, polymers have to be reprocessed in the solid state by drawing, rolling or extrusion. This is especially true for thick sheets, rods or objects of other shapes. The complementary nature of several techniques, including Raman, has been demonstrated in the characterization of segmental orientation in thick polyethylene samples [88]. In this work the original specimens, with dimensions of the order of a cm, were cut to obtain 1-mm thick platelets, and the crystalline and amorphous portions were compared. It was possible to measure crystalline orientation using X-ray diffraction or the 1894 cm^{-1} IR-active bands [88]. The orientation differs from the amorphous regions as indicated by the orientation behavior of the 1130 and 1060 cm^{-1} Raman-active skeletal stretching bands arising from vinyl end groups. The behavior for the first draw and changes in orientation of the second draw process were probed. Depending on the type of processing and the 3D objects produced, the degree of orientation between the center and the surface may differ. In some cases, the values obtained from Raman spectroscopy are lower than those calculated for the crystal phase from X-ray diffraction and from polarized IR, but are higher than the orientation determined for vinyl end groups. This intermediate result can be explained by the fact that the 1130 and 1060 cm^{-1} Raman bands are assignable to the all-trans conformers, found in both the amorphous and crystalline regions. The values obtained from X-ray and Raman methods at various locations in the samples are shown in Fig. 9.19.

Isotactic polypropylene (iPP) exists in a 31 helix. Its vibrations can be divided into 25 A modes and 26 degenerate E modes. The A modes are polarized and E modes depolarized in the Raman spectrum. Very different spectra are obtained for the uniaxially oriented sample. Here, the chain segment orientation lies somewhere between the random case for gases or liquids and the perfectly oriented case for macroscopic single crystals. In most cases, some degree of uniaxial orientation can be achieved with appropriate pro-

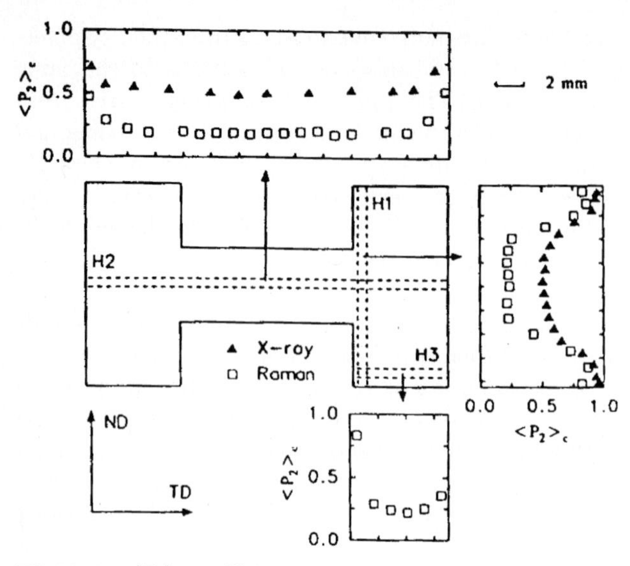

Fig. 9.19. $\langle P_2 \rangle_c$ coefficients measured by X-ray diffraction (110, 200 and 020 reflections) and by Raman spectroscopy at different positions for the H1, H2, and H3 platelets [88]

cessing techniques. The identification of the symmetries of normal vibrations is achieved by performing appropriate Raman scattering experiments [25,26], which are easier to conduct than the corresponding IR experiments. The assignments of various modes to A or E symmetry has been found to be important in structural consideration.

9.2.3 Definition of Orientation Function

It is important both to deduce the anisotropy achieved and to model the process as a function of draw ratio, at least in a semi-quantitative fashion. The orientation distribution function, which describes how chain segments are oriented to a reference axis usually defined by the processing method, is then employed. As mentioned above, even for highly deformed material, some imperfection in segmental orientation remains. Knowledge concerning the molecular mechanism and the degree of segmental orientation achievable by various drawing, extrusion or other deformation techniques is, in fact, quite limited. Two aspects describing the formation of anisotropic polymers need to be defined: a quantitative description of segment orientation achieved and the relationship between segment orientation and draw ratio.

The distribution of chain segments relative to the reference axis is defined as the fraction of chains in each angular volume element [89]:

$$\frac{\mathrm{d}n}{N} = f(\zeta, \phi)\,\mathrm{d}\zeta\,\mathrm{d}\phi \,, \tag{9.10}$$

where $\zeta = cos\theta$ with θ the angle measured with respect to the reference axis. In three dimensions, the distribution function is often expanded as [90,91]

$$f(\zeta,\phi) = \sum_{l=0}^{\infty} \sum_{m=-l}^{l} C_{lm} P_l^m(\zeta) \mathrm{e}^{\mathrm{i}m\phi} , \tag{9.11}$$

where P_l^m is the normalized associated Legendre polynomial and the coefficients C_{lm} are determined from

$$C_{lm} = \frac{1}{2\pi} \int_0^{2\pi} \int_{-1}^{1} f(\zeta,\phi) P_l^m(\zeta) \mathrm{e}^{-im\phi} \, \mathrm{d}\zeta \, \mathrm{d}\phi \, . \tag{9.12}$$

If the system has no angular dependence (uniaxially oriented with a symmetry plane perpendicular to the deformation axis), then all C_{lm} vanish except for $m = 0$. Thus the distribution function can be written as

$$f(\zeta,\phi) = \sum_{l=0}^{\infty} C_l P_l(\zeta) , \tag{9.13}$$

and then

$$C_l = \int_{-1}^{1} f(\zeta) P_l(\zeta) \mathrm{d}\zeta \, . \tag{9.14}$$

For an axially symmetric system, only even Legendre polynomials are retained in the expression:

$$\begin{aligned} P_0(\zeta) &= 1 \\ P_2(\zeta) &= \frac{1}{2}(3\zeta^2 - 1) \\ P_4(\zeta) &= \frac{1}{8}(35\zeta^4 - 30\zeta^2 + 3) \, . \end{aligned} \tag{9.15}$$

A full description of the segmental orientation function requires that the individual components be defined. The C coefficients represent the contribution from each Legendre polynomial. The second order coefficient is the well-known "Herman orientation function" [92] and is generally the only one used for structural analysis,

$$C_2 = \frac{3\left\langle \cos^2\theta \right\rangle - 1}{2} \, . \tag{9.16}$$

This expression represents an average taken over all orientations of the sample. Many methods can measure the second moment C_2, which ranges from -0.5 to 1.0. In one extreme the segment is oriented perpendicular and in the other parallel to the deformation axis. $C_2 = 0$ corresponds to random orientation. For IR spectroscopy, the second moment is usually given by:

$$C_2 = \frac{D-1}{D+2} \frac{D_0+2}{D_0-1} , \tag{9.17}$$

where D is the dichroic ratio, $D_0 = 2\cot 2\beta$, and β is the angle of the transition moment with respect to the chain axis.

Both Raman and NMR can measure higher moments such as C_4. In practice, experimental difficulties severely limit applications of the full theory [93]. Only X-ray diffraction is capable of accurately measuring higher moments of the distribution function [88,94], but the high quality diffraction data necessary for this are difficult to obtain. For small draw ratios, evaluation of C_2 is usually quite sufficient; higher moments are needed for highly deformed samples.

The expressions presented above measure the segment distribution relative to an axis. It is often desirable to relate this distribution function to the degree of deformation associated with the polymer chain transformation from isotropic to anisotropic states. This transformation is dependent on the size, perfection, and interconnectivity of the morphological units. In semi-crystalline polymers, the deformation behavior depends on the degree of crystallinity, crystal size, perfection of crystalline units, and, very importantly, the amorphous segments connecting these crystalline units. These expressions are highly dependent on the models used. The only case that can be solved in closed form, the Gaussian chain model, is a network consisting of totally flexible chains and chain segments. This model quantifies the angle between each chain segment and the end-to-end vector as this vector elongates and orients when the overall sample is deformed. The model gives a surprisingly weak dependence of the orientation on draw ratio [90,91]. However, if the number of connecting units between segments is small, the change in orientation function increases significantly with strain [95].

9.3 Long-Range Order and Disorder in Polymers

9.3.1 Initial Observations Made for Models and Polymers

The strength of vibrational spectroscopy lies in its ability to characterize the chemical composition and local structure of polymers. The intense longitudinal acoustic mode (LAM) in the extremely low-frequency ($< 50\,\mathrm{cm}^{-1}$) Raman spectra of semicrystalline polymers is sensitive to structure on a large length scale. Unlike other vibrational modes, this vibration is characteristic of polymer chain order over several hundred Å. Diffraction methods work best on long-range ordered structures. These techniques can accurately provide atomic placements and are capable of measuring the size, type, and perfection of crystalline structures. Raman spectra do not depend on long-range coherence. Therefore, this low-frequency vibration provides a different morphological probe to characterize structural units in semi-crystalline polymers.

The observation and characterization of LAM in n-paraffins and in polyethylene are well established. The earliest observation of such a Raman-active fundamental mode in n-paraffins, whose frequency is inversely proportional to the planar zigzag chain length, was made in 1949 [96]. In fact, a series of intense vibrations was observed in the low-frequency region of the Raman spectra. The main characteristics of the lowest frequency component of these vibrations were its high intensity and its frequency being inversely proportional to the straight chain segment length. It was hypothesized that this mode corresponds to the longitudinal acoustic vibration along the backbone of the entire chain segment, whose frequency is then given by:

$$\nu = \frac{m}{2cL}\sqrt{\frac{E}{\rho}} . \tag{9.18}$$

In this expression m is the order of the mode (only odd modes are Raman active), c is the speed of light, L is the straight chain segment, E is Young's modulus, and ρ is the single chain density. This expression is analogous to that for the sound velocity in an elastic rod. There are many other descriptions of this LAM, such as the frequency associated with the standing wave of an open organ pipe. Various continuum models have been proposed as an explanation for the origin of the mode. A full normal coordinate analysis provides the only complete explanation for this type of motion. A more detailed study showed higher orders of the LAM, whose frequencies agreed well with the calculated dispersion curves of an infinite planar zigzag chain [82].

This vibration was seen some years later in single crystals of polyethylene [45]. In that report a direct relationship was found between the central frequency of the mode and the long period in small angle X-ray scattering (SAXS). The assumption is that the straight chain segment is the segment within the crystalline region, which corresponds to the lamella thickness. It should be emphasized that these two techniques are very different in origin. The long period measured by SAXS originates from density differences between the crystalline and amorphous regions. SAXS measures the distance and how the individual lamellae stack relative to each other. Thus the measured long period includes contributions from the amorphous regions as well. An additional factor for consideration is the fact that chain tilt may occur in the crystalline regions [97]. The Raman technique measures the straight chain segment in the lamellae independent of the degree of tilt.

9.3.2 Other LAM Observations

Successful observations of LAM have been confined mainly to polymers possessing fairly simple chain conformations and chemical units such as polyethylene, which has a high single modulus but low density. The observation of LAM in other polymers proved elusive for quite some time, despite numerous attempts. Polymers such as the well-defined model polypeptides are among the notable failures [98–100]. Even in fairly similar systems such as iPP, LAM was not obtained until $\sim$ 25 years after the initial observation. Several unsuccessful attempts were made to observe the LAM in isothermally crystallized or cold drawn material [97]. The first successful LAM observation was for iPP prepared under special conditions of extrusion and orientation [87]. The sample was obtained by the "solid-state" extrusion technique. Both sample transparency and the preferential reduction in Rayleigh scattering, done with the iodine filter, made the observation possible [97]. The low-frequency Raman spectra observed at 300 and 100 K for a polypropylene sample extruded at 130 °C are shown in Fig. 9.20. Both Stokes and anti-Stokes shifts were measured. The mode was at 9.2 and 9.8 cm^{-1} at 300 K and 100 K, respectively. Two characteristics of the low-frequency Raman band observed in extruded polypropylene strongly indicate that it represents a LAM associated with lamellar components of the sample: (1) Its frequency is inversely proportional to chain length (shown below) and (2) The intensity is exceedingly strong, in fact much stronger than the skeletal deformation A mode at 398 cm^{-1}.

LAM has now been observed in a number of polymer systems such as polytetrafluoroethylene, POM [101–103], and poly(ethylene oxide) (PEO)

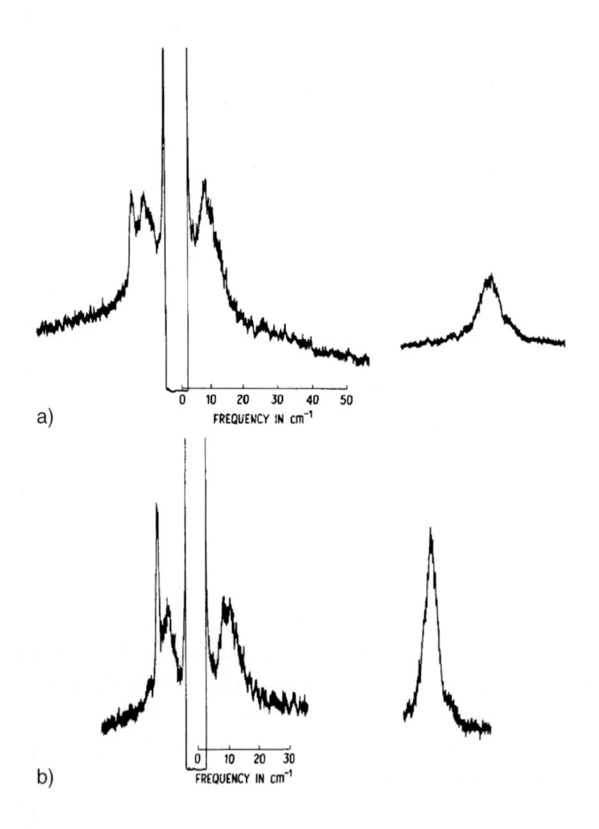

Fig. 9.20. Raman spectra of highly oriented transparent polypropylene (**a**) at 300 K and (**b**) at 110 K. Band at right is at 398 cm^{-1}

[104–106] as well as in some linear aliphatic polyesters [107–109] and a biodegradable thermoplastic, poly (β-hydroxybutyrate) (PHB) [110]. The inverse relationship between the long period measured by SAXS and band frequency was observed. In addition, the vibration was found to be extremely intense as is characteristic of longitudinal acoustic modes. The LAM observation in PHB is especially intriguing, since its chemical structure and the equilibrium conformation differ significantly from those of all other polymers known to exhibit LAM bands.

The fact that LAM is not easily found in other polymers can be understood from the fact that force as well as mass perturbations along the chain may lead to coupling of transverse and longitudinal motions. The effect of "perturbing" influences within chains such as methyl side-group placement on n-alkanes [111], double bond placement in trans-alkanes [112] and hydrogen bonds [113] has been studied and shown to affect the LAM to varying degrees in both intensity and frequency. In another study, simple linear aliphatic chains were modified by the placement of well-spaced ester groups along the polymer backbone. A LAM-like vibration was also observed for these polyesters [107–109]. Based on a series of normal vibrational analyses, the effect of masses placed off the main chain axis was investigated for this series of polyesters. Well-spaced ester groups act as unbalanced mass units, leading to the presence of several LAM-like modes, each with differing amounts of transverse and longitudinal character [108]. This may provide an explanation for the presence of the second unassigned low-frequency band centered around $9\,\mathrm{cm}^{-1}$ found for PHB. With increased structural complexity, these structural perturbations may become more significant, leading eventually to the total disruption of any longitudinal acoustic mode. Thus, in addition to its use as a morphological tool, LAM in PHB provides an opportunity to study the origin of LAM in a structurally more complicated polymer than those reported earlier. As such, it should aid in identifying factors that determine the appearance or non-appearance of LAM in specific polymer chains, leading to a better understanding of LAM in non-polyethylene-like structures.

In earlier studies on polyethylene-like chains, the strong intensity was found to be due to the large change in polarizability associated with the long chain motion of this mode and the Boltzmann factor associated with low lying vibrations [21,82,114]. The polarizability change of a number of polymers with non-carbon atoms has yet to be determined and, in fact, may be quite small. Interestingly, however, PHB, clearly exhibits LAM even though it has a large number of "perturbing" groups comprising a mix of closely spaced methyl units and ester groups as well as possibly an intermolecular effect between ester groups. The ester groups for the linear aliphatic polyesters were sufficiently well-spaced so that in the crystalline state the chains formed a nearly planar zigzag [108]. Compared with polymers such as iPP, POM, PEO and polytetrafluoroethylene in which LAM has also been observed, PHB is clearly the most unlike polyethylene.

Structural defects such as chain branching may affect the frequency of the LAM [111–113]. As intuition would suggest, placing the chain branch at the node, i.e. the center of the chain, produces little, if any, perturbation. In contrast, if this structural defect is placed at the antinode of the vibration, i.e. near the chain ends, substantial perturbation is observed. Following these earlier studies, end effects continue to attract interest. For example, the presence of an amorphous or loosely folded layer of material at the lamellae surface must perturb the effective frequency, intensity, and distribution of the LAM observed for semi-crystalline polymers. That the frequency-chain length relationship could not correspond exactly to the relationship derived from the n-paraffins was already evident from normal vibration calculations on n-paraffins and cyclic $C_{34}H_{68}$ [97]. Analysis of these perturbing effects can also be pursued by treating the LAM as the vibration of an elastic rod containing moduli associated with a crystalline portion capped by the amorphous regions at the ends [115]. This model was proposed in order to account for the anomalously high LAM frequency found for iPP [87]. End effects have also been observed for PEO, in which the specific interactions between hydroxy end groups strongly perturb both the frequency and relative intensity of the LAMs [83,87,104,116].

The effects of intermolecular interactions are also interesting to analyze. The earliest studies definitely established that the LAM of a planar zigzag chain conformation is a single chain phenomenon. Because of the planar zigzag structure, LAM in polymethylene systems is necessarily confined to the plane containing the skeletal backbone. The LAM is essentially independent of intermolecular interactions. N-alkanes are known to exhibit different crystalline phases. The LAM frequency remains unchanged when solid-solid phase transitions occur [117]. For fluoropolymers, low-frequency modes were found for both the solid and melt states [103]. Based on this observation, it was suggested that regular extended helical conformations exist in both states. The width of LAM relates to the existence of conformational defects known to exist in both n-alkanes and fluoro-alkanes [117,118]. A previous normal coordinate analysis of crystalline PEO has shown that, in contrast to all-trans polyethylene chains, in helical chains there is a significant interchain effect due to the large radial component present in the LAM atomic displacements for such polymers [106]. Unlike the planar symmetry case, the deformation of a helix must necessarily involve winding and unwinding of the entire chain segment. The temperature dependence of the frequency change suggests that intermolecular interactions may be more important for helical chains of low modulus than for those (such as polyethylene) of high modulus.

9.3.3 Applications of LAM to Polymer Structural Characterization

As mentioned previously, using the relation given by (9.18), the single chain elastic modulus within a crystalline lamella for various polymers can be cal-

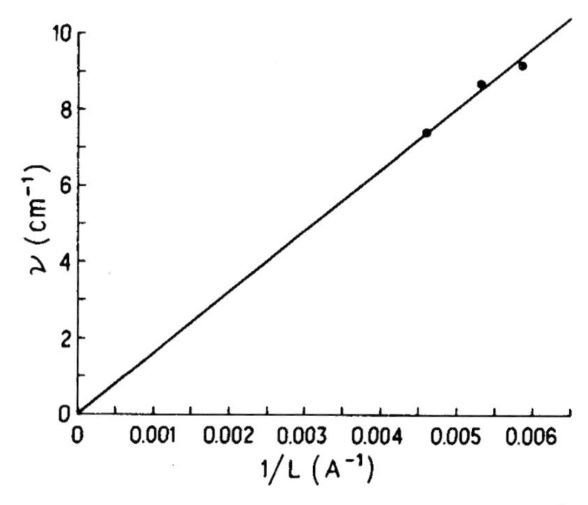

Fig. 9.21. Plot of Raman frequency versus L^{-1}. (L is the Bragg spacing of the SAXS maximum)

culated from the slope of the line fitted to the plot of ν versus L^{-1} (Fig. 9.21). This estimation requires knowledge of the single chain density, which is generally obtained from the unit cell parameters. For example, the modulus first calculated for polyethylene was 360 GPa, which was revised to $\sim$ 300 GPa when perturbing effects were considered [83]. For polypropylene, the LAM measurements yield a value of 36.6 GPa. This is to be compared with the measured elastic modulus along the chain direction, $\sim$ 41 GPa [86], and a theoretical value of 33 GPa obtained [119] using torsional force constants from a valence force field [18,120]. Given a crystal density 1.25 g/cm^3 [121], the single chain elastic modulus of PHB was calculated to be 29.6 GPa, a value consistent with that expected for a helix [110]. As stated previously, the fact that the modulus along the chain direction can be measured is particularly valuable when an estimate of processing efficiency is required.

LAM spectroscopy is useful for morphological studies, since it provides an independent measure of straight chain segment length in polymer systems. The central peak position corresponds to the most probable chain length, and the bandwidth reflects the chain length distribution. This often yields a chain length distribution when no other technique is available. In some instances the conversion from frequency space to chain-segment length space (weak broad LAM) requires special attention. The Raman intensity of the kth mode of a molecule of n chemical repeats is

$$I(n,k) \propto \frac{(\nu_0 - \nu_{n,k})^4}{\nu_{n,k}[1 - \exp(-h\nu_{n,k}/k_\text{B}T)]} S(n,k) \,, \tag{9.19}$$

where ν_0 and $\nu_{n,k}$ are the excitation and k^{th} vibrational frequencies, respectively, and S is the scattering activity, which can be related to the change in polarizability tensor element, $[\alpha'(n,k)]^2$. The resulting spectrum in terms of

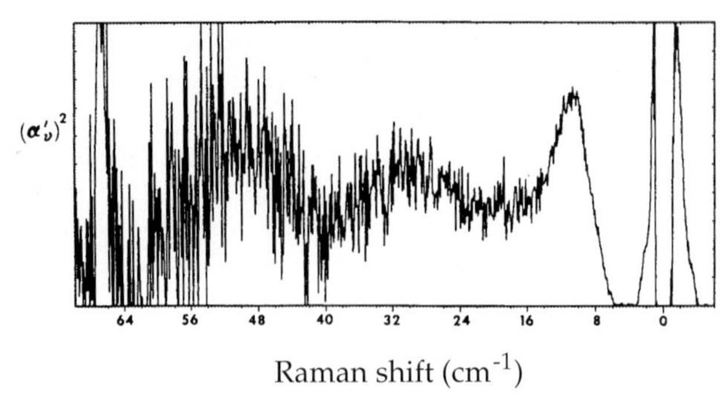

Fig. 9.22. Raman spectrum of annealed bulk-crystallized polyethylene plotted in terms of $(\alpha'_v)^2$[122]

the scattering activity is shown in Fig. 9.22. In this case, even LAM-5 can be seen above the noise. As mentioned above, the form of the longitudinal mode is clearly defined at the molecular level for a planar zigzag chain conformation. For large n, $[\alpha'(n,k)]^2$ is proportional to $1/n$. If the frequency is sufficiently low, the relative occupation term in the denominator is then proportional to $\nu(n,k)$. In order to convert the Raman spectrum to the chain segment length distribution, the observed intensity is effectively multiplied by ν^3 [122]. For sharp LAM bands, as observed for crystals containing narrow distributions, the peak frequency can be used to determine the most probable chain segment length. For broad LAM's the peak frequency can be considerably different from that for the most probable chain length in the calculated distribution [30,31,122].

The LAM observed for short oligomers is easily separated from the Rayleigh wings. Because of the longer chain segments in polymer systems, their LAM's are usually superimposed on the wings of the Rayleigh line making it difficult to obtain the true bandshape and intensity. It is possible to determine the difference between the polarized and depolarized spectra [122]. This applies to highly oriented samples, for which the LAM spectrum is polarized. Capaccio, Wilding, and Ward assumed the Rayleigh peak to be Gaussian in shape and removed it from the measured Raman spectra by a curve fitting routine [123]. One can simply approximate the LAM visually. It is also possible to record the Rayleigh profile under the same spectrometer settings used to obtain the Rayleigh contribution. In some cases a combination of Gaussian and Lorentzian functions is required to fully describe the Rayleigh line shapes ($0-40\,\mathrm{cm}^{-1}$) [30,31]. After the Rayleigh scattering is removed, the LAM peak is corrected for the Boltzmann factor and then converted to the chain length distribution.

Chain folding is a fundamental characteristic associated with crystallization behavior of semicrystalline polymers. From the kinetic theory of crystallization, the lamellar thickness is known to depend on the degree of supercooling [124]. The nature of chain folds and end groups may greatly influence the lamellar thickness and folding scheme in forming morphological subunits. DSC, wide angle X-ray diffraction, electron microscopy, and vibrational spectroscopy are all used to characterize the crystalline portion of semicrystalline polymers. LAM exhibits particular advantages in the characterization of morphological features. For narrow molecular weight fractions ($M_n = 6000-10{,}000$) a stepwise increase in SAXS lamellar spacings of PEO as a function of crystallization temperature has led to the concept of integer folds (IF) [125]. This phenomenon indicates that the chain ends are not located within the crystalline lamellae but are rejected into the thin amorphous layers separating the crystals [125]. All stems of a molecule are the same length with chain ends located at the lamellar surfaces. The initial LAM studies of PEO also provided evidence for such IF structures [104]. A Raman spectrum of PEO crystals is shown in Fig. 9.23. Three distinct peaks were obtained for the crystalline polymers, two of which are shown. None of these peaks is present in the melt spectrum. For samples of $M_n < 3000$, which are known to crystallize in a chain-extended form when melt-crystallized at 25 °C, the observed frequencies of the lowest component show a smooth shift to lower values corresponding to the gradual increase in lamellar thickness. Those peaks were therefore assigned to the fundamental LAM in PEO. The second series of peaks, which are broader and of lower intensity, have higher

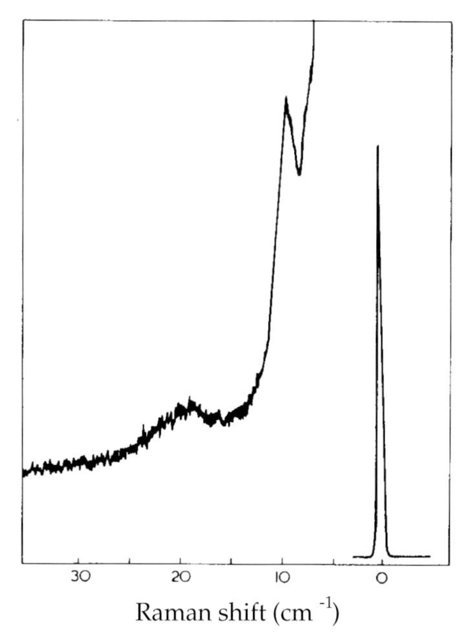

Fig. 9.23. Low-frequency Raman spectrum for crystalline poly(ethylene oxide) of average molecular weight 2000. The Rayleigh scattered peak is recorded on reduced scale [104]

frequencies by a factor of approximately 2.3 rather than 3.0 as would be expected for the second overtone. The Raman and X-ray evidence indicate that for a crystallization temperature of 25 °C, PEO of $M_n \leq 2000$ crystallizes in the chain-extended form and that of $M_n \geq 4000$ in the chain-folded form. Fraction 3000 contains both types of crystals. The Raman peak at 12.0 cm^{-1}, which is close to that of fraction 1500, indicates a once-folded crystal of fraction 3000. Unfortunately, the LAM frequency for the chain-extended crystal was too low to be observed in that work. Fraction 3000, crystallized at 35 °C, yielded a weak Raman peak at 12.0 cm^{-1}, reflecting a decrease in the fraction of once-folded material.

Subsequent LAM studies of PEO, together with DSC and SAXS data, suggest the existence of lamellae that contain a mixture of IF structures as well as lamellae comprised of chains with simple fractional-integer folds (Fig. 9.24) [105]. Such complex folding schemes are also suggested by results on ultralong n-paraffins, although the SAXS data were interpreted in terms of general noninteger folds [105]. For samples crystallized at lower temperatures, the 11.6 cm$^-$1 LAM band and the 107 Å SAXS spacing clearly indicate the presence of independent lamellae of once-folded molecules. On heating, such lamellae should be unstable and convert to bilayers, lamellae with observed $T_m = 58.8$ °C. The other LAM and SAXS spacings, i.e., 122 and 128 Å, respectively, are consistent with lamellae with more than one fold, since the length of the molecule is 211 Å. In this case, Raman contributes to a fundamental understanding of the polymer structure.

Annealing of semi-crystalline polymers is a difficult process to understand. It is known that both microscopic and macroscopic defects are reduced upon sample exposure to temperatures somewhat below the crystallization temperature. Polymer crystals are metastable, i.e. the lamellae thickness and lateral

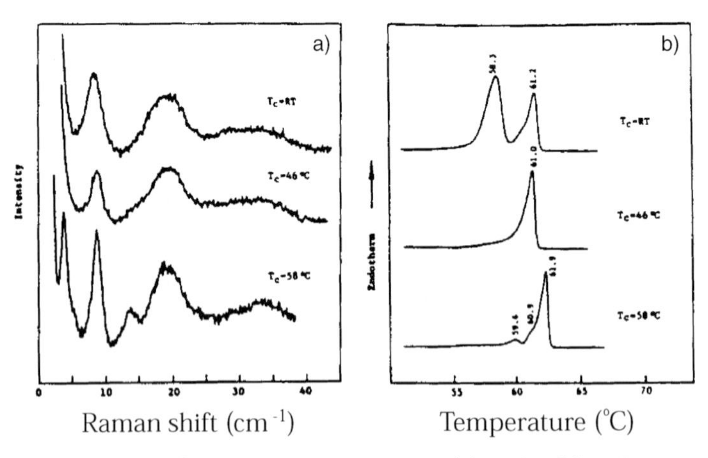

Fig. 9.24. Low-frequency Raman spectra (**a**) and DSC melting curves (**b**) of PEO 3000 crystallized at different crystallization temperatures (T_c) [105]

size are generally determined by the degree of supercooling balanced by the side and end free energies. Therefore, as a function of time, chain-folded crystals thicken on heat annealing, and the associated molecular process consists of refolding of chains to a greater fold length. The ultimate thickness is highly dependent on the initial thickness, and the process is known to be irreversible. Thus, thermodynamic equilibrium theory cannot be used to characterize this time-dependent behavior [126]. A kinetic theory first proposed by Sanchez seems to be the most successful [127]. The driving force is assumed to be the decrease in free energy caused by the decrease in the number of folds [127].

Molecular annealing mechanisms in polymers are difficult to study, since the initial and final states are ill defined. Many techniques can be employed to characterize structural changes during annealing. Traditionally thermal analysis can be used to follow changes in melting temperature when crystal volume increases. Small angle X-ray diffraction can be used to measure the long period when crystalline lamellae thicken. Density increases for annealed samples can also be measured. LAM has proven particularly useful in this area.

The change in lamellae thickness is intriguing. For some polymers such as polyethylene, on holding at a constant temperature the long spacing increases first by a rapid jump followed by a continuous fold-length increase proceeding logarithmically with time [127,128]. This increase can level off even on a logarithmic scale, an effect that depends on the molecular weight and, in particular, on the sharpness of its distribution. Other polymers, such as nylons, change in a discontinuous fashion. Chainfolded, solution-crystallized polyamides increase their fold lengths only with difficulty. When an increase does occur, it is in a discontinuous fashion. In nylon 66 an initial fold length of 54–55 Å increased in one step to 109 Å. The two reflections were visible simultaneously and the relative intensity of the former decreased gradually. In one instance the enlarged fold length increased discontinuously further to 225 Å, with the 109 Å reflection still visible. Clearly the lamellar thickness has increased by factors of two and four, the second corresponding to a repetition of the doubling. These data suggest a mechanism [128] illustrated in Fig. 9.25. This model was the basis for a detailed diffraction analysis of the annealing behavior of dilute solution-grown polyethylene crystals [129,130]. It was concluded that well-annealed samples ($> 10^5$ seconds) have reduced defect density, approaching that for unannealed crystals. Furthermore, for annealed samples, preferred chain stem lengths are multiples of the unannealed length. This evidence was cited as support for the crystal thickening model proposed earlier. The chain length distribution is plotted in Fig. 9.26.

A number of Raman studies have been carried out to follow the annealing behavior of semi-crystalline polymers [131]. The most detailed study of changes in segment length distribution employed the LAM technique [132]. As in most Raman studies, the time required to accumulate an adequate spec-

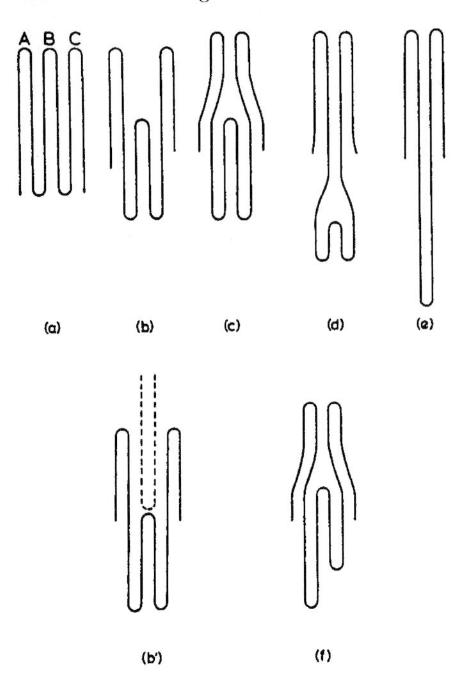

Fig. 9.25a–f. Schematic representation of various stages of the fold length increase to a doubled fold length by the Dreyfus scheme as described in the text [128]

trum is simply too long to follow the changes in segment length as samples are annealed. On the other hand, it is not entirely clear that high time resolution is an important issue. The distribution of lengths of straight-chain segments has been determined for dilute solution-grown polyethylene annealed at different temperatures (Fig. 9.27) [132]. Unannealed samples have a most probable chain length of 100 Å with a distribution of less than 20 Å. Samples annealed for a long time approach the well defined chain length distribution of the initial material. The measured distributions are interesting in that they are quite asymmetric. For the annealed sample, the band is asymmetric and is more intense on the low-frequency side. After the sample was annealed at 120 °C, the asymmetry was reversed so that the band became more intense on the high-frequency side. Bands associated with intermediate annealing temperatures appear to have shoulders. However, they do not appear to be consistent with the model proposed earlier [128].

In polyethylene crystals, the crystal thicknesses form in a continuous fashion as a function of supercooling [124]. Because of their chemical repeat, to preserve long-range order in the crystalline state, the lamellar thickness of nylons would have to be formed and change in a discontinuous fashion [128]. Due to the strong dipole–dipole interactions, one would expect that the formation and subsequent changes in polyester crystals would be different

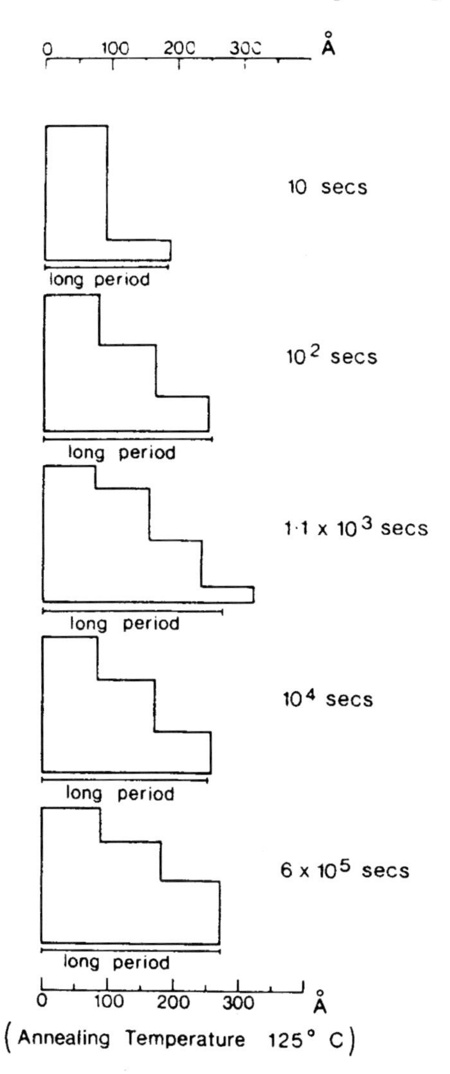

Fig. 9.26. Stepped crystal profiles [for (001) direction] calculated from the correlation functions of the specimens annealed at 125 °C. The bars under each profile represent the long period (to the same scale) as measured by SAXS [129,130]

from polyethylene and would possess some of the characteristics of nylons. The LAM observed for various as-crystallized and annealed linear aliphatic polyesters exhibit a fairly complex pattern (Fig. 9.28). In linear aliphatic polyesters, the changes in Raman or SAXS depend very much on the temperature and annealing time. As can be seen in Fig. 9.28, a straight line gives a poor fit to the data including different crystallization temperatures and relatively high annealing temperatures (58 °C). The line does not intercept the

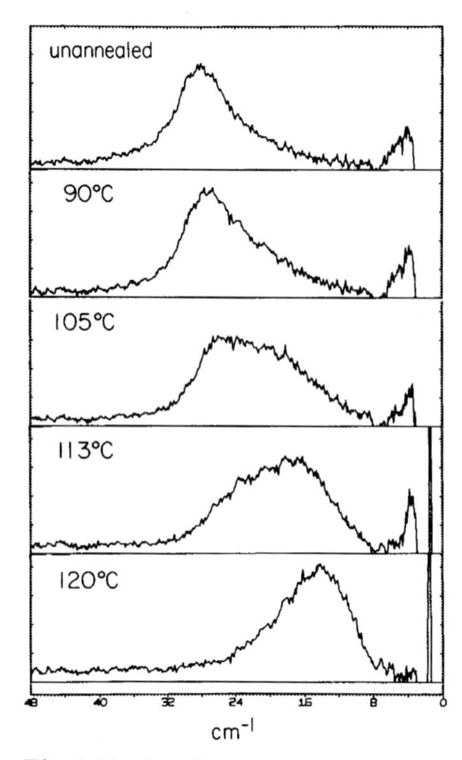

Fig. 9.27. Low-frequency Raman spectra of polyethylene single crystal at 28 °C after being annealed at successively higher temperatures. Annealing time at each temperature was 6 hours [132]

origin, as one would expect if the frequency were strictly inversely proportional to chain length. This apparent inconsistency may originate from a layer of amorphous material located along the crystal boundary. In addition, for samples annealed at lower temperature (50 °C), there can be changes in either Raman frequency or in the periodicity measured by SAXS, but both need not change simultaneously. For long annealing time these data can also be described by a linear function, again independent of the initial crystallization temperatures. This observation illustrates the complementary nature of the Raman and SAXS techniques. The differences revealed by the two techniques may be important in structural studies of polyester single crystals.

As mentioned in the previous section, a number of flexible polymers have been solid-state extruded or drawn to form ultra-oriented, high-modulus fibers. Information obtained from X-ray scattering, electron microscopy, vibrational spectroscopy, and calorimetry has been used to support a number of structural models, each differing in the description of the crystallite size and connectivity (Fig. 9.12). Poor mechanical properties are usually offered as evidence for a significant number of structural defects in the samples.

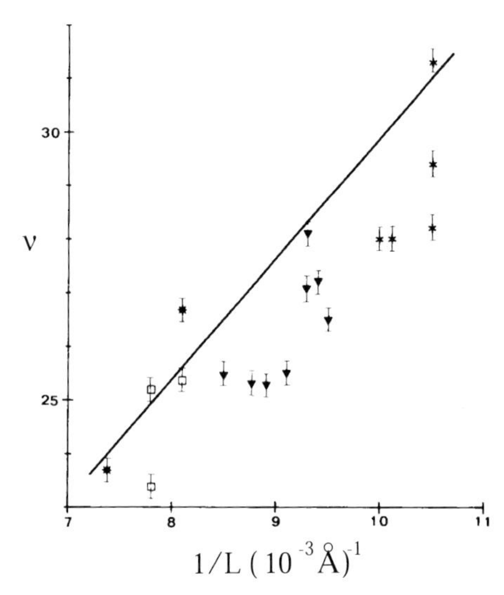

Fig. 9.28. Plot of Raman frequency vs. L^{-1}. (L is the Bragg spacing of the SAXS maximum.) (×) 9–8 Polyesters grown at 25 °C and annealed at 50 °C; (▼) samples grown at 35 °C and annealed at 50 °C; (∗) samples grown at 25 °C but annealed at 58 °C; (□) samples grown at 35 °C but annealed at 58 °C

Structural defects can be as high as one for every five chain stems of the lattice of the crystal block. The length distribution of straight-chain segments is therefore of great interest.

The deformation process was studied by measuring LAM spectra at various points on drawn polyethylene [123,131]. The analysis describes the disruption of the crystallites occurring at various draw ratios for solid-state extruded samples [30,31,133]. The all-trans segment length appears to be uniquely determined by extrusion temperature. This finding essentially agrees with earlier SAXS studies on drawn polymers. The LAM analyses are complicated by the fact that the true distribution of extended chain segments is sensitive to the observed half width and difficult to obtain. A split-billet method developed at the University of Massachusetts has proven effective in forming ultra-oriented high modulus samples [134]. The effects of temperature and draw ratio on chain orientation and extended chain length distribution during extrusion were studied by removing a partly extruded billet and examining points within the die region and along the extruded fiber, which revealed various stages of deformation [30,31].

In the solid extrusion process, during heating and application of pressure before extrusion, the distribution of chain lengths shifted to a higher L_{MAX}, by 20 Å when compared to the initial material. This change is not obvious when observed directly in frequency space but, as mentioned above, it shows up in the chain length distribution [30,31]. The shift is quite sensitive to the width of the observed LAM. This change in L_{MAX} is not purely a thermal effect. The initial material was annealed without pressure at 80 °C for 4 h. The L_{MAX} measured for this sample is 161 Å, a value less than that found for the undeformed polyethylene in the extrusion die. The fact that the chain length distribution can change for the 80 °C sample is somewhat surprising. From all indications, the chain mobility for polyethylene is limited at such a low temperature compared to the effective melting temperature (the melting point increases with pressure). One possibility is a preferential narrowing of the chain length distribution by the low molecular weight fraction, which would lead to a narrower LAM, as observed.

For samples prepared at a somewhat higher temperature, 130 °C, the change in L_{MAX} for the essentially undeformed sample increased even more dramatically. It is known that the morphology of the isotropic material can strongly influence its subsequent drawing behavior [135]. Formation of anisotropic materials always involves continuous plastic deformation of the spherulitic structure before necking can involve shear, slip, and rotation of stacked lamellae [136]. Phase transformations can also occur within individual lamellae. LAM studies indicate that, during the extrusion process, the deformed material actually might have originated from material morphologically different from that put into the die [30,31]. The LAM position and width can differ considerably for samples extruded at various temperature and draw ratios. For deformed samples, the chain length distribution usually contains a large number of short extended chain segments (Fig. 9.29).

Structure analyses based on LAM give information on the stability of highly deformed materials. As mentioned above, the most probable chain length and half-width in chain distribution have been established for a series of annealed polyethylene. Among the most interesting observations from LAM results on highly deformed materials is that the LAM bands exhibit considerable departure from the relationship found for annealed single crystals of polyethylene. This departure occurs only for samples extruded at low temperatures (Fig. 9.30). Samples extruded at 130 °C follow the same trend

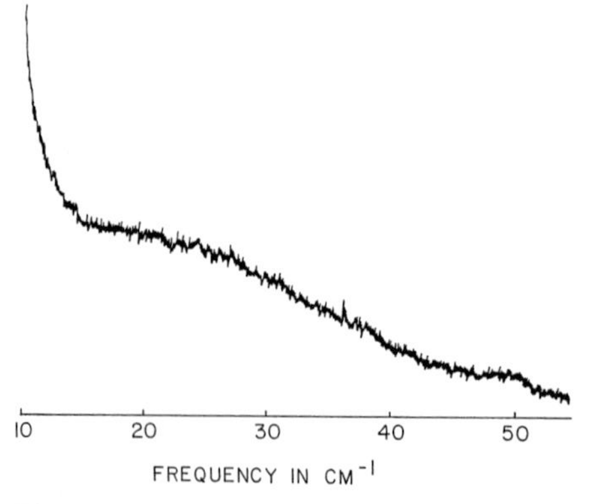

Fig. 9.29. LAM observed in drawn polyethylene; drawing temperature 75 °C

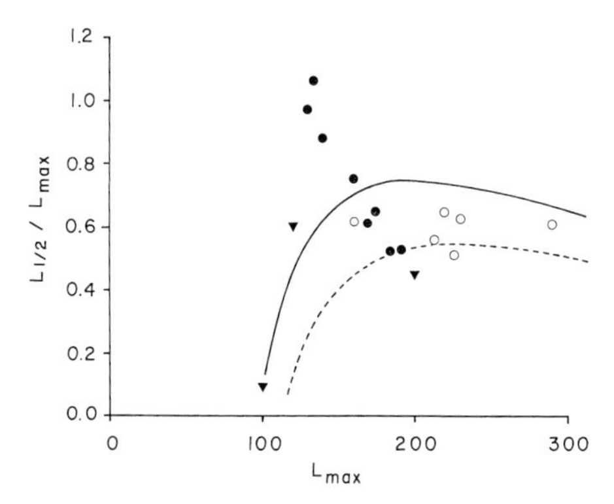

Fig. 9.30. Relationship for $L_{1/2}$ and L_{MAX} in variously prepared polyethylene; (•) 80 °C sample; (▼) 130 °C sample; (○) annealed single crystals [30,31]. The lines refer to the changes observed for annealed single crystals [132]

as annealed single crystals. Even though no fundamental reason exists to relate the half-width $L_{1/2}$ to L_{MAX}, there appears to be a relationship that occurs for particular processing methods or history [30,31]. The half-width $L_{1/2}$ is a useful quantity showing obvious differences in various polyethylene structures. The degree of chain mobility is an important factor in removing structural defects introduced during the mechanical deformation process. For samples extruded at low temperature, there is limited chain mobility. For samples extruded at high temperatures, more defects disappear, forming highly ordered structures that are dimensionally stable. Annealed polyethylene single crystals approach equilibrium and therefore contain fewer structural defects, resulting in smaller $L_{1/2}$ values. Low-frequency Raman data obtained from variously prepared polyethylene samples differ significantly not only in the straight-chain length, but also in the reduced half-width for the length distribution. Because of renewed interest in polyolefins, detailed analysis should provide information about the kind, number, and mobility of defects introduced into the straight chain segment distribution by various processing conditions.

9.4 Fermi Resonance Interaction and Its Application to Structural Analysis

Fermi resonance occurs when two different vibrations of the same symmetry type have nearly the same energy [137]. A classic example of molecular Fermi resonance is the case of CO_2, in which the ν_1 fundamental at $1337\,\mathrm{cm}^{-1}$ has nearly the same energy as the overtone (also called second harmonic) of the bending mode ν_2, whose fundamental is found at $667\,\mathrm{cm}^{-1}$ [16]. Due to the anharmonic terms in the Hamiltonian that couple the vibrations, the modes mix, share intensity, and repel each other. The degree of mixing and repulsion increase as the coupling terms in the Hamiltonian increase and as the frequency separation of the unperturbed levels decreases. This phenomenon affects certain levels observed in Raman studies of polymers, which in many cases has had a beneficial effect on understanding structural transitions.

The vibrational spectra of polymers provide many opportunities for observing Fermi resonances. In the $3300\,cm^{-1}$ region of proteins and polypeptides, the IR-active Amide A and B due to the fundamental NH stretching ($\sim 3300\,cm^{-1}$) are close to the second harmonic of the Amide II vibration near $1500\,cm^{-1}$ [138]. For synthetic polymers, the C–H stretching region is the most interesting. Knowledge of C–H stretching vibration changes in chain conformation and packing is quite limited. To first order, C–H stretching modes are highly localized and involve nearly pure hydrogen motion. They are usually insensitive to the molecular conformation and environment, in sharp contrast to skeletal modes at lower frequencies. This hypothesis appeared intuitively sound and was supported by a half century of evidence from IR spectroscopy. Until quite recently [29], C–H stretching modes were considered group frequencies and essentially ignored. It is therefore surprising that changes in the C–H stretching region of Raman spectra can be the first indications of changes in chain conformation and environment [139,140].

In trans-1,4-polybutadiene, at least 6 Raman lines are observed for the crystalline state at room temperature [75]. Given the inversion symmetry of the chain, at most one =C–H and two CH_2 stretching modes are expected. When the crystals are heated above 71 °C, a solid-solid phase transformation takes place, causing changes in both chain conformation and packing. The changes in the overall IR and Raman spectra are quite minimal at this transition. The most interesting observation is that sharp features in the C–H stretching region evolved into a profile containing at least 4 bands (Fig. 9.31). Attempts to explain this phenomenon were considered. Similarly in lipid bilayers, the skeletal bands in the $\sim 1100\,cm^{-1}$ region are indicative of structural transformation. The relative intensity of the $\sim 2900\,cm^{-1}$ CH_2 asymmetric stretching to the $2850\,cm^{-1}$ symmetric stretching band is one of the most sensitive indicators of structural change. Many models exist for the arrangements of protein and lipid in biomembranes. An important difference between these models is the degree to which lipid bilayers, with ordered fatty acid chains, play a role in the structure. Better understanding of such changes in structure would enable a better understanding of membrane functions [140,141]. One of the earliest reports of changes in the C–H stretching region was for the phospholipid-water mixture [140], in which the changes are related to the introduction of kinks in chain conformation.

The complexities observed in the Raman spectra of the C–H stretch region of polymethylene chains are the result of Fermi resonances between completely symmetric double excitations of fundamentals at lower frequencies with higher frequency C–H stretchings of the same symmetry. Complications of this nature are the rule rather than exception. The sensitivity of C–H

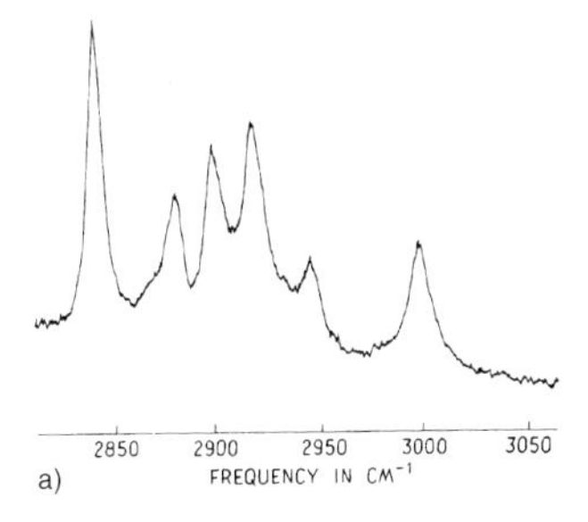

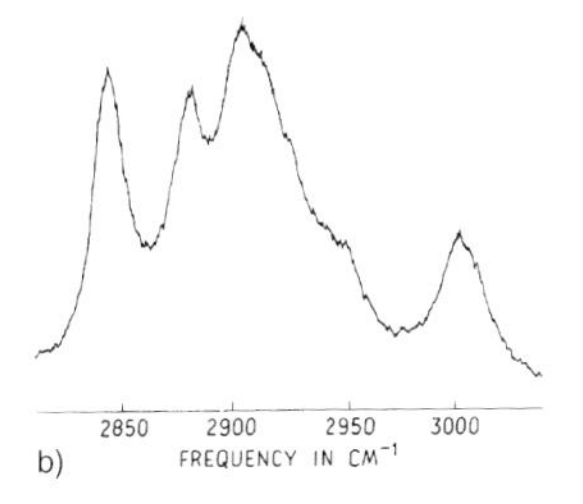

Fig. 9.31. Raman spectrum of TPBD, 5145 Å excitation, 1.5 cm^{-1} bandpass at 5100 Å; C–H region; (**a**) 23 °C; (**b**) 80 °C

stretching vibrations to structural change results from an inherent *amplification* factor. Low-lying vibrations are known to be sensitive to both chain conformation and chain packing differences. Because of the Fermi resonance, small frequency changes in the lower frequency modes can cause significant changes in the high-frequency spectrum.

Interpretations of the C–H region have long been plagued by unresolved contradictory assignments appearing in the literature. Around the CH_2 symmetric and asymmetric stretching, several seemingly broad and sometimes intense maxima exist. Most of the integrated intensity belongs to the broad bands. This is shown in Fig. 9.32 for n-alkanes. Contrary to expectation, the Raman-active bands observed for different intermolecular packings are quite different. Even though the overall spectra are similar, additional features exist for some crystalline unit cells. For example, a multiplet exists at 2850 cm^{-1} for triclinic packing. It is also clear that when chains disorder, as in the hexagonal phase, the relative intensities of the dominant components change.

Three additional features in the C–H stretching region are especially important. Consider first the Raman polarization measurements made on a solid-state extruded, uniaxially-oriented polyethylene shown in Fig. 9.33. Here, the direction of the incident radiation is perpendicular to the long axis of the sample, and the polarization of the incident and scattered radiation are parallel to this axis. This zz geometry gives the totally symmetric A_g species. Collection is at right angles. The intensity of the line at 2880 cm^{-1} is

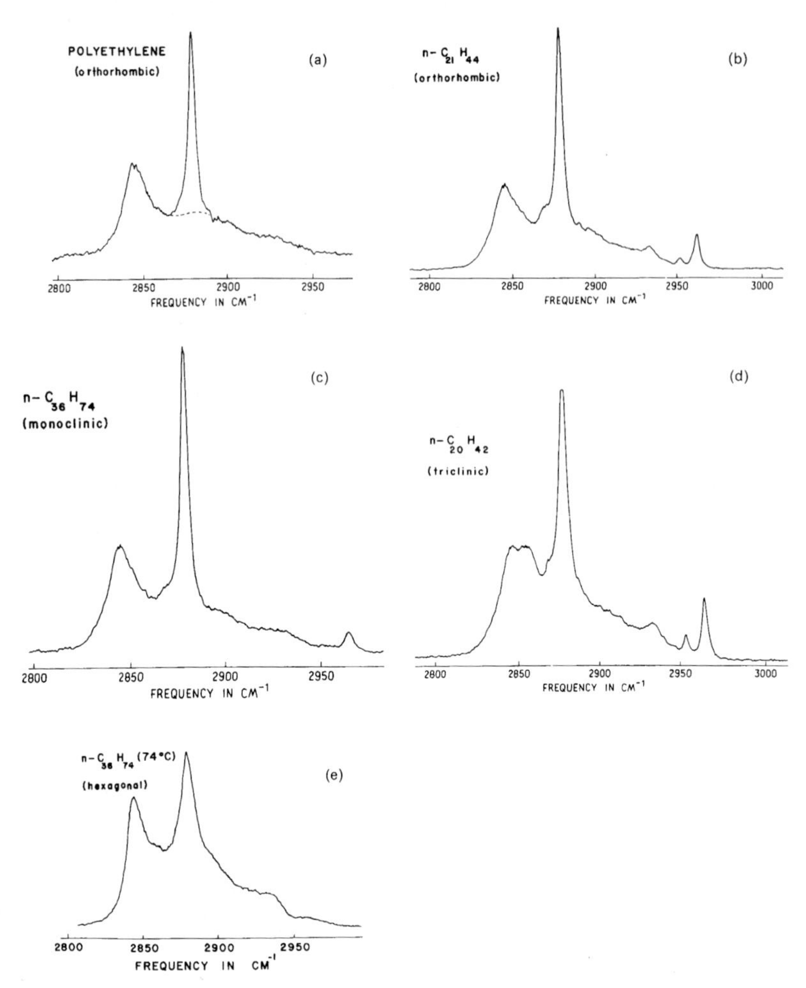

Fig. 9.32. C–H stretching region of the Raman spectra of (**a**) orthorhombic polyethylene at 110 K, (**b**) orthorhombic n-$C_{21}H_{44}$ at 110 K, (**c**) monoclinic n-$C_{36}H_{74}$ at 110 K, (**d**) triclinic n-$C_{20}H_{42}$ at 110 K, and (**e**) hexagonal n-$C_{36}H_{74}$ at 347.5 K. All spectra were taken with bandpass 1.5 cm^{-1} at 5100 Å. Laser power 200 mW at 5145.4 Å

greatly reduced, nearly absent, relative to the rest of the spectrum. (A residual intensity persists but can be eliminated by narrowing the collection angle or using more highly oriented samples.) Thus, the line at $\sim$ 2880 cm^{-1} can be confidently assigned to asymmetric stretching and the entire remaining broad profile to symmetric stretching. Associated with the latter, a previously hidden maximum is revealed near 2890 cm^{-1}. Next consider the Raman spectra for the isotopically isolated chain (n-$C_{36}H_{74}$ in n-C_{36}-D_{74}, 1:20) shown in Fig. 9.34. Unlike IR spectra from neat $C_{36}H_{74}$, the Raman spec-

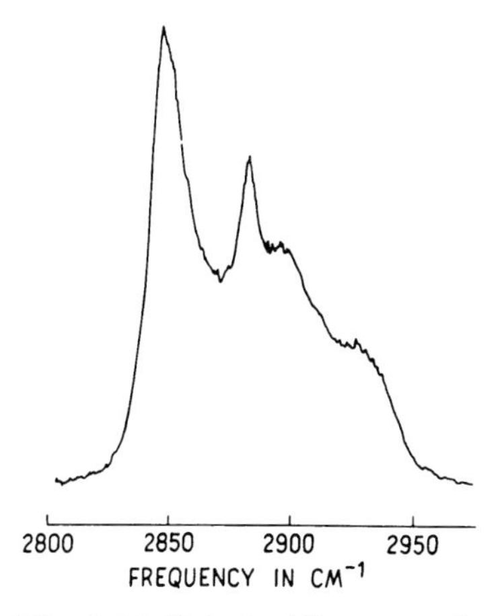

Fig. 9.33. Polarized Raman active bands in the C–H stretching region of uniaxially oriented polyethylene at 100 K; $y(zz)x$ geometry was employed with chain axes aligned with the z-axis. Bandpass is $2.5\,\mathrm{cm}^{-1}$ at 5100 Å; laser power 250 mW at 5145 Å

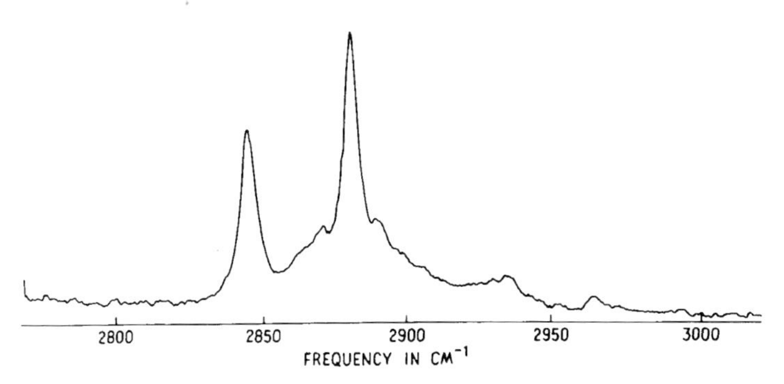

Fig. 9.34. Raman spectrum of isotopically isolated chains in the CH_2 stretching region; 1H/20D n-$C_{36}H_{74}$ at 110 K; bandpass is $2.5\,\mathrm{cm}^{-1}$ at 5100 Å; laser power 250 mW at 5145 Å

trum exhibits significant changes in the C–H stretching region. The line at $2845\,\mathrm{cm}^{-1}$ sharpens considerably, and its half-width reduces from the usual $22\,\mathrm{cm}^{-1}$ to about $10\,\mathrm{cm}^{-1}$. The $2880\,\mathrm{cm}^{-1}$ line is relatively unaffected; its half-width is still $\sim 7\,\mathrm{cm}^{-1}$. The peak heights of the two stretching bands are now nearly the same. There is considerable depletion in intensity on the low frequency side of the broad band underlying the $2880\,\mathrm{cm}^{-1}$ line. However, note that this broad band is still largely unaffected at frequencies higher than $2880\,\mathrm{cm}^{-1}$. The final feature is exemplified in the Raman spectrum of the n-$C_{36}H_{74}$ liquid state shown in Fig. 9.35. In this case the symmetric stretch mode persists as a relatively narrow, intense, and strongly polarized line near $2850\,\mathrm{cm}^{-1}$, that closely resembles the mode seen in the hexagonal

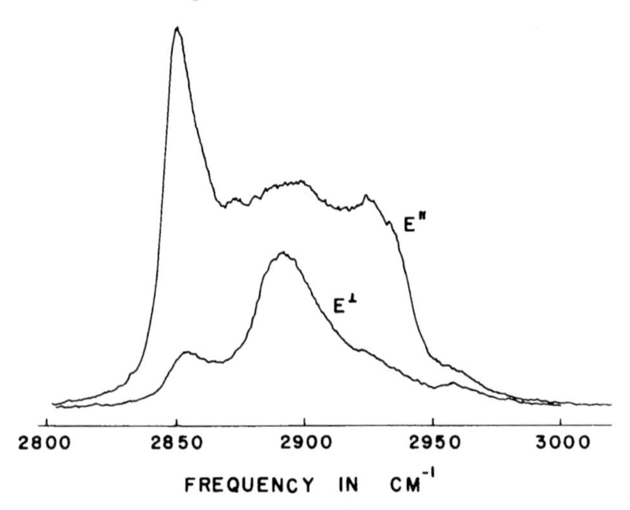

Fig. 9.35. Polarized Raman spectra of n-$C_{36}H_{74}$ in the melt. $E^{\perp}$ and $E^{\parallel}$ denote the polarization of scattered light relative to the polarization of the excitation; bandpass is $2.0\,cm^{-1}$ at 5100 Å; laser power 250 mW at 5145 Å

crystal. There is a second somewhat less intense and broader polarized line centered near $2925\,cm^{-1}$, which probably originates from Fermi resonance interaction between symmetric stretching and the harmonic of the bending component. In between these two lines, and obscured by them, is a weaker depolarized line near $2890\,cm^{-1}$. Its position, polarization and intensity identify it as the asymmetric stretch mode. This line is clearly revealed in the depolarized spectrum, where it is found to be quite symmetric in shape with a half width of about $25\,cm^{-1}$.

Two types of Fermi resonance interactions need be distinguished. The first is intramolecular and involves dispersion of the bending modes parallel to the chain axis. The second is intermolecular and involves perpendicular dispersion of the bending modes, and thus is crystal structure dependent. Much of the spectral structure can be explained by considering only the first type [142,143]. The analysis is complicated because of the involvement of three separate factors concerning the chain: conformation, mobility, and packing. Chains of planar zigzag conformation have been studied extensively; their normal modes are known for both polymers and oligomers. For such cases, the band structure in the C–H stretching region is determined in large part by intramolecular Fermi resonance interaction between the symmetric C–H stretching fundamental and the overtones of the HCH scissors modes [142–144]. This interaction also plays an important role in determining band shapes for the liquid and solid states where the resonance is affected by chain conformation and chain packing. The fact that the isotopically isolated spectrum is dramatically different suggests that interchain interactions are not insignificant and need careful consideration with regard to structural analysis. It is well known that intermolecular coupling can often be eliminated by isolating the molecule of interest in a matrix consisting of its deuterated analogue. Intramolecular Fermi resonance will, of course, persist.

Initially, an interpretation of this region suggested that Fermi resonance interactions occur between the symmetric C–H stretch fundamental and the overtone combinations of all scissors fundamentals, including those from infinite as well as all finite chains. All major features in the C–H stretch region of n-alkanes or polyethylenes can be accounted for in terms of Fermi resonance interactions between appropriate binary combinations of the methylene bending modes and the C–H stretching fundamentals [142,143]. The secondary bands are all broad, which is a consequence of the fact that the resonances involve a virtual continuum of binary combination states.

In [143], a Fermi resonance equation was set up to analyze the C–H stretch region in terms of two unknown parameters: the Fermi resonance interaction constant for an isolated methylene group and the frequency of the unperturbed symmetric C–H stretch fundamental. Solution of this equation yields a calculated spectrum having all the features of the observed spectrum. The observed spectrum can be quantitatively reproduced using a reasonable form of the CH_2 scissors dispersion curve, which permits adjustment of the values of the two unknown parameters.This calculation led to a modification of earlier views concerning the relation between critical points and band positions. That bands occur near or at critical point frequencies was interpreted to mean that they were associated with a high density of vibrational states. It is now clear that these bands represent unique states that occur near but not precisely at the critical points. Thus, an important factor is the density of binary states. For the isolated chain these can be approximated from the calculated bending mode dispersion curve [142,143]. The unexpected changes in the CH stretch regions produced by structural transitions result from the fact that lower vibrations are extremely sensitive to changes in chain packing. If the interchain environment changes, the Fermi resonance interactions amplify these changes resulting in significant changes in the relative band intensity.

A number of other points regarding the effects of Fermi resonance should be mentioned. Obviously, upon removing the effects of intrachain Fermi resonance, the remaining features must reflect interchain effects. The primary difference between the Raman spectrum of the isolated chain in a deuterated crystal and that in a normal crystal is the greater breadth of the $2845\,cm^{-1}$ line in the latter case (Fig. 9.34). There is also more intensity on the high frequency side of this line, which tends to fill in the trough between the lines at 2879 and $2845\,cm^{-1}$. The spectra of polyethylene and the orthorhombic and monoclinic forms of the n-paraffins are nearly identical as the chain packing is similar for all three cases. However, the triclinic form of the n-paraffins has a different structure from the hexagonal form, and these differences are reflected in the Raman spectra.

The relative intensity of the two C–H stretching modes in polyethylene is 5.1±0.5, as measured from polarized spectra of an oriented sample (Fig. 9.33). The much larger symmetric versus asymmetric intensity agrees with Raman intensity calculations based on bond polarizability parameters. By knowing the origins of the bands in this region, the changes observed during phase transformations can be assigned with more confidence. It is now understood that relative intensity changes result from changes in the underlying structure due to Fermi resonance interactions. Again, the apparent sensitivity results from the built-in *amplification* factor. Small changes in the frequency and intensity of the lower frequency vibrations can cause significant changes in the mixed higher frequency states. The C–H stretch intensity ratio is a useful guide for studying complex systems and has been used extensively in systems such as cholesteric liquid crystals and biomembranes.

9.5 Disordered States

An area of current interest is using Raman spectroscopy to analyze disordered chains, which may occur in polymers in solution, melt, or in the solid state. These chains lack long-range order but may contain short-ordered sequences, and they may adopt a specific conformational distribution depending on geometric constraints such as surfaces or interfaces. Other constraints may include the presence of junctions such as in networks. Long-range electrostatic forces along the chain, which occur in polymeric electrolytes or *polyelectrolytes*, may also perturb the structure. Chain disorder can affect macroscopic properties such as ionic conductivity, mechanical properties, solvation efficiency, corrosion inhibition, and stress build-up. Many characterization techniques are capable of measuring ordered structures, but few match the capability of vibrational spectroscopy, particularly Raman scattering, to quantify disordered chains. The frequency and relative intensity of Raman bands depend on the relative concentrations of specific localized structures. For samples involving well-defined structures and dependable band assignments, the method works well. However, when a broad distribution of a large variety of conformations is present, the method is difficult to apply with confidence. It is often hard to relate observed spectroscopic differences to well-defined structural changes. The bands of disordered structures are typically broad and weak and often overlapping. Yet it is precisely these disordered structures that are crucial in determining macroscopic properties. Raman spectroscopy is ideal for studies involving disordered structures because polarizability changes associated with different carbon-carbon backbone conformations are directly reflected in the spectra. Because it is a scattering phenomenon associated with a tensor, polarization data still contain considerable structural information for samples that are partially disordered or even those in melt or solution.

Unlike ordered chains with well-defined symmetry elements, disordered chains have a large number ($3N - 6$) of vibrational modes. It takes considerable effort to simulate Raman spectra for all possible conformations associated with chain disorder. Programs have been introduced that attempt to take structural defects into account [145]. From the dispersion curves, it is possible to obtain the vibrational density of states, which can lead to vibrational bands [146]. Attempts to predict band intensities, however, have proven extremely difficult. Molecular dynamics has also been employed [147–149] to calculate correlation functions that can be transformed to obtain vibrational frequencies. Lastly, direct quantum mechanical calculations have been carried out [150]. This method is seldom utilized in polymer structural studies because substantial computation power is required even for relatively small model compounds.

The marked increase in computing performance achieved over the past few years has served as a prime impetus for progress in spectroscopy research. In the past, large and challenging analyses of macromolecular systems were possible only on main-frame computers. The same calculations can be done today with greater speed on workstations or PC's available in virtually every laboratory. With increased computing capabilities, simulation methods have become much more useful tools in spectroscopy.

9.5.1 Normal Coordinate Approach

Vibrational modeling of polymers has usually been done using a normal coordinate analysis (NCA) based on Wilson's GF matrices and force constants transferred from small molecules [8,21]. Often, refinements in force field are needed to better fit the polymer system. Such analyses can only deal with a specific chain conformation generated with well-established bond lengths and valence and torsional angles [21]. The first computer programs, introduced in the 60's, were followed by many variations [17,151,152]. The generation of structures consistent with helical parameters found from diffraction data usually tests the structural parameters [17,18,151,152]. Until recently, calculation of vibrational spectra for all possible chain conformations, sufficiently large to describe disordered chains has not been feasible.

Since the Raman spectrum is sensitive to conformation, different spectra are observed for different chain conformations. The conformational distribution usually depends on the immediate molecular environment. A change in temperature, for example, will change the relative population of the various conformers. Lack of a quantitative definition of disorder has proven to be the major obstacle in analysis of spectra from disordered polymeric chains. Rather than analyzing one specific chain conformation for an ordered chain, the analysis requires the vibrational spectrum of a conformational distribution. This problem is complex, even in relatively short-chain molecules. For example, the number of possible conformers for a C_{12} hydrocarbon is of the order of 10^4, each with a unique spectrum often only slightly different from that of another. In order to treat polymers lacking long-range order, an approximate method was developed in [39]. The isotropic Raman spectrum $S(\nu)$ for the disordered state was simulated as a composite of contributions from the ensemble of chains generated by a *Monte Carlo* procedure that assigned both a conformation and a total probability for each chain. The spectrum of each conformation is calculated from force constant and intensity data transferred from smaller molecules. These individual spectra are then weighted by their probabilities and combined into a composite spectrum. With a surprisingly small number of structural and intensity parameters, it has been possible in this way to accurately reproduce experimental IR and Raman spectra of some n-alkanes [38,39,153,154]. The same method has been applied to more complicated polymers lacking long-range order [155]. The original code was designed for n-alkanes. Subsequent program modifications were introduced that allow non-equivalent bonds along the chain [41,155]. It should be mentioned that disordered chains may include configurational defects. For example, the degree of racemic content may significantly influence the crystallization rates of iPP and the associated physical properties. Generation of chains containing specific defect content is not as difficult as first envisioned. Flory developed methods for determining the probability of finding specific conformers along the chain, and these are incorporated into the programs. Polymeric chains containing both conformational and configurational defects can now be analyzed with confidence [40].

Since the validity of such an analysis depends on the ability to generate an accurate structure, conditional probabilities for each bond along the chain can be incorporated based on well-accepted rotational isomeric states. Highly accurate chain conformation distributions can be generated for chains with conformational as well as configurational defects. In practice, only rather large numbers (> 2000) of finite chains can be considered. The effects of conformational or configurational defects may extend over a few chemical repeats. Because of memory limitations, the size of finite molecules is generally limited to 24 chemical repeats. However, the use of a finite molecule of well-defined structure may introduce additional features in the regions of interest [40].

A requisite for accurate NCA analyses is the availability of a reliable force field. In general, only intramolecular potential energies are needed. With few exceptions, such as polyethylene and trans-1,4-polybutadienes, there is no explicit evidence of Raman-active intermolecular vibrations [42,46]. Therefore, interchain interactions based on small molecules are seldom taken into account, even though well-defined inter-atomic potentials can easily be incorporated. Because the number of force constants (both diagonal and off diagonal coupling) is much larger than the number of observations, few force constants can be determined with certainty. In both intra- and intermolecular interactions, the force fields are transferred from those established for small molecules. Among the many refined for specific polymers, probably the most used are those for n-alkanes and polypeptides [17–19,138].

It has recently been shown that only a few parameters are needed to calculate the isotropic Raman scattering intensities below $\sim 1000\,\mathrm{cm}^{-1}$ [38,39,156]. Additivity in bond polarizability is always assumed. The isotropic scattering activity for mode k can be represented as:

$$S_k(\mathrm{iso}) \propto \left(\sum_i \alpha_i L_{ik} \right)^2 , \tag{9.20}$$

where L_{ik} is the internal coordinate displacement amplitude of component i for eigenvector L_k, and α_i represents the mean polarizability derivative for internal coordinate i. The summation in (20) is a representation of the polarizability derivative, described in terms of the internal coordinates. The most important parameters for n-alkanes are associated with C–C stretch and C–C–C angle bending internal coordinates [154]:

$$S_k(\mathrm{iso}) \sim \left(D_R \sum_i L_{ik}^R + D_S \sum_j L_{jk}^S + D_\omega \sum_m L_{mk}^\omega + D_\zeta \sum_n L_{nk}^\zeta \right)^2 , \tag{9.21}$$

where the elements of the eigenvector matrix, L_{ik}^R and L_{mk}^ω, represent the backbone C–C stretch and C–C–C angle bending components, and L_{jk}^S and L_{nk}^ζ represent the C–C stretch and C–C–C angle bending components containing the methyl side groups for the normal mode k. The D coefficients represent the relative contribution to the total intensity of the four types of internal coordinates, for a unit amplitude of respective coordinate displacements. The relative ratio for these two parameters in n-alkanes is 0.295 [39]. Additional complexity is found for vinyl polymers when C–C bonds of the side group must be considered. Backbone and side group intensity contributions need not be differentiated [153], but this can depend on polymer tacticity [40]. The simulated isotropic Raman spectrum often needs to incorporate the relative population of each vibrational transition using:

$$I_{\mathrm{iso}} = \frac{S(\nu)}{\nu} \left[1 - \exp\left(-\frac{h\nu}{k_{\mathrm{B}}T} \right) \right] . \tag{9.22}$$

9.5.2 Molecular Dynamics Approach

The importance of simulation methods in the interpretation of vibrational spectra cannot be overstated [157]. Most such methods rely on the NCA approach, as discussed above, but with the tremendous computational power available today, methods using a molecular dynamics approach have proven to be a suitable alternative for simulating vibrational spectra [149]. In this technique the time trajectories of positions, velocities, and other parameters describing the polymer state are calculated. An autocorrelation function is introduced that describes the changes in these parameters as a function of time. The vibrational spectrum is then obtained by transforming the autocorrelation function into the frequency domain.

Molecular dynamics calculations incorporate both intra- and intermolecular interactions. The corresponding spectra thus contain information concerning polymer structure in the condensed state. The advantages of using molecular dynamics to simulate vibrational spectra include: (1) A broad frequency range can be included; (2) Not only individual molecules but molecular aggregates, ordered or disordered with weak and strong intermolecular interactions, can be treated; (3) Parameters pertaining to vibrational spectroscopic features, such as band frequency, intensity and shape, can all be calculated; and (4) Vibrational spectra for polymers under stress or at different temperatures can be examined.

Autocorrelation functions based on molecular dynamics have been applied to various spectroscopic techniques [158–162]. The early results were sufficiently encouraging to lead to more complicated studies of polyatomic molecules. For large systems, the most convenient procedure is to obtain the spectral density of vibrational states from classical trajectories [163]. However, this procedure does not easily give band intensities. A good illustration of the power of molecular dynamics is provided by hydrogen-bonded systems such as crystalline poly(p-phenylene terephthalamide), in which features associated with hydrogen bonds were better understood using the correlation function approach [163]. The crystalline structure and possible deformation modes of this material were simulated in [164]. Using periodic boundary con-

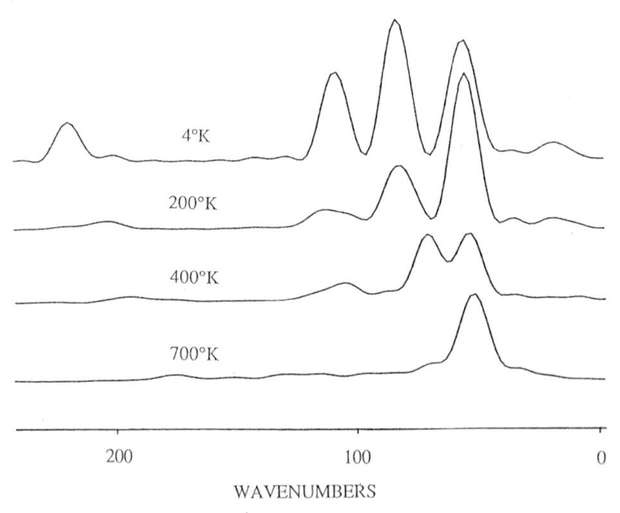

Fig. 9.36. Simulated spectra of vibrational density resulting from fluctuation of the cell pressure during molecular dynamics process at various temperatures

ditions, the crystalline unit cell is accurately described as an infinite 3D system [165]. The calculated vibrational spectra obtained from [164] are shown in Fig. 9.36. These spectra emphasize the assignment of the interchain hydrogen bond stretching vibrations, generally thought to be in the $100\,\mathrm{cm}^{-1}$ region [166]. The spectra below $250\,\mathrm{cm}^{-1}$ may contain external modes arising from intermolecular interactions. Internal modes, such as skeletal bending or torsional vibrations, may also exist in this region. Among the bands observed in the simulated spectrum, the vibration at $\sim 90\,\mathrm{cm}^{-1}$ has been assigned to hydrogen bond stretching. This assignment was based on two observations: (1) The $90\,\mathrm{cm}^{-1}$ feature is absent when secondary interactions due to hydrogen bonding are turned off. (2) The feature shifts down and is washed away as the temperature increases. Spectroscopic features known to be directly associated with hydrogen bonds are usually difficult to observe and can be clearly seen only at subambient temperatures [167]. Although molecular dynamics holds much promise, few successful examples are reported and most quantitative analyses still employ the NCA approach.

9.5.3 Examples

Conformational Distribution of PEO and its Model Compound in the Liquid State. The chain conformation distribution, related to the relative energy differences along each bond, is reflected in the relative intensity and, particularly, the bandshape of Raman-active skeletal bands. Based on the assigned Raman features it is possible to deduce the difference in relative energies between various rotational isomeric states. In the first study of this type, Snyder and coworkers analyzed the low-frequency (0–$600\,\mathrm{cm}^{-1}$) Raman spectra of n-alkanes in the liquid state [38]. This method was subsequently applied to the low-frequency Raman spectrum of molten state iPP. The vibrational analysis was successfully used to identify the correct model governing the chain [153], and it was extended to higher frequency vibrations (0–$1500\,\mathrm{cm}^{-1}$) of liquid n-alkanes [39].

The utility of Raman spectroscopy for deducing chain conformation distribution for polymers lacking long-range order is shown in the detailed study of PEO in both aqueous solution and molten state [41]. The structures of this polymer family, either as homopolymers or in copolymers, determine properties in a large number of applications, including thermoelastomers [168–170], polymeric electrolytes [171–174], and surfactants. Steric considerations alone suggest that the minimum energy conformer is all trans, designated as *ttt*, in a notation that successively references the backbone O–C, C–C and C–O bonds. However, the most stable conformation in the crystal includes a gauche conformation and is designated *tgt* [7]. Since the conformation present in the crystal results from the interaction of a number of factors including crystal-field forces, it is not possible to use the crystal structure alone to predict with certainty the dominant conformation present in the melt or solution. An early explanation of the introduction of gauche conformations suggested that they were due to electrostatic interactions between the net positive and negative charges that accumulate on the backbone due to introduction of the heteroatom into the chain [7].

The structure of PEO has continued to intrigue investigators over the years. The chain structure departs from *tgt* and changes to *ttt* when PEO is blended with PMMA [175]. The structure of PEO was recently studied in solution. Adsorbed chains in the air-solution interface weres found to be very different from the disordered structure expected for the solution [176–178]. In fact, the vibrational bands for these adsorbed chains have characteristics similar to those in the crystalline state [177]. Various theoretical studies have led to quite different results for chain conformations of PEO [7,179–190]. Under certain experimental conditions, the tgg' conformer, rather than the *tgt* sequence, can have the highest relative population [182]. It must be mentioned that the relative populations of various conformers, especially tgg', have yet to be confirmed experimentally.

Determining the conformational distribution for PEO either in solution or melt has proven to be difficult. Several liquid systems of PEO, including both molten and solution states, have been investigated using vibrational spectroscopy [191]. Raman spectra of PEO in the melt and both chloroform and aqueous solutions reveal that the conformations in aqueous solution retain the *tgt* conformation that occurs in the crystalline solid. Spectra of the chloroform solution, however, more closely resemble those of the melt than of the aqueous solution. The conformational distribution cannot be obtained from the spectra due to uncertainty in band assignments. Theoretical considerations met with mixed success. The characteristic ratio for PEO, defined as the ratio between the observed end-to-end chain length and that expected for completely flexible chains, ranges from 5.2 in aqueous solution to 6.9 in melt and solutions in organic solvents. As mentioned above, the relative energies of various rotational isomeric states differ considerably [184,192]. Because of differences in the local dielectric environment, identical sets of rotational isomeric states need not be considered [184,192]. In fact, it is more likely that different conformational distributions exist for PEO in different physical states.

The isotropic Raman spectra of PEO in melt and aqueous solutions shown in Fig. 9.37 suggest new indicators for characterizing conformational distribution. Considerable differences are evident in the $850\,\mathrm{cm}^{-1}$ band shape and the disordered LAM region near $365\,\mathrm{cm}^{-1}$ [41]. Although insignificant in the normal Raman spectrum, the shapes and positions of features in these two regions differ substantially in the isotropic spectrum. The strong bands near $850\,\mathrm{cm}^{-1}$ are almost mirror images in the two experiments. The most intense feature for the aqueous solution is at $880\,\mathrm{cm}^{-1}$, whereas that for the melt centers at $830\,\mathrm{cm}^{-1}$. The shapes of the skeletal deformations near $200\,\mathrm{cm}^{-1}$ are also quite different for the two samples. These correlate with differences in conformational distributions in the melt and aqueous solutions of PEO.

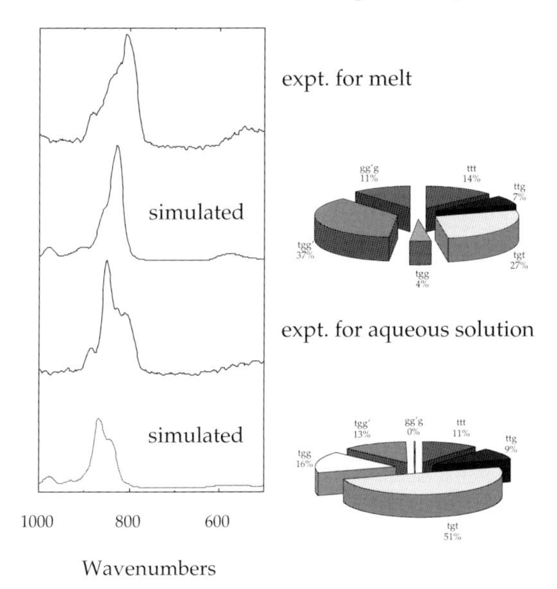

Fig. 9.37. Simulated and observed isotropic Raman spectra of poly(ethylene oxide). Different conformational distributions for melt and aqueous solution are described in text

Analyses of the experimental data are described in [41]. Upon removal of degenerate and high-energy states, the conformational distribution for PEO always contains the conformer sequences *ttt*, *tgt*, *tgg*, *tgg′*, and *ttg* [7,180,192]. A *Monte Carlo* method was used to generate three conformational distributions (1000 model chains) of PEO chains incorporating the energy terms of Flory's ($E_\sigma = -0.430$, $E_\rho = 0.900$ and $E_\omega = 0.338\,\text{kcal/mol}$) [7], Abe's ($E_\sigma = -0.5$, $E_\rho = 0.9$ and $E_\omega = 0.338\,\text{kcal/mol}$) [192], and Smith's models ($E_\sigma = 0.1$, $E_\rho = 1.4$ and $E_\omega = -1.3\,\text{kcal/mol}$) [180]. Differences between the models lie mainly in the percentages of *tgt* and *tgg′* in each conformational distribution [41]. Flory's and Abe's models produce the largest percentage of *tgt* conformer (lower panels in Fig. 9.37); Smith's simulation E is dominated by *tgg′* (upper panels). In order to complete a normal coordinate treatment of PEO and the model compound, a force field developed by Snyder and Zerbi for small ether molecules was transferred [193]. The isotropic Raman spectrum $S(\nu)$ for the liquid state was then simulated as a composite of contributions from the ensemble of chains, as discussed above.

As described previously and partially in Fig. 9.37 [41], the agreement between the simulated and observed spectra can be found not only in the $200\,\text{cm}^{-1}$ bands but also in the backbone bond stretching region. The *tgg′*-dominated distribution produces a disordered LAM with a maximum at $\approx 350\,\text{cm}^{-1}$ and a skeletal stretching vibration at $830\,\text{cm}^{-1}$ [41]. The distribution having mostly *tgt* conformers is associated with a disordered LAM centered at $\approx 282\,\text{cm}^{-1}$ and the skeletal band at $860\,\text{cm}^{-1}$. The simulated spectra from the two groups fit the two different observations, one for PEO in aqueous solution and the other in the melt. These results demonstrate that the conformational distribution of PEO in the melt is reasonably predicted by a set of statistical weights that include the so-called $O \cdots H$ attraction effect [180]. The results also indicate that for PEO in water the $O \cdots H$ attraction effect is lost. The simulated and experimental data suggest that by changing intermolecular interactions, it is possible to significantly change the conformational distribution in PEO.

Chain Conformation of Poly(Propylene Oxide) (PPO) Based on Polymeric Electrolytes. Polymer based electrolytes have generated great interest from both fundamental and applied points of view [194]. Linear polyether-based electrolytes have been widely investigated to understand the relationship between polymer structure and ionic conductivity. Polymer electrolytes possess better dimensional, thermal and chemical stability than inorganic and solution electrolytes. The host polymer used for polymer electrolytes usually contains highly polar atoms that can solvate various ions. In such systems, the structure of the polymer matrix is extremely important, since local segmental motions of the polymer chain and the number of free ions govern ionic conduction [195–200]. As various ions are highly soluble in polyethers such as PPO or PEO, they are often used as electrolytes. The solubility can be attributed to the specific interaction between the ether oxygen and cation [194,201]. The fact that that PEO exhibits much better solvating power for many salts than either poly (methylene oxide)[$-CH_2 - O-$] or poly (trimethylene oxide) [$-CH_2CH_2CH_2 - O-$] indicates that the solvation behavior can be altered by the polymer structure. The differences correlate with the spacing of polar atoms along the chain. The lower ionic conductivity of PPO- compared with PEO-based electrolytes has been attributed to steric hindrance caused by interference of the CH_3 side group with the interaction between ether oxygen and cations [202]. Undoubtedly, a better understanding of the chain conformational distribution of the host polymer will lead to better understanding of the ionic conduction mechanism.

Vibrational spectroscopy has been used to characterize conformational changes induced by polymer-salt interactions in PEO [203–207]. Although several spectroscopic studies have been conducted, very little quantitative structural analysis has been carried out for PPO-based electrolytes [208–212]. As discussed above, Raman-active bands are strongly dependent on the polarizability changes of C—C bonds along the backbone and can be extremely sensitive to chain conformational changes. Low frequency Raman bands in PPO-based electrolytes contain considerable contributions from CCO/COC skeletal bending modes and can be sensitive to chain conformational changes. These bands change significantly as a function of salt concentration [171–174,212,213]. The frequency shift has been interpreted in terms of changing chain stiffness [212,213]. As other new unexplained spectroscopic features are present when salts are introduced, a detailed interpretation of PPO structure in the presence of salts has yet to emerge. Vibrational spectra can be analyzed for a specific chain conformation. On the other hand, PPO used for polymer electrolytes has a disordered structure and the vibrational spectra generally consist of overlapping features associated with many different chain conformations.

As seen in Fig. 9.38, the bands in the 700–900 cm^{-1} region of the model compound, 1,2-dimethoxy-propane (DMP), are broad and ill defined in the absence of salt [171–174]. However, an obvious change in intensity at ca. 810 cm^{-1} is observed with increasing salt concentration. These band frequencies are independent of molecular weight and, therefore, they reflect a localized vibrational mode. Analysis of salt complex indicates the intensity change at 810 cm^{-1} is caused by interaction between the lithium cation and the ether oxygen. The calculated isotropic Raman spectra of a random chain and that of a sequence of $tg_\alpha t$ conformation along the $-O-C-C-O-$ bond of DMP

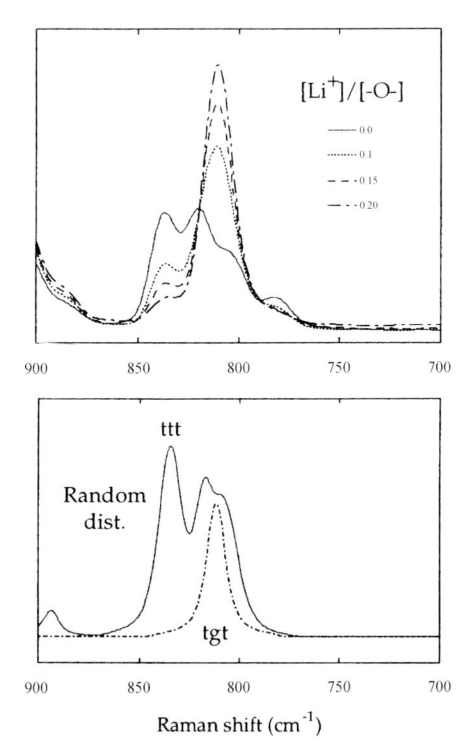

Fig. 9.38. *Top*: Experimental Fourier transform Raman spectra of DMP over the region between 700 and 900 cm^{-1} with various $[Li^+]/[-O-]$ concentrations; *Bottom*: The calculated spectra of DMP obtained for a random conformational distribution and for a regular sequence of $tg_\alpha t$ conformation in the 700–900 cm^{-1} region

are also shown in Fig. 9.38. The calculations show three bands at 835, 817, and 807 cm^{-1} for the random chain and a single band at 812 cm^{-1} for the $tg_\alpha t$ conformation. Independent of chain length, and based on a normal mode analysis, the lowest frequency component at 807 (or 812) cm^{-1} is attributed to trans(OC)-gauche α(C–C)-trans(C–O) ($tg_\alpha t$) conformation. The band centered at 817 cm^{-1} contains contributions from molecules associated with $g_\alpha g_\beta t$ and/or $g_\beta g_\alpha t$ structures. The main contributor to the band at 835 cm^{-1} is the trans(O–C)-trans(C–C)-trans(C–O) conformation of –O–C–C–O– bond. Based on the comparison of relative intensity of bands associated with different conformations, one finds $E_{g\alpha}$ associated with C–C has an extremely low value of –120 cal/mol relative to the trans conformer [171–174]. The energy difference is smaller than the measured value for DMP in the gaseous phase [214,215].

The experimental data in Fig. 9.38 are well represented by chains having a random conformational distribution. Based on the simulated spectra, observed bands centered at 836 and 807 cm^{-1} of DMP are assigned to ttt and $tg_\alpha t$ conformations, respectively. As mentioned above, the strong band at 811 cm^{-1} in the DMP/$LiClO_4$ complex can be definitely assigned to the $tg_\alpha t$ conformation, since the experimental and calculated spectra superimpose almost exactly. It follows that the intensity increase of this band upon introduction of salt is caused by the increase in the $tg_\alpha t$ conformation population induced by the interaction between the Li cations and ether oxygens. This model study serves as the basis to analyze the changes found for linear and crosslinked PPO's [171–174].

Because of their inherent advantages such as dimensional and thermal stability, network polymer based electrolytes have been of interest in spite of their somewhat lower ionic conductivity in comparison to the corresponding linear polymer based electrolytes [194,201]. The network derived from PPO with multi-functional isocyanates at junctions has been used as a model network polymer because atactic PPO has a narrow molecular weight distribution capable of yielding a homogeneous structure. The temperature dependence of the ionic conductivity of these network polymer electrolytes was described using the Williams–Landel–Ferry (WLF) equation [194]. This study suggests that the conductivity is mainly associated with local segmental motions of polymer chains. Further studies have shown two distinct dielectric relaxation processes, one related to the ether units and the other associated with the urethane groups at junctions [216,217]. The mechanism for ionic conductivity is not clearly established. It is generally believed that the lower ionic conductivity of network-based polymer electrolytes is due to their reduced chain segmental motion compared to that of the linear polymer [216,217]. The reduction in chain dynamics may be reflected in the conformational distribution of the chains [171–174].

The characteristic changes found in the linear chains (Fig. 9.38) are not observed in network polymers [171–174]. Since Raman active bands in the

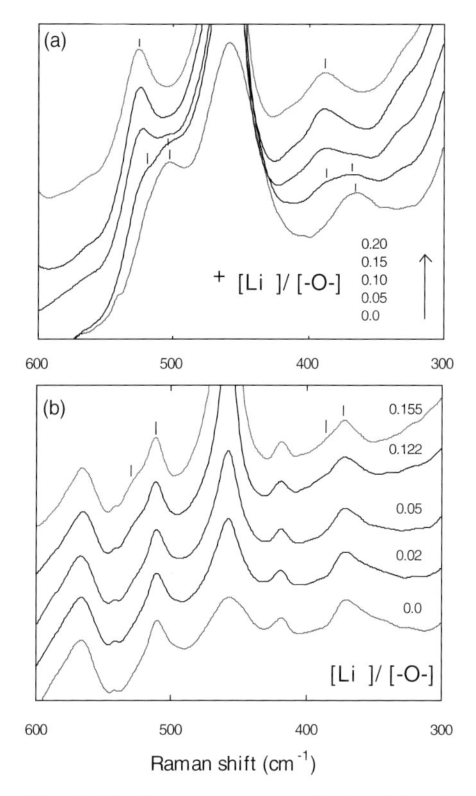

Fig. 9.39. Raman spectra obtained for linear and crosslinked polyethers as a function of salt concentration; 600–300 cm^{-1} region; (**a**) HPPO1000; (**b**) NPPO1000

300 and 800 cm^{-1} regions represent characteristic modes of PPO chains, relative intensity changes of these bands indicate the proportion of PPO chains interacting with the ions through changes in chain conformation by ion solvation. Therefore, the smaller changes observed in the networks (lower panel in Fig. 9.39) based on the short chain length (mol. wt. 1000) of PPO signify a low local concentration of Li cations interacting with ether oxygen. The total salt concentration based on the number of ether units is, however, the same as that in linear PPO-based electrolytes. From the crosslink density dependence of the spectral change in the 800 and 300–500 cm^{-1} regions, salt depletion associated with ether units in the network of low molecular weight PPO chains can be attributed to interaction of the Li cation with the urethane group of the crosslink point This is consistent with the intensity increase in the 1700 cm^{-1} region [171–174]. The molar content of the ether units to urethane groups, including the NPPO1000 sample, is at least 8:1. If Li cations are distributed between urethane and ether units based only on the statistical weight of the two components, the local content of Li cations should be much higher along the PPO chains, which should be reflected in

changes of the skeletal modes. As mentioned above, this is contrary to the spectroscopic observations [171–174]. These observations lend support to the greater solvation ability of the urethane group for the Li cation, compared with the ether unit [216,217].

Analysis of Polyelectrolyte Chain Conformation by Polarized Raman Spectroscopy. Global and local interactions can be separated for neutral polymers. The relationship between the global chain conformation for polyelectrolytes, in terms of a persistence length and an excluded volume parameter, and the chain conformation described by a rotational isomeric state model, is however, untested. Raman spectroscopy can partially close this gap in our understanding of polyelectrolytes by measuring conformational changes as chain stiffness is varied.

Most solution characterizations of polyelectrolytes employ methods such as viscometry and light scattering that provide only global structural information. These studies were interpreted to show that chain stiffness increases with ionization. It has been suggested that in the extreme case, the chain is fully extended. Polarized Raman spectroscopy is ideally suited to address issues of local structure and thus complements previous work [218,219]. As demonstrated above, for disordered chain conformations typically found in solution, the vibrational bands associated with specific chain conformation overlap, making it difficult to precisely characterize spectral features. Information provided by polarization studies, which are sensitive to changes in the conformational distribution, i.e., changes in the distribution of torsional angles among the various rotational states permitted in a vinyl chain backbone, can yield substantial structural information. Similar approaches have been pursued to study conformational changes in neutral polymer solutions, for example, to identify the θ condition or to quantify the enthalpy or entropy of helix formation [32,34–37]. A direct calculation of the persistence length from depolarization data has been hampered by the absence of polarizability derivative values for different conformations, even in small molecules. Recent results [220,221], in conjunction with a simplified form of the rotational isomeric state model, make it possible to calculate the persistence length using the C−H stretch depolarization ratio of poly(acrylic acid) (PAA) [218,219]. The need for a good signal to noise ratio is complicated by the fact that only very dilute concentrations are suitable for analysis. Here, multiple passes through the sampling volume are often necessary to obtain good data.

Focusing on the conformation-sensitive bands of a vinyl polymer, the depolarization ratio defined earlier can be equated to the number ratio of gauche and trans isomers that exhibit perpendicular and parallel polarized scattering:

$$\rho = \frac{xI_{\perp g} + I_{\perp \mathrm{t}}}{xI_{\| g} + I_{\| \mathrm{t}}} \, . \tag{9.23}$$

Here, x is the isomeric gauche to trans ratio, $I_{\perp \mathrm{g}}$ and $I_{\perp \mathrm{t}}$ are the scattered intensities for gauche and trans isomers analyzed by perpendicular polarization, respectively, and $I_{\| \mathrm{g}}$ and $I_{\| \mathrm{t}}$ are the intensities for parallel polarization. This expression can be reduced to a more useful form from which separate depolarization ratios for the gauche and trans isomers of PAA can be determined. For this we write the depolarization ratios in terms of each isomer's overall polarizability tensor α and its derivative α' with respect to the normal coordinate of vibration Q:

$$\rho = \frac{3\gamma'^2}{45\overline{\alpha}'^2 + 4\gamma'^2} , \tag{9.24}$$

where γ' is the derivative of the anisotropy of α, and $\overline{\alpha}'$ is the derivative of the trace of α, both with respect to Q [222].

To explicitly relate ρ to x, an additional relationship is required. It is then possible to relate the total intensity scattered by the gauche isomer I_{gauche} $(= I_{\perp \mathrm{g}} + I_{\| \mathrm{g}})$ to that scattered by the trans isomer I_{trans} $(= I_{\perp t} + I_{\| t})$. Using the calculated values of $\overline{\alpha}'$ and γ' for gauche and trans isomers of n-butane and the frequencies of the symmetric and asymmetric CH stretching vibrations for the same isomers, it is found that $0.648 I_{\mathrm{gauche}} = I_{\mathrm{trans}}$. This relationship, along with the values for ρ_{gauche} and ρ_{trans}, allows (9.23) to be rewritten in the more useful form,

$$x = \frac{0.0601 - 1.0887\rho}{\rho - 0.7718} , \tag{9.25}$$

which serves as a simple tool to analyze spectroscopic data for polymer conformations.

To relate the isomeric ratio x to the persistence length L_{t}, a rotational isomeric state model is used with one low-energy trans state of statistical weight one and two higher, equal energy gauche states of statistical weight σ [218,219]. The isomeric ratio is then

$$x = 2\sigma , \tag{9.26}$$

where

$$\ln \sigma = -\Delta E / RT , \tag{9.27}$$

ΔE is the energy change between the two states and R the gas constant. The persistence length is calculated from σ in the usual manner. We also need the average of $\cos\phi$, the cosine of the dihedral angle, given by

$$\langle \cos\phi \rangle = z^{-1} \sum \mu_\eta \cos\phi_\eta , \tag{9.28}$$

where z is the partition function or the sum of statistical weights, μ_η is the statistical weight of state η, and ϕ_η is the dihedral angle of that state. If the three rotation states are described by $\phi = 0$, 120 and -120 degrees, of statistical weights 1, σ, and σ, respectively, (9.28) becomes

$$\langle \cos\phi \rangle = (1-\sigma)/(1+2\sigma) . \tag{9.29}$$

The limiting characteristic ratio at high molecular weight, C_∞, can then be written

$$C_\infty = \left(\frac{1-\cos\theta}{1+\cos\theta}\right)\left(\frac{1-\langle\cos\phi\rangle}{1-\langle\cos\phi\rangle}\right) , \tag{9.30}$$

where θ is the valence angle between bonds [7]. Finally the persistence length is

$$L_t = \frac{L}{2}(C_\infty + 1) , \tag{9.31}$$

where L is the carbon–carbon bond length.

Raman studies reveal several interesting aspects of ionized structures [218, 219]. Based on polarization data, the ΔE, C_∞, and L_t values calculated from the depolarization data are given in Table 9.1. The energy difference between isomers grows by a surprisingly low value of $\approx 8\%$ as compared to its initial value of 1471 cal/mole for the unionized polymer to 1584 cal/mole for the fully ionized sample. In comparison, the energy change between the same isomers is 600 cal/mole for polyethylene at room temperature and 1100–2300 cal/mole for PTFE [7]. The small change in conformer energy upon PAA ionization indicates that the polymer does not stiffen appreciably as it ionizes in a salt-free, semidilute solution. The lengths calculated from spectroscopic data fall within the range of persistence lengths determined via SAXS [223]. According to the present analysis, L_t changes from 14.6 to 17.6 Å upon ionization. Muroga et al. studied persistence length changes for PAA in similar semidilute solutions using SAXS and found $L_t = 8$–15 Å [223]. Such analyses demonstrate the potential of Raman scattering for studying the interplay of local and non-local polyelectrolyte forces.

Table 9.1. Conformational analysis of depolarized raman data

Ionization	ρ	ΔE (cal/mole)	C_∞	L_t(Å)
0	0.1438	1471	18.0	14.6
0.29	0.1418	1486	18.5	15.0
0.50	0.1332	1553	20.7	16.7
0.72	0.1303	1578	21.6	17.4
0.93	0.1296	1584	21.8	17.6

Structure of Syndiotactic Polypropylenes. Polyolefins are often considered as commodity materials with production capacity speculated to be in excess of usage. Actually, the opposite is true. Because of recent advances in catalyst design [224], synthesis of highly-ordered polyolefins with fewer defects and better control of the defect distribution is now possible. These advances in chemistry have raised an even greater interest in the study of polyolefin structure and development of more diverse applications. For example, the ability to precisely control configuration has sparked interest in structural studies of sPP. Even though iPP is one of the most used polymers, sPP does not enjoy the same degree of success, and far fewer studies have been undertaken to examine the structure of the syndiotactic isomer. Many questions regarding their structure and associated changes remain unanswered. Raman studies have been carried out to elucidate the spectra obtained for sPP with various conformational distributions and configurations [40]. Based on the force field and rotational isomeric state models developed previously, the conformational distributions associated with the amorphous phase and their change as a function of time and deformation have now been assigned. As stated earlier, only 2 polarizability parameters associated with C$-$C stretching and CCC angle bending are required to duplicate the observed intensities. The conformation sensitive bands at 400 and 800 cm^{-1} are of great interest. Features in these regions have been correlated with the distribution of chain conformation and configuration.

For sPP, the number of seemingly stable chain conformations and packing is intriguing. The relative volume fraction of each state is highly dependent on thermal history, time, processing, and, in many cases, type and number of configurational defects. There are at least 3 different crystalline forms associated with sPP. The helical form I, with ggtt chain conformation, is the most common [225,226]. Form II with an all trans sequence can be obtained by quenching the polymer from the melt and then stretching [226]. Form III, with conformational features of both forms I and II, has also been proposed with the conformation $(g_2t_2g_2t_6)_n$ [227–229]. It was recently found that the planar zigzag structure forms spontaneously over an extended length of time by quenching from melt into ice water and storing at or below 0 °C [230,231]. Form III has also been observed when a drawn sample of form II is exposed to benzene, toluene or xylene vapor [227,228]. Both forms II and III are stable at room temperature, then revert to form I spontaneously with heating. Although conformational analysis has been carried out for polypropylenes, direct verification of the conformation distribution associated with the disordered phase has not been verified. Characterization of disordered structures of sPP may be divided into two interrelated analyses: one deals with the disorder associated with chain conformations different from long regular sequences of *ggtt* or *tttt*, and the second deals with configurational defects, racemic or meso content [40].

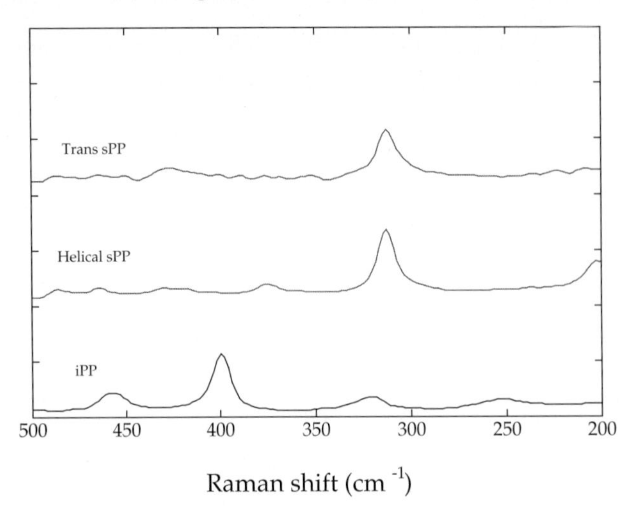

Fig. 9.40. Fourier transform Raman spectra obtained for various syndiotactic and isotactic polypropylenes

The earliest method used to characterize sPP configurations was IR spectroscopy, which suggested a structural parameter referred to as "syndiotacticity index" [232–234]. The usefulness of Raman scattering was first demonstrated when the skeletal bending modes in the 300 cm^{-1} region (Figs. 9.3 and 9.40) of polypropylene were found to be sufficiently sensitive to differentiate between racemic and meso sequences for characterizing the overall chain configuration [40,235]. Similarly, the broad, ill-defined bands in the 800 cm^{-1} region (Fig. 9.4) were suggested to be extremely sensitive to processing differences, configurational differences and thermal history. The differences suggest that structures in various sPP are complex and encompass all the various chain conformations found earlier. Raman studies, including both experimental and theoretical normal vibrational analyses, were applied to the ensemble of chain conformations for sPP containing known configurational and thermal history (Figs. 9.3, 9.4, 9.40–9.42) [40].

The structural parameters and force fields for polyolefins are well defined [18,151,236]. The chain conformational distribution can be obtained by using the rotational isomeric states generated previously [163,237,238]. Energy contours from experimental and theoretical studies have yielded low energy minima around the *gg* and *tt* pairs. For the isotactic and atactic polymer, C_∞ has been accurately calculated using the 3-state model with suitable parameterization [239]. The simplicity and convenience of this 3-state model outweighs the possible advantage of higher accuracy associated with the 5-state model [238]. The statistical weight matrices and the generation of conditional probability of each bond along the chain follow the procedure first presented by Flory [7]. The statistical weight matrices for the sPP containing configurational defects are given in [7,240,241].

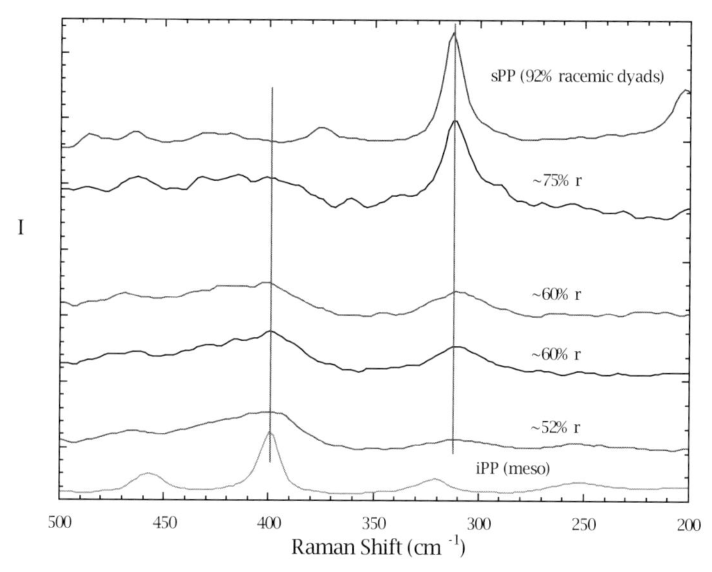

Fig. 9.41. Fourier transform Raman spectra obtained for various syndiotactic polypropylenes with known configurational defect

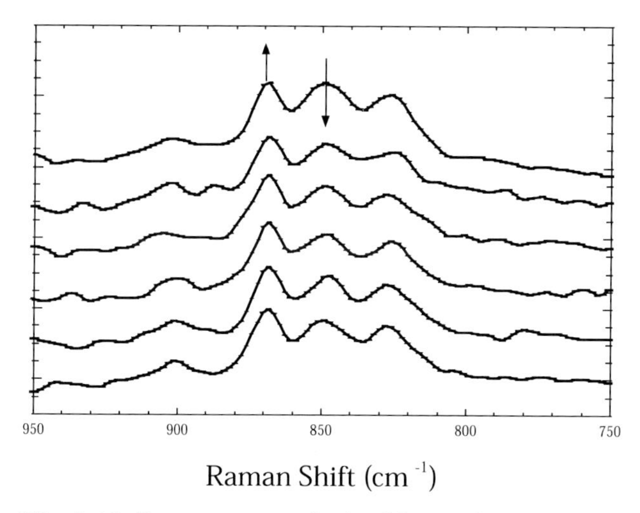

Fig. 9.42. Raman spectra obtained for syndiotactic polypropylene as a function of time at room temperature. The initial sample is obtained by quenching a molten sample into liquid nitrogen then kept at $-5\,^{\circ}$C for 140 hours

Raman spectra obtained from polypropylenes of different conformational distributions can be accurately simulated [40]. Identification of specific features such as those centered at 800 and 400 cm^{-1} enable an understanding of the change of polypropylenes from a completely disordered state to a more or-

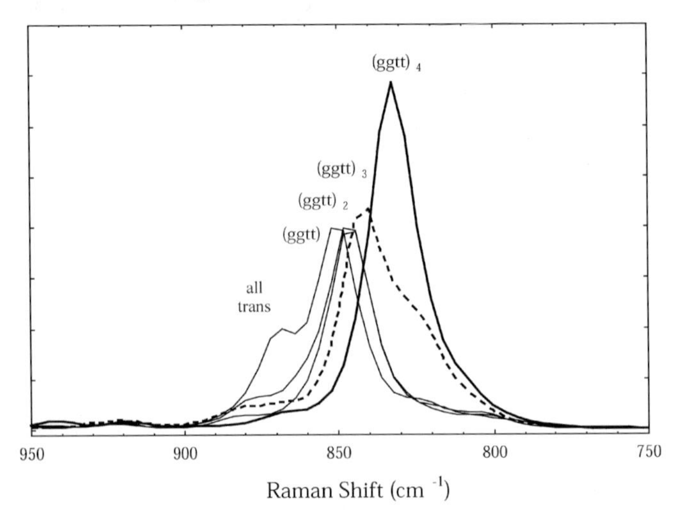

Fig. 9.43. Simulated Raman spectra of syndiotactic polypropylenes in the 800 cm^{-1} region with conformational defects. The helical segments are as specified in the text

dered state or states including crystallization. Although the bands found near 800 cm^{-1} are generally referred to as regularity bands, they have never been accurately assigned. The bands near 826 and 845 cm^{-1} are suggested to be associated with form I and the amorphous chains, respectively [235,236]. The high frequency component at 865 cm^{-1} has never been assigned. The broad band at 845 cm^{-1} is the dominant feature of a polypropylene melt. With increasing configurational disorder, the 826 and 865 cm^{-1} bands decrease in intensity relative to the 845 cm^{-1} peak.

Based on simulation results, the 830 cm^{-1} band can only be associated with long helical *ggtt* chain conformations (Fig. 9.43). All extremely short helical segments situated between either conformational or configurational defects contribute to the intensity in the middle 850 cm^{-1} region. The calculated spectra for all possible conformers of isolated helical segments embedded between sequences of random conformations, including the 2 global minima found in the conformational search, with the exception of the $tggttg'g't$, all exhibit intensity in the region around 850 cm^{-1}. The band at 845 cm^{-1} is then attributed to the amorphous structure, and indeed it does not appear in any normal coordinate analysis of long ordered chains.

The simulated molten state possesses at least 39% of sequences of forms *ggtt* or $ttg'g'$. The *tttt* sequence accounts for 26% of the distribution. The remaining conformers have an insignificant contribution [40]. A simulation of iPP in the molten state has been carried out [153]. The molten spectra for sPP and iPP compare well with simulated spectra. sPP spectroscopic features below 500 cm^{-1} are significantly different from those of iPP in both the solid and melt. The major difference is the presence of the peak around 400 cm^{-1}

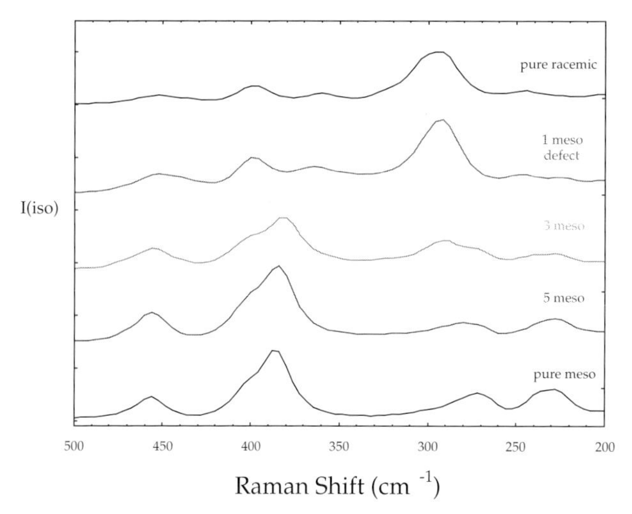

Fig. 9.44. Simulated Raman spectra of syndiotactic polypropylenes in the 300 cm^{-1} region with various number of configurational defects

in the isotactic polymer and the peak around 300 cm^{-1} in the syndiotactic form, crystalline or amorphous. This difference is verified by disordered chain calculations (Fig. 9.44) in which the numbers of configurational defects are systematically varied and their placement along the chain is based on Monte Carlo methods. It can be seen that two dominant bands in the 300–400 cm^{-1} region are truly configurational bands, i.e. their presence depends only on configurational content and is independent of conformation. The simulated results compare well with experimental data (Fig. 9.41). The relative intensity is a true syndiotacticity index, as shown in Fig. 9.3, and it compares well with all reported values.

Simulations for an all trans chain of different chain lengths ranging from $n = 6$ to $n = 11$ are shown in Fig. 9.45. Two components at 850 and 870 cm^{-1} are found. If the relative intensity of the 2 components varies as shown, the higher frequency component has equal or higher intensity than the lower component at $n = 13$. This higher frequency component is then assigned to an all trans structure of fairly long length. With conformational defects along the chain, it is natural to expect that the intensity of the observed 845 cm^{-1} component would increase as observed. Based on these experimental and simulated Raman features, it is possible to infer that sPP crystallizes into a helical structure at crystallization temperatures above 0 °C (Fig. 9.4). Conversely, only all trans structures are formed in sPP crystallized below 0 °C, (Fig. 9.42). Raman data are clear regarding the structures formed, but the mechanisms responsible for such behavior are still unknown.

Acknowledgements. For a project such as this, it is hard to overestimate the tremendous assistance offered by my students and research associates. I wish to extend particular thanks to my Chemistry colleague, Professor Howard Stidham; my associates, Amanda Chaparro, Dr. Tom Hahn, Amy Heintz, Shuhui Kang, and Dr. Wu Suen; and to Sophia for going over the manuscript page by page. Funding from Materials Research Science and Engineering Center, and 3M are especially appreciated.

References

1. C.V. Raman: Indian J. Phys. **2**, 387 (1928)
2. C.V. Raman, U.S. Krishnan: Nature **121**, 501 (1928)
3. G. Landsberg, L. Mandelstam: Naturwissenschaften **16**, 557 (1928)
4. S.P.S. Porto: J. Opt. Soc. Am. **56**, 1585 (1966)
5. K. Nakamoto: *Infrared and Raman Spectra of Inorganic and Coordination Compounds* (Wiley, New York 1986)
6. F. Gornick, L. Mandelkern: J. Appl. Phys. **33**, 907 (1962)
7. P.J. Flory: *Statistical Mechanics of Chain Molecules* (Interscience Publishers, New York 1969)
8. P. Painter, M. Coleman, J.L. Koenig: *The theory of vibrational spectroscopy and its application to polymeric materials* (Wiley, New York 1982)
9. J.L. Koenig: *Spectroscopy of Polymers* (American Chemical Society, Washington DC 1992)
10. G. Zerbi: *Modern Polymer Spectroscopy* (Wiley-VCH, New York 1999)
11. N.B. Colthup: J. Opt. Soc. Am. **40**, 397 (1950)
12. N.B. Colthup, L.H. Daly, S.E. Wiberly: *Introduction to Infrared and Raman Spectroscopy* (Academic Press, Boston 1990)

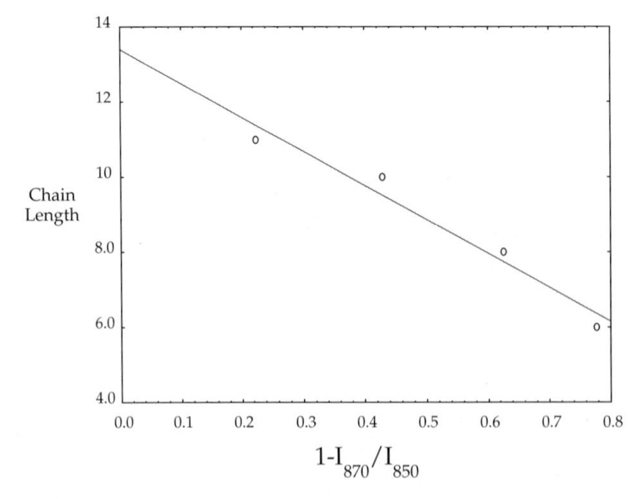

Fig. 9.45. Relative intensity of the 2 components associated with all trans structure

13. H. Tadokoro: *Structure of Crystalline Polymers* (Wiley-Interscience, New York 1979)
14. R. Zbinden: *Infrared Spectroscopy of High Polymers* (Academic Press, New York 1964)
15. S. Krimm: Fortschr. Hochpolym.-Forsch. **2**, 51 (1960)
16. G. Herzberg: *Infrared and Raman Spectra*, Vol. 2 (Van Nostrand, Princeton 1945)
17. R.G. Snyder, J.H. Schachtschneider: Spectrochim. Acta **19**, 85 (1963)
18. R.G. Snyder, J.H. Schachtschneider: Spectrochim. Acta **20**, 853 (1964)
19. R.G. Snyder, J.H. Schachtschneider: Spectrochim. Acta **21**, 169 (1965)
20. G. Turrell: *Infrared and Raman Spectra of Crystals* (Academic Press, New York 1972)
21. E.B. Wilson Jr., J.C. Decius, P.C. Cross: *Molecular vibrations* (McGraw-Hill, New York 1955)
22. C.W. Bunn: Trans. Faraday Soc. **35**, 483 (1939)
23. B. Wunderlich: *Macromolecular Physics*, Vol. 1 (Academic Press, New York 1973)
24. B. Wunderlich: *Macromolecular Physics*, Vol. 2 (Academic Press, New York 1976)
25. R.T. Bailey, A.J. Hyde, J.J. Kim: Adv. Raman Spectrosc. **1**, 296 (1972)
26. R.T. Bailey, A.J. Hyde, J.J. Kim: Spectrochim. Acta **A30**, 91 (1974)
27. R.G. Snyder: J. Mol. Struct. **36**, 222 (1971)
28. R.G. Snyder: J. Mol. Struct. **37**, 353 (1971)
29. R.G. Snyder, S.L. Hsu, S. Krimm: Spectrochim. Acta **A34**, 395 (1978)
30. Y.K. Wang, D.A. Waldman, R.S. Stein, S.L. Hsu: J. Appl. Phys. **53**, 6591 (1982)
31. Y.K. Wang, D.A. Waldman, R.S. Stein, S.L. Hsu: Macromolecules **15**, 1452 (1982)
32. I.W. Shepherd: Biochem. J. **155**, 543 (1976)
33. I.W. Shepherd: Rep. Prog. Phys. **38**, 565 (1975)
34. P.W. Thornley, I.W. Shepherd: J. Polym. Sci., Polym. Phys. Ed. **15**, 97 (1977)
35. P.W. Thornley, I.W. Shepherd: J. Polym. Sci., Polym. Phys. Ed. **15**, 1339 (1977)
36. R. Speak, I.W. Shepherd: J. Poly. Sci. Symp. **44**, 209 (1974)
37. A.J. Hartley, Y.K. Leung, J. Mc Mahon, C. Booth, I.W. Shepherd: Polymer **18**, 336 (1977)
38. R.G. Snyder, Y. Kim: J. Phys. Chem. **95**, 602 (1991)
39. R.G. Snyder: J. Chem. Soc. Faraday Trans. **88**, 1823 (1992)
40. T.D. Hahn: in *Polymer Science and Engineering Department*, (University of Massachusetts, Amherst 1998)
41. X. Yang, Z. Su, D. Wu, S.L. Hsu, H.D. Stidham: Macromolecules **30**, 3796 (1997)
42. S.L. Hsu, W.H. Moore, S. Krimm: J. Appl. Phys. **46**, 4185 (1975)
43. D.E. Williams: J. Chem. Phys. **43**, 4424 (1965)
44. D.E. Williams: J. Chem. Phys. **45**, 3770 (1966)
45. W.L. Peticolas, G.W. Hibler, J.L. Lippert, A. Peterlin, H. Olf: Appl. Phys. Lett. **18**, 87 (1971)

46. C.K. Wu, M. Nicol: J. Chem. Phys. **58**, 5150 (1973)
47. S.L. Hsu, J.P. Sibilia, K.P. O'Brien, R.G. Snyder: Macromolecules **11**, 990-995 (1978)
48. H. Tadokoro, S. Yasumoto, S. Murahashi, I. Nitta: J. Polym. Sci. **44**, 266 (1960)
49. M. Shimomura, M. Iguchi, M. Kobayashi: Polymer **29**, 351 (1988)
50. T. Uchida, H. Tadokoro: J. Polym. Sci. (A-2) **5**, 63 (1967)
51. J.L. Koenig: in *Applied Spectroscopy Reviews*, Vol. 4, ed. by E.G. Brame (Marcel Dekker, New York 1971) pp. 233–306
52. J.P. Luongo, R. Salovey: J. Polym. Sci. B **3**, 513 (1965)
53. S.W. Cornell, J.L. Koenig: J. Polym. Sci. A **7**, 1965 (1969)
54. S.L. Hsu, W.H. Moore, S. Krimm: Biopolymers **15**, 1513 (1976)
55. S.L. Hsu: in *Physics*, (University of Michigan, Ann Arbor 1975) p. 206
56. F. Viras, T.A. King: Polymer **25**, 1411 (1984)
57. V.V. Tarasov: *New Problems in the Physics of Glass* (Oldbourne Press, Jerusalem 1963)
58. B. Wunderlich: Pure Appl. Chem. **67**, 1019 (1995)
59. C. Kittel: *Introduction to Solid State Physics* (Wiley, New York 1996)
60. F.J. Deblase, M.L. McKelvy, M. Lewin, B.J. Bulkin: J. Polym. Sci.: Lett. Ed. **23**, 109 (1985)
61. J.D. Savage, Y.K. Wang, M. Corbett, H.D. Stidham, S.L. Hsu: Macromolecules **25**, 3164 (1992)
62. N. Ishihara, M. Kuramoto: in *Catalyst Design for Tailor-Made Polyolefins*, Vol. 89, pp. 339–350 (Elsevier 1994)
63. A. Immirzi, F. de Candia, P. Iannelli, A. Zambelli: Makromol. Chem., Rapid Commun. **9**, 761 (1988)
64. V. Vittoria, F. de Candia, P. Iannelli, A. Immirzi: Makromol. Chem., Rapid Commun. **9**, 765 (1988)
65. N.M. Reynolds, J.D. Savage, S.L. Hsu: Macromolecules **22**, 2867 (1989)
66. M. Kobayashi, T. Nakaoki, N. Ishihara: Macromolecules **22**, 4377 (1989)
67. R.A. Nyquist: Appl. Spectrosc. **43**, 440 (1989)
68. G. Guerra, V.M. Vitagliano, C. De Rosa, V. Petraccone, P. Corradini: Macromolecules **23**, 1539 (1990)
69. M. Gomez, A.E. Tonelli: Macromolecules **23**, 3385 (1990)
70. D.C. Doherty, A.J. Hopfinger: Macromolecules **22**, 2472 (1989)
71. S.J. Spells, I.W. Shepherd, C.J. Wright: Polymer **18**, 905 (1977)
72. L.A. Carreira, T.G. Towns: J. Chem. Phys. **63**, 5283 (1975)
73. W.G. Fateley, R.K. Harris, F.A. Miller, R.E. Witkowski: Spectrochim. Acta **21**, 231 (1965)
74. S. Tsuzuki, K. Tanabe, E. Osawa: J. Phys. Chem. **94**, 6175 (1990)
75. S.L. Hsu, S. Krimm: J. Polym. Sci., Phys. Ed. **14**, 521 (1976)
76. G. Natta, P. Corradini: Nuovo Cimento Suppl. **1**, 9 (1960)
77. S. Iwayanagi, I. Sakurai, T. Sakurai, T. Seto: J. Macromol. Sci.: Phys. Ed. B **2**, 163 (1968)
78. K. Suehiro, M. Takayanagi: J. Macromol. Sci: Phys. Ed. B **4**, 39 (1970)
79. T. Tatusumi, T. Fukushima, K. Imada, M. Takayanagi: J. Macromol. Sci.: Phys. Ed. B **1**, 450 (1967)

80. J.M. Stellman, A.E. Woodward, S.D. Stellman: Macromolecules **6**, 330 (1973)
81. M. Mackley: MRS Bulletin **22**, 47 (1997)
82. R.F. Schaufele, T. Shimanouchi: J. Chem. Phys. **47**, 3605 (1967)
83. G.R. Strobl, R. Eckel: J. Polym. Sci., Phys. Ed. **14**, 913 (1976)
84. P. Smith, P.J. Lemstra: J. Mat. Sci. **15**, 505 (1980)
85. A. Zwijnenburg, A.J. Pennings: Colloid Polym. Sci. **253**, 452 (1975)
86. I. Sakurada, T. Ito, K. Nakamae: J. Polym. Sci. C **15**, 75 (1966)
87. S.L. Hsu, S. Krimm: J. Appl. Phys. **47**, 4265 (1976)
88. C.P. Lafrance, P. Chabot, M. Pigeon, R.E. Prud'homme, M. Pezolet: Polymer **34**, 5029 (1993)
89. W. Kuhn, F. Grun: Kolloid Z. **101**, 248 (1942)
90. R.J. Roe, W.R. Krigbaum: J. Chem. Phys. **40**, 2608 (1964)
91. R.J. Roe, W.R. Krigbaum: J. Appl. Phys. **35**, 2215 (1964)
92. P.H. Hermans, J.J. Hermans, D. Vermaas, A. Weidinger: J. Polym. Sci. **3**, 1 (1947)
93. D.I. Bower: J. Polym. Sci., Phys. Ed. **10**, 2135 (1972)
94. C.P. Lafrance, M. Pezolet, R.E. Prud'homme: Macromolecules **24**, 4948 (1991)
95. H.J. Marrinan: J. Polym. Sci. **39**, 461 (1959)
96. S.I. Mizushima, T. Shimanouchi: J. Am. Chem. Soc. **71**, 1320 (1949)
97. H.G. Olf, A. Peterlin, W.L. Peticolas: J. Polym. Sci., Phys. Ed. **12**, 359 (1974)
98. V. Renugopalakrishnan, T.W. Collette, L.A. Carreira, R.S. Bhatnagar: Macromolecules **18**, 1786 (1985)
99. P.C. Painter, L.E. Mosher, C. Rhoads: Biopolymers **21**, 1469 (1982)
100. B. Fanconi: Biopolymers **12**, 2759 (1973)
101. J.F. Rabolt, B. Fanconi: J. Polym. Sci. Polym. Lett. Ed. **15**, 121 (1977)
102. J.F. Rabolt: Polymer **22**, 890 (1981)
103. J.F. Rabolt, B. Fanconi: Polymer **18**, 1258 (1977)
104. A. Hartley, Y.K. Leung, C. Booth, I.W. Shepherd: Polymer **17**, 354 (1976)
105. K. Song, S. Krimm: Macromolecules **22**, 1505 (1988)
106. K. Song, S. Krimm: J. Polym. Sci., Polym. Phys. Ed. **28**, 63 (1990)
107. M.J. Folkes, A. Keller, J. Stejny: Colloid Polym. Sci. **253**, 354 (1975)
108. C. Chang, Y.K. Wang, D.A. Waldman, S.L. Hsu: J. Polym. Sci., Polym. Phys. Ed. **22**, 2185 (1984)
109. Y.K. Wang, P.H.C. Shu, R.S. Stein, S.L. Hsu: J. Polym. Sci., Polym. Phys. Ed. **18**, 2287 (1980)
110. B. Chien, H.D. Stidham, S.L. Hsu: Macromolecules **29**, 4247 (1996)
111. B. Fanconi, J. Crissman: J. Polym. Sci., Polym. Lett. Ed. **13**, 421 (1975)
112. J.F. Rabolt, R. Twieg, C. Snyder: J. Chem. Phys. **76**, 1646 (1982)
113. J.F. Rabolt: J. Polym. Sci. Polym. Phys. Ed. **17**, 1457 (1979)
114. G.R. Strobl, R. Eckel: Colloid Polym. Sci. **258**, 570 (1980)
115. A. Peterlin, H.G. Olf, W.L. Peticolas, G.W. Hibler, J.L. Lippert: J. Polym. Sci. B **9**, 359 (1971)
116. S. Krimm, S.L. Hsu: J. Polym. Sci., Polym. Phys. Ed. **16**, 2105 (1978)
117. J.D. Barnes, B.M. Fanconi: J. Chem. Phys. **56**, 5190 (1972)
118. S.L. Hsu, N. Reynolds, S.P. Bohan, H.L. Strauss, R.G. Snyder: Macromolecules **23**, 4565 (1990)

119. S.L. Hsu, S. Krimm, S. Krause, G.S.Y. Yeh: J. Polym. Sci., Polym. Lett. Ed. **14**, 195 (1976)
120. M. Asahina, S. Enomoto: J. Polym. Sci. **59**, 101 (1962)
121. M. Yokouchi, Y. Chatani, H. Tadokoro, K. Teranishi, H. Tani: Polymer **14**, 267 (1973)
122. R.G. Snyder, S.J. Krause, J.R. Scherer: J. Polym. Sci., Polym. Phys. Ed. **16**, 1593 (1978)
123. G. Capaccio, M.A. Wilding, I.M. Ward: J. Polym. Sci., Polym. Phys. Ed. **19**, 1489 (1981)
124. A. Keller: Rep. Prog. Phys. **31**, 41 (1968)
125. J.P. Arlie, P. Spegt, A. Skoulios: Makromol. Chem. **104**, 212 (1967)
126. A. Peterlin: J. Polym. Sci. B **1**, 279 (1963)
127. I. Sanchez: J. Appl. Phys. **44**, 4332 (1973)
128. P. Dreyfus, A. Keller: J. Polymer Science, B **8**, 253 (1970)
129. A.H. Windle: J. Mat. Sci. **10**, 252 (1975)
130. A.H. Windle: J. Mat. Sci. **10**, 1959 (1975)
131. G.V. Fraser, P.J. Hendra, M.E.A. Cudby, H.A. Willis: J. Mat. Sci. **9**, 1270 (1974)
132. R.G. Snyder, J.R. Schere, D.H. Reneker, J.P. Colson: Polymer **23**, 1286 (1982)
133. C.J. Farrell, A. Keller: J. Mat. Sci. **12**, 966 (1977)
134. A.E. Zachariades, W.T. Mead, R.S. Porter: Chem. Rev. **80**, 351 (1980)
135. G. Capaccio, T.A. Compton, I.M. Ward: Polym. Eng. Sci. **18**, 533 (1978)
136. A. Peterlin: Polym. Eng. Sci. **18**, 488 (1978)
137. E. Fermi: Z. Physik **71**, 250 (1931)
138. W.H. Moore, S. Krimm: Biopolymers **15**, 2439 (1976)
139. B.P. Gabor, W.L. Peticolas: Biochim. Biophys. Acta **465**, 260 (1977)
140. B.J. Bulkin, A. Krishnamachan: J. Am. Chem. Soc. **94**, 1109 (1972)
141. J. Thomas, S.L. Hsu, D.A. Tirrell: New J. Chem. **18**, 407 (1994)
142. R.G. Snyder, S.L. Hsu, S. Krimm: Spectrochim. Acta A **34**, 395 (1978)
143. R.G. Snyder, J.R. Scherer: J. Chem. Phys. **71**, 3221 (1979)
144. S. Abbate, G. Zerbi, S.L. Wunder: J. Phys. Chem. **86**, 3140 (1982)
145. B. Fanconi: Ann. Rev. Phys. Chem. **31**, 265 (1980)
146. A. Rubcic, G. Zerbi: Macromolecules **7**, 754 (1974)
147. B.G. Sumpter, D.L. Thompson: J. Chem. Phys. **86**, 2805 (1987)
148. G. Cardini, V. Schettino: J. Chem. Phys. **94**, 2502 (1990)
149. J.A. McCammon, S.C. Harvey: *Dynamics of proteins and nucleic acids* (Cambridge University Press, Cambridge 1987)
150. A.P. Scott: J. Phys. Chem. **100**, 16502 (1996)
151. J.H. Schachtschneider, R.G. Snyder: Spectrochim. Acta **21**, 1527 (1965)
152. J.H. Schachtschneider, R.G. Snyder: Spectrochim. Acta **19**, 117 (1963)
153. V.K. Hallmark, S.P. Bohan, H.L. Strauss, R.G. Snyder: Macromolecules **24**, 4025 (1991)
154. D.A. Cates, H.L. Strauss, R.G. Snyder: J. Phys. Chem. **98**, 4482 (1994)
155. H.J. Tao, W.J. MacKnight, K.D. Gagnon, R.W. Lenz: Macromolecules **28**, 2016 (1995)
156. J. Tang, A.C. Albrecht: in *Raman Spectroscopy: Theory and Practice*, Vol. 2, ed. by H.A. Szymanski (Plenum Press, New York 1970) p. 33

157. D.J. Lacks: J. Phys. Chem. **99**, 14430 (1995)
158. L.A. Nafie, W.L. Peticolas: J. Chem. Phys. **57**, 3145 (1972)
159. R.G. Gordon: J. Chem. Phys. **40**, 1973 (1964)
160. R.G. Gordon: J. Chem. Phys. **41**, 1819 (1964)
161. R.G. Gordon: J. Chem. Phys. **42**, 3685 (1965)
162. R.G. Gordon: J. Chem. Phys. **43**, 1307 (1965)
163. X. Yang, S.L. Hsu: Macromolecules **26**, 1465 (1993)
164. X. Yang, S.L. Hsu: Macromolecules **24**, 6680 (1991)
165. M.G. Northolt: Eur. Polym. J. **10**, 799 (1974)
166. W.F.X. Frank, H. Fiedler: Infrared Phys. **19**, 481 (1979)
167. D.Y. Shen, S.K. Pollack, S.L. Hsu: Macromolecules **22**, 2564 (1989)
168. H.S. Lee, Y.K. Wang, S.L. Hsu: Macromolecules **20**, 2089 (1987)
169. H.S. Lee, S.L. Hsu: Macromolecules **22**, 1100 (1989)
170. H.S. Lee, Y.K. Wang, W.J. MacKnight, S.L. Hsu: Macromolecules **21**, 270 (1988)
171. S. Yoon, K. Ichikawa, M.W.J., S.L. Hsu: Macromolecules **28**, 4278 (1995)
172. S. Yoon, K. Ichikawa, W.J. MacKnight, S.L. Hsu: Macromolecules **28**, 5063 (1995)
173. S. Yoon, W.J. MacKnight, S.L. Hsu: J. Appl. Polym. Sci. **21**, 197 (1997)
174. S. Yoon: in *Polymer Science and Engineering*, (University of Massachusetts, Amherst 1995)
175. X. Li, S.L. Hsu: J. Polym. Sci., Polym. Phys. Ed. **22**, 1331 (1984)
176. Y. Ren, C.W. Meuse, S.L. Hsu, H.D. Stidham: J. Phys. Chem. **98**, 8424 (1994)
177. Y. Ren, M.S. Shoichet, T.J. McCarthy, H.D. Stidham, S.L. Hsu: Macromolecules **28**, 358 (1995)
178. Y. Ren: in *Physics*, (University of Massachusetts, Amherst 1995)
179. K. Inomata, A. Abe: J. Phys. Chem. **96**, 7934 (1992)
180. R.L. Jaffe, G.D. Smith, D.Y. Yoon: J. Phys. Chem. **97**, 12745 (1993)
181. A. Abe, K. Tasaki: J. Mol. Struct. **145**, 309 (1986)
182. G.D. Smith, R.L. Jaffe, D.Y. Yoon: J. Am. Chem. Soc. **117**, 530 (1995)
183. S. Tsuzuki, T. Uchimaru, K. Tanabe, T. Hirano: J. Phys. Chem. **97**, 1346 (1993)
184. G.D. Smith, D.Y. Yoon, R.L. Jaffe, R.H. Colby, R. Krishnamoorti, L.J. Fetters: Macromolecules **29**, 3462 (1996)
185. S. Neyertz, D. Brown, J.O. Thomas: J. Chem. Phys. **101**, 10064 (1994)
186. S. Neyertz, D. Brown: J. Chem. Phys. **102**, 9725 (1995)
187. H. Liu, F. Muller-Plathe, W.F. van Gunsteren: J. Chem. Phys. **102**, 1722 (1995)
188. K. Tasaki, A. Abe: Polym. J. **17**, 641 (1985)
189. K. Tasaki: J. Am. Chem. Soc. **118**, 8459 (1996)
190. K. Tasaki: J. Am. Chem. Soc. **118**, 8459 (1996)
191. J. Koenig, A.C. Angood: J. Polym. Sci. (A-2) **8**, 1787 (1970)
192. A. Abe, K. Tasaki, J.E. Mark: Polym. J. **17**, 883 (1985)
193. R.G. Snyder, G. Zerbi: Spectrochim. Acta **A23**, 391 (1967)
194. J.R. MacCallum, C.A. Vincent: *Polymer Electrolyte Reviews*, Vol. 1 (Elsevier Applied Science, London 1987)
195. C.C. Lee, P.V. Wright: Polymer **23**, 681 (1982)

196. D.R. Payne: Polymer **23**, 690 (1982)
197. M.G. McLin, C.A. Angell: J. Phys. Chem. **95**, 9464 (1991)
198. L.M. Torell, S. Schantz: J. Non-Cryst. Solids **131**, 981 (1991)
199. M. Watanabe, J. Ikeda, I. Shinohara: Polym. J. **15**, 175 (1983)
200. M. Watanabe, J. Ikeda, I. Shinohara: Polym. J. **15**, 65 (1983)
201. M. Watanabe, N. Ogata: Brit. Polym. J. **20**, 181 (1988)
202. C. Vachon, M. Vasco, M. Perrier, J. Prud'homme: Macromolecules **26**, 4023 (1993)
203. H. Takahashi, T. Kyu, Q. Tran-Cong, O. Yano, T. Soen: J. Polym. Sci. B **29**, 1419 (1991)
204. B.L. Papke, M.A. Ratner, D.F. Shriver: J. Phys. Chem. Solids **42**, 493 (1981)
205. J. Maxfield, I.W. Shepherd: Polymer **16**, 505 (1975)
206. H. Matsuura, K. Fukuhara: J. Mol. Struct. **126**, 251 (1985)
207. J.L. Koenig, A.C. Angood: J. Polym. Sci. (A-2) **8**, 1787 (1970)
208. R. Frech, J. Manning, D. Teeters, B.E. Black: Solid State Ionics **28–30**, 954 (1988)
209. R. Frech, J. Manning, B.E. Black: Polymer **60**, 1785 (1989)
210. J. Manning, R. Frech, E. Hwang: Polymer **31**, 2245 (1990)
211. J. Manning, R. Frech: Polymer **33**, 3487 (1992)
212. R. Bergman, C. Svanberg, D. Andersson, A. Brodin, L.M. Torell: J. Non-Cryst. Solids **235**, 225 (1998)
213. S. Schantz, L.M. Torell, J.R. Stevens: J. Appl. Phys. **64**, 2038 (1988)
214. T. Hirano, T. Miyajima: J. Mol. Struct. **126**, 141 (1985)
215. T. Miyajima, T. Hirano, H. Sato: J. Mol. Struct. **125**, 97 (1984)
216. K. Ichikawa, L.C. Dickinson, W.J. MacKnight, M. Watanabe, N. Ogata: Polymer **33**, 4699 (1992)
217. K. Ichikawa, W.J. MacKnight: Polymer **33**, 4693 (1992)
218. W.J. Walczak, D.A. Hoagland, S.L. Hsu: Macromolecules **25**, 7317 (1992)
219. W. Walczak, D.A. Hoagland, S.L. Hsu: Macromolecules **29**, 7514 (1996)
220. K.M. Gough: J. Chem. Phys. **91**, 2424 (1989)
221. W.F. Murphy, J.M. Fernandez-Sanchez, K. Raghavachari: J. Phys. Chem. **95**, 1124 (1991)
222. D. Steele: *Theory of Vibrational Spectroscopy* (W.B. Saunders Co., Philadelphia 1971)
223. Y. Muroga, I. Noda, M. Nagasawa: Macromolecules **18**, 1576 (1985)
224. J.A. Ewen, R.L. Jones, A. Razavi, J.D. Ferrara: J. Am. Chem. Soc. **110**, 6255 (1988)
225. G. Natta, P. Corradini: Nuovo Chim. Suppl. **15**, 40 (1960)
226. G. Natta, M. Peraldo, G. Allegra: Makromol. Chem. **75**, 215 (1964)
227. T. Asakura, A. Aoki, D.T., M. Demura, T. Asanuma: Polym. J. **28**, 24 (1996)
228. Y. Chatani, H. Maruyama, T. Asanuma, T. Shiomura: J. Polym. Sci. B Polym. Phys. **29**, 1649 (1991)
229. P. Sozzani, R. Simonutti, A. Comotti: Mag. Res. Chem. **32**, S45 (1994)
230. T. Nakaoki, Y. Ohira, H. Hayashi, F. Horii: Macromolecules **31**, 2705 (1998)
231. T.D. Hahn, W. Suen, H.D. Stidham, A. Seidle, S.L. Hsu: Polymer (in press)
232. M. Peraldo, M. Cambini: Spectrochim. Acta **21**, 1509 (1965)
233. A. Zambelli, G. Natta, I. Pasquon: J. Polym. Sci. C **4**, 411 (1963)

234. J.J. Boor, E.A. Youngman: J. Polym. Sci (A-1) **4**, 1861 (1966)
235. G. Masetti, F. Cabassi, G. Zerbi: Polymer **21**, 143 (1980)
236. J.M. Chalmers: Polymer **18**, 681 (1977)
237. G. Allegra, P. Ganis, P. Corradini: Makromol. Chem. **61**, 225 (1963)
238. U.W. Suter, P.J. Flory: Macromolecules **8**, 765 (1975)
239. U. Biskup, H.-J. Cantow: Macromolecules **5**, 546 (1972)
240. W.L. Mattice, U.W. Suter: *Conformational Theory of Large Molecules. The Rotational Isomeric State Model in Macromolecular Systems* (Wiley, New York 1994)
241. P.J. Flory, J.E. Mark, A. Abe: J. Am. Chem. Soc. **88**, 639 (1966)
242. G.S. Cieloszyk, M.T. Cruz, G.L. Salinger: Cryogenics **13**, 718 (1973)
243. R.B. Stephens: Phys. Rev. B **13**, 852 (1976)
244. D.C. Prevorsek, P.J. Harget, R.K. Sharma, A.C. Reimscheussel: J. Macromol. Sci.: Phys. Ed. B **8**, 127 (1973)
245. E.W. Fisher, H. Goddar: J. Polym. Sci. C **16**, 4405 (1969)
246. E.S. Clark, L.S. Scott: Polym. Eng. Sci. **14**, 682 (1974)
247. J. Clements, R. Jakeways, I.M. Ward: Polymer **19**, 639 (1978)
248. A.G. Gibson, G.R. Davies, I.M. Ward: Polymer **19**, 683 (1978)
249. A.E. Zachariades, W.T. Mead, R.S. Porter: in *Ultra-high modulus polymers*, ed. by A. Ciferri, I.M. Ward (Applied Sciences Publishers, London 1977)
250. A.H. Windle: in *Developments in Oriented Polymers*, Vol. 1, ed. by I.M. Ward (Applied Sciences Publishers, London 1982)

IX C.V. Raman: A Personal Note

Samuel Krimm

As I stepped off the plane in Bangalore in the early morning of September 20, 1962, there, much to my surprise, was Raman himself standing beside his chauffeured car to welcome me on my visit to the Raman Research Institute. I was on my way from presenting an invited paper at the International Symposium on Molecular Structure and Spectroscopy in Tokyo to spend a year-long sabbatical in Cambridge, England, and my request to visit with him was warmly responded to by Raman, not only with his invitation but also with the comment that he "had the pleasure of seeing" my 1960 review article on "Infrared Spectra of High Polymers." In fact, his hospitality was warm and gracious, including lunch (prepared by his charming wife) and the tour of his (largely empty) Institute and its well-groomed gardens.

His tone cooled markedly when I mentioned that I had arranged to visit with S. Bhagavantum at the Indian Institute of Science that afternoon (relations between the two Institutes, I later found out, being somewhat strained). But he became quite heated the next day when, in my ignorance of the prior interactions between the two, I asked his opinion of Max Born's ideas on lattice dynamics! The temperature returned to normal as he discussed the importance of applying "My Effect" to the study of the vibrational spectra of polymers, to which I was trying to bring a more fundamental understanding.

The forthcoming advent of the laser light source initiated a new era in the Raman spectral studies of polymers, both synthetic and biological, since such spectra provided important information complementary to that given by infrared spectra. Together with the detailed insights into vibrational modes resulting from normal mode analyses, polymer spectroscopy began approaching the maturity achieved in the spectroscopic analysis of small molecules.

A recent vibrational study of alpha-helical poly (L-alanine) provides a good example of the results of such developments: polarized Raman spectra have led to the clear identification of E_2-species (only Raman-active) modes [1], whose assignment is crucial to a complete spectral analysis; and an *ab initio* based normal mode analysis has resulted in the reproduction of all observed (and well-assigned) bands above $200\,cm^{-1}$ to less that $5\,cm^{-1}$ [2]. With the introduction of spectroscopically-derived potential energy functions to provide accurate normal mode frequencies [3], and the possibility that appropriate electrostatic models can predict reliable intensities [4], it is clear that the Raman spectroscopy of polymers will flourish in its role of contributing to a deeper understanding of the physical properties of these materials.

References

1. S.-H. Lee, S. Krimm: J. Raman Spectrosc. **29**, 73 (1998)
2. S.-H. Lee, S. Krimm: Biopolymers **46**, 283 (1998)
3. K. Palmo, N.G. Mirkin, S. Krimm: J. Phys. Chem. A **102**, 6448 (1998)
4. K. Palmo, S. Krimm: J. Comput. Chem. **19**, 754 (1998)

10 Raman Scattering in Perovskite Manganites

V.B. Podobedov and A. Weber

Abstract. Raman spectroscopy is shown to be a useful tool for the study of phase transitions in lanthanum manganese compounds of the perovskite type. In this work doped and undoped $La_{1-x}R_xMnO_3$ (R = Ca, Sr) single crystals and thin films deposited on $LaAlO_3$ and $NdGeO_3$ substrates are studied in polarized Raman scattering as functions of doping ($0 < x < 1$) and temperature ($5 < T < 423\,K$). Raman spectra of single crystals exhibit a significant dependence on the amount and kind of the dopant R, and they are sensitive to structural and magnetic transitions driven by both doping and temperature. Raman spectra of the doped $La_{1-x}R_xMnO_3$ crystals have weakly-resolved features and intensities comparable to those of second-order spectra in cubic crystals. The small distortion of the nearly cubic lattice of doped $La_{1-x}R_xMnO_3$ single crystals results in structured phonon Raman spectra composed of two parts. These are attributed 1) to the distorted non-cubic perovskite lattice and are consistent with the selection rules for a tetragonal structure, and 2) to the density of vibrational states manifested as second-order Raman scattering. An effect of spin-lattice interaction corresponding to a paramagnetic to canted anti-ferromagnetic phase transition was found in the Raman spectra of undoped $LaMnO_3$. A correlation of lattice transformation as well as magnetic ordering was found in thin films of $La_{1-x}Ca_xMnO_3$. Structural changes, preferential orientation of films as well an influence of deposition conditions can be ascertained from the phonon Raman spectra. The effect of spin-lattice interaction due to spin ordering below T_c was observed in spectra of $La_{1-x}Ca_xMnO_3$ films. A simple frequency dependence of the totally symmetric external mode with doping x was found for $La_{1-x}Sr_xMnO_3$ crystals. This dependence is demonstrated to be a useful diagnostic tool for determining the doping value in $La_{1-x}Sr_xMnO_3$ crystals. The results of the present work may be extended to other related manganite structures, in particular $La_{1-x}Sr_xMnO_3$ films.

The magnetoresistance effect has been known for over 100 years. The discovery of "Giant Magneto Resistance" (GMR) in 1986 [1] started a new era of research in magnetic materials in view of their potential incorporation in efficient technological devices, particularly those used in magnetic recording heads [2]. More recently it was found that another class of materials exhibits much greater magnetoresistance properties, and the effect was accordingly dubbed "Colossal Magneto Resistance" (CMR) [3]. Most representative of the CMR materials are the manganite perovskites $La_{1-x}R_xMnO_3$ where the lanthanide cation is trivalent (La^{3+}) and R is a divalent alkaline earth cation (R = Ca, Sr, Ba). The end members of this series, with $x = 0$ ($LaMnO_3$) and $x = 1$ (e.g., $CaMnO_3$) are antiferromagnetic insulators in their ground states. The CMR effect is accompanied by a ferromagnetic -to-paramagnetic phase transition and is greatest near the transition temperature T_c. The various aspects of CMR are described in a variety of publications and reviews [4]. While devices based on the GMR effect are already entering the market place, there are many obstacles that must be overcome before CMR materials are found that render them useful for commercial exploitation. There is, accordingly, a great amount of research underway to better understand and ultimately fabricate magnetic recording devices based on the CMR effect [5]. While much of the work is driven by potential applications, there is also considerable intrinsic interest in the study of the basic physics and chemistry of the doped perovskite manganites. Many different experimental techniques are used alongside theoretical models in this undertaking. Among these, the technique of Raman spectroscopy has proven itself to be of considerable importance in the study of solid-state materials in general. It is the purpose of this chapter to report on a series of Raman spectroscopic experiments on doped and undoped lanthanum manganite crystals as well as thin films.

The microstructural characteristics of the $La_{1-x}R_xMnO_3$ lattice, which may be effectively changed by the doping value [6] or by oxygen/lanthanum stoichiometry [7,8], essentially define the macroscopic properties of $La_{1-x}R_x$ MnO_3 compounds [9]. Both processing and annealing conditions, the film thickness, and film-substrate lattice mismatch introduce additional dependencies into the macroscopic and microscopic behaviors of thin-film structures. Different studies of $La_{1-x}R_xMnO_3$ compounds over a wide range of doping and temperature have demonstrated a variety of structural transitions (cubic, rhombohedral, tetragonal, orthorhombic and monoclinic crystal symmetries) [9–13]. Various magnetic phases (para- and ferromagnetic insulators and metals, canted antiferromagnetic insulators) have been intensively studied in the lanthanum manganese compounds by different techniques [9–12]. In particular, several studies [10,14] have demonstrated a strong dependence of the main lattice parameters and bond lengths over a wide range of temperature and doping. These studies found that due to related changes in the lattice [9], the principal parameters of the CMR materials, the Curie temperature T_c and magnetoresistance, strongly correlate with a microstructural Mn–O–Mn bond angle. Asamitsu et al. [15] have shown that the variation of an external magnetic field can control the structural phase transition in $La_{0.83}Sr_{0.17}MnO_3$ crystals when the magnetic moments and charge carriers are coupled to the crystal structure. An important role for the understanding of the bulk metallic and electronic properties of $La_{1-x}Sr_xMnO_3$ system is a strong electron–phonon coupling related to a Jahn–Teller splitting of the energy levels of the Mn^{3+} ion [16].

In addition to a basic physical interest related to electron–phonon, Coulomb and magnetic interactions, potential practical applications of $La_{1-x}R_x$ MnO_3 compounds, in particular, film structures [17–19], have stimulated the study of their properties by various techniques. The structure of materials, their lattice distortion, charge carrier concentration, and oxygen stoichiometry control the CMR properties of $La_{1-x}R_xMnO_3$ compounds. Due to their known polarization and spatial resolution capabilities, nondestructive optical spectroscopies (infrared, Raman) may play an important role for optimizing the properties of magnetoresistive materials. Most of the known infrared (IR) studies were carried out with ceramic samples and with small spectral resolution [20,21]. Raman scattering (RS) studies of the $La_{1-x}Sr_xMnO_3$ systems [13,22,23], as well as studies of closely related perovskite-like ferroelectrics [24–26], provide information that complements that obtained from IR investigations.

Optical spectra of phonons, including those observed in RS are known to be sensitive to many of the above lattice effects. In view of the great attention given to this system, any information about optical phonons in the $La_{1-x}R_xMnO_3$ materials is of current interest. However, in perfect cubic perovskite structures the first-order RS is forbidden by the selection rules. On the other hand, small distortions of the nearly-cubic LSMO/LCMO structures lower the symmetry and first-order RS becomes allowed, although its intensity is still strongly dependent on the amount of distortion of the perovskite unit cell. As a rule, RS in such doped lanthanum manganese materials is extremely weak. In our studies of single crystals [13,23], it was found that Raman spectra of manganites usually contain three different comparable contributions: sharp features that follow the selection rules for first-order RS, a relatively broad second-order RS, and electronic RS.

In this chapter we concentrate on Raman studies of the $La_{1-x}R_xMnO_3$ (R = Sr, Ca) systems. These are designated LSMO or LCMO for R = Sr or Ca, respectively. The experimental data shown are mainly for Sr-doped single crystals and Ca-doped films, although samples doped with Pb as well as the spectra for $Nd_{1-x}Sr_xMnO_3$ crystals were found to have some common Raman features as well. This chapter is organized as follows. In Sect. 10.1 we briefly describe the Raman technique used in the experiments and present the basic considerations for obtaining the selection rules for the Raman spectra of the $La_{1-x}R_xMnO_3$ system. Section 10.2 considers Raman scattering in $La_{1-x}Sr_xMnO_3$ single crystals ($x = 0.1, 0.2$ and 0.3). At these levels of doping the LSMO materials have a nearly cubic structure. However, as mentioned above, the existing lattice distortion gives rise to first-order Raman scattering. Section 10.3 deals with the undoped $LaMnO_3$ single crystal which undergoes a paramagnetic to canted-antiferromagnetic transition at the Néel temperature $T_N = 140\,K$. Here we show that some Raman features of this material can also be found in the spectra of doped LSMO and LCMO compounds. In Sect. 10.4 we emphasize the use of Raman spectroscopy for nondestructive optical characterization of LCMO films. A correlation between Raman data and resistivity and magnetic measurements is demonstrated.

10.1 Manganite Structure and Selection Rules for Optical Vibrational Modes

The perovskite-type materials may be described by the general formula ABO_3. Many perovskite materials were previously the subject of extensive spectroscopic studies in Raman and IR, mostly due to their ferroelectric properties [24–26]. The structure of these materials involves a network of corner sharing oxygen octahedra. The metal ion B is located in the center of an octahedron, while the octahedral oxygens themselves are located in the edges of a cube consisting of A cations. In the case of the colossal magnetoresistive $La_{1-x}R_xMnO_3$ manganites, the manganese occupies the B-site while the rare earth cation (La^{3+} and doping material R) occupies the A-site. Two well-known perovskite structures are shown in Fig. 10.1: (a) the ideal structure with a perfect cubic symmetry O_h^1 and (b) the distorted orthorhombic structure D_{2h}^{16}. Since the parent material $LaMnO_3$ is an antiferromagnetic insulator at room temperature, the paramagnetic-to-ferromagnetic transition can be induced by proper doping of the system. Usually this is accomplished by replacing the La^{3+} ion by alkaline earth elements, mostly Sr or Ca. The values of the CMR are, typically, largest for doping values in the range from $x = 0.2$ to $x = 0.4$, whereas the end members of the $La_{1-x}R_xMnO_3$ series (with $x = 0.0$ and $x = 1.0$) remain antiferromagnetic insulators at low temperatures [6].

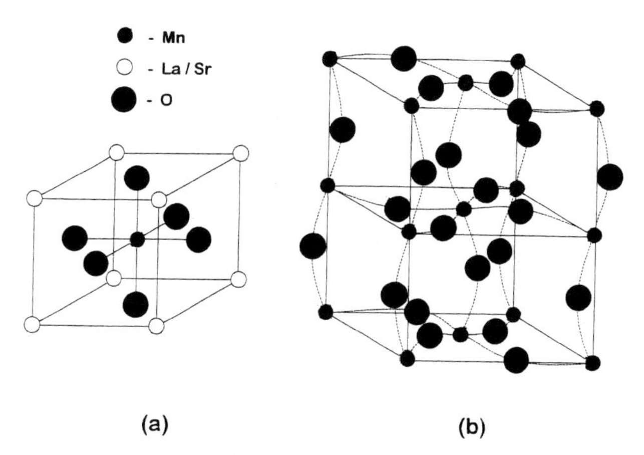

Fig. 10.1. Ideal **(a)** and distorted orthorhombic **(b)** perovskite structures of $La_{1-x}Sr_xMnO_3$

The $La_{1-x}R_xMnO_3$ system may exist in one of several phases: paramagnetic and ferromagnetic insulators and metals, canted antiferromagnetic insulator and antiferromagnetic insulator phases, depending on the doping value x, and the temperature. A graphic depiction of these phases and the possible transitions among them are captured in a phase diagram. The magnetic phase diagrams for $La_{1-x}R_xMnO_3$ have been reported in [4c] and [9]. Moreover, early structural studies of manganite and related systems, mostly powders or ceramics, have shown that this material may have cubic [21,27], rhombohedral [10,15,28] or orthorhombic [29–31] symmetries depending on the temperature and value of doping x. The activities of vibrational modes in optical spectra are obtained by the methods of group theory. For the three commonly reported symmetries of $La_{1-x}R_xMnO_3$ materials these are summarized below.

The perfect perovskite structure belongs to the cubic space group O_h^1 ($m3m$), and a unit cell contains one formula unit. The optical vibrations are classified as

$$\Gamma = 3F_{1u}(\mathrm{IR}) + F_{2u} \,,$$

where the three F_{1u} species are "allowed" (active) in infrared (IR) absorption while the F_{2u} species are inactive ("forbidden") in both IR and Raman Scattering (RS). The IR-active modes of $La_{1-x}R_xMnO_3$ materials were analyzed within this assumed symmetry [20,21]. No Raman modes are allowed in the first-order spectra. However, the crystal lattice of most real $La_{1-x}R_xMnO_3$ materials is distorted, for example due to the Jahn–Teller (JT) and other effects. The existing lattice distortions therefore permit the appearance of a first-order Raman spectrum.

At moderate values of doping, e.g., $x \approx 0.3$, the rhombohedral symmetry D_{3d}^6 ($R\overline{3c}$) is expected. For this structure the unit cell contains two formula units. Therefore the following activity of optical vibrational modes is expected:

$$\Gamma = A_{1g}(\mathrm{RS}) + 4E_g(\mathrm{RS}) + 3A_{2g} + 2A_{1u} + 3A_{2u}(\mathrm{IR}) + 5E_u(\mathrm{IR}) \,.$$

So far Raman spectra for the rhombohedral structure of $La_{1-x}R_xMnO_3$ single crystals have been reported only in [51]. However no significant difference with the spectra from the orthorhombic phase was found.

The symmetry of the parent $LaMnO_3$ material, as well as those lightly doped ($x < 0.125$) is usually described by the space group $D_{2h}^{16}(mmm)$. In this case the unit cell has four formula units and the optical vibrations are classified as

$$\begin{aligned}\Gamma = {} & 7A_g(\mathrm{RS}) + 5B_{1g}(\mathrm{RS}) + 7B_{2g}(\mathrm{RS}) + 5B_{3g}(\mathrm{RS}) \\ & + 8A_u + 9B_{1u}(\mathrm{IR}) + 7B_{2u}(\mathrm{IR}) + 9B_{3u}(\mathrm{IR}) \,.\end{aligned}$$

Accordingly, the Raman spectrum is expected to be composed of seven A_g, five B_{1g}, seven B_{2g}, and five B_{3g} modes. Since the early study by Geller [30],

$LaMnO_3$ is known to possess a distorted perovskite orthorhombic structure with both $\sqrt{2} \times \sqrt{2}$ doubling of the unit cell in the a-c plane and additional doubling along the b axis. This structure was studied by Raman spectroscopy in [22] and [23].

The $La_{1-x}R_xMnO_3$ single crystals used in this study were prepared by the floating-zone technique. Dispersive X-ray analysis, X-ray diffraction and microwave absorption studies [32] were used to characterize the samples. The crystallographic axes of the samples were determined from the above X-ray study as well as by scattering polarization properties in the present Raman measurements. The film samples studied had a thickness $d \approx 270\,\mathrm{nm}$ and were prepared by the laser ablation technique by sputtering of the LCMO ceramic target, followed by an annealing in oxygen at $800\,°\mathrm{C}$ [33]. Film samples were deposited on two types of substrates, (001) : $LaAlO_3$ (LAO) and (001) : $NdGaO_3$ (NGO).

The samples were placed in the center of a superconducting magnet/cryostat operating over the ranges of temperature from $T = 4.3\,\mathrm{K}$ to $350\,\mathrm{K}$ and magnetic field from $H = 0\,\mathrm{T}$ to $8\,\mathrm{T}$ [34]. Except when mentioned otherwise, the Raman spectra were excited by the radiation from an Ar^+ ion laser ($\lambda = 514.5\,\mathrm{nm}$). The excitation power was kept below $50\,\mathrm{mW}$ to avoid both possible sample damage and the uncertainty in its temperature due to heating. It was found from the Stokes/anti-Stokes Raman intensity measurements that at this power the local temperature increase did not exceed $10\,°\mathrm{C}$ for well-polished $La_{1-x}R_xMnO_3$ crystals or films. Different polarization geometries were produced by means of a half-wave silica plate and a wide-band polarizer. A triple-stage multichannel spectrometer equipped with a CCD detector operating at $140\,\mathrm{K}$ was used to obtain the Raman spectra. The spectral resolution was typically set at about $4\,\mathrm{cm}^{-1}$. Although most of the spectra were obtained in a back-scattering geometry, the weakest Raman spectra were detected using grazing incidence excitation and a 90°-scattering geometry to reduce Raman contribution from the cryostat quartz windows to the recorded signal. In nearly all experiments (see, however, Sect. 10.3, Fig. 10.15) the laser beam was first reflected off a plane diffraction grating, which spatially dispersed the radiation emitted by the laser plasma discharge. Thus only the selected line, usually the Ar laser line at 514.5 nm, was allowed to irradiate the sample and produce Raman spectra. The spectra thus generated were free from any false lines due to Rayleigh scattering of the incoherent plasma radiation.

10.2 Doped Crystals ($x > 0$)

As noted above, a small distortion from the cubic perovskite crystal lattice results in a complex structure of the Raman spectra of $La_{1-x}R_xMnO_3$ single crystals. We note three different contributions to the intensity of the Raman spectra, namely those due to first-order RS, second-order RS, and electronic

RS [35,36]. As will be shown below, the first part follows the selection rules for vibrational transitions (phonons). It depends on the symmetry of the distorted crystal lattice. The second contribution is due to the density of vibrational states and usually exhibits broad-band features. The third contribution, however, is an electronic RS that usually accompanies the phonon spectra and contributes to the total intensity as a continuous background. For more details on electronic RS in $La_{1-x}R_xMnO_3$ materials, we address the reader to [35] and [36].

In Fig. 10.2 the Raman spectrum of a $La_{1-x}Sr_xMnO_3$ single crystal ($x = 0.2$, $T = 293\,K$) is represented by curve 1. The xx scattering geometry was chosen since this spectrum contains all the features typical for Raman spectra of $La_{1-x}Sr_xMnO_3$ single crystals. The abrupt intensity drop below $50\,cm^{-1}$ in spectrum 1 is due to the chosen wavelength setting of the premonochromator. The possible contribution of parasitic elastically scattered light to the low-frequency part of Raman spectrum may be then estimated as follows. Since below $50\,cm^{-1}$ this contribution is found to be less than 5% of the maximum intensity, most of the spectrum above this frequency has to be due to inelastic scattering in the $La_{0.8}Sr_{0.2}MnO_3$ crystal. Curve 2, which represents the imaginary part of the Raman susceptibility, $\mathrm{Im}\,\chi$, is obtained from spectrum 1 by normalization with the thermal factor $[1 + n(\omega)] = [1 - \exp(\hbar\omega/k_BT)]^{-1}$. Note that the features of this spectrum were found to be typical for many other $La_{1-x}R_xMnO_3$ samples (Fig. 10.3) as well as for $La_{1-x}Ca_xMnO_3$ films (Sect. 10.4). From Figs. 10.2 and Fig. 10.3 it follows that the Raman spectrum of a $La_{1-x}Sr_xMnO_3$ single crystal may be fitted by a superposition of three components: few relatively sharp and weak peaks, an intense wide band spanning a region of $\sim 700\,cm^{-1}$, and a relatively flat background extending into the high-frequency region. As mentioned above, this third component should be assigned to electronic RS in $La_{1-x}Sr_xMnO_3$ materials [35] and is presented in Fig. 10.2 by the horizontal dotted line 3.

The sharp peaks located at the top of the wide band have different polarizations and may be separated in different scattering geometries [13]. These sharp peaks are located in three frequency regions: $180\,cm^{-1}$ to $300\,cm^{-1}$, $400\,cm^{-1}$ to $520\,cm^{-1}$, and $580\,cm^{-1}$ to $680\,cm^{-1}$ (Fig. 10.4). The frequency positions of the Raman bands are somewhat higher than those of the IR F_{1u} modes [20,37]. However for three different dopants R = Ca, Sr and Pb, the main Raman features are very similar. A full set of polarized Raman spectra was obtained in order to define a possible mode assignment. Polarized spectra for a $La_{0.7}Sr_{0.3}MnO_3$ crystal are shown in Fig. 10.5. The bands located at $190\,cm^{-1}$ and $435\,cm^{-1}$ were found to be the most representative for the spectrum of doped $La_{0.7}Sr_{0.3}MnO_3$ single crystals. These bands can be well separated in the $x'y'$ and $y'y'$ scattering geometries (the prime symbol $'$ denotes a 45° rotation of the sample in the xy-plane) as may be seen from Fig. 10.5. The observed polarization dependence of the two main bands

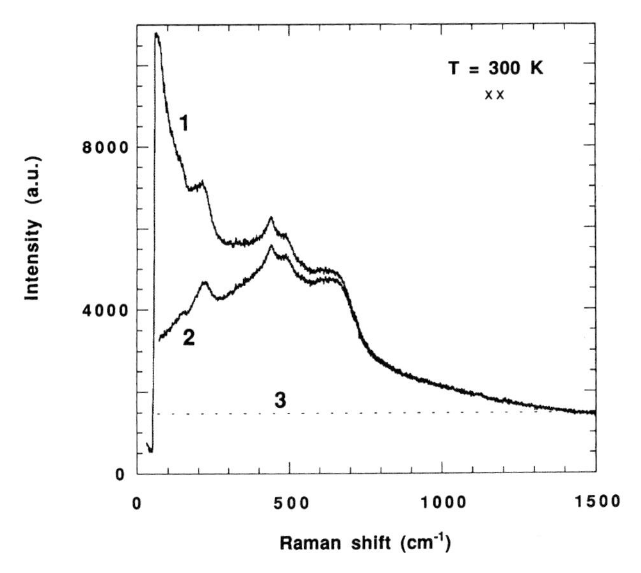

Fig. 10.2. Raman spectrum (1), Raman susceptibility (2) of $La_{0.8}Sr_{0.2}MnO_3$ single crystal. The *dotted line* (3) indicates the contribution of electronic scattering

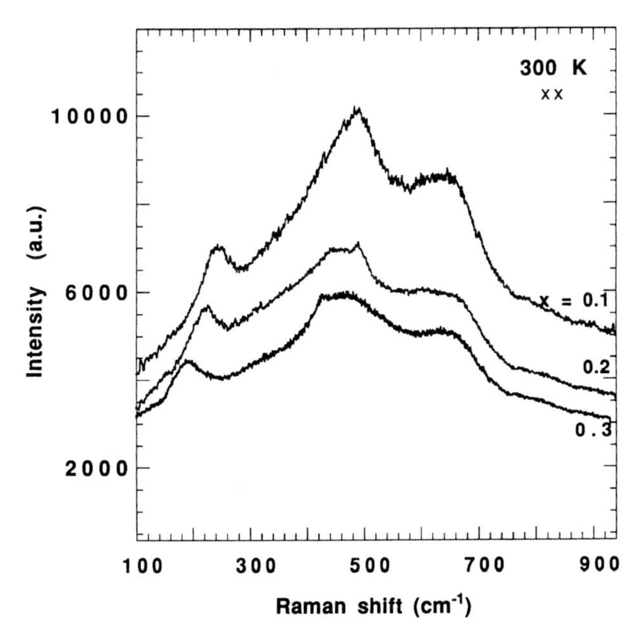

Fig. 10.3. Raman spectra of $La_{1-x}Sr_xMnO_3$ single crystals with different values of doping, x. The spectra are corrected by the thermal factor $[1 + n(\omega)] = [1 - \exp(\hbar\omega/k_BT)]^{-1}$

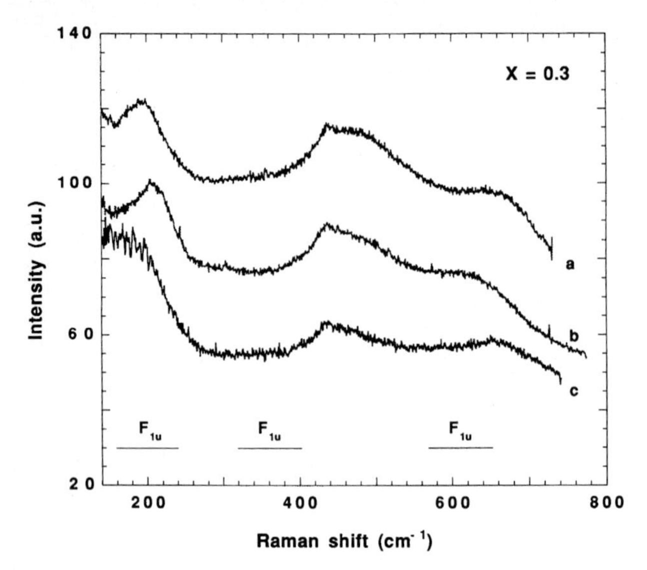

Fig. 10.4. Raman spectra of manganite single crystals with the same value of doping, $x = 0.3$: a – $La_{0.7}Sr_{0.3}MnO_3$, b – $La_{0.7}Ca_{0.3}MnO_3$, c – $La_{0.7}Pb_{0.3}MnO_3$. The positions of the three IR-active vibrational modes, F_{1u}, are indicated by horizontal lines in the bottom of the figure

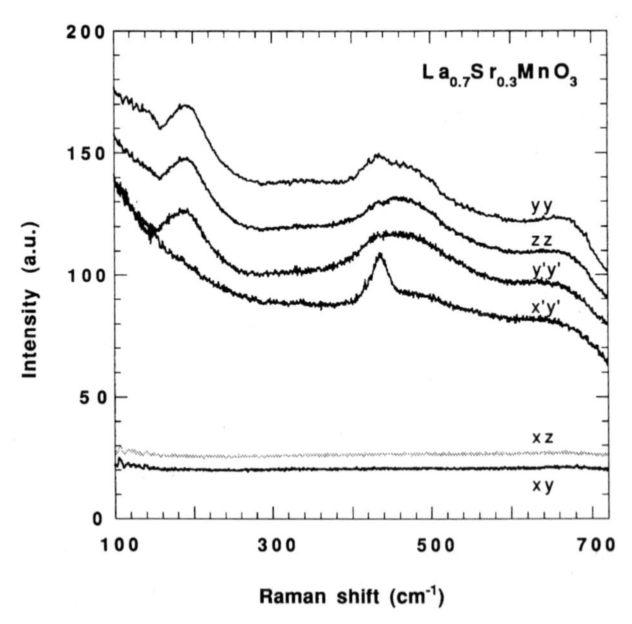

Fig. 10.5. Polarized Raman spectra of $La_{0.7}Sr_{0.3}MnO_3$ single crystal in different scattering geometries

demonstrates their different symmetries. Another band at about 450 cm^{-1} usually is more intense in parallel scattering geometries whereas the remaining two features at 145 cm^{-1} and 650 cm^{-1} exhibit weak polarization dependence in the xy-plane. In the xy and xz geometries the Raman spectra have the smallest intensity. According to previous neutron studies [9,10,28], $La_{1-x}Sr_xMnO_3$ compounds must possess the rhombohedral (D_{3d}^6) symmetry at a doping value of $x \approx 0.3$. From a factor group analysis one would therefore expect one A_{1g} and four E_g vibrational modes to be Raman active (Sect. 10.1). The observed spectra, however, are more consistent with a tetragonally-distorted structure, which would account for the known polarization properties of the corresponding Raman tensor components. It is interesting to note that in the closely related study of a $BaCeO_3$ perovskite compound [38,39] the Raman spectra were also found to be more consistent with a tetragonal (D_{4h}^5) space group, although a disagreement with neutron diffraction data was noted in [40]. Within the tetragonal symmetry, the bands at 190 cm^{-1} and 435 cm^{-1} definitely exhibit the polarization properties of A_{1g} and B_{1g} vibrational modes, respectively. On the other hand, the number of observed spectral peaks and their polarization properties are not consistent with E_g modes, which are typical for rhombohedral D_{3d}^6 symmetry. A similar polarization behavior of the vibrational modes typical for tetragonal lattice symmetry was also observed in a $Nd_{0.7}Sr_{0.3}MnO_3$ single crystal (Fig. 10.6). The preference of selecting a tetragonal structure against a rhombohedral one is motivated by the presence of the B_{1g} mode (435 cm^{-1}), the absence

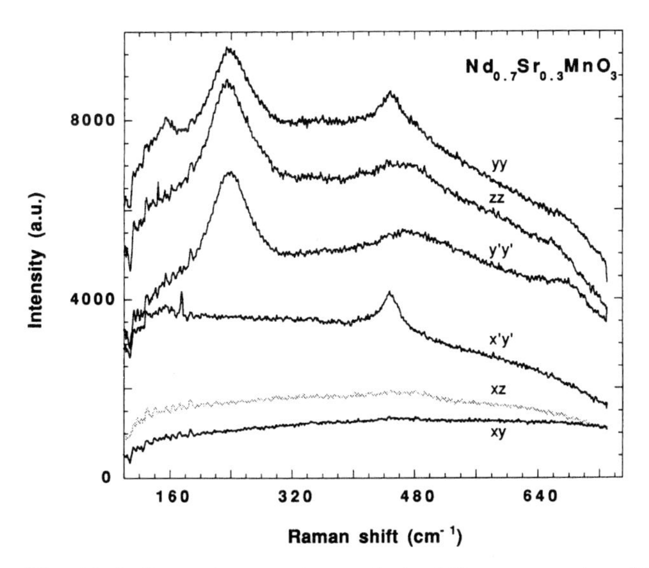

Fig. 10.6. Room temperature polarized Raman spectra of $Nd_{0.7}Sr_{0.3}MnO_3$ single crystal in different scattering geometries

of E_g modes, as well as by the specific polarization dependence of spectra in different scattering geometries (Figs. 10.5 and 10.6).

The main contribution to the total RS intensity in $La_{1-x}Sr_xMnO_3$ crystal is the broad band covering a range of about 700 cm^{-1} as seen from Figs. 10.2 and 10.3. It is very likely that this band is mainly due to second-order RS arising in the rhombohedral (pseudo-cubic [10] or cubic O_h^1) perovskite disordered structures typical for doped $La_{1-x}Sr_xMnO_3$ materials. The first-order scattering discussed above may therefore be explained by a small distortion of a lattice structure, one that leads to comparable intensities of the relatively weak first- and second-order spectra in nearly cubic perovskite compounds. For this reason, therefore, we consider the structure of phonon Raman spectra of doped $La_{1-x}Sr_xMnO_3$ single crystals to be composed of two parts. The first one is related to the distorted non-cubic perovskite structure and follows the selection rules for the first-order RS in tetragonal structure. The second part is mainly due to the density of vibrational states and is classified as second-order RS in a nearly cubic medium. The coexistence of both first- and second order RS contributions and their possible separation by hyper-Raman spectroscopy was described in [41].

Low-frequency Raman spectra of three $La_{1-x}Sr_xMnO_3$ samples with different values of doping are presented in Fig. 10.7. For optimum display, all

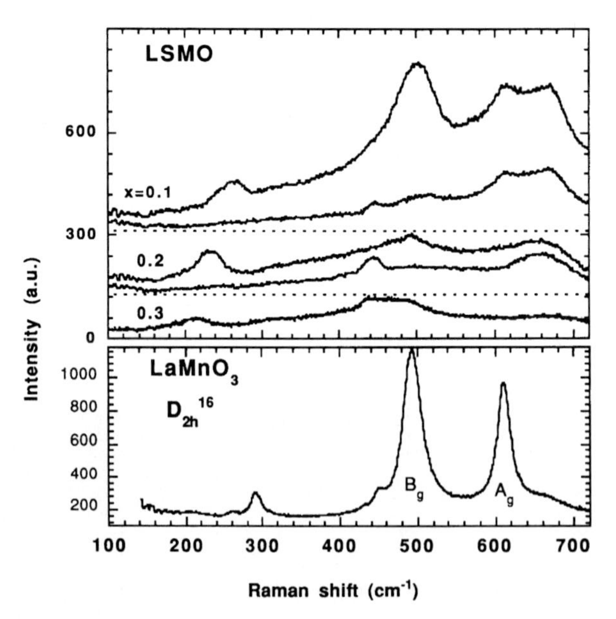

Fig. 10.7. Raman spectra of $LaMnO_3$ single crystal (*bottom panel*) and polarized spectra of $La_{1-x}Sr_xMnO_3$ single crystal with different value of doping, x. (*upper panel*). Upper and bottom spectra for each value of x correspond to the $x'x'$ and $x'y'$ scattering geometries, respectively. For $x = 0.3$ the spectra for the yy scattering geometry is shown

spectra are shifted in the vertical direction and have the same intensity scale. The Raman spectrum of a $LaMnO_3$ crystal is shown together with a set of spectra of doped samples for comparison. Note that the features related to the D_{2h}^{16} orthorhombic structure (the bands at $493\,cm^{-1}$ and $609\,cm^{-1}$), typical for undoped systems, are gradually decreasing in intensity with increasing value of the doping x. We attribute this phenomenon to the gradual relative decrease of the orthorhombic lattice distortion due to reduction of the JT effect in doped materials. The smaller intensity of RS at higher values of x indicates that $La_{1-x}Sr_xMnO_3$ compounds tend toward a cubic structure. In the spectrum of $La_{0.9}Sr_{0.1}MnO_3$ (Fig. 10.7, upper panel) the intensities of the bands at $493\,cm^{-1}$ and $609\,cm^{-1}$, typical for the D_{2h}^{16} orthorhombic symmetry, are weak, but still stronger than those in highly doped crystals. No new bands typical only for the rhombohedral phase, for example, E_g modes, were detected in the experimental spectra covering the transition from the orthorhombic ($x = 0$) to the proposed rhombohedral one in doped $La_{1-x}Sr_xMnO_3$ crystals. This may be expected if the corresponding rhombohedral distortion of the structure were not big enough. As was found from the neutron study [10], the lengths of the Mn–O bonds in this case become almost equal. Nevertheless, Raman spectra at higher doping still display a reduction of the total scattering intensity, a change of the band shape and an anomalously large shift of the band associated with the A_{1g} mode with both temperature and doping (Figs. 10.8 and 10.9). At this point, we note that

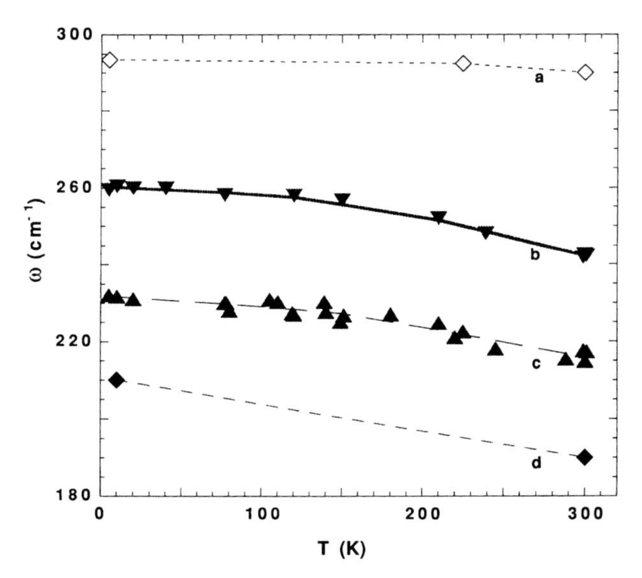

Fig. 10.8. Temperature dependence of the position of A_{1g} mode in Raman spectra of $La_{1-x}Sr_xMnO_3$ single crystals: curves a, b, c and d correspond to $x = 0, 0.1, 0.2$ and 0.3, respectively

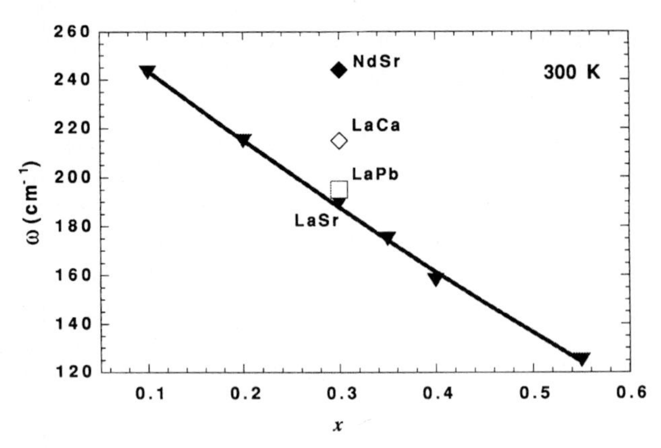

Fig. 10.9. Raman shift of the A_{1g} mode in $La_{1-x}Sr_xMnO_3$ single crystals as a function of doping value x. Positions of the same modes in other manganites with $x = 0.3$ are shown for comparison

the frequency of this band (Fig. 10.9) may be used for a fast non-destructive measurement of the doping value x in a $La_{1-x}Sr_xMnO_3$ system (see below).

As known from the classification of the phonon modes in cubic perovskite structures [20,21,24], the low-energy part of the vibrational spectrum is usually assigned to external vibrations. In the case of the $La_{1-x}R_xMnO_3$ system, these vibrations involve the motion of the La and Sr (or other doping ions) with respect to MnO_6 octahedra. In [38] a similar conclusion was drawn in a Raman study of the $BaCeO_3$ perovskite system. The main features observed in the present experiments are consistent with the same assignment of the A_{1g} Raman mode since its position was found to be sensitive to both the amount (Fig. 10.8) and kind (Fig. 10.9) of the dopant. Nevertheless a complete lattice dynamics calculation of the $La_{1-x}R_xMnO_3$ system would be necessary for an unambiguous assignment of the Raman spectrum.

As shown in Fig. 10.9, the frequency change of the above A_{1g} mode, $[243\,\mathrm{cm}^{-1}$ (at $x = 0.1$) – $125\,\mathrm{cm}^{-1}$ (at $x = 0.55)] = 118\,\mathrm{cm}^{-1}$, has the opposite sign to what might be expected from the difference in the atomic weights of Sr(87) and La(139). Such behavior is unusual in terms of considerations appropriate to the mixed crystals [42]. A possible reason for the $118\,\mathrm{cm}^{-1}$ softening of the A_{1g} mode may be given in terms of the specific interaction between La and Sr ions and the MnO_6 octahedron. A strong force constant describing this interaction may overcompensate the mass effect and thus explain the anomalous shift. The La and Sr ions occupy the sites between the oxygen octahedra and the ionic radius of this site, $\langle r_A \rangle$, increases with Sr doping in $La_{1-x}Sr_xMnO_3$ compounds [43]. The overlap of the Mn-site d orbitals and oxygen p orbitals was found to be strongly affected by the internal pressure generated by the La/Sr-site, and a wide-ranging dependence of the Curie temperature T_C on the change of ionic radius of a La site $\langle r_A \rangle$ was demonstrated [9]. We suggest that this phenomenon is also responsible for the decrease of the force constant in the La/Sr–MnO_6 system as well as for the corresponding shift of the related Raman mode with doping.

We now consider the large temperature shift of this band, which is essentially higher than what would be expected from the usual temperature-dependent anharmonic effect. For all doped crystals this change is approximately $20\,cm^{-1}$ in the temperature range from 5 K to 300 K (Fig. 10.8). Assuming the validity of the above argument, the shift of $\sim 20\,cm^{-1}$ can also be explained by a temperature dependence of the ionic radius $\langle r_A \rangle$. When the value of doping x is a constant, an additional anharmonic effect is also present and this results in the stronger reduction of the Raman shift with temperature.

A suggested practical application of the foregoing may be a fast optical determination of the doping value x in the $La_{1-x}Sr_xMnO_3$ system. This involves the measurement of the Raman shift of the low-frequency A_{1g} mode. This method is based on the observed change in the Raman displacement, $\Delta\omega = 118\,cm^{-1}$, which corresponds to a change in the doping value $\Delta x = 0.45$. For this purpose, the curve in Fig. 10.9 may be used as a calibration guide. The position of the A_{1g} mode, at least over the range $0.1 < x < 0.55$, may be determined from the Raman spectrum with an uncertainty $\delta\sigma < 2\,cm^{-1}$. Therefore, the estimated uncertainty in the measured value of x should be of the order of $\Delta x(\delta\sigma/\Delta\omega)$, which is better than 7% even for $x = 0.1$. For the determination of x by this method it is necessary that the calibration curve be obtained from samples with well-defined x. The stoichiometric composition must be also well known since it may affect the vibrational spectrum. The influence of additional factors can be accounted for in the calibration procedure.

Raman spectroscopy as a non-destructive optical technique has another advantage that is often used for spatially-resolved analysis. The laser radiation may be focused into a diffraction-limited spot of a diameter $D = 4\lambda f/\pi d$ where λ is the wavelength, f is the focal length of the lens, and d is the diameter of the (collimated) laser beam. In the present case, the diameter of a laser beam was $d = 3$ mm and the focal length of focusing lens, $f = 100$ mm. Therefore, D is approximately 0.02 mm for $\lambda = 514.5$ nm. Since Raman spectra are collected exactly from the illuminated laser spot, this value corresponds to an in-plane spatial resolution. In Figs. 10.10 and 10.11 some Raman spectra from the surface of a $La_{0.8}Sr_{0.2}MnO_3$ single crystal are shown as a function of position of the laser beam on the crystal. Both sets of spectra correspond to a displacement of ± 1.5 mm from an almost invisible boundary line on the surface of the crystal. All spectra have the same intensity scale but were shifted for clarity in the vertical direction within each top and bottom sets. The obvious difference between both sets is the intensity of the peak at about $500\,cm^{-1}$ (Fig. 10.10). It was found that this peak is usually stronger in the orthorhombic phase of the $La_{1-x}Sr_xMnO_3$ system (see also Figs. 10.12 and 10.13). In

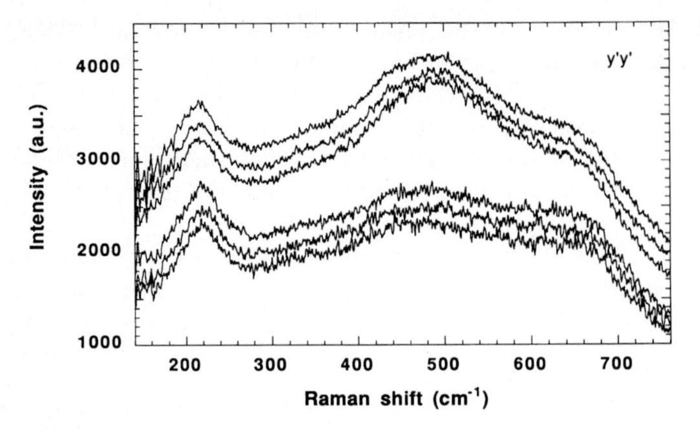

Fig. 10.10. Spatial dependence of Raman scattering from $La_{0.8}Sr_{0.2}MnO_3$ single crystal. Lower and upper sets of spectra correspond to a maximum displacement of ±1.5 mm from a selected boundary on the surface of crystal

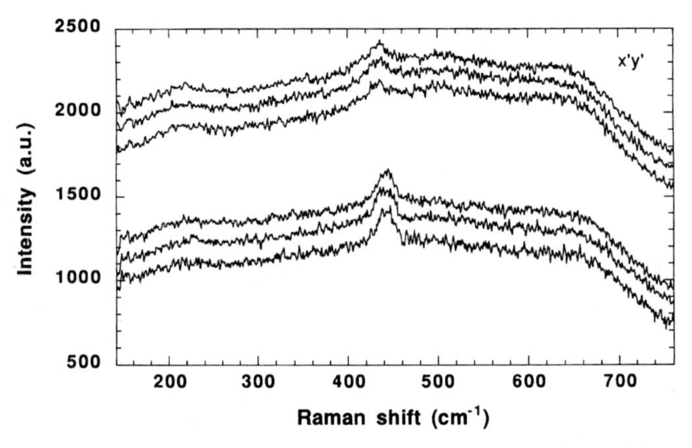

Fig. 10.11. The same notation as in Fig. 10.10. The $x'y'$ scattering geometry is shown

the crossed polarization shown in Fig. 10.11, one can observe the change in the intensity of the 435 cm^{-1} peak, which is a signature of the tetragonal phase. Accounting for the relatively small signal accumulation time (2 min), this example demonstrates the principal possibilities of Raman spectroscopy as a tool for a fast diagnostic of the nonuniformity of $La_{1-x}Sr_xMnO_3$ crystal surfaces or films.

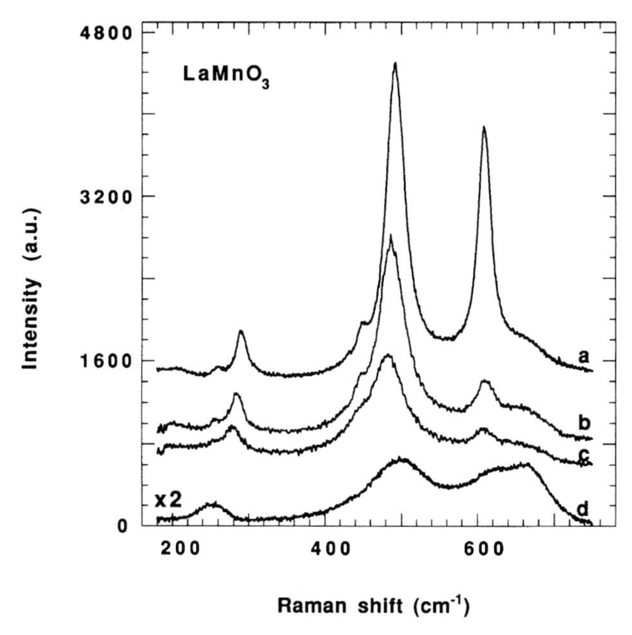

Fig. 10.12. Temperature dependent Raman spectra of $LaMnO_3$ single crystal. Spectra a, b and c correspond to $T = 293\,K$, $373\,K$ and $423\,K$, respectively. The spectrum of a doped $La_{0.9}Sr_{0.1}MnO_3$ sample at $T = 293\,K$ is shown for comparison

10.3 Undoped Crystals ($x = 0$)

The stoichiometric compound $LaMnO_3$, the parent material for the $La_{1-x}R_x$ MnO_3 series, is an antiferromagnetic insulator at room temperature and does not exhibit CMR behavior. It was found that $LaMnO_3$ possesses the $GaFeO_3$ structure [30], and the D_{2h}^{16} (mmm) orthorhombic symmetry of $La_{1-x}R_xMnO_3$ crystal was reported for both undoped [22,23,29] and lightly doped ($x < 0.125$) samples [10]. As follows from a group theoretical analysis (Sect. 10.1), the following activity of the Raman modes is expected for different scattering geometries: $A_{1g} - xx,\ yy,\ zz$; $B_{1g} - xy$; $B_{2g} - xz$; $B_{3g} - yz$. Here the x and y axes are located in the ac plane while z axis coincides the b direction.

As shown in Sect. 10.2, the Raman spectra of $La_{1-x}R_xMnO_3$ crystals contain features that can be interpreted as first-order scattering from tetragonal structures. We show below that some of these features have their origin from a parent $LaMnO_3$ material. A set of spectra from an undoped sample at different temperatures is shown in Fig. 10.12, together with that of $La_{0.9}Sr_{0.1}MnO_3$. It was found that with an increase of the sample temperature the features related to the D_{2h}^{16} orthorhombic structure gradually decrease in intensity. We explain this as a relative reduction of the orthorhombic lattice distortion, which comes about by the stronger thermal motion of the ions. Many previous studies have indicated that at high temperatures perovskite-like compounds approach a cubic structure. At the same time the intensities of the bands at 493 cm^{-1} and 609 cm^{-1}, related to the D_{2h}^{16} orthorhombic symmetry, are weak in the spectrum of $La_{0.9}Sr_{0.1}MnO_3$ (Fig. 10.12). It follows from Figs. 10.4–10.7, that a further drop of these intensities occurs in highly doped crystals. Two main conclusions may be drawn from the above observation. First, the RS from $La_{1-x}R_xMnO_3$ single crystals, at least in the range of $0.1 < x < 0.3$, contains the features of both structural phases (tetragonal and orthorhombic). Secondly, this behavior is also in qualitative agreement with the decrease of the Jahn–Teller distortion with increasing doping. As indicated in Figs. 10.4–10.7 and 10.12, the reduction of the JT distortion in $La_{1-x}R_xMnO_3$ crystals at higher values of x has the same effect on Raman spectra as the change of the crystal symmetry due to the temperature-induced phase transition. This observation is also confirmed by polarized Raman spectra of doped crystals [13]. Jahn–Teller distortion is known to be responsible for the orthorhombic distortion of the nearly-cubic perovskite $La_{1-x}R_xMnO_3$ structure. The JT splitting of the Mn^{3+} ion energy levels plays an important role in the understanding of metallic and electronic properties of these and similar systems.

Polarized Raman spectra of an undoped $LaMnO_3$ crystal in different scattering geometries are shown in Fig. 10.13. The main maxima were observed

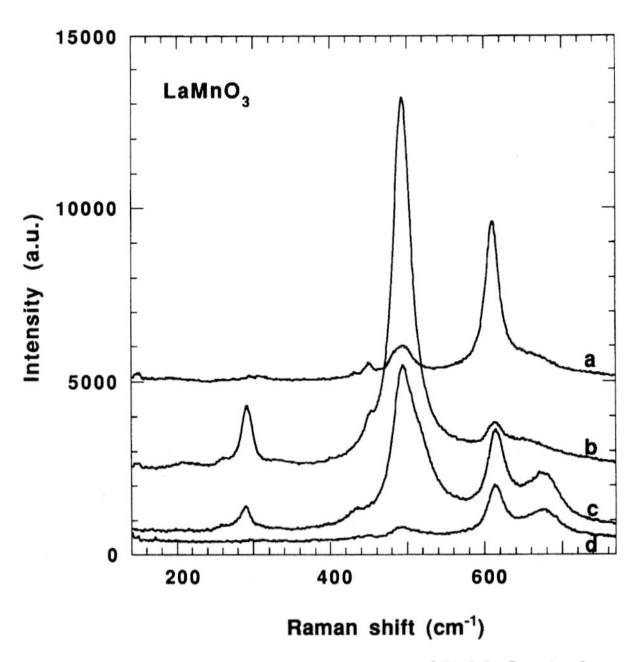

Fig. 10.13. Polarized Raman spectra of $LaMnO_3$ single crystal in different scattering geometries. Spectra a, b, c, and d correspond to $x'y'$, $y'y'$, zz and xz scattering geometries, respectively

at 96 (not shown in Fig. 10.13), 142, 210, 260, 291, 333, and 493 cm^{-1} (A_{1g} modes), and 149, 309, 450, 484, and 609 cm^{-1} (B_{1g} modes). These modes can be easily separated and identified by polarized Raman scattering. We also note, however, the presence of some additional bands whose assignments becomes more problematic because their intensities do not correlate with those predicted for D_{2h}^{16} orthorhombic symmetry (Sect. 10.1). For example, the wide band at 675 cm^{-1} exhibits a weak polarization dependence while the band at 430 cm^{-1} appears for almost all scattering geometries. The strong peak at 660 cm^{-1} was usually present in RS from unpolished faces of the crystal (Fig. 10.14). Since this peak was never observed from freshly-polished crystals, we consider it to be due to degraded surface compounds, which very likely may be lanthanum or manganese oxides. The band at 675 cm^{-1} was also observed in the spectra of doped $La_{1-x}R_xMnO_3$ materials and may be explained in part as a contribution of the second-order Raman process. Both a relatively small distortion of a nearly cubic lattice and its disorder in the perovskite structures results in comparable intensities of the first- and second-order RS.

At the Néel temperature $T_N \approx 140$ K, undoped $LaMnO_3$ compounds undergo a phase transition from paramagnetic insulator to the canted-antiferromagnetic phase [10,11]. We studied the temperature dependence of the Raman spectra of $LaMnO_3$ crystal in order to find the RS features related to this transition. Over the extended temperature range from 5 K to 423 K, most of the observed phonon peaks show both ordinary and anomalous shifts with temperature. Some irregularity in the Raman shift was found at T close to $T_N \approx 140$ K. In particular, the strongest effect was observed for the B_{1g} mode at 609 cm^{-1}. As known from IR spectra [24], the high-frequency vibrational mode in $La_{1-x}R_xMnO_3$ crystal may be assigned to the internal vibration related to the mutual Mn–O motion within the oxygen octahedron.

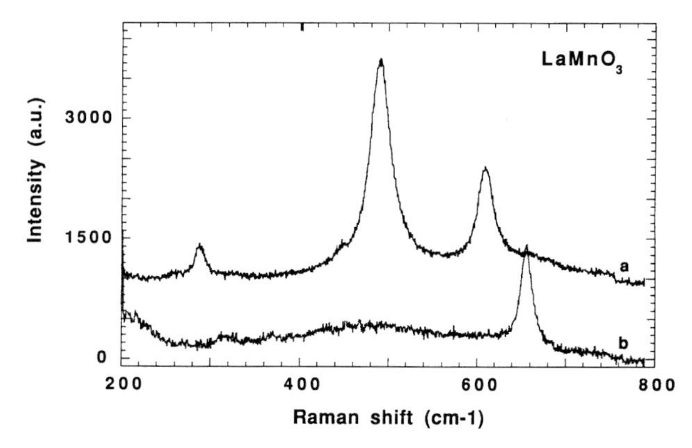

Fig. 10.14. Comparison of Raman spectra from a freshly polished surface of $LaMnO_3$ single crystal (*a*) and from an unpolished surface of the same crystal (*b*)

It is known also that this high-frequency mode should be sensitive to the Mn–O bond lengths. In the orthorhombic D_{2h}^{16} phase ($x = 0$), Mn ions occupy the c_i sites. Since these ions do not undergo displacements in vibrational modes that are active in RS, the Raman-active Mn–O mode involves only the motion of the oxygen atoms. We classify, therefore, the B_{1g} mode at 609 cm^{-1} as a stretching Mn–O vibration, due to both its location and the relatively strong intensity. The same assignment of the high-frequency B_{1g} mode in the similarly structured compounds $BaCeO_3$ and $SrCeO_3$ was done in [38] and was later elaborated upon for $LaMnO_3$ [22].

Polarized Raman spectra involving this mode are shown in Fig. 10.15. In this case the spatial dispersion filter (see Sect. 10.1) was not used and a sharp (Rayleigh scattered) Ar^+ plasma emission line shows up in all spectra. The position of this line serves as a fiduciary mark and demonstrates that over the whole range of temperatures from 120 K to 278 K the temperature generated shifts of the Raman lines are genuine and cannot be attributed to instrumental drifts or instabilities. Thus the observed shift of the B_{1g} mode is not an artifact and will be discussed below. The abrupt frequency shift of the B_{1g} mode at T close to $T_N \approx 140$ K is shown in the Fig. 10.16. In the IR study of the LCMO system [21], a similar phenomenon was explained by the strong electron–phonon coupling near the metal-insulator transition [16,21] resulting in a polaron-related effect on the vibrational mode. Here we observe the shift of the Raman B_{1g} mode in a $LaMnO_3$ insulating phase. On the other hand, in the recent study of the orthorhombic $LaMnO_3$ phase [44], the variations of

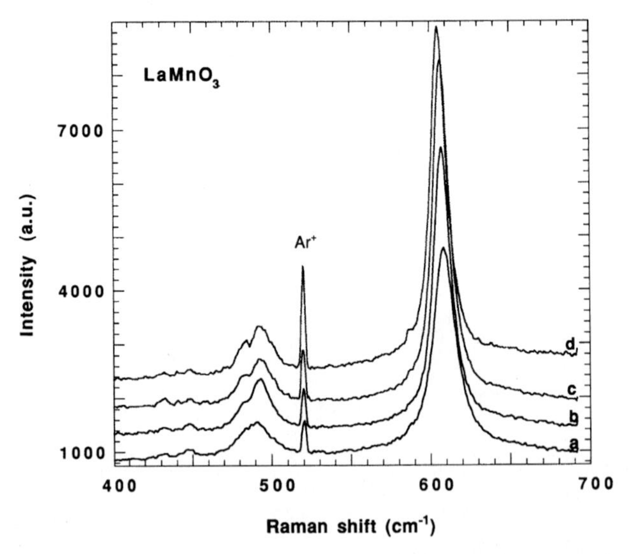

Fig. 10.15. Raman spectra of a $LaMnO_3$ single crystal in course of paramagnetic insulator-to-canted antiferromagnetic metal transition. Spectra a, b, c and d correspond to $T = 120$ K, 140 K, 180 K and 278 K, respectively

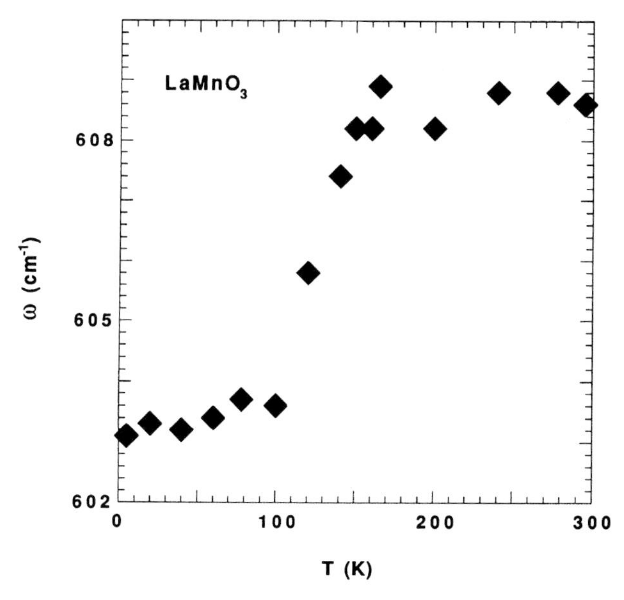

Fig. 10.16. Temperature dependence of the Raman shift of the high-frequency B_{1g} mode in $LaMnO_3$ single crystal

the lattice parameters, unit-cell volume, and sublattice magnetization were reported, and their irregular behavior in the temperature range from 100 K to 150 K was clearly indicated. We found that the Raman shift of the B_{1g} mode at 609 cm^{-1} observed in the present experiment (Fig. 10.16), qualitatively correlates well with the value of the magnetization [44]. For this reason, a spin-lattice interaction may serve as a reasonable explanation of the observed irregular shift at temperatures close to $T_N \approx 140$ K. A spin-lattice interaction arises from the stabilization energy required to bring the MnO_6 octahedra to a particular structural configuration. This stabilization energy is determined by the relative orientation between the Mn spins of near neighbor MnO_2 planes. The MnO_6 octahedra change accordingly in response to the antiferromagnetic ordering of the Mn spins as $LaMnO_3$ undergoes a paramagnetic to canted-antiferromagnetic transition.

10.4 Films

Before dealing with the analysis of Raman scattering from $La_{1-x}Ca_xMnO_3$ (LCMO) films we consider the Raman spectra of the LCMO ceramic target material used for preparing the films. In Fig. 10.17 such Raman spectra of ceramics with $x = 0.3$ are shown for both parallel and crossed polarizations. We note that the Raman spectra of the target material are consistent with spectra of doped manganite single crystals (Sect. 10.2) [13,23], if allowance is made for an intensity redistribution due to the arbitrary orientation of

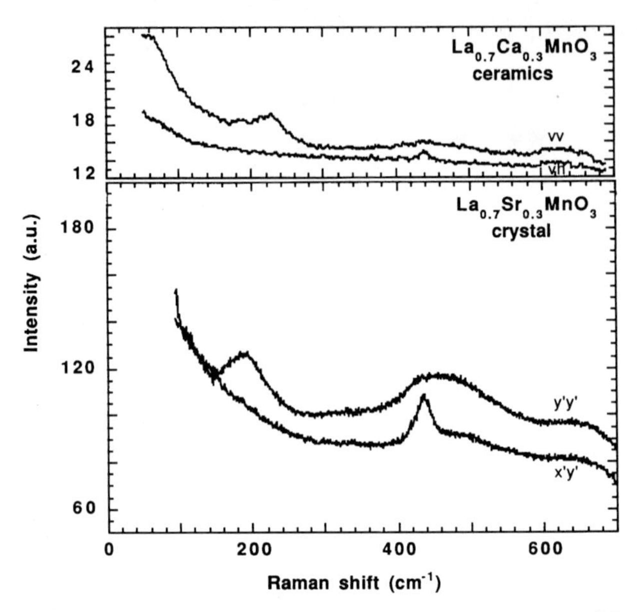

Fig. 10.17. Comparison of polarized Raman spectra of $La_{0.7}Ca_{0.3}MnO_3$ ceramics (*upper panel*) and $La_{0.7}Sr_{0.3}MnO_3$ single crystal (*lower panel*)

microcrystals in ceramics. As was shown above, the two vibrational modes of $A_{1g}(\sim 200\,\mathrm{cm}^{-1})$ and $B_{1g}(435\,\mathrm{cm}^{-1})$ symmetries characterize the Raman spectra of these crystals. The polarization properties of the components of the scattering tensor in the ceramic are consistent with a tetragonal unit cell. The Raman spectra of the films prepared from the ceramic by laser ablation were, however, strongly modified. Polarized Raman spectra of these films on $LaAlO_3$ (LAO) substrates are presented in Fig. 10.18 (see also Fig. 10.23). In addition to the peaks already present in the spectra of the ceramics, we note also two additional peaks below $200\,\mathrm{cm}^{-1}$. A better resolved structure, clearly seen at lower temperature (Fig. 10.23), was observed in the range from $400\,\mathrm{cm}^{-1}$ to $500\,\mathrm{cm}^{-1}$. Other notable differences are the softening of the low frequency A_{1g} mode and the higher intensity of the peak at $660\,\mathrm{cm}^{-1}$ for the films annealed in oxygen.

Among the possible reasons for the change of Raman spectra in films, and by implication, therefore, the microscopic symmetry, we consider first the mechanical strain, arising from the mismatch of the lattice parameters between the substrate and the LCMO materials. As found from this Raman experiment, there was no essential difference between the spectra of the films ($d \sim 270\,\mathrm{nm}$) on $NdGeO_3$ (NGO) and LAO substrates (Fig. 10.19). However, from [17] it is known that the lattice mismatch in LCMO-LAO (1.8%) is about an order of magnitude higher than that in the LCMO-NGO system (0.2%). The absence of observable differences in the Raman spectra of the two films may therefore be better understood in terms of a limited pene-

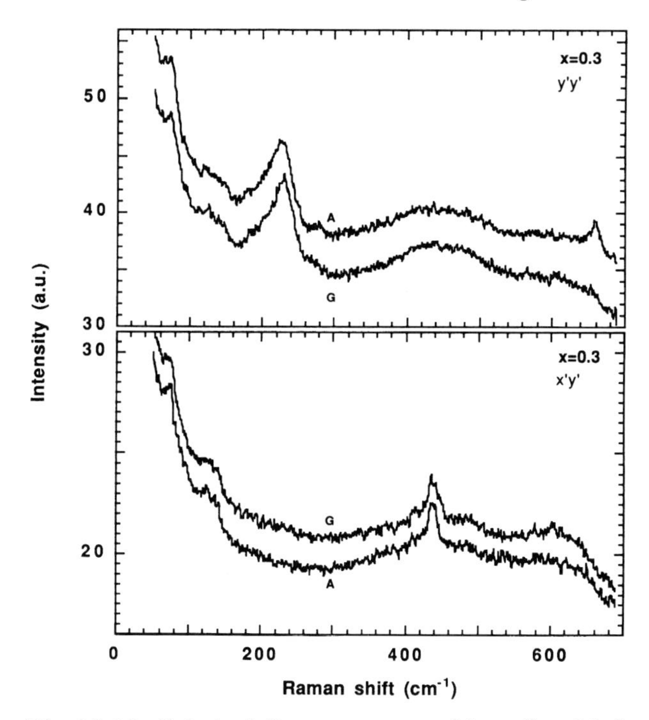

Fig. 10.18. Polarized Raman spectra of $La_{0.7}Ca_{0.3}MnO_3$ as-grown (G) and annealed (A) films

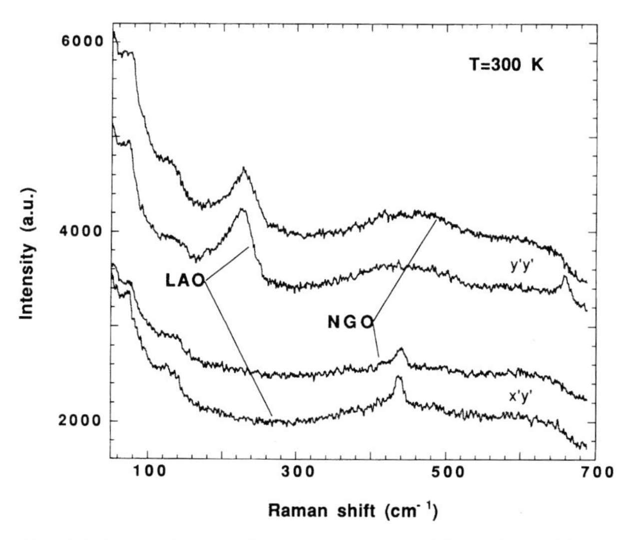

Fig. 10.19. Polarized Raman spectra of $La_{0.7}Ca_{0.3}MnO_3$ films on $LaAlO_3$ (LAO) and $NdGaO_3$ (NGO) substrates. For each film, two spectra in the $y'y'$ and $x'y'$ scattering geometries are shown

tration depth ($\sim$ 10 nm) of the excitation radiation. While in the LCMO-substrate system the strains should be more effective near the substrate-film interface [45], they are expected to be smaller on the other (film-air) side of the film, which contributes primarily to the RS. The appearance of additional peaks in the spectra of the films at 87, 145, 410, and 470 cm^{-1} (Fig. 10.23) may therefore be the signature of a rhombohedral (D_{3d}^6) phase of $La_{1-x}Ca_xMnO_3$. This phase was observed in some neutron diffraction studies of $La_{1-x}Sr_xMnO_3$ materials [9,10]. As defined by the selection rules (Sect. 10.1), the Raman spectra from the rhombohedral (D_{3d}^6) LCMO phase should contain $A_{1g}+4E_g$ modes. Moreover, the E_g modes are expected to be present in both parallel and crossed geometries and this was experimentally observed. From the changes observed in the Raman spectra of the target and the film, we may conclude that the film's structure is affected mainly by the deposition process. The lattice mismatch between the film and the substrate would be expected to be more important for Raman spectra from thinner films.

Raman spectra of films with different doping are presented in Fig. 10.20. The main change in the intensity was found for the broad band at about 600 cm^{-1}. However, taking into account the data from other films and LSMO single crystals, this change may be mainly attributed to the oxidation (anneal-

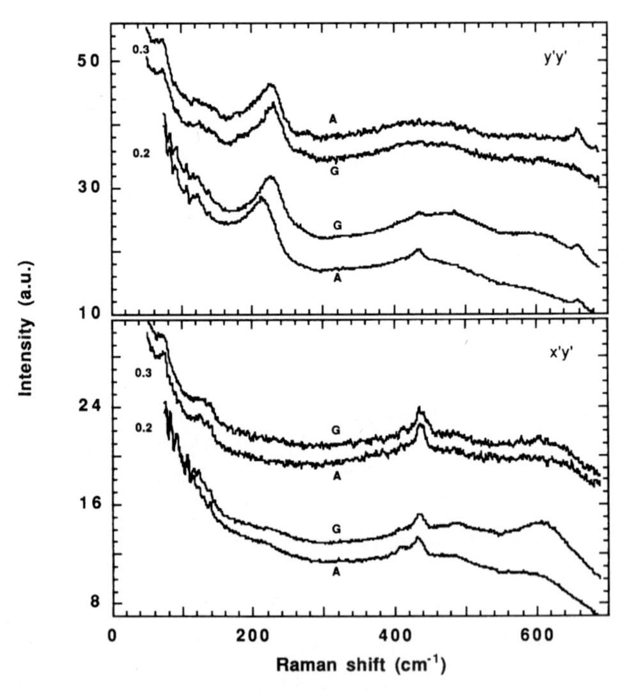

Fig. 10.20. Comparison of polarized Raman spectra of $La_{1-x}Ca_xMnO_3$ films ($x =$ 0.2 and 0.3) in the $x'y'$ and $y'y'$ scattering geometries

ing) process, rather than to a doping dependence. The Raman study of LSMO single crystals [13,23] has shown that this band cannot be assigned to a fundamental lattice vibration due to its temperature behavior, large linewidth and weak polarization dependence. This band may therefore be a disorder-induced Raman feature, and its origin is expected to reside in a lattice defect due to an existing oxygen deficit. It is known that a deficit of oxygen introduces defect-disorder features in perovskite structures while appropriate annealing reduces local disorder and leads to relaxation of the Mn–O–Mn bond angle [46,47]. The disorder-induced nature of the band at $600\,\mathrm{cm}^{-1}$ receives additional support in the present Raman spectra of $La_{1-x}Ca_xMnO_3$ films, since this band is less intense in annealed structures (Fig. 10.20) in which the oxygen deficit is smaller. The relative intensity of this disorder-induced Raman band may, therefore, serve as a good indicator of the oxygen deficit in LCMO films. Furthermore, the comparable intensities of the $600\,\mathrm{cm}^{-1}$ bands in LSMO single crystals and annealed films indicate also a comparable structural disorder in these two systems and emphasizes a relatively high crystallinity of particular films prepared by the laser ablation technique.

As seen from the experimental data, the Raman spectra of some LCMO film samples also contain a peak at about $660\,\mathrm{cm}^{-1}$. This peak appears more often in the spectra of annealed samples though less intense features were detected also from as-grown samples. Due to its intensity dependence this peak may be considered as a Raman signature of the film annealed in oxygen. However a peak at about $660\,\mathrm{cm}^{-1}$ was also observed in an undoped $LaMnO_3$ (LMO) single crystal provided the Raman scattered light was collected from a previously unpolished face of the crystal (see Fig. 10.14). Upon polishing, this peak disappeared and only the well-known spectrum of the LMO structure was observed [23]. To verify that this peak is due to the near-surface area of a film, i.e., closest to its film-air boundary, we compared the Raman spectra obtained with different excitation wavelengths. Using the set of wavelengths available from Ar^+ and Kr^+ lasers, we found that the normalized (to the A_{1g} mode) intensity of the $660\,\mathrm{cm}^{-1}$ peak strongly depended on the excitation wavelength, λ_i (Fig. 10.21). The observed large intensity change, about 15-fold in the range from 458 nm to 647 nm, cannot, however, be explained only by the change in penetration depth of the exciting light. An additional intensity dependence due to a resonance Raman effect may likely be present. However, if a resonance Raman feature contributes only to the intensity of one selected peak ($660\,\mathrm{cm}^{-1}$), different ground electronic states for the LCMO material should be expected, including the one that is responsible for the mode at $660\,\mathrm{cm}^{-1}$. Thus, whatever contributes to the trends in the intensity, the surface effect, or resonance Raman, the $660\,\mathrm{cm}^{-1}$ peak cannot be assigned to a vibration mode of the LCMO lattice. More likely, this mode is related to one of the simple oxides of lanthanum or manganese and probably describes a long-term degradation of the sample surface area.

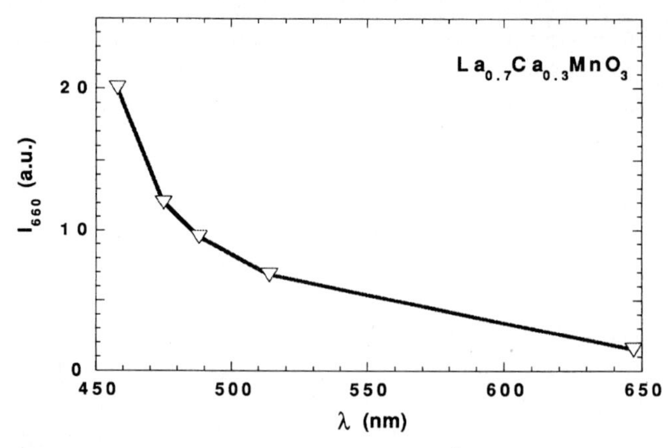

Fig. 10.21. Dependence of the $660\,\mathrm{cm}^{-1}$ peak intensity in the Raman spectra of a $La_{0.7}Ca_{0.3}MnO_3$ film on the excitation wavelength. These intensities were normalized to that of A_{1g} mode for each λ

Raman spectra of LCMO films with different values of doping x are shown in Fig. 10.22 ($x = 0.2$) and Fig. 10.23 ($x = 0.3$). The unpolarized spectra of a $La_{0.8}Ca_{0.2}MnO_3$ film, shown in Fig. 10.22, contain all the A_{1g}-, B_{1g}- and E_g-like modes. Moreover, the A_{1g} and B_{1g} modes may be separated by polarization as shown in Fig. 10.23 for a $La_{0.7}Ca_{0.3}MnO_3$ film. In the latter case, the x' and y' axes, and therefore the orientation of the LCMO film, were found from the highest intensity contrast between A_{1g} and B_{1g} modes. To illustrate the behavior at low temperatures, we point to the disappearance of both broad bands at about $460\,\mathrm{cm}^{-1}$ and $600\,\mathrm{cm}^{-1}$, while the band

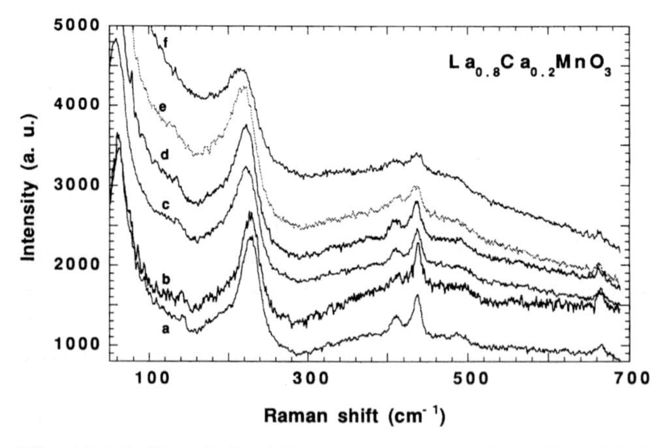

Fig. 10.22. Unpolarized Raman spectra of a $La_{0.8}Ca_{0.2}MnO_3$ film at different temperatures. Spectra a, b, c, d, e and f correspond to $T = 120$, 140, 170, 200, 230 and 260 K, respectively

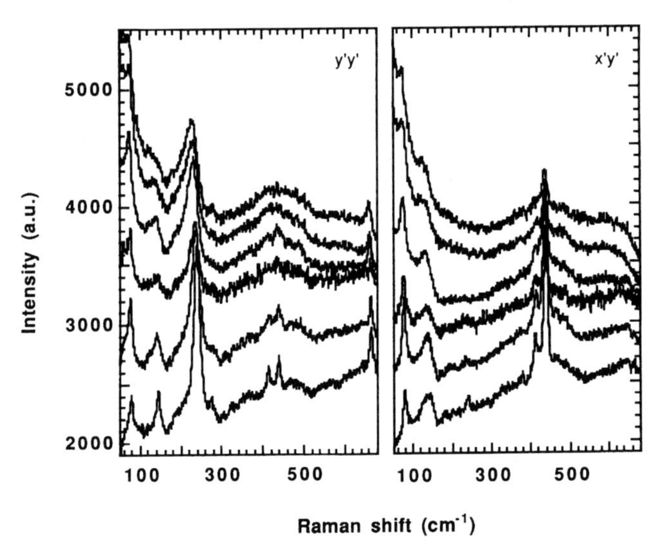

Fig. 10.23. Polarized Raman spectra of an annealed $La_{0.7}Ca_{0.3}MnO_3$ film in the $x'y'$ and $y'y'$ scattering geometries. The spectra in each panel correspond to temperatures (from *top* to *bottom*) 300, 250, 220, 180, 130, 70 and 7 K, respectively

at $460\,cm^{-1}$ evolves into a clearly resolved structure. Besides these changes, there are new peaks at $410\,cm^{-1}$ and $470\,cm^{-1}$ that exhibit the typical polarization behavior of the E_g modes in the $y'y'$ and $x'y'$ scattering geometries. Therefore, the band at $460\,cm^{-1}$, which is unresolved at higher temperatures, should be assigned to second-order Raman scattering rather than to a particular phonon mode. For the reason discussed above, the band at $600\,cm^{-1}$ was assigned as a disorder-induced Raman feature. Note that the strongest $A_{1g}(493\,cm^{-1})$ and $B_{1g}(609\,cm^{-1})$ modes in undoped LMO materials should also be considered as possible contributors to the intensity of the broad bands at $460\,cm^{-1}$ and $600\,cm^{-1}$ (see Sect. 10.3), since the second-order Raman spectrum is usually related to the density of vibrational states.

Compared to as-grown films, a small $(3-5\,cm^{-1})$ softening of the A_{1g} mode in the annealed samples is always present in Raman spectra (Figs. 10.16 and 10.20). In an inverse process, i.e. upon deoxygenation of LCMO films, a hardening of the related $225\,cm^{-1}$ Raman peak was observed [48]. The annealing of as-grown manganite films in oxygen usually reduces the number of oxygen vacancies and increases the T_c of the sample. This leads simultaneously to an effective lattice change, i.e. a compression of the unit cell along the crystallographic axes b and c [49]. As known from the high pressure Raman studies, such compression, however, should lead to a hardening of the A_{1g} mode, i.e., to the opposite effect. Therefore, the reason for the softening of the A_{1g} mode in the Raman spectra of LCMO films may be the same as in the case of doping $LaMnO_3$ with strontium [13]. Besides, the increase of the ionic radius $\langle r_A \rangle$ of the La/Ca site is able to overcompensate the expected hardening of a vibrational mode induced by compression of the unit cell.

With a decrease of temperature from 300 K to 7 K, smooth hardening of the A_{1g} mode was observed in Raman scattering from $La_{1-x}Ca_xMnO_3$ films (Fig. 10.24). However, the linewidth of this mode exhibits a relatively steep decrease below 250 K (Fig. 10.25). The LCMO compounds with $x = 0.3$ have a paramagnetic insulator-to-ferromagnetic metal phase transition at $T_c \sim$ 250 K [6]. In analyzing the factors affecting the linewidth of the A_{1g} mode at 227 cm^{-1}, we consider three main causes: electron–phonon interaction, anharmonicity, and electron spin-lattice coupling. Usually, an electron–phonon interaction without a spin effect leads mainly to the broadening of the phonon mode [50], which is opposite to the observed effect. The temperature dependent Raman behavior of the A_{1g} mode (Fig. 10.24) also suggests that the steep linewidth change in the temperature range from 250 K to 210 K cannot be explained by an anharmonic effect. The observed narrowing of the 227 cm^{-1} Raman peak was therefore attributed to the spin-lattice interaction affected by spin ordering below T_c when LCMO undergoes the paramagnetic insulator-to-ferromagnetic metal phase transition. Note that the character of the 227 cm^{-1} peak narrowing in the temperature range 7 – 300 K correlates satisfactorily with the magnetization data as seen from Fig. 10.25. A similar correlation was found also with resistivity data.

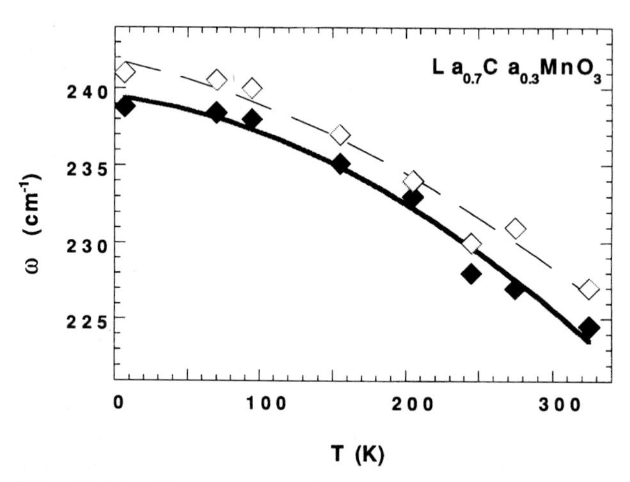

Fig. 10.24. Temperature dependence of Raman shift of the A_{1g} mode in $La_{0.7}Ca_{0.3}MnO_3$ film. *Open* and *closed diamonds* are related to as-grown and annealed samples, respectively

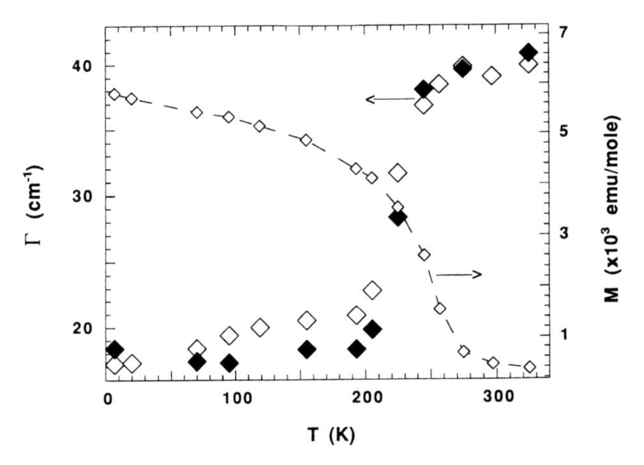

Fig. 10.25. Temperature dependence of Raman linewidth of the A_{1g} mode in as-grown (*open diamonds*) and annealed (*closed diamonds*) $La_{0.7}Ca_{0.3}MnO_3$ films. The magnetization data for annealed sample (*small open diamonds*) are shown for comparison

10.5 Summary

Raman spectroscopy is shown to be a useful tool for the study of the phase transitions in lanthanum manganese compounds. The observed temperature-dependent polarized Raman spectra of doped and undoped $La_{1-x}R_xMnO_3$ single crystals have shown a significant dependence of scattering by optical phonons on both the kind and amount of the dopant. They are sensitive to structural and magnetic transitions driven by both doping and temperature.

Polarized Raman spectra of undoped and doped $La_{1-x}R_xMnO_3$ single crystals have been studied in the temperature range from 5 K to 423 K (R = Sr, Ca). Raman spectra of the doped $La_{1-x}R_xMnO_3$ crystals have weakly-resolved features and intensities comparable to those of the second-order spectra in cubic media. The spectral intensity is mainly related to scattering by optical phonons at $q \sim 0$ (for $x \approx 0$) and the density of vibrational states for $x > 0.1$. The small distortion of the nearly cubic lattice of doped $La_{1-x}Sr_xMnO_3$ single crystals results in a structured phonon Raman spectrum that is composed of two parts. The first part is assigned to the distorted non-cubic perovskite lattice and follows the selection rules for first-order scattering from a tetragonal structure. This indicates a predisposition toward a tetragonal structure in doped $La_{1-x}Sr_xMnO_3$ single crystals. The second part is attributed to second-order Raman scattering, and its shape reflects the density of vibrational states.

An effect of the spin-lattice interaction was found in Raman spectra of undoped $LaMnO_3$. The most intense B_{1g} mode in $LaMnO_3$ crystal has an abrupt frequency shift near $T_N = 140$ K, which corresponds to a paramagnetic to canted-antiferromagnetic phase transition. A simple frequency dependence of the A_{1g} external mode upon x was found. This dependence may be used for optical diagnosis of the doping value x in the $La_{1-x}Sr_xMnO_3$ crystals. Although the frequency of the A_{1g} mode was found to be sensitive to doping, the observed shift is opposite to that expected from the difference in the atomic weights of Sr and La. The anomalies of phonon Raman scattering from doped crystals were explained by both the temperature and doping dependence of the ionic radius of a La/Sr-site.

A correlation of lattice transformation as well as magnetic ordering in $La_{1-x}Ca_xMnO_3$ films with Raman scattering data was found. It was shown that the structural changes, preferable orientation of films as well as an influence of deposition conditions can be ascertained from phonon Raman spectra. The effect of spin-lattice interaction due to spin ordering below T_c was observed in Raman spectra of $La_{1-x}Ca_xMnO_3$ films. The results of this work may be extended to other related manganite structures, in particular, the $La_{1-x}Sr_xMnO_3$ films. These results prove that Raman spectroscopy can be a useful probe for optical diagnostics of LCMO films.

Acknowledgement. The authors gratefully acknowledge Drs. Raju V. Datla for fruitful discussions and support of this work; D.B. Romero, J.P. Rice, H.D. Drew for fruitful cooperation, and J. Hougen for helpful discussions. We also thank Drs. S. Bhagat, R. Schreekala, M. Rajeswari, R. Ramesh and T. Venkatesan from the University of Maryland for samples of $La_{1-x}R_xMnO_3$ single crystals and films.

References

1. P. Grünberg, R. Schreiber, Y. Pang, M.B. Brodsky, H. Sowers: Phys. Rev. Lett. **57**, 2442 (1986)
2. S.S.P. Parkin: Annu. Rev. Mater. Sci. **25**, 357 (1995); See also J.L. Simonds: Phys. Today **48**, 26 (1995)
3. a) S. Jin, T.H. Tiefel, M. McCormack, R.A. Fastnacht, R. Ramesh, L.H. Chen: Science **264**, 413 (1994); b) M. McCormack, S. Jin, T.H. Tiefel, R.M. Fleming, J.M. Phillips, R. Ramesh: Appl. Phys. Lett. **64**, 3045 (1994)
4. See, for example: a) A.P. Ramirez: J. Phys.: Condens. Matter **9**, 8171 (1997); b) J. Fontcuberta: Phys. World February 1999, p. 33; c) C.N.R. Rao, A.K. Cheetham, R. Mahesh: Chem. Mat. **8**, 2421 (1996); d) F. Damay, N. Nguyen, A. Maignan, M. Hervieu, B. Raveau: Solid State Commun. **98**, 997 (1996); e) topical papers on "Understanding and utilizing colossal magnetoresistance materials", special issue on Research Frontiers in: Phil. Trans. R. Soc. London A **356**, 1469 (1998)
5. The magnitude of magnetoresistance, MR, is expressed quantitatively by the relation $MR = \Delta R/R'$, where $\Delta R = R - R_H$ is the change in the resistance when the material is placed in a magnetic field H and R' is either the zero-field resistance R or the resistance R_H in the presence of the magnetic field. Both definitions are found in the literature

6. P. Schiffer, A.P. Ramirez, W. Bao, S.-W. Cheong: Phys. Rev. Lett. **75**, 3336 (1995)
7. A. Maignan, C. Michel, M. Hervieu, B. Raveau: Solid State Commun. **101**, 277 (1997)
8. S. Pignard, H. Vincent, J.P. Senateur, J. Pierre, A. Abrutis: J. Appl. Phys. **82**, 4445 (1997)
9. H.Y. Hwang, S.W. Cheong, P.G. Radaelli, M. Marezio, B. Batlogg: Phys. Rev. Lett. **75**, 914 (1995)
10. H. Kawano, R. Kajimoto, M. Kubota, Y. Yoshizawa: Phys. Rev. B **53**, R14709 (1996)
11. E.O. Wollan, W.C. Koehler: Phys. Rev. **100**, 545 (1955)
12. R. Mahendrian, S.K. Tiwary, A.K. Raychaudhuri, T.V. Ramakrishnan, R. Mahesh, N. Rangavittal, C.N.R. Rao: Phys. Rev. B **53**, 3348 (1996)
13. V.B. Podobedov, A. Weber, D.B. Romero, J.P. Rice, H.D. Drew: Solid State Commun. **105**, 589 (1998)
14. P.G. Radaelli, D.E. Cox, M. Marezio, S-W. Cheong, P.E. Schiffer, A.P. Ramirez: Phys. Rev. Lett. **75**, 4488 (1995)
15. A. Asamitsu, Y. Moritomo, Y. Tomioka, T. Arima, Y. Tokura: Nature **373**, 407 (1995)
16. A.J. Millis, P.B. Littlewood, B.I. Shraiman: Phys. Rev. Lett. **74**, 5144 (1995)
17. M.E. Hawley, C.D. Adams, P.N. Arendt, E.L. Brosha, F.H. Garzon, R.J. Houlton, M.F. Hundley, R.H. Heffner, Q.X. Jia, J. Neumeier, X.D. Wu: J. Crystal Growth **174**, 455 (1997)
18. K.B. Li, Z.Z. Qi, X.J. Li, J.S. Zhu, Y.H. Zhang: Thin Solid Films **304**, 386 (1997)
19. V.B. Podobedov, D.B. Romero, A. Weber, J.P. Rice, R. Shreekala, M. Rajeswari, R. Ramesh, T. Venkatesan, H.D. Drew: Appl. Phys. Lett. **73**, 3217 (1998)
20. T. Arima, Y. Tokura: J. Phys. Soc. Jpn. **64**, 2488 (1995)
21. K.H. Kim, J.Y. Gu, H.S. Choi, G.W. Park, T.W. Noh: Phys. Rev. Lett. **77**, 1877 (1996)
22. M.N. Iliev, M.V. Abrashev, H.-G. Lee, V.N. Popov, Y.Y. Sun, C. Thomsen, R.L. Meng, C.W. Chu: Phys. Rev. B **57**, 2872 (1998)
23. V.B. Podobedov, A. Weber, D.B. Romero, J.P. Rice, H.D. Drew: Phys. Rev. B **58**, 43 (1998)
24. M.D. Fontana, G. Metrat, J.L. Servoin, F. Gervais: J. Phys. C **17**, 483 (1984)
25. C.H. Perry, N.E. Tornberg: In *Light Scattering in Solids* (Springer, New York 1969) p. 467
26. J.L. Verble, E. Gallego-Lluesma, S.P.S. Porto: J. Raman Spectrosc. **7**, 7 (1978)
27. M.C. Martin, G. Shirane, Y. Endoh, K. Hirota, Y. Moritomo, Y. Tokura: Phys. Rev. B **53**, 14 285 (1996)
28. A. Asamitsu, Y. Moritomo, R. Kumai, Y. Tomioka, Y. Tokura: Phys. Rev. B **54**, 1716 (1996)
29. J.B.A.A. Elemans, B. van Laar, K.R. van der Veen, B.O. Loopstra: J. Solid State Chem. **3**, 238 (1971)
30. S. Geller: J. Chem. Phys. **24**, 1236 (1956)
31. W.E. Pickett, D.J. Singh: Phys. Rev. B **53**, 1146 (1996)
32. S.E. Lofland, V. Ray, P.H. Kim, S.M. Bhagat, M.A. Manheimer, S.D. Tyagi: Phys. Rev. B **55**, 2749 (1997)

33. A. Goyal, M. Rajeswari, R. Shreekala, S.E. Lofland, S.M. Bhagat, T. Boettcher, C. Kwon, R. Ramesh, T. Venkatesan: Appl. Phys. Lett. **71**, 2535 (1997)
34. V.B. Podobedov, J.P. Rice, A. Weber, H.D. Drew: J. Superconductivity **10**, 205 (1997)
35. R. Gupta, A.K. Sood, R. Mahesh, C.N.R. Rao: Phys. Rev. B **54**, 14 899 (1996)
36. S. Yoon, H.L. Liu, G. Schollerer, S.L. Cooper, P.D. Han, D.A. Payne, S.-W. Cheong Z. Fisk: Phys. Rev. B **58**, 2795 (1998)
37. Y. Okimoto, T. Katsufuji, T. Ishikawa, T. Arima, Y. Tokura: Phys. Rev. B **55**, 4206 (1997)
38. T. Scherban, R. Villeneuve, L. Abello, G. Lucazeau: Solid State Commun. **84**, 341 (1992); J. Raman Spectrosc. **24**, 805 (1993)
39. F. Genet, S. Loridant, G. Lucazeau: J. Raman Spectrosc. **28**, 255 (1997)
40. K.S. Knight, N. Bonanos: Solid State Ionics **77**, 189 (1995)
41. V.B. Podobedov: J. Raman Spectrosc. **27**, 731 (1996)
42. A.S. Barker, Jr., A.J. Sievers: Rev. Mod. Phys. **47**(2), S1 (1975)
43. W. Archibald, J.-S. Zhou, J.B. Goodenough: Phys. Rev. B **53**, 14 445 (1995)
44. Q. Huang, A. Santoro, J.W. Lynn, R.W. Erwin, J.A. Borchers, J.L. Peng, R.L. Greene: Phys. Rev B **55**, 14 987 (1997)
45. A.J. Millis, A. Goyal, M. Rajeswari, K. Ghosh, R. Shreekala, R.L. Greene, R. Ramesh, T. Venkatesan: (unpublished)
46. J. Nowotny, M. Rekas: J. Am. Ceram. Soc. **81**, 67 (1998)
47. D.C. Worledge, G.J. Snyder, M.R. Beasley, T.H. Geballe, R. Hiskes, S. DiCarolis: J. Appl. Phys. **80**, 5158 (1996)
48. N. Malde, P.S.I.P.N. de Silva, A.K.M. Akther Hossain, L.F. Cohen, K.A. Tomas, J.L. MacManus-Driscoll, N.D. Mathur, M.G. Blamire: Solid State Commun. **105**, 643 (1998)
49. A.M. DeLeon-Guevara, P. Berthet, J. Berthon, F. Millot, A. Revcolevschi, A. Anane, C. Dupas, K. LeDang, J.P. Renard, P. Veillet: Phys. Rev. B **56**, 6031 (1997)
50. L.A. Falkovsky, W. Knap, J.C. Chervin, P. Wisniewski: Phys. Rev. B **57**, 11 349 (1998)
51. E. Granado, N.O. Moreno, A. Garcia, J.A. Sanjurjo, C. Rettori, I. Torriani, S.B. Oseroff, J.J. Neumeier, K.J. McClellan, S.W. Cheong, Y. Tokura: Phys. Rev. B **58**, 11435 (1998)

X Raman Scattering from Perovskite Ferroelectrics

R.S. Katiyar

Raman scattering has proven to be extremely useful for studies of phase transitions in ferroelectric materials and, in particular, as a tool to characterize soft modes [1] and features related to order-disorder phenomena [2]. The dynamical behavior of phase transitions and their sensitivity to temperature and doping can be easily elucidated by Raman scattering measurements [3–5]. The perovskite (ABO_3) ferroelectric materials in thin-film form are of considerable interest, particularly for non-volatile ferroelectric random access memory and microwave devices. The ferroelectric properties depend strongly on film preparation, chemical composition and microstructures (crystal structures, orientation, grain size, domain structure, and surface roughness). Micro-Raman scattering from ferroelectric thin films provides a useful link between the phonon spectrum and the microstructure of the film.

Micro-Raman investigations on nanocrystalline $PbTiO_3$, $(PbBi_2)Ta_2O_9$, $(PbLa)TiO_3$, $Pb(Sc_{1/2}Ta_{1/2})O_3$ and $(PbZr)TiO_3$ (PZT) reveal that the low-frequency modes are grain size and thickness dependent [6,7]. A distinct soft-mode-induced phase transition appears in each material and the transition temperature decreases with decreasing grain size. Figure X.1 shows softening of the E(1TO) mode as a function of Zr concentration x for PZT powder with grain size of 60 nm. The extrapolated phase boundary appears at $x = 0.4$, whereas in the corresponding bulk samples the boundary inferred from Raman data is at $x = 0.52$ [4]. The temperature dependence of the frequency and damping of the E(1TO) mode in $PbTiO_3$ with varying film thickness are plotted in Fig. X.2. As for bulk samples, the temperature dependence of the soft-mode frequency is in accordance with the Curie–Weiss behavior.

Relaxor ferroelectrics, such as $PbMg_{1/3}Nb_{2/3}O_3$ (PMN) and $PbSc_{1/2}Ta_{1/2}O_3$, are a special class of complex oxide materials with large dielectric response extending over a wide range of temperature that do not follow the Curie–Weiss law. The crystallographic space group in relaxors is different from that of proper ferroelectrics in that the former exhibit nanoscale ordered regions dispersed in a three dimensional disordered matrix. Relaxors with $1:1$ B site ions show the possibility of order–disorder transformations, while members with $1:2$ B site ionic arrangements show intrinsically disordered behavior. The size of the ordered regions in relaxors is typically in the range 2–5 nm which makes Raman scattering an ideal means to probe their structural features [8]. Unlike proper perovskite ferroelectrics, such as

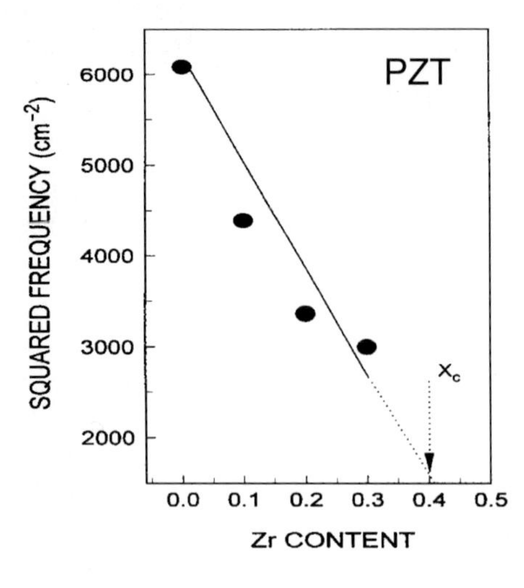

Fig. X.1. ω^2 vs. x for PTZ powder with 60 nm grain size; ω is the frequency of the E(1TO) mode [6]

$BaTiO_3$ and $PbTiO_3$, soft modes have not been observed in relaxor ferroelectrics.

However, Raman scattering has demonstrated potential usefulness to study static and dynamical aspects of the B-site ordering, to distinguish different types of substitution in the A and B sublattices and to probe selection rule deviations. One of the major conclusions drawn from Raman studies is that the space group of the relaxor nanoregions is $Fm3m$ and that this symmetry remains as the samples go through the maximum of the dielectric constant at T_m [8]. This study also confirms an important common structural aspect of complex perovskites in that the nanoscale clusters exist with 1:1 B-site order, regardless of whether the stoichiometric composition for the B ions is 1:1 or

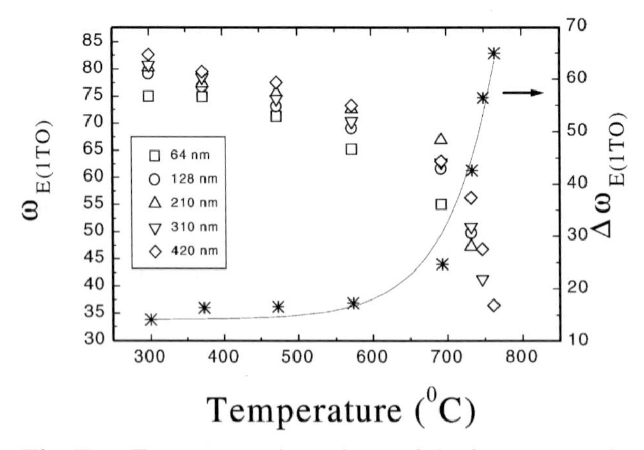

Fig. X.2. Temperature dependence of the frequency and ping of the E(1TO) mode in $PbTiO_3$ for various film ness [7]

1:2. These findings establish a relationship between Raman spectra and crystal symmetry in materials consisting of nanoscale regions, thus showing that Raman scattering serves as a complementary tool of direct microstructural characterization that has been developed in the last decade.

As the temperature increases above T_{m}, one observes a gradual loss of individuality in the dynamics of the B′ and B″ octahedra. Finally, above 700 K and for most materials, the nanoregions transform to a primitive cubic lattice ($Pm3m$) where all first-order Raman modes disappear. In PMN, the dynamic relaxation of translational symmetry correlates with the appearance of a broad central peak reflecting competing interactions between two frustrated phase transitions [8].

References

1. W. Cochran: Phys. Rev. Lett. **3**, 412 (1959)
2. P. da R. Andrade, R.S. Katiyar, S.P.S. Porto: Ferroelectrics **8**, 637 (1974)
3. G. Burns, B.A. Scott: Phys. Rev. Lett. **25**, 1191 (1970)
4. A. Pinczuk: Solid State Commun. **12**, 1035 (1973)
5. R. Merlin, J.A. Sanjurjo, A. Pinczuk: Solid State Commun. **16**, 931 (1975)
6. J.F. Meng, R.S. Katiyar, G.T. Zou, X.H. Wang: Phys. Status Solidi A **164**, 851 (1997)
7. P.S. Dobal, S. Bhaskar, S.B. Majumder, R.S. Katiyar: J. Appl. Phys. **86**, 828 (1999)
8. I.G. Siny, R.S. Katiyar, A.S. Bhalla: Ferroelectrics Review **2**, 1 (2000)

Index

Printing: Mercedes-Druck, Berlin
Binding: Stürtz AG, Würzburg

Springer Series in

MATERIALS SCIENCE

1 **Chemical Processing with Lasers***
By D. Bäuerle

2 **Laser-Beam Interactions with Materials**
Physical Principles and Applications
By M. von Allmen and A. Blatter
2nd Edition

3 **Laser Processing of Thin Films and Microstructures**
Oxidation, Deposition and Etching of Insulators
By. I. W. Boyd

4 **Microclusters**
Editors: S. Sugano, Y. Nishina, and S. Ohnishi

5 **Graphite Fibers and Filaments**
By M. S. Dresselhaus, G. Dresselhaus, K. Sugihara, I. L. Spain, and H. A. Goldberg

6 **Elemental and Molecular Clusters**
Editors: G. Benedek, T. P. Martin, and G. Pacchioni

7 **Molecular Beam Epitaxy**
Fundamentals and Current Status
By M. A. Herman and H. Sitter 2nd Edition

8 **Physical Chemistry of, in and on Silicon**
By G. F. Cerofolini and L. Meda

9 **Tritium and Helium-3 in Metals**
By R. Lässer

10 **Computer Simulation of Ion-Solid Interactions**
By W. Eckstein

11 **Mechanisms of High Temperature Superconductivity**
Editors: H. Kamimura and A. Oshiyama

12 **Dislocation Dynamics and Plasticity**
By T. Suzuki, S. Takeuchi, and H. Yoshinaga

13 **Semiconductor Silicon**
Materials Science and Technology
Editors: G. Harbeke and M. J. Schulz

14 **Graphite Intercalation Compounds I**
Structure and Dynamics
Editors: H. Zabel and S. A. Solin

15 **Crystal Chemistry of High-T_c Superconducting Copper Oxides**
By B. Raveau, C. Michel, M. Hervieu, and D. Groult

16 **Hydrogen in Semiconductors**
By S. J. Pearton, M. Stavola, and J. W. Corbett

17 **Ordering at Surfaces and Interfaces**
Editors: A. Yoshimori, T. Shinjo, and H. Watanabe

18 **Graphite Intercalation Compounds II**
Editors: S. A. Solin and H. Zabel

19 **Laser-Assisted Microtechnology**
By S. M. Metev and V. P. Veiko
2nd Edition

20 **Microcluster Physics**
By S. Sugano and H. Koizumi
2nd Edition

21 **The Metal-Hydrogen System**
By Y. Fukai

22 **Ion Implantation in Diamond, Graphite and Related Materials**
By M. S. Dresselhaus and R. Kalish

23 **The Real Structure of High-T_c Superconductors**
Editor: V. Sh. Shekhtman

24 **Metal Impurities in Silicon-Device Fabrication**
By K. Graff 2nd Edition

25 **Optical Properties of Metal Clusters**
By U. Kreibig and M. Vollmer

26 **Gas Source Molecular Beam Epitaxy**
Growth and Properties of Phosphorus Containing III–V Heterostructures
By M. B. Panish and H. Temkin

* The 2nd edition is available as a textbook with the title: *Laser Processing and Chemistry*